SPECIAL PAPERS IN PALAEONTOLOGY NO. 77

EVOLUTION AND PALAEOBIOLOGY OF EARLY
SAUROPODOMORPH DINOSAURS

EDITED BY

PAUL M. BARRETT *and* DAVID J. BATTEN

with 107 text-figures and 24 tables

THE PALAEONTOLOGICAL ASSOCIATION
LONDON

March 2007

CONTENTS

[Special Papers in Palaeontology 77, 2007, pp. 5–7]

FOREWORD

by PAUL M. BARRETT* *and* TIM J. FEDAK†

*Department of Palaeontology, The Natural History Museum, Cromwell Road, London SW7 5BD, UK; e-mail: P.Barrett@nhm.ac.uk
†Department of Biology, Dalhousie University, Life Sciences Centre, Halifax, Nova Scotia, B3H 4J1, Canada; e-mail: tfedak@dal.ca

SAUROPODA includes many of the most familiar dinosaurs, such as *Apatosaurus* (formerly '*Brontosaurus*'), *Diplodocus* and *Brachiosaurus* (the latter finding fame in the *Jurassic Park* movies). The unusual appearance of these animals, with their extraordinarily elongate necks, proportionally small skulls and enormous size (with body masses of up to 70 tonnes in some taxa), has fascinated generations of museum visitors and professional palaeontologists alike. Over 100 valid sauropod genera are recognized currently, with representatives recovered from almost every continent (excluding Antarctica), from deposits of Late Triassic to latest Cretaceous age (Upchurch *et al.* 2004). Work over the past 30 years has greatly advanced our knowledge of sauropod phylogeny, biomechanics, physiology and ecology, offering insights into the limits of life at large body size and contributing to more general debates on dinosaur diversity, palaeobiology and palaeoecology (Upchurch *et al.* 2004; Curry Rogers and Wilson 2005; Tidwell and Carpenter 2005). Most work on sauropods to date has concentrated on eusauropods; however, more basal taxa, such as *Vulcanodon*, remained poorly known owing to the fact that they were represented by incomplete specimens, thereby frustrating attempts to investigate the early evolutionary history of the group. This has severely hampered our understanding of the sequence of character acquisitions that ultimately resulted in the specialized feeding, locomotory and axial systems seen in derived sauropods. Similarly, the evolutionary origins of gigantism in the group were obscure.

However, the past few years have witnessed several revolutions in our knowledge of early sauropod evolution. These revolutions have been driven by two main factors. The first of these is the discovery of new basal sauropod taxa from deposits of Late Triassic–Early Jurassic age, an interval that had previously yielded almost no sauropod material. The emergence of some of these taxa (e.g. *Isanosaurus* and *Gongxianosaurus*) resulted from continuing fieldwork (He *et al.* 1998; Buffetaut *et al.* 2000), whereas others (e.g. *Antetonitrus*, the 'Lufeng sauropod' and *Chinshakiangosaurus*) were 'found' as a result of re-evaluating material already present in museum collections (Barrett 1999; Yates and Kitching 2003; Upchurch *et al.* in press). All of these taxa present novel combinations of primitive and derived character states and their inclusion in sauropod phylogenies has led to the recognition of novel tree topologies and has started to illuminate the sequence of character acquisition that occurred in basal sauropods (e.g. Yates and Kitching 2003; Upchurch *et al.* in press). The second major factor has been an unprecedented increase in interest in a group of dinosaurs that had been severely neglected for the past century: the prosauropods.

Prosauropods were among the first dinosaurs to be described scientifically, with *Thecodontosaurus*, from the Late Triassic fissure fills of the Bristol Channel region of the UK, being the first to be named (Riley and Stutchbury 1836). Other early discoveries of prosauropods in Germany (reviewed in Moser 2003) and South Africa (Owen 1854) provided numerous well-preserved specimens. However, associations of prosauropod material with teeth from carnivorous archosaurs led to much taxonomic confusion and for many years some prosauropods were allied with the faunivorous theropod dinosaurs. Nevertheless, some early workers (e.g. von Huene 1920, 1932) did recognize that at least some prosauropods were closely related to sauropods, leading to proposal of the name Sauropodomorpha von Huene, 1932 for a group consisting of sauropods and some prosauropods. New prosauropod taxa continued to be described and named during the late 19th and 20th centuries, particularly from Germany (von Huene 1932), China (reviewed in Young 1951) and Argentina (e.g. Bonaparte 1972). However, only a handful of specialists have worked on this material (F. von Huene in Germany, C.-C. Young in China, J. F. Bonaparte in Argentina and P. M. Galton in the USA being foremost among them) and prosauropods have generally been regarded as 'poor relations' of the larger and more spectacular sauropods by the majority of dinosaur palaeontologists. Possible reasons for this might include, but are probably not limited to, the relatively conservative anatomy of prosauropods, which lack the horns, frills, armour and other appendages often found in other, more charismatic, dinosaur clades, and the discovery of new material in regions remote from research institutions in North America and Europe. This is unfortunate as large amounts of excellent material were available for study that could have contributed to the debates that shaped our current model of dinosaur biology. Moreover, prosauropods dominated Late Triassic and Early Jurassic terrestrial ecosystems (comprising 95 per cent of the standing vertebrate biomass in some faunas), attained a global distribution, and achieved a reasonably high level of species richness (Galton 1990; Galton and Upchurch 2004). They ranged in size from under 1 m in length (*Thecodontosaurus*) to over 10 m (*Riojasaurus*), embraced diets encompassing omnivory to obligate high-browsing herbivory, and lived in diverse environments that included sand-seas and broad, well-watered flood-plains. By all measures they were successful animals, yet work on prosauropods has lagged behind that on other dinosaur clades.

Early non-numerical cladistic studies on dinosaurs confirmed the monophyly of Sauropodomorpha (Gauthier 1986), a conclusion now supported by computer-based analyses (e.g. Sereno 1999; Langer 2004). However, there has been little consensus on whether prosauropods form a monophyletic clade that is the sister-group of Sauropoda or a paraphyletic grade consisting of taxa that represent successively distant outgroups to the latter clade (Gauthier 1986; Galton 1990; Sereno 1999; Benton *et al.* 2000; Yates 2003, 2004; Yates and Kitching 2003; Galton and Upchurch 2004; Langer 2004). Although there is still a lack of consensus over this issue, many of these analyses demonstrated that at least some 'prosauropod' taxa (e.g. *Blikanasaurus*) were in fact more closely related to sauropods than to other 'prosauropods', reinvigorating debates over early sauropod phylogeny, which in turn are having an important impact on studies dealing with feeding, locomotion, ecology and diversity in Sauropodomorpha. As shown by some of the papers in this volume, a large number of animals that were formerly 'prosauropods' (e.g. *Melanorosaurus*) are now thought to be part of early sauropod history. Work on early sauropodomorph dinosaurs is now proceeding at an unprecedented rate, with new work on prosauropod biology even gracing the pages of *Science* (Reisz *et al.* 2005; Sander and Klein 2005).

This volume arose from a symposium entitled 'The evolution of giants: tracing the early evolution of sauropodomorph dinosaurs' that was held as a part of the 65th Annual Meeting of the Society of Vertebrate Paleontology, in Mesa, Arizona, on 22 October 2005. The aim of this symposium (co-convened by Paul M. Barrett and Tim J. Fedak) was to allow dinosaur workers specializing on prosauropod and basal sauropod dinosaurs to meet and present results on varied projects that ranged from phylogeny to functional morphology to taphonomy. This breadth of topics is reflected in the diversity of papers presented in this volume.

Many of the contributions provide anatomical descriptions that form the basis for new palaeobiological analyses. Langer *et al.* give the first detailed account of the forelimb structure of *Saturnalia*, the most primitive sauropodomorph from the Carnian of Brazil, and use this as a basis for reconstructing the forelimb musculature of the animal and commenting on its locomotory biology. Bonnan and Senter also concentrate on forelimbs, providing a novel analysis of limb function in the prosauropods *Massospondylus* and *Plateosaurus*. A third paper, by Bonnan and Yates, examines the forelimb anatomy of *Melanorosaurus*, a taxon now regarded by many workers as an early member of the sauropod lineage, and its implications for the evolution of the characteristic pronated limb posture of sauropod dinosaurs.

Upchurch *et al.* propose a new sauropodomorph phylogeny, which identifies a series of basal taxa, a monophyletic group of 'core' prosauropods and a pectinate array of taxa leading to Sauropoda. This phylogeny underpins the paper by Barrett and Upchurch that examines the evolution of sauropodomorph feeding mechanisms and their possible ecological significance. In addition, Yates provides a new phylogenetic study, in which the majority of 'prosauropod' taxa form successive outgroups to Sauropoda: this paper also contains a detailed description of the skull of *Melanorosaurus*, which takes on new significance as the earliest known skull of a sauropod dinosaur. Wedel discusses the evolution and possible functional roles of postcranial pneumaticity in dinosaurs, with special reference to the situation in basal sauropods and 'prosauropods', while Klein and Sander present an exceptionally detailed study of bone histology in *Plateosaurus* that yields many new insights on growth patterns and growth rates in early dinosaurs. Pol and Powell provide much needed information on the anatomy of the early sauropod *Lessemsaurus* from the Late Triassic of Argentina, while Fedak and Galton add new details on the braincase of the American taxon *Anchisaurus*. The vexed taxonomic questions relating to the validity and synonymy of *Anchisaurus* and *Ammosaurus* are addressed by Sereno, who also reviews work on sauropodomorph systematics to date.

We hope that this volume will serve as a benchmark for future work on prosauropods and basal sauropods. Although they have been neglected in the past, we hope that this selection of new work proves that they are worthy of interest and are contributing to some of the most active and constructive debates in dinosaur studies.

Acknowledgements. Our sincere thanks go to all of the participants who took part in 'The evolution of giants' symposium, especially to those who contributed to this volume. We thank the following individuals for their help in reviewing manuscripts: M. Benton, M. Bonnan, M. Carrano, A. Chinsamy, L. Claessens, G. Erickson, J. Farlow, J. Harris, D. Henderson, D. Hone, J. Hutchinson, M. Langer, D. Pol, P. Senter, P. Upchurch, M. Vickaryous, D. Weishampel, J. Wilson and A. Yates. S. Modesto acted as the Handling Editor for papers co-authored by PMB. We thank Professor D. Batten, the Senior Editor of *Special Papers in Palaeontology*, for his support and technical assistance throughout the production of this volume and the Palaeontological Association for its interest in this project. Finally, we offer our grateful thanks to the Society of Vertebrate Paleontology for hosting the symposium at their 65th Annual Meeting in Mesa.

REFERENCES

BARRETT, P. M. 1999. A sauropod dinosaur from the Lower Lufeng Formation (Lower Jurassic) of Yunnan Province, People's Republic of China. *Journal of Vertebrate Paleontology*, 19, 785–787.

BENTON, M. J., JUUL, L., STORRS, G. W. and GALTON, P. M. 2000. Anatomy and systematics of the prosauropod dinosaur *Thecodontosaurus antiquus* from the Upper Triassic of southwest England. *Journal of Vertebrate Paleontology*, 20, 77–108.

BONAPARTE, J. F. 1972. Los tetrapodos del sector superior de la Formacion Los Colorados, La Rioja, Argentina (Triásico Superior). *Opera Lilloana*, 22, 1–183.

BUFFETAUT, E., SUTEETHORN, V., CUNY, G., TONG HAI-YAN, LE LOEUFF, J., KHANSUBHA, S. and JONGAUTCHARIYAKUL, S. 2000. The earliest known sauropod dinosaur. *Nature*, 407, 72–74.

CURRY ROGERS, K. A. and WILSON, J. A. 2005. *The sauropods: evolution and paleobiology*. University of California Press, Berkeley, CA, 349 pp.

GALTON, P. M. 1990. Basal Sauropodomorpha–Prosauropoda. 320–344. *In* WEISHAMPEL, D. B., DODSON, P. and OSMÓLSKA, H. (eds). *The Dinosauria*. University of California Press, Berkeley, CA, 733 pp.

—— and UPCHURCH, P. 2004. Prosauropoda. 232–258. *In* WEISHAMPEL, D. B., DODSON, P. and OSMÓLSKA, H., (eds). *The Dinosauria*. Second edition. University of California Press, Berkeley, CA, 861 pp.

GAUTHIER, J. 1986. Saurischian monophyly and the origin of birds. *Memoirs of the California Academy of Sciences*, 8, 1–55.

HE XIN-LU, WANG CHANG-SHENG, LIU SHANG-ZHONG, ZHOU FENG-YUN, LIU TU-QIANG, CAI KAI-JI and DAI BING 1998. A new sauropod dinosaur from the Early Jurassic in Gongxian County, south Sichuan. *Acta Geologica Sichuan*, 18, 1–6, pl. 1. [In Chinese, English abstract].

HUENE, F. von 1920. Bemerkungen zur systematic und stammesgeschichte einiger reptilien. *Zeitschrift für Inducktive Abstammungs- und Vererbungslehre*, 24, 162–166.

—— 1932. Die fossile Reptil-Ordnung Saurischia, ihre Entwicklung und Geschichte. *Monographie Geologie Paläontologie*, 1, 1–361.

LANGER, M. C. 2004. Basal Saurischia. 25–46. *In* WEISHAMPEL, D. B., DODSON, P. and OSMÓLSKA, H. (eds). *The Dinosauria*. Second edition. University of California Press, Berkeley, CA, 861 pp.

MOSER, M. 2003. *Plateosaurus engelhardti* Meyer, 1837 (Dinosauria: Sauropodomorpha) aus dem Feuerletten (Mittelkeuper; Obertrias) von Bayern. *Zitteliana (Reihe B, Abhandlungen der Bayerischen Staatssammlung für Paläontologie und Geologie)*, 24, 3–186.

OWEN, R. 1854. *Descriptive catalogue of the fossil organic remains of Reptilia and Pisces contained in the museum of the Royal College of Surgeons of England*. Taylor and Francis, London, 184 pp.

REISZ, R. R., SCOTT, D., SUES, H.-D., EVANS, D. C. and RAATH, M. A. 2005. Embryos of an Early Jurassic prosauropod dinosaur and their evolutionary significance. *Science*, 309, 761–764.

RILEY, H. and STUTCHBURY, S. 1836. A description of various remains of three distinct saurian animals, recently discovered in the Magnesian Conglomerate near Bristol. *Proceedings of the Geological Society of London*, 2, 397–399.

SANDER, P. M. and KLEIN, N. 2005. Unexpected developmental-plasticity in the life history of an early dinosaur. *Science*, 310, 1800–1802.

SERENO, P. C. 1999. The evolution of dinosaurs. *Science*, 284, 2137–2147.

TIDWELL, V. and CARPENTER, K. 2005. *Thunder-lizards: the sauropodomorph dinosaurs*. Indiana University Press, Bloomington, IN, 495 pp.

UPCHURCH, P., BARRETT, P. M. and DODSON, P. 2004. Sauropoda. 259–322. *In* WEISHAMPEL, D. B., DODSON, P. and OSMÓLSKA, H. (eds). *The Dinosauria*. Second edition. University of California Press, Berkeley, CA, 861 pp.

—— —— ZHAO XI-JIN and XU XING in press. A re-evaluation of *Chinshakiangosaurus chunghoensis* Ye *vide* Dong 1992 (Dinosauria, Sauropodomorpha): implications for cranial evolution in basal sauropod dinosaurs. *Geological Magazine*.

YATES, A. M. 2003. A new species of the primitive dinosaur *Thecodontosaurus* (Saurischia: Sauropodomorpha) and its implications for the systematics of early dinosaurs. *Journal of Systematic Palaeontology*, 1, 1–42.

—— 2004. *Anchisaurus polyzelus* (Hitchcock): the smallest known sauropod dinosaur and the evolution of gigantism among sauropodomorph dinosaurs. *Postilla*, 230, 1–57.

—— and KITCHING, J. W. 2003. The earliest known sauropod dinosaur and the first steps towards sauropod locomotion. *Proceedings of the Royal Society of London, Series B*, 270, 1753–1758.

YOUNG CHUNG-CHIEN 1951. The Lufeng saurischian fauna. *Palaeontologica Sinica, Series C*, 13, 1–96, pls 1–12.

[Special Papers in Palaeontology 77, 2007, pp. 9–55]

THE FIRST COMPLETE SKULL OF THE TRIASSIC DINOSAUR *MELANOROSAURUS* HAUGHTON (SAUROPODOMORPHA: ANCHISAURIA)

by A D A M M. Y A T E S

Bernard Price Institute for Palaeontological Research, University of the Witwatersrand, Private Bag 3, Johannesburg, WITS 2050, South Africa; e-mail: yatesa@geosciences.wits.ac.za

Typescript received 1 February 2006; accepted in revised form 20 July 2006

Abstract: The skull and atlas-axis complex of an articulated sauropodomorph skeleton (NM QR3314) from the Late Triassic lower Elliot Formation of South Africa is described. The specimen is identified as the poorly known taxon *Melanorosaurus readi* Haughton. A revised diagnosis for this taxon is proposed and NM QR3314 is referred to it on the basis of several sacral autapomorphies, most notably the presence of five sacral vertebrae. The skull and atlas-axis complex reveal several autapomorphic characters that can be used to diagnose the taxon further. These are: a transversely broad internarial bar; an enlarged premaxilla; a short ridge on the dorsolateral surface of the posterior end of the maxilla; loss of the anteroventral process of the nasal; an elongate post-erolateral process of the nasal; a dorsoventrally expanded anterior end of the jugal; an elongate vomer; an anteroposteriorly shortened fossa on the posterior ventral margin of the basiparasphenoid plate and a broad, dorsoventrally shallow anterior neural canal. The skull also displays an intriguing mix of derived characters, several of which support a position close to Sauropoda. Other characters support a closer relationship to *Plateosaurus* than to Sauropoda but these are interpreted as a mix of convergences and symplesiomorphies following the phylogenetic relationships presented herein.

Key words: *Melanorosaurus*, Sauropodomorpha, skull, Late Triassic, South Africa, Elliot Formation.

T H E Late Triassic lower Elliot Formation of South Africa and Lesotho has been known to contain dinosaurs for almost as long as dinosaurs have been recognized as a distinct group. The first to be named, *Euskelosaurus browni* Huxley, 1866, was described from postcranial fragments sent back to England by the pioneering South African collector Alfred Brown (Van Heerden 1979; Cooper 1980; Galton and Van Heerden 1998). Despite this early start, the dinosaurs of the lower Elliot Formation have remained frustratingly difficult to characterize and classify, owing to a paucity of reasonably complete specimens and the almost total absence of skulls from the unit. Thus, the discovery of a largely complete and articulated skeleton (NM QR3314) with a complete skull in the lower Elliot Formation is of considerable importance. The discovery is made all the more significant by its identification as *Melanorosaurus readi* Haughton, 1924 (Galton *et al.* 2005).

Melanorosaurus readi is a well-known dinosaur in the sense that the name is frequently used. However, like most other dinosaurs from the lower Elliot Formation, its anatomy remains poorly known and its relation-ships with other dinosaurs are contentious (Yates 2003*a*; Galton *et al.* 2005). Nevertheless, it has been prominent in discussions on the origin of the gigantic sauropod dinosaurs for many years (e.g. Romer 1956; Charig *et al.* 1965; Cruickshank 1975; Yates 2003*b*; Yates and Kitching 2003). Although it is usually classified as a prosauropod, several workers have suggested that *Melanorosaurus* shares a more recent common ancestry with sauropods than with typical 'prosauropods' such as *Plateosaurus* and *Massospondylus*. If true it would therefore represent an early stage on the lineage leading to the highly specialized sauropods. Both sauropod (Yates 2003*b*, 2004; Yates and Kitching 2003) and prosauropod (Galton and Upchurch 2004) affinities for *M. readi* have been supported in recent cladistic analyses.

In this paper the complete skull and atlas-axis complex of NM QR3314 are described (Text-figs 1–18). The skull characters that diagnose the species are discussed, as is its position in sauropodomorph phylogeny, in the light of these new data. However, it is necessary first to discuss the morphology of the proximal caudal vertebrae and the sacrum in order to justify the referral of NM QR3314 to *M. readi*.

Institutional abbreviations. AM, Amherst College Museum, Amherst, Massachusetts; BMNH, Natural History Museum, London; BP, Bernard Price Institute for Palaeontological Research, University of the Witwatersrand, Johannesburg; BRSMG, Bristol City Museum and Art Galleries, Bristol; BRSUG, Department of Earth Sciences, University of Bristol, Bristol; CM, Carnegie Museum, Pittsburgh; GPIT, Institut und Museum für Geologie und Paläontologie der Universität Tübingen, Germany; IVPP, Institute for Vertebrate Paleontology and Paleoanthropology, Beijing; LV, Museum of Lufeng Dinosaurs, Jingshan; MACN, Museo Argentino de Ciencias Naturales, Buenos Aires; MB, Museum für Naturkunde der Humboldt Universität, Berlin; MCP, Museu de Ciências e Tecnologia, Pontifícia Universidade Católica do Rio Grande do Sul, Porto Alegre; MOR, Museum of the Rockies, Bozeman; NM, National Museum, Bloemfontein; PULR, Museo de Paleontología, Universidad Provincial de La Rioja, La Rioja; PVL, Fundación Miguel Lillo, Universidad Nacional de Tucumán, San Miguel de Tucumán; QG SAM, South African Museum, Iziko Museums, Cape Town; SMNS, Staatliches Museum für Naturkunde, Stuttgart; TATE, Tate Geological Museum, Casper College, Casper, Wyoming; TM, Transvaal Museum, Pretoria; UCMP, Univeristy of California Museum of Paleontology, Berkeley; YPM, Peabody Museum of Natural History, Yale University, New Haven.

Anatomical abbreviations. af, articulating facet; amf, anterior maxillary foramen; amp, anteromedial process of the maxilla; an, angular; aoc, articular surface for the occipital condyle; aof, antorbital fossa; aofen, antorbital fenestra; ar, articular; atr, atlantal rib; aur, auricular recess; axi, axial intercentrum; bo, basioccipital; bpt, basipterygoid process; bs-ps, basiparasphenoid complex; bt, basal tuber; ci, crista interfenestralis; cp, crista prootica; ct, crista tuberalis; d, dentary; ds, dorsal surface; dt, dentary tooth; ec, ectopterygoid; emf, external mandibular fenestra; ep, epipterygoid; epi, epiphysis; ex-op, exoccipital opisthotic complex; f, frontal; fm, foramen magnum; fo, fenestra ovalis; II–XII, foramina for cranial nerves II–XII; inf, internasal fenestra; itf, infratemporal fenestra; j, jugal; jf, foramen for jugular vein; lac, lachrymal; lp, lateral process of the laterosphenoid; ls, laterosphenoid; mf, metotic fissure; mx, maxilla; n, nasal; nc, neural canal; ncs, neurocentral suture; nf, narial fossa; np, neurapophysis; ns, neural spine; o, orbit; oc, occipital condyle; od, odontoid; os, sulcus for the ophthalmic nerve (V1); p, parietal; pa, prearticular; pal, palatine; pd, perilymphatic duct; pe, pedicel of neurapophysis; plp, posterolateral process of the premaxilla; pls, posterolateral spur of the nasal; pmx, premaxilla; po, postorbital; pop, paroccipital process; poz, postzygapophysis; pp, parapophysis; ppf, postparietal fenestra; pr, prootic; pra, proatlas; prf, prefrontal; prz, prezygapophysis; psp, parasphenoid process; pt, pterygoid; ptf, post-temporal fenestra; q, quadrate; qf, quadrate foramen; qj, quadratojugal; rap, retroarticular process; sa, surangular; sc, semicircular canal; se, symphyseal expansion; sg, stapedial groove; snf, subnarial foramen; so, supraoccipital; sp, splenial; spf, splenial foramen; sq, squamosal; st, stapes; stf, supratemporal fossa; vcmf, foramen for mid cerebral vein; ve,

vestibule; vl, anteroventral lappet; vo, vomer; vp, ventral process of proatlas.

MATERIAL

A team from the National Museum, Bloemfontein, excavated NM QR3314 in 1994 from the farm Damplaats in the Ladybrand District, Free State, South Africa. The excavation site is well below the disconformity that marks the boundary between the lower and upper Elliot Formation. This places the specimen in the Norian Stage of the Late Triassic (Lucas and Hancox 2001). The specimen was buried in red overbank mudstones typical of the Elliot Formation. Although articulated the skeleton is poorly preserved. The bones are heavily cracked and have undergone a small degree of explosive deformation caused by the swelling of clay minerals that have infiltrated the cracks. If this cracking was caused by exposure to the sun then it is likely that the individual died during a severe drought when there were no scavengers active in the area to disturb the articulation of the skeleton. The specimen may have been a juvenile when it died because the posterior dorsal vertebrae have open neurocentral sutures. This is supported by its size (the estimated femur length is 530 mm), which is smaller than other specimens of this species, including the paralectotype, SAM-PK-K3450 (femur length 604 mm) and NM 1551 (femur length 645 mm).

The skull has also undergone dorsoventral compression, which seems to have caused dislocation of several bones and localized deformation of the vertical pillars of the skull (Text-figs 1–2). It has also caused the ventral ends of the quadrates to rotate in towards the midline (Text-fig. 4), thus narrowing the posterior palate and forcing the pterygoids into midline contact (Text-fig. 5).

The specimen has received only cursory attention in the literature. Welman (1999) described its basicranium, referring to it as an unidentified prosauropod, while Galton *et al.* (2005) listed it as a referred specimen of *Melanorosaurus readi*.

REFERRAL OF NM QR3314 TO *MELANOROSAURUS READI*

As stated above, the syntype series of *Melanorosaurus readi* includes no cranial material. Therefore, NM QR3314 can only be identified as this species on the basis of its postcranial skeleton. However, the poor preservation and incomplete preparation of the postcranium of NM QR3314 hampers direct comparison with the syntypes.

The syntype series consists of a number of postcranial bones catalogued under two numbers (SAM-PK-K3449 and 3450). The accessions catalogue at SAM records that SAM-PK-K3450 was collected 20 yards (*c.* 18 m) east of the collection site for SAM-PK-K3449. In addition Haughton (1924) wrote that the femur (which is part of SAM-PK-K3450) was derived from the sandstone overlying the red mudstone unit that contained the other bones of both SAM-PK-K3449 and 3450. Unfortunately, since this time a lot of extraneous material has been accessioned under both of these numbers, including a dentigerous jaw fragment of the cynodont *Cynognathus*, presumably from the Early Triassic Burgersdorp Formation, and sauropodomorph bones with a clearly different style of preservation, some of which are recognizable as originally belonging to SAM-PK-K3532. The latter is a poorly preserved composite specimen of uncertain systematic position (Galton *et al.* 2005). Compounding the problem is the lack of a precise list of the bones included in the original collections. A thorough description and evaluation of the composition of the syntype series will be the subject of a future work (Yates, Galton and Van Heerden in prep.). It is sufficient to point out here that a core sample of bones can be recognized as belonging to the original lot of SAM-PK-K3449 because they can be matched to bones that were figured and described by Haughton (1924), or share the distinctive style of preservation with these bones. This core group of bones may represent a single incomplete skeleton and should be treated as the lectotype of *Melanorosaurus readi*. It should be noted that this excludes the femur (and other bones catalogued as SAM-PK-K3450). This was clearly the intention of Haughton who, when erecting the taxon, stated that the femur was 'in doubtful association with the other remains' (Haughton 1924, p. 433). Several authors have stated that the only diagnostic bone in the syntype series is the femur (Van Heerden 1979; Galton 1998; Yates 2003*a*); however, the proximal caudal vertebrae of SAM-PK-K3449 are diagnostic. They display the following unique combination of character states: proximodistally compressed centra with tall ovoid faces; median ventral furrows; and short hyposphenal ridges. Unfortunately these characters cannot be determined in NM QR3314 because the proximal caudal vertebrae are poorly preserved, are preserved in articulation and have their ventral surfaces embedded in plaster in the display case. Thus, NM QR3314 cannot be referred to *M. readi* by direct comparison with SAM-PK-K3449, although it should be noted that the ulnae of the two specimens are virtually identical (Bonnan and Yates 2007). Nevertheless NM QR3314 can be referred to *M. readi* with confidence by comparison with NM 1551, which is also referred to *M. readi*.

Van Heerden and Galton (1997) first referred NM 1551 to *M. readi* on the basis of similarities between the femur of this specimen and that of the syntype series. Both femora are almost straight in anterior view, gently sinuous in lateral view, and have the fourth trochanter on the

medial margin of the shaft in posterior view. In addition, the lesser trochanter is located on the lateral margin of the shaft so that it is visible in posterior view, and the shaft has an elliptical cross-section below the fourth trochanter in each specimen. This combination of characters is unusual, but not unique, among early sauropodomorphs and can also be found in *Antetonitrus ingenipes* (Yates and Kitching 2003, fig. 2c) and probably *Lessemsaurus sauropoides* (Pol and Powell 2007). Furthermore, the femur of SAM-PK-K3450 does not belong to the lectotype of *M. readi* as discussed above. Nevertheless, it is clear that NM 1551 has the same distinctive type of proximal caudal vertebrae as SAM-PK-K3449 and this confirms the identification of NM 1551 as *M. readi*. NM 1551 also has a distinctive sacrum that has been used to diagnose *M. readi* (Galton and Upchurch 2004). The sacrum of NM QR3314 is largely complete, moderately well preserved, well exposed and can be compared with that of NM 1551.

The sacrum of NM 1551 has been described as consisting of four vertebrae: a dorsosacral, two primordial sacrals and a caudosacral (Van Heerden and Galton 1997). I disagree with this interpretation and believe that the preserved sacral vertebrae consist of two dorsosacral vertebrae followed by two primordials. This reinterpretation is based on the morphology of the last two sacrals in the sequence (first and second primordials in this paper, previously thought to be the second primordial and a caudosacral) and comparisons with the sacrum of NM QR3314. The second primordial is a distinctive sacral element that can be recognized in other early saurischians by its posterolaterally projecting transverse processes in dorsal view, the *en echelon* shape of the articular surface of the sacral rib in lateral view, and the projecting anterior margin of the transverse process that partly roofs the intercostal space between the first and second primordial ribs. The fourth sacral in NM 1551 displays all of these features and is identified as the second primordial sacral. Moser (2003) has proposed that the presence of paramedian depressions on the ventral surface of the centrum and the presence of a geniculation at the suture between the sacral rib and the centrum are diagnostic of the second primordial sacral, but these features do not apply in the case of *M. readi*. None of the sacral centra of NM 1551 has paramedian depressions, while both the third and the fourth sacrals in the series display geniculations at the sutural contact of the sacral rib to the centrum.

Thus, the sacrum of NM 1551 is interpreted as consisting of two dorsosacrals in front of the two primordials. Nevertheless, a caudosacral that was not preserved or collected is also believed to have been present. Prominent scars on the dorsolateral corners of the posterior surface of the second primordial sacral rib show where the ribs of a caudosacral would have attached to the sacrum. Fitting the ilium to the assembled sacrum shows that the second sacral rib extended for a considerable distance along the postacetabular process leaving little space for the caudosacral to contact the ilium. Thus, the caudosacral probably attached largely, or wholly, to the second primordial sacral. Five sacral vertebrae is a unique condition among basal sauropodomorphs and it is regarded as an autapomorphy of *M. readi*. It is convergently developed in derived eusauropods (*Patagosaurus* + (Omeisauridae + (*Jobaria* + Neosauropoda))) (Wilson 2002), but there is abundant evidence that *M. readi* is neither part of, nor especially closely related to, this clade (see below).

A further autapomorphy of the sacrum of NM 1551 relates to the first primordial sacral. In this sacral the articular surface that contacted the ilium is divided into two parts, a dorsal section and a more extensive ventral section that joins with the articular surfaces of the second dorsosacral in front and the second primordial sacral behind. The sacral rib is emarginated between these two surfaces so that there is a small elliptical fenestra visible in anterior or posterior view when the rib is articulated with the ilium. When the non-articulating emarginated surface is viewed laterally a deep pit bordered anteriorly and posteriorly by thin laminae of bone, can be seen penetrating the sacral rib. The presence of a non-articulating gap between the dorsal and ventral sections of the articular surface of the first primordial sacral is not unique among basal sauropodomorphs and can also be seen in *Riojasaurus incertus* (PVL 3805). However, the deep lateral pit has not been seen in any other taxon and is an autapomorphy of *M. readi*.

The sacrum of NM QR3314 clearly displays both of the autapomorphies discussed above and we can be confident of its referral to *M. readi*. In this specimen the caudosacral is present and clearly articulates with the posterior surface of the second primordial without making contact with the ilium. Only three sacral vertebrae are preserved anterior to the caudosacral (the second dorsosacral and the two primordial sacrals) on the sacral block, but the anterolateral surfaces of the second dorsosacral are clearly fresh breaks indicating that the neural arch of the first dorsosacral has broken away. Its centrum is present on display with the rest of the skeleton in the galleries of the National Museum (though it is disassociated from the sacral block) but the neural arch is missing.

Revised diagnosis for Melanorosaurus readi

Few explicit diagnoses for *Melanosaurus readi* have been proposed. Gauffre (1993) noted that a fourth trochanter placed at the midlength of the femoral shaft was an apomorphic character that linked *Melanorosaurus readi* with his *Melanorosaurus thabanensis* (a doubtfully valid

taxon of basal sauropod, or near-sauropod grade sauropodomorph) and so is an implicit diagnosis for the genus. However, most sauropods also display this condition (e.g. *Antetonitrus ingenipes*: Yates and Kitching 2003, fig. 2c; *Cetiosaurus oxoniensis*: Upchurch and Martin 2003, fig. 12a) indicating that this character probably diagnoses a clade containing *Melanorosaurus* and Sauropoda. Galton and Upchurch (2004) diagnosed *Melanorosaurus readi* on the basis of its four sacral vertebrae. As explained above, I disagree with Van Heerden and Galton's (1997) interpretation of the sacrum of *Melanorosaurus readi* but agree that the sacrum is unique and include several characters from it in the following, revised diagnosis:

A sauropodomorph displaying the following autapomorphies: a transversely broad internarial bar; an enlarged premaxilla; a short ridge on the dorsolateral surface of the posterior end of the maxilla; loss of the anteroventral process of the nasal; an elongate posterolateral process of the nasal; anterior end of the jugal dorsoventrally expanded relative to the suborbital bar; an elongate vomer (over 30 per cent of the total skull length); an anterposteriorly shortened fossa on the posterior ventral margin of the basiparasphenoid plate; a broad, dorsoventrally shallow anterior neural canal; five sacral vertebrae with two dorsosacrals, two primordial sacrals and a caudosacral; a deep pit bounded by laminae penetrating the lateral surface of the first primordial sacral rib; caudosacral attaching largely, or wholly, to the posterior surface of the second primordial sacral vertebra; ovoid face of the proximal caudal vertebra at least 1·42 times higher than wide.

DESCRIPTION

Skull roof

Premaxilla. The premaxilla consists of a rectangular main body with a dorsal internarial process and a large posterolateral process. The pair meet in an acute anterior symphysis, forming a pointed snout in dorsal view. The premaxillary body is relatively large compared with other early sauropodomorphs with its dentigerous portion occupying 32 per cent of the total length of the upper tooth row. The premaxillary bodies form a projecting muzzle that is offset from the dorsal process of the premaxilla by a gentle inflection. The dorsal ends of the dorsal processes are poorly preserved, but are clearly broad and dorsoventrally flattened. By contrast, the base of the dorsal process is transversely compressed. The cross-section of each dorsal process at its base is close to three times longer anteroposteriorly than they are mediolaterally wide.

Despite the large size of the premaxillae their teeth are not enlarged relative to the anterior maxillary teeth. The premaxillae also retain the primitive sauropodomorph number of four teeth. Consequently, the premaxillary tooth row contains large gaps between the teeth, which are at least equal to the length of one

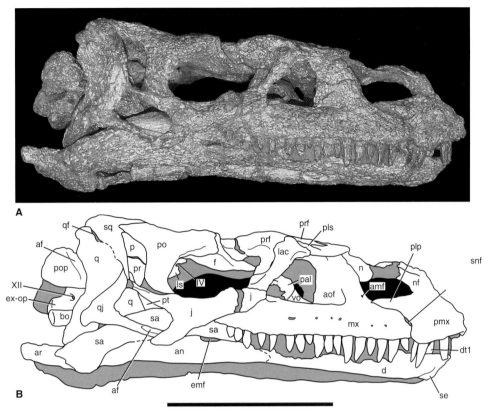

TEXT-FIG. 1. *Melanorosaurus readi*, NM QR3314. Skull and mandible in right lateral view. A, photograph. B, interpretive drawing. Grey areas represent matrix and bones from the left side of the skull. For explanation of abbreviations, see text. Scale bar represents 100 mm.

alveolus. There is also a short toothless gap between the first tooth and the symphysis. A pair of small ventral projections arises from the alveolar margin, one on each side of the midline, in the symphyseal gap. The lateral surface of the premaxillary body is pierced by a foramen, set in a depression, above the second tooth. The narial fossa is poorly impressed onto the premaxilla and a distinct margin is only visible on the dorsal process. The posterolateral process is an elongate tapering process that forms the ventral floor of the external naris. Its long axis is horizontal and lies on the dorsal surface of the anterior ramus of the maxilla, terminating at the base of the ascending process of the maxilla. A foramen exits from the base of the posterolateral process on its medial margin. The right-angled junction of the posterolateral process with the posterior margin of the premaxillary body creates an L-shaped suture between the premaxilla and maxilla. The subnarial foramen is an elliptical gap centred just below the right-angled bend of this L-shaped suture.

Medially a pair of posterior processes is in midline contact and visible in palatal view. They extend underneath the anteromedial processes of the maxilla to contact the anterior tip of the vomers. Similar processes are also present in *Plateosaurus engelhardti* (MB XXIV), *Massospondylus carinatus* (Barrett and Yates 2006, fig. 3) and *Yunnanosaurus huangi* (IVPP V505).

Maxilla. The maxilla is the typical elongate, triradiate element seen in other non-eusauropod sauropodomorphs. The right side

has a full complement of 19 imbricated teeth that extend to the posterior tip of the maxilla, below the orbit. The short anterior ramus is offset from the ascending process by a strong inflection in the anterior profile of the maxilla. Unlike many basal sauropodomorphs, such as *Efraasia minor* (Galton 1985*a*, pl. 4, fig. 1), *Plateosaurus engelhardti* (Galton 1984*a*, fig. 3a), *Riojasaurus incertus* (PULR 56), *Coloradisaurus brevis* (PVL 3967), *Lufengosaurus huenei* (Barrett *et al.* 2005*a*, fig. 2b) and *Jingshanosaurus xinwaensis* (LV 3), the anterior ramus is as deep as it is long. In this respect it matches the condition in *Yunnanosaurus huangi* (IVPP V505), some specimens of *Massospondylus carinatus* (e.g. BP/1/4934) and eusauropods (e.g. *Camarasaurus lentus*; Madsen *et al.* 1995, fig. 10). The anterodorsal corner is curved slightly towards the medial side so that it faces dorsolaterally. This region forms part of the ill-defined narial fossa. An anterior maxillary foramen opens on the anterior ramus of the maxilla, medial to the tip of the posterolateral process of the premaxilla, well inside the narial fossa.

The ascending ramus is unusually broad owing to a wide medial wall of the antorbital fossa. Most early sauropodomorphs have a narrow, crescent-shaped antorbital fossa on the ascending process of the maxilla, but in NM QR3314, as in *Unaysaurus tolentinoi* (Leal *et al.* 2004, fig. 2b), *Plateosaurus engelhardti* (Galton 1985*a*, fig. 4b) and *Coloradisaurus brevis* (PVL 3967), the antorbital fossa is broad and subtriangular with a nearly straight posterior margin. However, in *Plateosaurus engelhardti* and

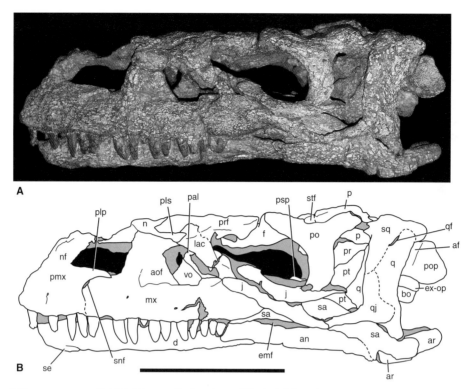

TEXT-FIG. 2. *Melanorosaurus readi*, NM QR3314. Skull and mandible in left lateral view. A, photograph. B, interpretive drawing. Grey areas represent matrix and bones from the left side of the skull. For explanation of abbreviations, see text. Scale bar represents 100 mm.

Unaysaurus tolentinoi the anterior rim of the antorbital fossa remains sharp and raised above the medial wall along its entire length, whereas in NM QR3314 (and *Coloradisaurus brevis*) the rim is a rounded ridge that fades out a short distance above the anteroventral corner of the antorbital fossa. Above that level the anterior margin of the antorbital fossa is marked by nothing more than a gentle change in slope of the lateral surface of the ascending ramus. Owing to the lack of an anteroventral process of the nasal the ascending ramus of the maxilla forms the entire posterior margin of the external naris. A narrow, pointed process extends posteriorly from the posterodorsal corner of the ascending process to underlap the anterior ramus of the lachrymal. Distortion of the skull on the right side has separated the lachrymal and the maxilla, revealing a sulcus on the lateral side of this posterodorsal process for reception of the anterior ramus of the lachrymal.

The posterior ramus of the maxilla forms the main body of the bone. In lateral view, the dorsal and ventral margins are parallel until it reaches the jugal, at which point it abruptly tapers beneath the jugal. A short dorsolateral ridge projects forward from the anterior end of the jugal on both sides. The ventral margin of the internal antorbital fenestra is no higher than that of the external antorbital fenestra (*sensu* Witmer 1997); consequently, there is no expression of the antorbital fossa, posterior to the ascending ramus. The lateral neurovascular foramina are difficult to discern but each appears to be associated with a shallow depression. These show that there were six (possibly seven) linearly arranged foramina, with the posteriormost one being enlarged relative to the others.

Medially there is a narrow maxillary shelf that broadens anteriorly to become confluent with the subtriangular anteromedial process. These processes approach each other but fail to make midline contact. The posterolateral process of the premaxilla articulates with the notch between the anterior ramus and the anteromedial process.

Nasal. The nasals are among the most unusual and autapomorphic elements of the skull roof. Most strikingly there is no triangular anteroventral process (which would normally form part of the posterior margin of the external naris on the lateral surface of the skull). Instead, the nasals are restricted entirely to the dorsal surface of the skull. The lateral margins of the nasals are strongly inflected towards the midline at the posterodorsal corners of the external nares. Each inflection point creates a rounded corner that projects anterior to the ascending ramus of the maxilla and overhangs the posterodorsal corner of the external naris. Anterior to these rounded corners the nasals form the posterior part of the internarial bar. This bar is unusually wide, especially towards the posterior end where it is approximately 9 per cent of the total skull length. Unfortunately poor preservation makes it impossible to determine the shape of the premaxilla-nasal suture, or to recognize where along the length of the internarial bar this suture occurs.

Behind the level of the posterior narial margin there is a longitudinal, elliptical depression on the midline. Although the nasals are thin and poorly preserved in this region it appears

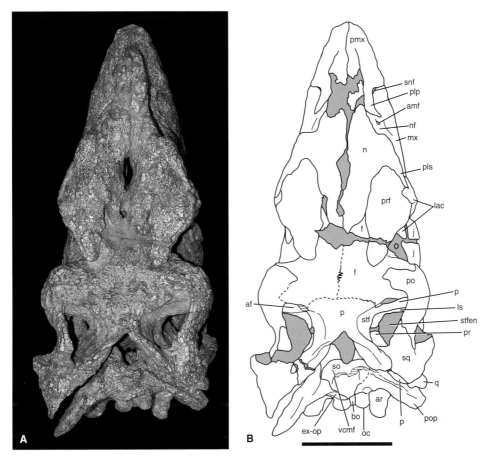

TEXT-FIG. 3. *Melanorosaurus readi*, NM QR3314. Skull in dorsal view. A, photograph. B, interpretive drawing. Grey areas represent matrix and hatched areas represent broken bone surfaces. For explanation of abbreviations, see text. Scale bar represents 100 mm.

that, like some basal sauropodomorphs (e.g. *Efraasia minor*: Galton 1985*a*, pl. 3, fig. 1; *Plateosaurus engelhardti*: Galton 1984*a*, pl.1, fig. 3; *Massospondylus carinatus*: Sues *et al.* 2004, fig. 7c), they failed to form a midline contact in the region of the depression, thus creating an elliptical internasal fenestra. The posterior margin of the conjoined nasals forms a posteriorly projecting subtriangular process that separates the prefrontals and is incised into the anterior end of the frontal pair. As in other saurischians, each nasal has a posterolateral spur that overlaps the lachrymal-prefrontal complex. In *Melanorosaurus readi*, however, this spur, which extends along the lateral margin of the prefrontal and the dorsal end of the lachrymal, is much longer than in other saurischians (almost 20 per cent of the width of the skull roof at the posterior tip of the spur).

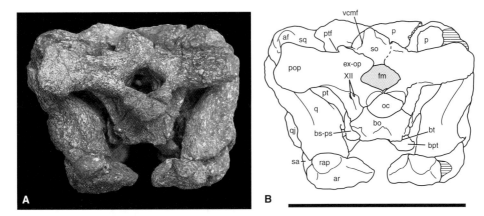

TEXT-FIG. 4. *Melanorosaurus readi*, NM QR3314. Skull and mandible in occipital view. A, photograph. B, interpretive drawing. Grey areas represent matrix and hatched areas represent broken bone surfaces. For explanation of abbreviations, see text. Scale bar represents 100 mm.

Lachrymal. As in prosauropods and stem sauropodomorphs, the lachrymal has a strong anterodorsal inclination in lateral view, although this may have been accentuated by dorsoventral compression of the skull. The ventral ramus of the lachrymal (which forms the bulk of the preorbital bar) flares anteroposteriorly at its dorsal end. As in *Anchisaurus polyzelus* (Yates 2004, fig. 7) and sauropods (Wilson and Sereno 1998) the antorbital fossa was not impressed into the anteroventral corner of the lachrymal. A short, bluntly triangular anterior ramus projects from the dorsal end of the lachrymal to fit into a lateral sulcus developed at the posterodorsal tip of the ascending ramus of the maxilla. The anterior ramus of the lachrymal is itself partly overlapped by the posterolateral spur of the nasal. A small, shallow, semicircular fossa along the ventral side of the dorsal ramus is the only impression the antorbital fossa makes upon the lachrymal. The lateral surface of the dorsal end of the lachrymal bears a small boss similar to the one seen in *Massospondylus carinatus* (Sues *et al.* 2004), immediately posterior to the tip of the posterolateral spur of the nasal. A small dorsomedial extension of the lachrymal is exposed on the skull roof, incised into the lateral margin of the prefrontal.

Prefrontal. The prefrontals are exceptionally large with a broad exposure on the skull roof. Dorsally the prefrontals are shaped roughly like a teardrop. The anterior end is bluntly rounded and fits into the embayment between the main body of the nasal and its posterolateral spur. There is a more attenuated posterior process that extends back along the dorsal orbital margin and almost reaches the level of the postorbitals. Thus, the frontal contribution to the margin of the orbit is severely restricted. The prefrontals are much wider than in most sauropodomorphs, except *Plateosaurus*, and occupy 60 per cent of the width of the skull roof at their widest point. About halfway along the length of the prefrontal there is a descending process that extends down the medial side of the lachrymal. As with the dorsal end of the lachrymal this process was transversely expanded and the dorsal end of the preorbital strut is unusually broad (three times wider transversely than anteroposteriorly at the level just below the semicircular fossa on the dorsal end of the lachrymal). Owing to the severe fracturing of the preorbital bar, caused by dorsoventral deformation, it is particularly difficult to discern the path of the suture between the lachrymal and the prefrontal. Assuming that the elongate foramen exposed on the posterior surface of

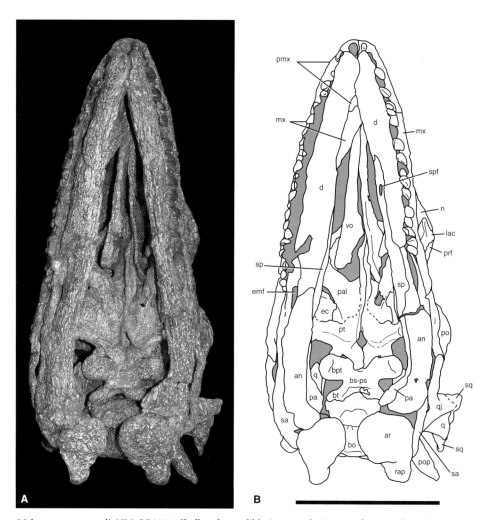

TEXT-FIG. 5. *Melanorosaurus readi*, NM QR3314. Skull and mandible in ventral view. A, photograph. B, interpretive drawing. Grey areas represent matrix and bones from the skull roof. For explanation of abbreviations, see text. Scale bar represents 100 mm.

the preorbital bar lay on this suture (as it does in other sauropodomorphs), the descending process of the prefrontal extended ventrally for at least half the length of the preorbital bar.

Frontal. The frontal pair forms the dorsal skull roof in the orbital region. The midline and frontoparietal sutures are difficult, if not impossible, to discern in most places, perhaps indicating that these sutures were starting to close and that these bones were fused in adults. Nevertheless, some short stretches of suture can be observed and poor preservation may be responsible for the apparent absence of these sutures along most of the length of these contacts. A short length of tightly interlocking midline suture is visible in the centre of the frontal pair. Due to the difficulty of tracing the sutures, the posterior extent of the frontals is uncertain. Nevertheless, it is likely that the frontal reached as least as far as the margin of the supratemporal fossa. The bones that comprise the anterior wall of the supratemporal fossa are difficult to interpret. On the left side they are largely missing (possibly because of over-preparation) leaving a gap between the lateral process of the laterosphenoid and the anterior rim of the supratemporal fossa. On the right side, there is a smaller gap. A short length of suture extending medially from this gap seems to mark the ventral limit of a descending flange from the skull roof. This flange is interpreted as part of the frontal contribution to the wall of the supratemporal fossa. Lateral to this flange is the slot-like gap, which may represent the position of an over-prepared ovoid fossa within the supratemporal fossa. Such fossae are present in several other basal sauropodomorphs (e.g. *Unaysaurus tolentinoi*: Leal *et al.* 2004, pl. 1; *Massospondylus carinatus*, BP/1/4779, 5241; *Jingshanosaurus xinwaensis*, LV 3) and are located in the frontal contribution to the supratemporal fossa. Regardless of how the bones in this region are interpreted, the frontal contribution to the supratemporal fossa could not have been as extensive as it is in *Plateosaurus engelhardti*.

Laterally, a process of the frontal is sutured to the anterior margin of the postorbital. In dorsal view, this process is set at a sharp angle to the orbital margin anterior to it, resulting in a dorsal orbital rim that is notched, like that of *Plateosaurus engelhardti* (von Huene 1926, pl. 1, fig. 4), rather than gently concave as in most other sauropodomorphs (e.g. *Massospondylus carinatus*: Sues *et al.* 2004, fig. 7c; *Anchisaurus polyzelus*: Galton 1976, fig. 13a; *Jingshanosaurus xinwaensis*: Zhang and Yang 1994, pl. 3; *Shunosaurus lii*: Chatterjee and Zheng 2002, fig. 4b). The orbital margin itself is very short (approximately 25 per cent of the total length of the frontal) owing to the great posterior extension of the prefrontal, which also closely resembles the condition in *Plateosaurus engelhardti* more closely than in any other sauropodomorph.

Anteriorly each frontal tapers to a pointed, triangular process that would have wedged between the nasal and the prefrontal, but deformation has caused them to project below the level of the skull roof into the interior of the snout.

Jugal. The jugal is a triradiate bone with two slender posterior processes (the postorbital and the quadratojugal processes) that define the anteroventral corner of the infratemporal fenestra and a deeper, anterior, main body that forms the suborbital bar. It has a bluntly subtriangular anterior end that is excluded from participating in the antorbital fenestra by a lachrymal-maxilla contact. The suborbital bar itself is moderately deep, with its minimal height being equal to just over one-quarter of the distance between the anterior tip of the infratemporal fenestra and the anterior tip of the jugal. In most basal sauropodomorphs the dorsal margin of the suborbital bar is approximately straight and the jugal tapers continuously toward the anterior end. In NM QR3314, however, the dorsal margin is concave with a suborbital constriction and a noticeable anterior expansion.

The postorbital process extends backwards at an angle of approximately 45 degrees to the quadratojugal process on the right side of the skull. The two rami have been forced closer together on the left side, resulting in a large longitudinal fracture running through the entire jugal. On the right side the junction between the two processes forms a rounded emargination that represents the anteroventral corner of the infratemporal fenestra. The fenestra probably underlapped the orbit by a short distance prior to deformation. The degree of underlap was certainly not as large as it is in *Jingshanosaurus xinwaensis* (Zhang and Yang 1994, fig. 3), *Anchisaurus polyzelus* (Yates 2004) or sauropods (Wilson and Sereno 1998). Deformation of the skull has caused the postorbital process to separate from the postorbital itself, and to project into the infratemporal fenestra on both sides of the skull. In the undeformed skull the postorbital process would have been more erect and articulated along the posterior margin of the postorbital to the midlevel of the orbit. The quadratojugal process was shorter and deeper than the postorbital process. It also displays a slight upward curvature and tapers to a single point, as in Massospondylidae (e.g. *Massospondylus carinatus*: Sues *et al.* 2004, fig. 5a) and Sauropoda (e.g. *Brachiosaurus brancai*: Janensch 1935–36, fig. 21). This is in contrast to most other saurischian taxa where the posterior tip is forked (Sereno *et al.* 1993). Furthermore, a prominent articular facet on the anterior end of the jugal process of the right quadratojugal indicates that the jugal overlapped the dorsolateral surface of the quadratojugal as it does in Massospondylidae and Sauropoda, rather than underlapping it as in *Plateosaurus engelhardti* (von Huene 1926, pl. 1, fig. 2). The quadratojugal process extends for less than half the length of the infratemporal bar.

Postorbital. The robust postorbital is a triradiate bone similar to that of other non-sauropod sauropodomorphs. The anterior ramus is broad in dorsal view and curves medially to articulate with the frontal. Its lateral margin forms a rounded, horizontal shelf that projects over the posterodorsal corner of the orbit. The ventral ramus curves anteriorly and articulates along the anterior margin of the posterodorsal process of the jugal. The degree of anterior curvature has been exaggerated by dorsoventral compression of the skull. Unlike *Anchisaurus polyselus* (Yates 2004, fig. 8) and advanced eusauropods (Wilson 2002) the ventral ramus is transversely narrower than its anteroposterior dimension. The posterior ramus, together with the squamosal, forms the supratemporal bar. Unlike the situation in eusauropods (Wilson and Sereno 1998), the posterior ramus is longer than it is deep. Correlated with the primitively elongate temporal bar is the subtriangular shape of the supratemporal fenestra with its long axis orientated anteroposteriorly. The base of the posterior ramus is directed posteromedially but it curves

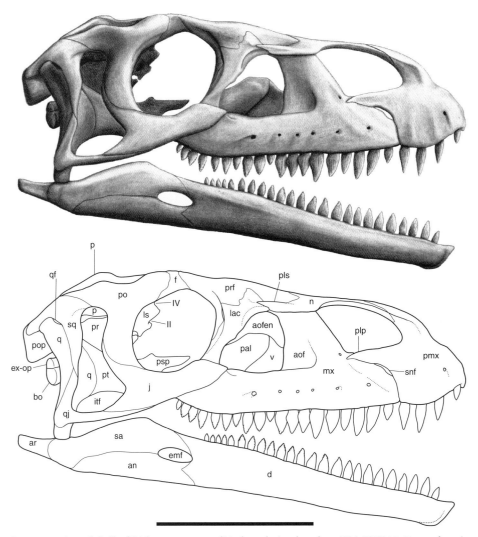

TEXT-FIG. 6.   Reconstruction of skull of *Melanorosaurus readi* in lateral view based on NM QR3314. For explanation of abbreviations, see text. Scale bar represents 100 mm.

laterally at its posterior tip, resulting in a laterally concave margin to the supratemporal bar in dorsal view. The anterior ramus is directed slightly dorsally resulting in a concave dorsal margin of the postorbital when viewed laterally.

Squamosal. The squamosal is a tetraradiate bone with three short horizontal processes arising from the dorsal end and a longer quadratojugal ramus extending anteroventrally. The short, triangular anterior and anteromedial processes form the posterolateral margin of the supratemporal fenestra. The anterior process is directed slightly medially, thus adding to the bowed appearance of the temporal bar in dorsal view. It articulates with the medial side of the posterior process of the postorbital. The anteromedial process contacts the posterolateral surface of the parietal. A stout, triangular process projects posteriorly and receives the paroccipital process on its posteromedial surface. This articular surface is gently concave and tilted slightly ventrally. Unlike most other basal sauropodomorphs the quadratojugal ramus is broadly triangular in lateral view, rather than

slender and splint-like as it is in most basal sauropodomorphs including *Jingshanosaurus xinwaensis* (LV 3) and basal anchisaurians (e.g. *Yunnanosaurus huangi*: Young 1942, fig. 2; *Anchisaurus polyzelus*, YPM 1883).

Quadratojugal. The two rami of the quadratojugal diverge at an angle of approximately 50 degrees, which is similar to the angle seen in *Riojasaurus incertus* (Bonaparte and Pumares 1995, fig. 3), *Massospondylus carinatus* (Sues *et al.* 2004, fig. 7a–b), *Jingshanosaurus xinwaensis* (LV 3) and *Yunnanosaurus huangi* (Young 1942, figs 1–2) and much wider than the angle in *Plateosaurus engelhardti* (Galton 1984*a*, pl. 2, fig. 1) or *Coloradisaurus brevis* (PVL 3967). The squamosal ramus contacts the posterior ventral edge of the squamosal, excluding the quadrate from the margin of infratemporal fenestra. It is broader than the jugal ramus, but only 77 per cent as long. The jugal ramus is a slender prong that extends forward to a level close to the posterior margin of the orbit. As in *Massospondylus carinatus* (Sues *et al.* 2004, fig. 5a), the articulated jugal ramus would have extended

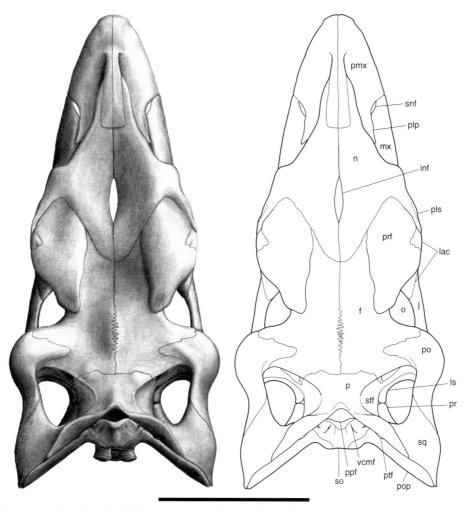

TEXT-FIG. 7.   Reconstruction of skull of *Melanorosaurus readi* in dorsal view based on NM QR3314. For explanation of abbreviations, see text. Scale bar represents 100 mm.

ventral and medial to the quadratojugal process of the jugal. This can be determined from the clear jugal facet on the dorsolateral surface of the jugal ramus on the right quadratojugal.

Quadrate. The quadrate shaft extends from the mandibular condyle to the dorsal quadrate head. It is gently sinuous in posterior view with the lower portion bowed medially, while the upper portion is bowed laterally. The dorsal two-thirds of the shaft are narrower, where it forms a prominent posterior ridge. The quadrate head forms a transversely expanded, bar-like condyle that is received by the ventral surface of the squamosal and is orientated anteromedially–posterolaterally. The posterolateral end of the quadrate head is exposed in lateral view. Two wing-like laminae protrude anterolaterally and anteromedially, respectively, from the upper two-thirds of the shaft. The posterolateral surface of the anterolateral lamina is concave, forming a vertically orientated trough. However, the lateral margin does not protrude posteriorly, so this trough is not enclosed to form a posteriorly facing fossa as it does in eusauropods (Wilson and Sereno 1998). The lateral margin was in contact with the quadratojugal and squamosal for most of its length. Thus, it would

appear to display the derived, sauropod-like, condition of having no quadrate foramen (Yates 2004). However, a narrow, elliptical gap between the quadrate and the squamosal, at the dorsal end of the quadrate, immediately below the quadrate head represents a dorsally shifted quadratojugal foramen. It is important to note that sauropods also display a short gap between the quadrate head and the lateral articular surface for the squamosal (e.g. *Camarasaurus* sp.: Madsen *et al.* 1995, figs 20k, o, s, w; *Apatosaurus* sp., TATE 099). Consequently the character described by Yates (2004) needs to be modified to reflect that the quadrate foramen is not absent in sauropods but has shifted dorsally and has changed shape. Nevertheless, it is clear that *Melanorosaurus readi* has the derived, sauropod-like condition.

The posteromedial lamina overlaps the anterolateral surface of the quadrate wing of the pterygoid. Other bones largely obscure this lamina but nonetheless it is clear that it is dorsally placed, with the ventral end placed at 64 per cent of the length of the shaft from the dorsal end.

Parietal. The parietal pair seems to be indistinguishably fused and there is no sign of the parietal foramen. It is also difficult to

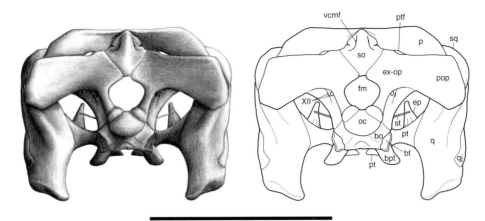

TEXT-FIG. 8. Reconstruction of skull of *Melanorosaurus readi* in occipital view based on NM QR3314. For explanation of abbreviations, see text. Scale bar represents 100 mm.

distinguish the fronto-parietal suture but a short section appears on the right side, adjacent to the supratemporal fossa. Dorsally the parietal pair forms most of the medial margin of the supratemporal fenestra and the walls of the surrounding supratemporal fossa. Within the supratemporal fossa an anterolateral process of the parietal can be seen on the right side. It extends along the dorsal surface of the lateral process of the laterosphenoid to contact the postorbital. The anterolateral process contacts the postorbital at the lateral edge of the fossa unlike *Massospondylus carinatus* and *Yunnanosaurus huangi* where the parietal-postorbital suture appears midway along the supratemporal fossa (Young 1942, fig. 3; Sues *et al.* 2004, fig. 3a).

The posterior margin of the parietal pair is strongly concave in dorsal view. The dorsal surface is humped, at the vertex of this embayment, between the supratemporal fossae. On each side of this hump there is a weak ridge separating the dorsolaterally facing walls of the supratemporal fossa from the dorsal skull roof. This ridge becomes a strong, shelf-like structure as the margin of the supratemporal fossa curves anterolaterally. The minimum distance between the supratemporal fossae is small (21 mm) compared with the width of the skull at this level (96 mm). Posterolaterally there are two wing-like processes with vertically orientated transverse axes. These processes articulate with the posteromedial side of each squamosal. They are dorsally expanded and rise above the dorsal surface of the squamosal where they join. In posterior view the posterolateral processes are relatively shallow (maximum depth of 15·6 mm) and are similar to the depth of the foramen magnum. This is a plesiomorphic trait among sauropodomorphs (Wilson 2002).

Braincase

Supraoccipital. In occipital view the supraoccipital is a tall, diamond-shaped plate that is as high as it is wide. Its occipital surface bears a broad median ridge. This ridge is flanked on each side by sulci, in which lie the foramina for the vena cerebralis media. The lateral margin of the supraoccipital is notched on each side, at the level of these foramina. Internally, this notch continues as a sulcus that extends medially to meet the internal foramen for the vena cerebralis media.

The dorsal margin of the supraoccipital is slightly concave anteriorly, so that when the supraoccipital is articulated with the parietal, which has a strongly posteriorly concave dorsal margin, a large gap is present between the two bones. This gap, or postparietal fenestra, opens directly into the endocranial cavity. Despite the seeming biological implausibility of having such an opening directly over the brain it is a widespread feature among basal sauropodomorphs (e.g. *Plateosaurus engelhardti*, SMNS 12949; *Massospondylus carinatus*, BP/1/4779, 5241; *Anchisaurus polyzelus*, YPM 1883) and is not thought to be the result of deformation.

The internal, or anterior, side of the supraoccipital shows a deep median sulcus that forms the posterior wall of the endocranial cavity. On each side of the cavity are triangular projections that contact the prootic, on their ventral surface, and the parietal, dorsally.

Exoccipital-opisthotic complex. As in almost all dinosaurs the exoccipital and opisthotic are indistinguishably fused. The complex forms the posteriormost lateral wall of the braincase, the paroccipital process and most of the rim of the diamond-shaped foramen magnum. Its contribution to the dorsolateral rim of the foramen magnum projects slightly posteriorly and is weakly raised to form the articulation for the proatlas.

The posterolateral wall of the braincase has a posterior projection from its ventral end that forms the dorsolateral part of the occipital condyle. Anterior to the condyle the wall is pierced by two hypoglossal foramina (cranial nerve XII). Anterior to these foramina is the metotic fissure (Text-fig. 10B). This is a deep, elliptical fossa that extends from the dorsal end of the basal tubera to the base of the paroccipital process. It is bordered anteriorly by the crista interfenestralis, which forms a narrow, sharp-edged septum. Within the right metotic fissure, there is a separate anteroventral foramen for the jugular vein with two slit-like foramina for cranial nerves IX–XI posterior to it. The floor of the left metotic fissure is damaged, uniting these foramina into a single large opening.

The paroccipital processes are directed posterolaterally, making the occiput V-shaped in dorsal view. In posterior view they are directed laterally, with a slight ventral deflection. This is not readily visible in the occipital view of the skull, owing to the

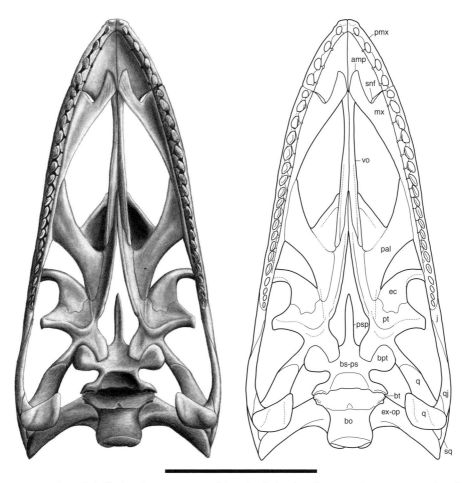

TEXT-FIG. 9.   Reconstruction of skull of *Melanorosaurus readi* in palatal view based on NM QR3314. For explanation of abbreviations, see text. Scale bar represents 100 mm.

angle that the occipital plate articulates against the deformed skull (Text-fig. 4), but is apparent when the occipital plate is viewed strictly posteriorly (Text-fig. 10A). The process maintains a constant depth for most of its length but there is a moderate ventral expansion at its distal end (preserved on the left side). A shallow step on the dorsal margin, near its base, marks the position of a small, slit-like, post-temporal fenestra. Facing laterally into the post-temporal fenestra is a foramen, presumably for the vena capitis dorsalis.

The anterolateral face of the paroccipital process is scarred by longitudinal striations for the basal half of its length. This scar is the sutural surface for the posterolateral process of the prootic. Extending along the ventral margin of this scar is a low thin crest, which is the posterolateral extension of the crista prootica. The crest defines the dorsal margin of the stapedial groove. On the left side a short section of the stapes lies in place within this groove. The groove terminates anteromedially at the anteriorly facing articular facet that fitted against the posterior wall of the prootic. At its anteromedial termination, the stapedial groove forms the posterior margin of the fenestra ovalis. The anterior margin of the fenestra ovalis was formed by the prootic.

The anterior articular facet is excavated by the posterior part of the vestibular recess, a small pocket just 4·9 mm across (Text-

fig. 10C). The posterior wall of the perilymphatic groove extends ventrally from the vestibule until it reaches the suture between the basioccipital and the exoccipital-opisthotic complex. Immediately medial to the perilymphatic duct is the internal opening of the jugular foramen. A small pillar separates this foramen from the internal openings of the two foramina for cranial nerves IX–XI.

Basioccipital. The basioccipital forms the ventral part of the occipital plate and the majority of the reniform occipital condyle. Extending anteriorly from the occipital condyle, it contributes a narrow median strip to the floor of the endocranial cavity. At the anterior end of this strip there is a low central keel, as in most other sauropodomorphs (e.g. *Thecodontosaurus caducus*; Yates 2003*b*, fig. 4f). The main body of the basioccipital is ventrally directed, lowering the basal tubera to a level well below that of the ventral margin of the occipital condyle. The basal tubera are broad and anteroposteriorly flattened. They are angled dorsolaterally from the ventral margin of the basioccipital. The occipital face of the basioccipital, adjacent to each tuber, is slightly concave, on each side of a central raised area. In contrast to *Anchisaurus polyzelus* (Yates 2004, fig. 3) and eusauropods (e.g. *Camarasaurus* sp.; Madsen *et al.* 1995,

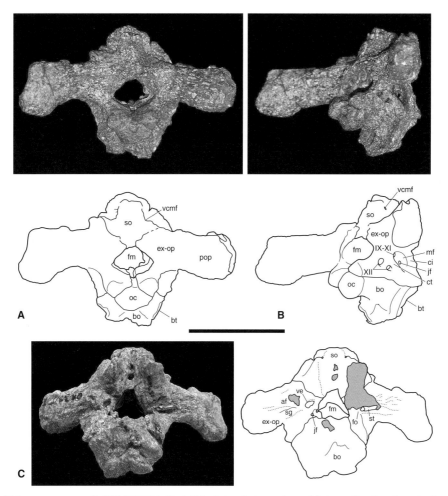

TEXT-FIG. 10. *Melanorosaurus readi*, NM QR3314. Occipital plate, photographs and interpretive drawings, in A, posterior, B, posterolateral and slightly ventral, and C, anterior views. For explanation of abbreviations, see text. Scale bar represents 50 mm.

fig. 26b–c), where there is a deep ventral notch between the basal tuberae, a median sheet of bone connects the basioccipital components of the basal tubera of *Melanorosaurus readi*. This median sheet creates a sharp posteroventrally directed ridge that connects the basal tubera. The posterior surface of this ridge is rather flat while there is a small medial notch on the antero-ventral surface. Unlike more basal sauropodomorphs (e.g. *Massospondylus carinatus*, SAM-PK-K1314; *Coloradisaurus brevis*, PULR unnumbered; *Jingshanosaurus xinwaensis*, LV 3) the basi-occipital component of the basal tuber lies strictly posterior to the basiparasphenoid component, rather than partly medial to it.

Basisphenoid-parasphenoid complex. In ventral view this complex is formed from a cross-shaped basiparasphenoid plate, with the basipterygoid processes and the basal tuberae forming the 'arms' of the cross, and an anteriorly projecting parasphenoid rostrum. Extending across the middle of the ventral surface of the basiparasphenoid plate is a ventrally projected, transverse ridge. This ridge separates the well-defined posterior fossa from the antero-ventrally facing anterior area. The posterior fossa is located between the widely separated basiparasphenoid components of

the basal tuberae. It is similar to the fossa seen in other anchi-saurs (e.g. *Anchisaurus polyzelus*; Yates 2004, fig. 3b) in that it is open posteriorly, but differs from them in being wider than it is anteroposteriorly deep. The basiparasphenoid components of the basal tuberae are simple, laterally projecting processes that lack finished bone on their posterior surfaces.

Anterior to the transverse ridge the basiparasphenoid plate slopes dorsally toward the parasphenoid rostrum. The plate is transversely concave between the basipterygoid processes. The ventrolaterally, and slightly anteriorly, directed basipterygoid processes are unusually large, and expanded anteroposteriorly, making them appear paddle-shaped in ventral view.

The horizontal parasphenoid rostrum is about as wide trans-versely as it is deep dorsoventrally as it is in other basal saurop-odomorphs that are more closely related to sauropods than to *Plateosaurus* (e.g. *Massospondylus carinatus*, SAM-PK-K1314; *Anchisaurus polyzelus*, YPM 1883; *Jingshanosaurus xinwaensis*, LV 3), and bears a longitudinal dorsal sulcus. It also exhibits the derived condition of being raised well above the level of basal tub-erae and the bases of the basipterygoid processes, producing a V-shaped ventral margin of the braincase. A raised parasphenoid rostrum is a synapomorphy of *Jingshanosaurus xinwaensis* (Zhang and Yang 1994, fig. 5) and Anchisauria (e.g. *Anchisaurus polyzelus*; Yates 2004, fig. 9d; *Camarasaurus* sp.; Madsen *et al.* 1995, fig. 22).

Welman (1999, p. 229) described a small raised area at the back of the basiparasphenoid plate as a 'posteromedian extension of the basiparasphenoid into the braincase floor'. He also reconstructed two foramina on either side of this raised area and interpreted them as the openings for the palatine ramus of the facial nerve (VII: Welman 1999, fig. 1). No other dinosaur shows paired foramina in this position and it appears likely that the foramina are simply notches between the ventral surface of the basiparasphenoid plate and the 'raised area'. The raised area itself appears to be a small flake of bone, possibly from the disrupted posterior margin of the basiparasphenoid on the left side, which is sitting in the posterior fossa. The irregular shape of the raised area and its asymmetrical position supports this interpretation. There are many other flakes of bone that have become slightly dislocated from their original position, scattered over the skull surface.

Prootic. The poorly exposed prootics are visible posteriorly, once the rear section of the braincase has been removed (Text-fig. 11), and through the infratemporal fenestra.

The rear view exposes the medial side of the prootic (the internal braincase wall and the posterior articular surface that contacts the opisthotic-exoccipital complex) while the view through the infratemporal fenestra shows a small portion of the lateral surface of the prootic and its contact with the parietal.

The internal wall of the prootic descends ventromedially from the ventral margin of the parietal to meet its antimere in a midline contact on the anterodorsally sloping braincase floor. The internal foramina for the abducens nerves (VI) can be seen near the anterior end of the braincase on either side of the midline. The bones of the palate obscure the external opening of the abducens nerve. When the right internal wall of the prootic is observed in posteromedial view, the internal opening of the foramen for the trigeminal nerve (V) is visible. It takes the form of a deep notch in the anterior margin of the prootic that is closed off anteriorly by the laterosphenoid. Also visible in this view is a single internal opening for the facial nerve (VII), which is located posterior to the trigeminal foramen and posteroventral to the shallow, concave, prootic portion of the auricular recess. It heads laterally, anteroventral to the vestibule.

The articular surface for the contact with the exoccipital-opisthotic complex lies posteriorly and is extended posterolaterally by a stout process. The posterolateral process is shallowly grooved along its length. Between the posterolateral process and the internal wall, on the right side, the anterior part of the vestibule is exposed. Curving posteroventrally from the vestibule is a groove that forms part of the perilymphatic duct. Extending lateral from the vestibule is a deep, narrow canal, marking the passage of the horizontal semicircular canal. Underneath the semicircular canal is an embayment in the margin of the posterior articular facet, forming the anterior margin of the fenestra ovalis. The crista prootica extends ventrally from the fenestra ovalis, becoming deeper ventrally until it terminates in a rounded lappet that projects posterolaterally, immediately above the basiparasphenoid component of the basal tuber. Nestled on the anterior side of the crista prootica is the external opening of the foramen for the facial nerve (VII).

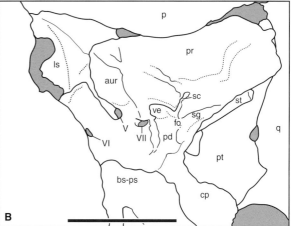

TEXT-FIG. 11. *Melanorosaurus readi*, NM QR3314. Prootic and surrounding bones in oblique view looking posteromedially and slightly dorsally after the occipital plate has been removed. A, photograph. B, interpretive drawing. For explanation of abbreviations, see text. Scale bar represents 20 mm.

Laterosphenoid. This bone can be observed through the orbits (Text-fig. 12) and through the supratemporal fenestra. It lies in front of the prootic and dorsally is in contact with the ventral surface of the frontal. It forms a roughly triangular plate that is angled ventromedially, although the two antimeres do not make midline contact. The posterodorsal corner is produced into a stout lateral process that is in contact with the ventral surface of the anterolateral process of the parietal, along its length, and the medial surface of the postorbital, at its distal end. Anteriorly there are two notches separated by an anterior process. The dorsal notch is the deeper of the two and forms part of the margin of the opening for the trochlear nerve (IV). The ventral notch seems to mark the pathway for the oculomotor nerve (III).

A thin posteroventral process divides the dorsal opening for the mid cerebral vein from the trigeminal foramen ventrally. Ventral to this process is a deep notch that forms the anterior margin of the trigeminal foramen. Confluent with the trigeminal foramen is a groove that extends across the lateral surface of the ventral end of the laterosphenoid that marks the path of the

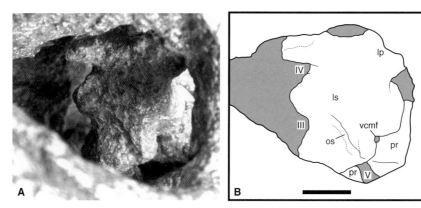

TEXT-FIG. 12.   *Melanorosaurus readi*, NM QR3314. Left laterosphenoid in oblique view, looking anterolaterally and slightly ventrally through the left orbit. A, photograph. B, interpretive drawing. For explanation of abbreviations, see text. Scale bar represents 10 mm.

ophthalmic branch (V1) of the trigeminal nerve. Internally the laterosphenoid has a simple, smooth concave surface.

Orbitosphenoid.   This bone is missing and given the presence of most other cranial bones, including the stapes, it is likely that it was unossified.

Stapes.   Both stapes are preserved. The right stapes is the more complete one of the pair but is preserved out of its natural position, lying on the ventral margin of the posterolateral process of the prootic (Text-fig. 11). It is a slender rod, 13 mm long (as preserved) and 2 mm in diameter. The footplate is not visible.

Palate

Pterygoid.   There are four principal rami of this bone, all radiating from the articulation with the basipterygoid processes. The vertically orientated quadrate ramus is situated posterolaterally: it is a thin sheet-like structure that articulates with the anteromedial lamina of the quadrate. The quadrate ramus flares posteriorly and has a concave posterior margin, giving it a shape vaguely reminiscent of a fish tail. Posteromedially is a short, flattened, subrectangular process that wraps around the basipterygoid process. The notch between the quadrate ramus and the posteromedial process forms the articular surface for the basipterygoid process. In NM QR3314 the posteromedial processes of the pterygoid contact each other medially, but this is probably an artefact owing to the narrowing of the posterior palate caused by the inward rotation of the quadrates. Extending laterally from the junction of the quadrate ramus and posteromedial process is the transverse process. This process also curves ventrally towards its distal end, which braces the medial surface of the mandible. It is supported by a thickened ridge that extends along its posterior margin, while anteriorly it forms a thin sheet, which articulates with the ectopterygoid.

The posterior ridge of the transverse process crosses the ventral surface of the pterygoid, curving anteriorly as it reaches the medial margin of the bone, just anterior to the base of the posteromedial process. At this point the ridge rises into a tall, thin, ventrally projecting crest. This crest, which is in midline contact with its antimere, extends for the full length of the anterior ramus of the pterygoid. A deep sulcus lies between the crest and the medial margin of the palatine. The anterior tips of these crests insert between the posterior ends of the vomers.

Epipterygoid.   A disarticulated epipterygoid lies on the dorsal surface of the right pterygoid flange. It is a flat, subtriangular plate, which is similar to the epipterygoid of *Massospondylus carinatus* (Sues *et al.* 2004; Barrett and Yates 2006) with an expanded base and a grooved basal lateral surface. The slightly curved, and tapering, dorsal end also resembles the epipterygoid of *M. carinatus* except that the tip is broader in NM QR3314.

Ectopterygoid.   The ectopterygoid is a simple flat triangular plate with an anterolateral process. The triangular plate is attached to the anterior margin of the transverse flange of the pterygoid and has a simple flat ventral surface, which lacks any form of fossa. Its posteromedial margin contacts the posterior end of the palatine. The slender, rod-like, anterolateral process curves dorsally and laterally as it extends towards the medial side of the upper jaws. The distal tip of the right process can be observed through the right orbit, showing that it largely contacts the medial side of the jugal, although a small contact with the maxilla cannot be ruled out.

Palatine.   The palatine (Text-fig. 13) is one of the most transformed bones of the skull, in comparison with other basal sauropodomorphs. As in eusauropods the posterolateral process is absent (e.g. *Camarasaurus* sp.; Madsen et al. 1995, fig. 38). Ventrally the palatine forms an anterolaterally orientated strap that runs from the lateral side of the pterygoid to the medial margin of the maxilla at the level of the anterior margin of the antorbital fenestra, resembling the narrow anterolateral process of eusauropods (Wilson and Sereno 1998). An anteromedial process arises from the dorsal surface of the palatine to join to the dorsal margin of the vomer, at its posterior end. It forms a tongue-shaped process that faces laterally, and is best seen through the antorbital fenestra. The process is convex laterally, and hollowed medially. The lateral wall of this process is thin and has largely broken away, leaving the margins as a dorsally ascending arch.

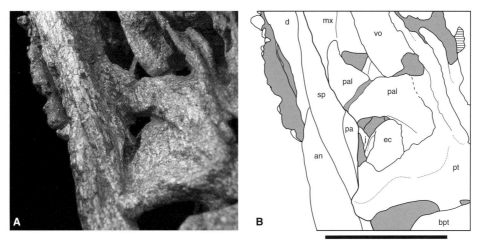

TEXT-FIG. 13.   *Melanorosaurus readi*, NM QR3314. Right palatine and surrounding bones in oblique view looking ventromedially and slightly posteriorly. A, photograph. B, interpretive drawing. For explanation of abbreviations, see text. Scale bar represents 50 mm.

Vomer. The vomers are elongate, sheet-like bones that are angled ventrolaterally away from the midline, forming a narrow, tent-like structure. The vomers are separated for the posterior quarter of their length but are otherwise in midline contact. The anterior ends of the pair are pointed and shallow, but the dorso-ventral depth increases posteriorly. The dorsal margin rises sharply to form a peak at its posterior end, where it contacts the dorsally directed anteromedial process of the palatine.

Like *Plateosaurus engelhardti* (Galton 1984*a*, pl. 1, fig. 4) and *Lufengosaurus huenei* (Barrett *et al.* 2005*a*, fig. 4b) the vomers are narrow in ventral view and lack the broad semilunate shape seen in some other sauropodomorphs (e.g. *Massospondylus carinatus*; Barrett and Yates 2006).

Mandible

Dentary. The lower jaws are tightly appressed, and the lateral surface of the dentary is partly obscured by the overbite of the upper jaws. Each dentary is elongate, roughly measuring five times longer than its maximum depth (measured medially). The dentary pair meet to form a narrow V-shaped arrangement, as in other basal sauropodomorphs. It cannot be determined if a posterior lateral ridge was present or absent, but its presence is likely as one occurs in the broad-jawed basal sauropod *Chinshakiangosaurus chunghoensis* (Barrett *et al.* 2005*b*). Anteriorly, there is a weak longitudinal ventral flange, along the symphysis, creating a small ventral expansion in lateral view. The first dentary tooth is inset from the pointed anterior tip by a distance of approximately 15 mm, which is about three times the width of the first dentary tooth.

Splenial. The splenial is a flattened trapezoidal bone on the medial surface of the mandible, similar in shape to that of *Plateosaurus engelhardti* (Galton 1984*a*, fig. 5). The left splenial is more easily observed. It shows a small, fully enclosed splenial foramen located in the anterior half of the bone, near its ventral margin. An elongate posterior spur projects from the postero-ventral corner of the splenial to overlap the ventromedial margin

of the angular, to the level of the posterior rim of the external mandibular fenestra.

Surangular. The postdentary regions of both mandibular rami are dorsoventrally crushed, and partially obscured by the infra-temporal bar, making a full description of the bones in this region impossible. The surangular does not appear to have differed significantly from those of *Plateosaurus engelhardti*, *Massospondylus carinatus* or *Jingshanosaurus xinwaensis*. It had a low rounded coronoid eminence. The mandible is approximately 40 mm deep (14 per cent of total mandibular length) at the peak of the coronoid eminence, but dorsoventral crushing has reduced this height. Between the right surangular and the angular is the external mandibular fenestra, which is a small, ovoid opening with a maximum diameter of 17·5 mm orientated longitudinally. Such a small external mandibular fenestra (6·2 per cent of total mandibular length) is a derived condition (Upchurch 1995) shared with *Jingshanosaurus xinwaensis* (LV 3), *Yunnanosaurus* (Young 1942, fig. 2) and sauropods (Russell and Zheng 1993, fig. 1). Posteriorly the surangular overlaps the lateral surface of the articular to a point posterior to the mandibular glenoid.

Angular. Both angulars are a little crushed, and their articular contacts are obscure making a detailed description impossible. It appears to have been about 85 mm long and 14 mm deep under the external mandibular fenestra. Anteriorly it is overlapped by the dentary on the lateral side and the splenial on the ventral side. Posterior to the splenial the angular curves medially to form the ventral margin of the mandible. Posteriorly it overlaps the ventrolateral surface of the surangular.

Prearticular. This bone is largely obscured on both sides. What little can be seen indicates that this bone does not differ from that of other basal sauropodomorphs.

Articular. The articular (Text-figs 4–5) has a peculiar and aut-apomorphic shape. In lateral view it is shallow, but in dorsal or

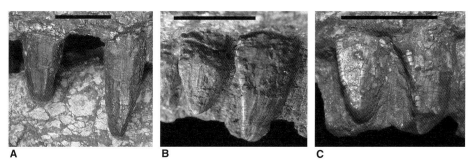

TEXT-FIG. 14. *Melanorosaurus readi*, NM QR3314. Dentition. A, left premaxillary teeth 3 and 4. B, right maxillary teeth 3 and 4. C, right maxillary teeth 14 and 15. Scale bars represent 10 mm.

ventral view, the main body is greatly expanded medially by a large semilunate process. The semilunate medial expansion has a sharp ventromedial margin. The dorsal surface slopes downwards from the mandibular glenoid to the ventromedial margin. Extending posteriorly from the lateral side is a stout retroarticular process. The retroatricular process extends 25 mm back from the posterior lip of the mandibular glenoid. It is shallow (about 10 mm deep at its midlength), giving the appearance of being elongate in lateral view. However, in dorsal view that part of the process projecting beyond the medial semilunate flange is as broad as it is long. Both the medial and the lateral surfaces of the retroarticular process bear indistinct, shallow fossae. As in other basal sauropodomorphs there is a strong inflection on the medial margin in dorsal view, between the retroarticular process and the main body of the articular.

Dentition. The teeth of NM QR3314 are distinctive amongst early sauropodomorphs in that they are fluted, have relatively finely serrated marginal carinae and narrow crowns with their widest point close to the base (Text-fig. 14). The crowns are elongate, with a slenderness index (the ratio of crown height to the maximum crown width; Upchurch 1998) of up to 2·13. The long axis of each crown is not recurved in the mesiodistal plane but the apical end is slightly inclined towards the lingual side in many teeth. There is a weak mesiodistal expansion a short distance above the base in all teeth that are fully erupted and well exposed. The transverse cross-section of the teeth cannot be determined but it is D-shaped in isolated tooth crowns that are referable to *M. readi* (BP/1/5334). These teeth are identified as *M. readi* on the basis of their elongate shape, weak basal expansion, poorly developed serrations, strong fluting and occurrence in the lower Elliot Formation. In these isolated teeth the lingual face has a low central raised area extending along its length that is flanked on each side by a shallow concavity.

Sharp carinae extend along the mesial and distal margins of each crown. The serrations are finer than in most other sauropodomorphs with similarly sized teeth, with approximately two serrations per millimetre. The serrations do seem to be upwardly angled as they are on other sauropodomorph teeth (Galton 1985*b*).

The teeth are sculpted with two sets of ornament. The strongest of these are a series of longitudinal ridges ('fluting') that run along the length of the tooth crowns. There are approximately four such ridges per face. The second set of ornament consists of smaller scale, discontinuous wrinkles that are similar to those seen in *Anchisaurus polyzelus* (Yates 2004, fig. 6b). Given that the teeth are thickly covered in glue, these wrinkles are hard to see in NM QR3314, but are clearly visible in the isolated teeth.

The second and third premaxillary teeth are the tallest and most slender tooth crowns, while the maxillary teeth become progressively shorter and proportionately wider. The premaxillary teeth and the first two or three maxillary teeth lack any serrations on their carinae (both mesial and distal). Thereafter, they possess a few serrations that are restricted to the apical region of the crown. The posteriormost teeth have more extensively serrated carinae, with the serrations extending along most of the height of the crown.

Of all described sauropodomorph teeth, those of *Jingshanosaurus xinwaensis* most closely resemble the teeth of *Melanorosaurus readi*. Both taxa have fluted elongate crowns with poorly developed basal expansions and serrations in the premaxillary and anterior maxillary teeth (Zhang and Yang 1994). Nevertheless *Jingshanosaurus xinwaensis* differs from *Melanosaurus readi* in having elliptical cross-sections of the crowns and weakly developed fluting on the labial surface of the maxillary teeth (LV 3).

Hyoid elements

Ceratohyal. A pair of ceratohyals was recovered. These are similar to the ceratohyals of other sauropodomorphs, being long (half the length of the skull) and strongly bowed with a slightly expanded posterior end (Text-fig. 15).

Anterior vertebral column

Proatlas. Both proatlantal elements are preserved as part of the same block containing the atlantal neurapophyses (Text-fig. 16). In dorsal view, each element presents a smooth flat surface that is similar in shape to the sole of a shoe, with the toe anterior and the heel posterior. Laterally, a strong ridge separates the sole-shaped surface from a ventrally projecting, hump-like process that is situated near the anterior end. The area between the peak of the hump and the anterior margin of the proatlas defines an oblique elliptical articulating facet that faces anteriorly, medially and ventrally. It would have lain over the dorsal margin of the foramen magnum. The dorsoventrally flattened heel

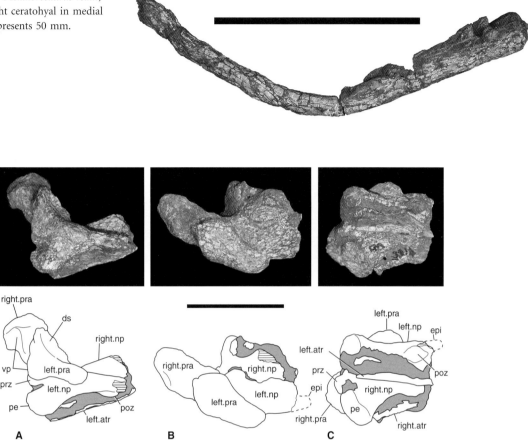

TEXT-FIG. 15. *Melanorosaurus readi*, NM QR3314. Right ceratohyal in medial view; scale bar represents 50 mm.

TEXT-FIG. 16. *Melanorosaurus readi*, NM QR3314. Atlas and proatlas. A, left proatlas and atlantal neurapophysis in lateral view. B, left proatlas and atlantal neurapophysis in dorsal view. C, left proatlas and atlantal neurapophysis in ventral view. Note that the right side elements and the atlantal ribs have been displaced and lay at oblique angles to the left elements. The left epiphysis is copied from photographs of casts made before the process was lost. Grey areas represent matrix and hatched areas represent broken bone surfaces. Scale bar represents 50 mm.

overlapped the anterior dorsal surface of the atlantal neurapophyses.

Atlas. Two separate neurapophyses, the atlantal intercentrum and two atlantal ribs are preserved (Text-figs 16–17). The left neurapophysis and rib are the more completely preserved and exposed of the pair. The neurapophysis (Text-fig. 16) resembles that of *Riojasaurus incertus* (Bonaparte and Pumares 1995, fig. 5c–d) and *Camarasaurus* sp. (Madsen *et al.* 1995, fig. 51c–f) in having a transversely broad and flat dorsal surface, and in lacking a dorsolateral crest seen in other basal sauropodomorphs (e.g. *Thecodontosaurus antiquus*, uncatalogued BRSUG material; *Plateosaurus engelhardti*, MB 1927.19.1; *Massospondylus carinatus*, BP/1/5241). At the anterior end of the bone on its lateral side there is the rectangular pedicel, which projects anteroventrally to contact the atlantal intercentrum. The postzygapophysis is an oval, ventromedially facing facet on the posterior medial margin of the bone. The lateral margin once projected posterior to the postzygapophysis, forming a short, blunt-ended epiphysis, much like that seen in

Riojasaurus incertus (Bonaparte and Pumares 1995, fig. 5c–d). The epiphysis was only preserved on the left side. Unfortunately, it has since been broken off and lost: nevertheless, it is recorded in an earlier cast of the block. The left atlantal rib is a simple, slender, straight rod of bone that is approximately as long as the atlantal neurapophysis (Text-fig. 16).

The atlantal intercentrum is a small dumb-bell-shaped bone (Text-fig. 17). The median constriction is caused by the crescentic bevels, both anteriorly and posteriorly, for the reception of the occipital condyle and the odontoid, respectively. The ventral surface is mostly flat with upwardly curved lateral sides and two small lappets on the anteroventral margin, either side of the midline.

Axis. The axis is only moderately well preserved (Text-fig. 18). The odontoid and axial intercentrum are attached; the latter is apparently fused while an open suture remains at the junction of the axial centrum and the odontoid. The neurocentral suture of the axis is also open and this is a further indication that the specimen was immature at the time of death. The total length of

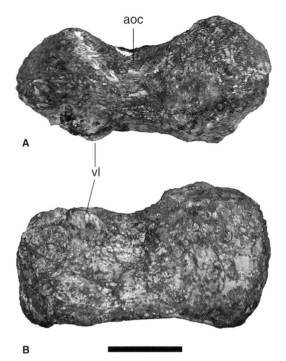

TEXT-FIG. 17. *Melanorosaurus readi*, NM QR3314. Atlantal intercentrum in A, anterior, and B, ventral views. For explanation of abbreviations, see text. Scale bar represents 10 mm.

the axis (80 mm) is shorter than all postaxial cervicals. The axial centrum is stout for a basal sauropodomorph: minus the axial intercentrum it is about 2·5 times as long as the posterior height of the centrum. The lateral surfaces are flattened and rather slab-sided, while the ventral surface is too damaged to determine if a ventral keel was present. The parapophysis is a small raised tubercle located near the ventral margin of the anterior end of the centrum. There is no apparent diapophysis, nor is there a trace of a transverse process on the neural arch. Anteriorly the neural canal is dorsoventrally low, transversely broad, and sub-rectangular. The foramen magnum is also wider than deep so this appears to be a real feature of the anterior spinal canal of *M. readi*. The posterior opening of the neural canal is obscured by matrix and it is not possible to determine if this unusual shape continues through the axis. The middle cervical vertebra of NM 1551 (Galton *et al.* 2005, fig. 1,3a–d) shows a normal, ovoid neural canal, indicating that the unusual shape of the neural canal was confined to the anterior end of the spinal column. The prezygapophyses are low tubercles that are set a short distance back from the anterior edge of the neural arches. The transverse width between the prezygapophyses (42 mm) is less than the width between the postzygapophyses (60 mm). In lateral view, the neural arch is surmounted by a long, low, rectangular neural spine. The anterior end of the neural spine is level with the prezygapophyses but the posterior end terminates at the base of the postzygapophyses. The neural spine widens from a simple transversely compressed lamina at its anterior end to a broad, tent-like structure formed by two laminae that diverge from the summit. Tall, ridge-like epipophyses are confluent with the two tented laminae that form the posterior end of the neural spine. These extend onto the dorsal surface of the paddle-like postzygapophyses but terminate abruptly well before the posterior end of the postzygapophyses. The postzygapophyses do not overhang the axial centrum, with the postzygapophyseal tips being level with the posterior face of the centrum.

The odontoid is a dorsoventrally flattened structure that is semicircular in dorsal view. It is as wide as the neural canal and located immediately below it. Surrounding the ventral margin of the odontoid is the reniform articular surface of the axial intercentrum. This articular surface is gently convex transversely and dorsoventrally and is wider (*c.* 43 mm) than the posterior width

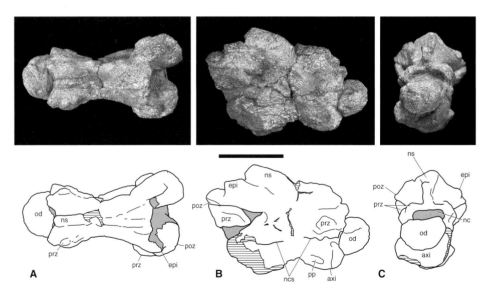

TEXT-FIG. 18. *Melanorosaurus readi* (NM QR3314). Axis in A, dorsal, B, right lateral, and C, anterior views. Grey areas represent areas of matrix; hatched areas represent broken bone surfaces. For explanation of abbreviations, see text. Note that the prezygapophyses of cervical vertebra 3 remain articulated with the postzygapophyses. Scale bar represents 50 mm.

of the axial centrum (*c*. 39 mm). There is no midline, anterior projection under the odontoid.

AUTAPOMORPHIES OF THE SKULL

1. *Transversely broad internarial bar.* The internarial bar of most saurischians is a narrow strut with a transverse width (measured at the anteriormost extent of the nasals) that is less than 4 per cent of the total length of the skull. It is 1·8 per cent in the basal saurischian *Herrerasaurus ischigualastensis* (Sereno and Novas 1993, fig. 1c), 3·8 per cent in the neotheropod *Allosaurus fragilis* (Madsen 1976, pl. 2a), 3·1 per cent in the sauropod *Camarasaurus lentus* (Madsen *et al.* 1995, fig. 5c) and 1·8 per cent in the basal sauropodomorph *Plateosaurus engelhardti* (SMNS 13200). In contrast, the width of the internarial bar of NM QR3314 is apparently close to one-tenth of the length of the skull (*c*. 9·0 per cent). Although the midsection of the internarial bar of this specimen is poorly preserved, its bases are intact enough to establish that the bar was unusually broad (Text-fig. 3).

2. *Enlarged premaxillae.* The dentigerous portion of the each premaxilla occupies 32 per cent of the total length of the upper tooth row (Text-figs 1–2). In other early sauropodomorphs this fraction is less than 25 per cent. For example it is 18 per cent in *Plateosaurus engelhardti* (Galton 1984*a*, pl. 2, fig. 1) and 22 per cent in *Shunosaurus lii* (Chatterjee and Zheng 2002, fig. 4). The premaxillae of NM QR3314 maintain a shape that is similar to that of other early sauropodomorphs, indicating that the difference is related to an overall enlargement of the premaxillae as opposed to a simple anteroposterior elongation.

3. *Short ridge on the dorsolateral surface of the posterior end of the maxilla.* Both maxillae bear a short dorsolateral ridge that extends forward from the anterior end of the jugal (Text-figs 1, 6). A longer, more strongly pronounced ridge is present in a similar position on the maxilla of *Lufengosaurus huenei* (Barrett *et al.* 2005*a*). This is interpreted as an example of convergence in the phylogeny presented below.

4. *Nasals lack an anteroventral process.* The external naris of other early saurischians forms an embayment on the anterolateral margin of the nasals. This embayment separates the narrow anterodorsal process (which forms part of the internarial bar) from the triangular anteroventral process. This process forms the posterodorsal margin of the external naris and extends along the anterior margin of the posterolateral process of the premaxilla or, in the case of derived saurischians that have a reduced posterolateral process of the premaxilla, the anterior margin of the ascending ramus of the maxilla. NM QR3314 is unique among sauropodomorphs in lacking the antero-

ventral process of the nasal so that the ascending ramus of the maxilla forms the entire posterior margin of the external naris (Text-figs 1–2).

5. *Nasal with a long, slender posterolateral process.* Basal eusaurischians are characterized by a short, pointed posterolateral process of the nasal that overlaps the lateral surface of the lachrymal. This process has been documented in coelophysids, ceratosaurs, tetanurans and early sauropodomorphs. In NM QR3314 it is far longer than in any other eusaurischian, reaching at least 17 per cent of the length of the nasal (Text-figs 1–3).

6. *Transversely broad prefrontals.* In most early dinosaurs the prefrontals form a relatively small proportion of dorsal skull roof. In many sauropodomorphs more derived than *Thecodontosaurus* the prefrontal is elongated but remains transversely narrow. The prefrontals of NM QR3314 are unusually broad, with each occupying 60 per cent of the distance between the midline and the lateral margin of the dorsal roof at their widest point (Text-fig. 3). *Plateosaurus* is the only other sauropodomorph to display similarly broad prefrontals but the distribution of other character states indicates that this similarity is convergent.

7. *Dorsoventrally expanded anterior end of the jugal.* Basal archosauriforms have a dorsal process that arises from the anterior end of the jugal and forms the base of the preorbital bar, below the lachrymal (e.g. *Euparkeria capensis*; Ewer 1965, fig. 2). This condition is retained in *Herrerasaurus ischigualastensis* (Sereno and Novas 1993, fig. 1a) but is lost in more derived saurischians which have a simply tapering anterior end of the jugal (e.g. *Eoraptor lunensis*: Sereno *et al.* 1993, fig. 1b; *Coelophysis rhodesiensis*: QG 165; *Plateosaurus engelhardti*: Galton 1984*a*, pl. 2, fig. 1). NM QR3314 has partly reverted to the primitive condition and has a dorsoventrally expanded anterior end of the jugal which is deeper than the section of the jugal under the orbit. Rauhut (2003*a*) noted that many derived neotheropods also have a dorsoventrally expanded anterior end of the jugal.

8. *Elongate vomers.* The vomers of most crurotarsans are less than one-quarter of the length of the skull: for example, it is 10 per cent in *Parasuchus hislopi* (Chatterjee 1978, fig. 2b), 23 per cent in *Stagonolepis robertsoni* (Walker 1961, fig. 3a) and 18 per cent in *Erpetosuchus granti* (Benton and Walker 2002, fig. 2c). This would appear to be the basal archosaur condition and it is present in dinosaurs primitively. It is retained in ornithischians (e.g. 25 per cent in *Hypsilophodon foxi*; Galton 1974, fig. 4b), most other basal sauropodomorphs (e.g. 19 per cent in *Massospondylus carinatus*, SAM-PK-K1314) and derived eusauropods (e.g. 22 per cent in *Camarasaurus lentus*; Wilson and Sereno 1998, fig. 7d). In contrast, the vomers of NM QR3314 are far more elongate, extending for 36 per cent of skull length (Text-fig. 5). The vomers

of neotheropods are similarly elongate: for example, they are 35 per cent in *Sinraptor dongi* (Currie and Zhao 1993, fig. 3c), and *Plateosaurus engelhardti* (Galton 1984*a*, pl. 1, fig. 5) also has unusually long vomers that closely resemble those of NM QR3314. These instances are interpreted as convergences in the phylogeny presented below. The vomers of *Lufengosaurus huenei* are, like those of *Melanorosaurus* and *Plateosaurus*, narrow and straight-sided in ventral view (Barrett *et al.* 2005*a*) so this taxon might be thought of as having the derived condition as well. Unfortunately the length of the vomers cannot be determined precisely because the rostral tips are missing. However, given that the right vomer contacts the virtually complete right maxilla, and that vomers do not extend much beyond this contact, if at all, in other sauropodomorphs (e.g. *Melanorosaurus readi*, NM QR3314; *Massospondylus carinatus*, SAM-PK-K1314; *Plateosaurus engelhardti*, MB XXIV), it seems that very little of the vomers are missing and that they occupied no more than 25 per cent of the total skull length using the unusually abbreviated skull reconstruction of the type specimen. In reality the skull was probably longer with the vomers occupying even less of its length.

9. *Anteroposteriorly short, crescentic fossa at the posterior end of the basiparasphenoid plate.* Wilson (2002) listed a basipterygoid/basisphenoid recess as a synapomorphy of Neosauropoda. Presumably he was referring to the prominent U-shaped fossa that is impressed upon the ventral surface of the basiparasphenoid plate that widely separates the basal tuberae. Yates (2004) noted a similar fossa on the ventral surface of the parabasisphenoid plate of *Anchisaurus polyzelus* and suggested that the feature was a synapomorphy of all known sauropods (a clade that is labelled Anchisauria in the present work). NM QR3314 has a similar fossa, but in this skull it is far wider transversely than it is anteroposteriorly long, making it crescentic rather than U-shaped (Text-fig. 5). Thus, the presence of the fossa is interpreted as a synapomorphy of Anchisauria but its unusual shape in NM QR3314 is interpreted as an autapomorphy.

10. *Articular forms a broad, flat plate that is wider than the mandibular ramus.* The subglenoid part of the articular of most dinosaurs is a stout piece of bone that is as deep as it is wide transversely. In some basal sauropodomorphs it is flattened as it is in NM QR3314 (e.g. *Massospondylus carinatus*, SAM-PK-K1314), but in these taxa it is no wider than the mandibular ramus immediately in front of it. In the case of NM QR3314, however, the plate-like articular is flared medially, extending well beyond the glenoid socket, making the articular much wider than the mandibular ramus in front of it (Text-figs 4–5).

11. *Anterior neural canal wider than deep.* In most saurischians the anterior neural canal is subcircular, or

sightly higher than wide (e.g. *Herrerasaurus ischigualastensis*: Sereno and Novas 1993, fig. 11e; *Camarasaurus* sp.: Madsen *et al.* 1995, fig. 53b, h, p, u), when observed both at the foramen magnum and at the axis in anterior view. In contrast, the foramen magnum of NM QR3314 is shaped like a dorsoventrally flattened diamond (Text-fig. 4), while the neural canal of the axis in anterior view is subrectangular and much wider than deep (Text-fig. 18C).

PHYLOGENETIC ANALYSIS

A new cladistic analysis of early sauropodomorph relationships was performed. An effort was made to include all diagnostic species of sauropodomorph basal to Eusauropoda. *Chinshakiangosaurus chunghoensis* and *Yimenosaurus youngi* were excluded because the specimens had not been examined first-hand and the published descriptions were too brief for adequate coding. *Mussaurus patagonicus* was excluded because the currently available specimens are juveniles and new data based on more mature specimens will be available soon (D. Pol, pers. comm. 2005). *Kotasaurus yamanpalliensis* was excluded because it is a chimaera based upon two sauropod taxa of different phylogenetic position (O. Rauhut, pers. comm. 2005).

Data from NM QR3314 and NM 1551 were used to complete the scoring for *Melanorosaurus*. Contrary to previous analyses, *Melanorosaurus* was among the more completely coded taxa in the new analysis (it could be scored for 88·2 per cent of the characters). The analysis included 42 other dinosauriform taxa (Table 1), 33 of which were sauropodomorphs, and three outgroup taxa (*Euparkeria*, Crurotarsi and *Marasuchus*). These taxa were scored for 353 characters (see Appendix).

Results

The character-taxon matrix (see Appendix) was analysed using PAUP 4.0b (Swofford 2002) with the following settings: heuristic search; random addition sequence with 20 replicates; and TBR branch-swapping algorithm. All characters were unweighted and 36 multistate characters were ordered (8, 13, 19, 23, 40, 57, 69, 92, 102, 108, 117, 121, 134, 144, 147, 149, 150, 157, 167, 170, 171, 177, 205, 207, 222, 227, 242, 251, 254, 277, 294, 299, 336, 342, 349, 353). Uncertainties in multistate characters were treated as polymorphisms: for example, if a character state may be 1 or 2 but is certainly not 0, the character would be coded as 1, 2. The search found 60 equally parsimonious trees with a tree length of 1094 steps. The strict consensus of these trees is well resolved (Text-fig. 19). The analysis

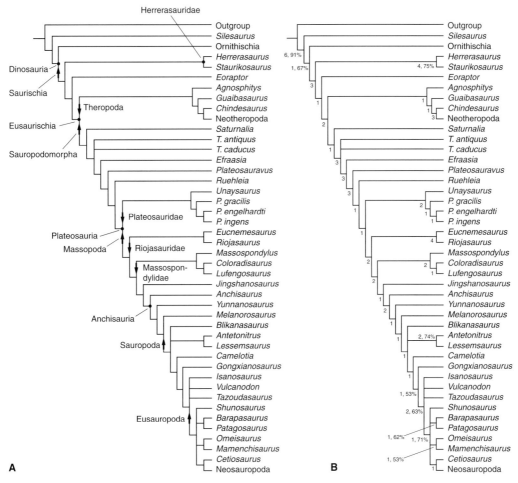

TEXT-FIG. 19. Strict consensus tree of 60 most parsimonious trees (tree length = 1094) produced by the analysis of 42 dinosauriform taxa with a particular emphasis on basal sauropodomorphs. A, cladogram applied with phylogenetic taxonomy; arrows represent stem-based taxa, dots represent node-based taxa. B, cladogram with robustness measures; decay index on the left and bootstrap percentage (when over 50 per cent) on the right.

contained a range of dinosauriform taxa outside of Sauropodomorpha (*Silesaurus*, Ornithischia, *Herrerasaurus*, *Staurikosaurus*, *Eoraptor*, *Agnosphitys*, *Guaibasaurus*, *Chindesaurus* and Neotheropoda) so that the precise order of outgroups to the Sauropodomorpha would not be determined *a priori*. The relationships of these taxa that were found can be gleaned from the cladogram (Text-fig. 19), but only the ingroup relationships of the Sauropodomorpha will be discussed here.

The following sauropodomorph taxa are found to be serially closer outgroups to the least inclusive clade containing *Plateosaurus* and Sauropoda: *Saturnalia*, *Thecodontosaurus* (the two species of *Thecodontosaurus* form a trichotomy with the clade of all more derived sauropodomorphs), *Efraasia*, *Plateosauravus* and *Ruehleia*. The name Plateosauria is available for the least inclusive clade that contains *Plateosaurus* and *Massospondylus*, which, in this analysis, also contains Sauropoda. Sereno's (1998) defini-

tion of Sauropodomorpha also conforms to this clade, but I prefer to use Sauropodomorpha in a stem-based sense as the sister group of Theropoda (Galton and Upchurch 2004). Relatively few species of this clade [*Unaysaurus*, *Plateosaurus gracilis* and *Plateosaurus* (= *Gresslyosaurus*) *ingens*] are found to be more closely related to *Plateosaurus engelhardti* than to Sauropoda, so I refer to them as Plateosauridae and decline to use its heterodefinitional synonym, Prosauropoda. The sister group of Plateosauridae would be Sauropoda, according to the currently widely used definition of the clade (Sereno 1998). However, the content of Sauropoda has expanded beyond all recognition in the topology found here. Thus, a re-definition of Sauropoda is warranted and a new clade name is required for the most inclusive clade that contains *Saltasaurus loricatus* but not *Plateosaurus engelhardti*. Massopoda (Yates 2006) is used here for this clade. Massopod characters of the skull of NM

QR3314 include a short antorbital fossa that has a rostro-caudal length that is less than that of the orbit and the extension of the infratemporal fenestra underneath the orbit. The most basal massopods were found to be a small clade of just two taxa: *Riojasaurus* and *Eucnemesaurus*. The next clade to branch away from the sauropod stem was Massospondylidae (including *Massospondylus*, *Lufengosaurus* and *Coloradisaurus*). Cranial characters supporting the clade of Massospondylidae and all other remaining massopods that can be seen in NM QR3314 include a wide exposure of the maxilla in the margin of the external naris, a slot-shaped subnarial foramen (reversed in sauropods) and an extremely weak impression of the antorbital fossa onto the dorsal end of the ascending ramus of the maxilla. Above the Massospondylidae there is a series of singleton taxa (*Jingshanosaurus*, *Anchisaurus*, *Yunnanosaurus* and *Melanorosaurus*) that form serially closer outgroups to Sauropoda. Above *Jingshanosaurus*, the content of the clade of all remaining massopods matches those of Sauropoda in Yates and Kitching (2003) and Yates (2004). This clade also conforms to the recently defined Anchisauria (the least inclusive clade containing *Melanorosaurus* and *Anchisaurus*; Galton and Upchurch 2004). Anchisaurian synapomorphies that are present in the skull of *Melanorosaurus* include: the shape of the braincase in lateral view, where the ventral surface is lowered to a level well below the ventral rim of the occipital condyle, while the parasphenoid rostrum is raised above it; the presence of wrinkled enamel on the tooth crowns; failure of the antorbital fossa to impress upon the ventral end of the lachrymal; a reduced quadrate foramen located immediately below the quadrate head and bordered laterally by the squamosal; and a wide separation of the basal tubera by a caudally open U-shaped fossa on the basiparasphenoid plate.

Sauropoda as defined in this paper is the most inclusive clade containing *Saltasaurus* but not *Melanorosaurus* (Yates 2006). Thus, *Melanorosaurus* is, by definition, at least part of the sister-group of Sauropoda. What this analysis has found is that *Melanorosaurus* is a singleton taxon that is more closely related to Sauropoda than any other non-sauropod sauropodomorph. The derived cranial characters that support the clade of *Melanorosaurus* + Sauropoda are: an inflection at the base of dorsal process of the premaxilla, creating a premaxillary muzzle; posterior margin of the external naris formed almost entirely by the maxilla; a reversal to a broad quadratojugal ramus of the squamosal; and the loss of the posterolateral process of the palatine, creating a narrow, anterolaterally directed ramus that contacts the maxilla.

Given that the analysis contained a number of poorly known, and thus poorly constrained, taxa, it is unsurpris-

ing that the bootstrap and decay index values are low. In order to evaluate better the strength of the hypothesis presented here one of the 66 most-parsimonious trees was compared with the shortest trees that conform to three competing hypotheses of sauropodomorph relationships: partial, or 'core' prosauropod monophyly (e.g. Yates and Kitching 2003; Yates 2004); fully inclusive prosauropod monophyly (e.g. Galton and Upchurch 2004); and fully inclusive melanorosaurid monophyly, regardless of its wider relationships (e.g. Galton and Upchurch 2004).

The first of these trees was obtained by searching for the shortest tree when the following sauropodomorph topology was enforced as a 'backbone' constraint: (*Saturnalia* + ((*Riojasaurus* (*Plateosaurus* + *Massospondylus*)) + (*Vulcanodon* + (*Shunosaurus* + Neosauropoda)))). The most parsimonious trees when this constraint was enforced were 1106 steps long (12 steps longer than the MPTs). The second tree was obtained by searching for the shortest tree that contains *Saturnalia*, *Thecodontosaurus antiquus*, *T. caducus*, *Efraasia*, *Plateosauravus*, *Ruehleia*, *Unaysaurus*, *Plateosaurus gracilis*, *P. ingens*, *P. engelhardti*, *Eucnemesaurus*, *Riojasaurus*, *Massospondylus*, *Coloradisaurus*, *Lufengosaurus*, *Jingshanosaurus*, *Yunnanosaurus*, *Anchisaurus*, *Camelotia* and *Melanorosaurus* in a clade exclusive of all other ingroup taxa. This gathers all non-sauropod sauropodomorphs into a single monophyletic group. Although *Camelotia* was found to be a true sauropod in the present analysis it is also included here because it was specifically included in the monophyletic Prosauropoda of Galton and Upchurch (2004). The shortest trees that contain such a clade were 1155 steps long (61 steps longer than the MPTs). The third tree was found by searching for the shortest tree that contains *Riojasaurus*, *Eucnemesaurus*, *Camelotia* and *Melanorosaurus* in an exclusive clade, following the content included in Galton and Upchurch (2004). Although *Eucnemesaurus* was not included in their analysis it is included here because of its robust, close relationship to *Riojasaurus*. The shortest trees that conform to this constraint were 1119 steps long (25 steps longer than the MPTs).

When compared with the first tree ('core' prosauropod monophyly), using a Templeton test, the most parsimonious tree was not found to be a significantly better explanation of the data ($P = 0.138$) at $P < 0.05$. On the other hand, the Templeton test result for the comparison with the second tree (fully inclusive prosauropod monophyly) is highly significant ($P < 0.0001$) and the result for the comparison with the third tree (inclusive melanorosaurid monophyly) was also significant ($P = 0.025$). In conclusion, the hypothesis that all non-sauropod sauropodomorphs form a single large clade as suggested by Galton and Upchurch (2004) can be confidently rejected but it is

TABLE 1. Sources of data (literature and specimens) for the ingroup taxa used in the cladistic analysis. OTU, operational taxonomic unit; (c) following a specimen number indicates that only a cast of this specimen was examined. Additional data for *Staurikosaurus*, *Eoraptor*, *Chindesaurus*, *Unaysaurus*, *Plateosaurus* and *Riojasaurus* were gleaned from detailed photographs of the type specimens (referred specimens in the case of *Plateosaurus*).

OTU	Specimens
Silesaurus opolensis	Dzik 2003
Ornithischia	BMNH R1111, R196, RU.B17, B23; BP/1/4885; SAM-PK-K337, K1332; UCMP 130580
	Sereno 1991; Thulborn 1972; Santa Luca 1980; Butler 2005; Galton 1974; Norman *et al.* 2004
Herrerasaurus ischigualastensis	PVL 2566
	Novas 1989, 1993; Sereno 1993; Sereno and Novas 1993
Staurikosaurus pricei	Colbert 1970; Galton 1977, 2000
Eoraptor lunensis	Sereno *et al.* 1993; Rauhut 2003*a*; Langer 2004
Agnosphitys cromhallensis	Fraser *et al.* 2002
Guaibasaurus candelariensis	Bonaparte *et al.* 1999
Chindesaurus bryansmalli	Long and Murry 1995
Neotheropoda	CM 31374 (c); BMNH RU.P76/1; BP/1/5243, 5278; QG165 (currently held at the BP); MB.R.2175.7.4, unnumbered (holotype of *Elaphrosaurus bambergi*); UCMP 37302, 37303, 47721, 77270; MOR 693 (c)
	von Huene 1934; Raath 1969; Rowe 1989; Welles 1984; Currie and Zhao 1993; Madsen 1976; Madsen and Welles 2000; Gilmore 1920; Rauhut 2003*a*
Saturnalia tupiniqum	MCP 3844-PV, 3845-PV
	Langer *et al.* 1999; Langer 2003
Thecodontosaurus caducus	BMNH RU.P24, RU.P24/3, RU.P77/1
	Kermack 1984; Yates 2003*b*
Thecodontosaurus antiquus	YPM 2192, 2195; numerous BRSUG and BRSMG specimens
	von Huene 1908; Benton *et al.* 2000
Anchisaurus polyzelus	YPM 208, 209, 1883; AM 41/109(c)
	Galton 1976; Yates 2004
Riojasaurus incertus	PVL 3526, 3662, 3663, 3805, 3808, PULR 56
	Bonaparte 1972, 1999; Bonaparte and Pumares 1995
Efraasia minor	SMNS 12354, 12667, 12668, 12684, 12843, 14881
	Galton 1973, 1984*b*, 1985a; Galton and Bakker 1985; Yates 2003*c*
Plateosauravus cullingworthi	SAM-PK-K3341, 3342, 3344–3348, 3350, 3351, 3356, 3607, 3609, UCMP 42858
	Haughton 1924; Van Heerden 1979
Ruehleia bedheimensis	MB RvL 1, 2, 3
	Galton 2001
Eucnemesaurus fortis	TM 121, BP/1/6107, 6110–6115, 6220
	Van Hoepen 1920; Van Heerden 1979; Galton and Van Heerden 1998; Yates 2006
Unaysaurus tolentinoi	Leal *et al.* 2004
Plateosaurus gracilis	GPIT 18318a; YPM 2679 (c); SMNS 5715
	von Huene 1908, 1915; Galton 1985a; Yates 2003*c*
Plateosaurus engelhardti	GPIT skelett 1, skelett 2, SMNS 12950, 13200, MB.R. 1937, skelett 25, XXIV
	von Huene 1926; Galton 1984*a*, 1985*c*; Moser 2003
Plateosaurus (= *Gresslyosaurus*) *ingens*	Moser 2003; Galton 1986
Massospondylus carinatus	BP/1/4376, 4693, 4779, 4930, 4934, 4955, 5238, 5241, SAM-PK-K1314
	Van Hoepen 1920; Cooper 1981; Gow 1990; Sues *et al.* 2004; Barrett and Yates 2006
Coloradisaurus brevis	PULR unnumbered, PVL 3967, unnumbered (field no. 6)
	Bonaparte 1978
Lufengosaurus huenei	IVPP V15
	Young 1941; Barrett *et al.* 2005*a*

TABLE 1. Continued.

OTU	Specimens
Yunnanosaurus huangi (= *robustus*)	IVPP V94, V505
	Young 1942, 1951
Jingshanosaurus xinwaensis	LV 3
	Zhang and Yang 1994
Melanorosaurus readi	SAM-PK-K3449, 3450, NM R1551, NM QR3314
	Haughton 1924; Van Heerden and Galton 1997; Galton *et al.* 2005;
	Bonnan and Yates 2007
Antetonitrus ingenipes	BP/1/4952, NM QR 1545
	Yates and Kitching 2003
Lessemsaurus sauropoides	PVL 4822
	Bonaparte 1999; Pol and Powell 2007
Blikanasaurus cromptoni	SAM-PK-K403
	Galton and Van Heerden 1998
Camelotia borealis	Galton 1998
Isanosaurus attavipachi	Buffetaut *et al.* 2000
Gongxianosaurus shibeiensis	He *et al.* 1998
Vulcanodon karibaensis	Raath 1972; Cooper 1984
Tazoudasaurus naimi	Allain *et al.* 2004
Shunosaurus lii	Dong *et al.* 1983; Zhang 1988; Chatterjee and Zheng 2002
Barapasaurus tagorei	Jain *et al.* 1979; Wilson and Sereno 1998
Patagosaurus fariasi	PVL 4170, 4076; MACN CH933
	Bonaparte 1986*a*; Rauhut 2003*b*
Cetiosaurus oxoniensis	Upchurch and Martin 2002, 2003
Mamenchisaurus spp.	Young and Chao 1972; Russell and Zheng 1993;
	Ouyang and Ye 2002; Ye *et al.* 2001
Omeisaurus spp.	Dong *et al.* 1983; He *et al.* 1988; Tang *et al.* 2001
Neosauropoda	MB.R2181, R2223.2.2, R222.2.3, R2249; TATE 099(c); YPM 1980, 1225
	Hatcher 1903; Calvo and Salgado 1995; Osborn and Mook 1921;
	Gilmore 1936; Janensch 1935–36; Madsen *et al.* 1995

not yet possible to reject intermediate hypotheses where a smaller group of 'core' prosauropods (riojasaurids, plateosaurids and massospondylids) form a monophyletic group. Furthermore, the hypothesis that the content of Melanorosauridae is monophyletic, as suggested by Galton and Upchurch (2004), can be rejected.

DISCUSSION

Although the skull of *Melanorosaurus* displays a number of cranial character states that nest it close to the sauropods (as defined here) it is remarkably *Plateosaurus*-like in general appearance and, presumably, function. It also displays two derived character states that are otherwise found only in *Plateosaurus* among sauropodomorphs. These are transversely expanded prefrontals and elongate vomers. Nevertheless, the weight of evidence strongly favours the interpretation that *Melanorosaurus* is quite distant from *Plateosaurus* and is a close relative of the Sauropoda.

Previous classifications of Sauropodomorpha have placed *Melanorosaurus* with *Riojasaurus* in the low-diversity taxon Melanorosauridae. However, these taxa were found to be quite distant from one another in several phylogenetic analyses based on their postcranial anatomy (Yates 2003*b*, 2004; Yates and Kitching 2003). The distant placement of these two taxa is further supported by a comparison of the skulls. *Riojasaurus* differs from *Melanorosaurus* in the following features: a pointed rostrum with strong parasagittal premaxillary ridges; a rounded (vs. slot-shaped) subnarial foramen; lack of an inflection at the base of the dorsal process of the premaxilla; a slender quadratojugal process of the squamosal; a jugal process of the quadratojugal that is subequal in length to the squamosal process and articulates with the lateral side of the jugal; transversely expanded basal tubera with the basioccipital component placed medial to the basiparasphenoid component; a narrow midline notch between the basal tubera; a transverse septum between the basipterygoid processes; a tubercle on the ventrolateral margin of the braincase between the basal tuber and the basipterygoid process; teeth with smooth enamel and coarse serrations extending into the basal half of the crown in anterior maxillary teeth (PULR 56). Many of these features indicate that *Riojasaurus* is less closely related to sauropods than *Melanorosaurus*. As described above, a Templeton test of the shortest tree that gathers the traditional melanorosaurids into a clade was found to be a significantly poorer explanation of the data than the most parsimonious tree, where the component taxa are widely scattered. The Melanorosauridae has been defined as the most inclusive clade including *Melanorosaurus readi* but not *Anchisaurus polyzelus* (Galton and Upchurch 2004). In the phylogeny presented here it would include all Sauropods as well as *Melanorosaurus*. Thus, it is necessary to modify the definition slightly by adding *Saltasaurus loricatus* as a second, exclusive anchor taxon. This restricts Melanorosauridae to a low-diversity clade including only those sauropodomorphs more closely related to *M. readi* than to sauropods or any other well-known sauropodomorph that is the eponym of a higher taxon (Yates 2006). No other named sauropodomorph currently fits this definition.

Cranial specializations that diagnose Sauropoda or the less inclusive Eusauropoda seem to be linked to two main changes. One suite of characters seems to form a complex related to anteroposterior shortening of the skull concentrated in the supratemporal and infratemporal regions (Wilson and Sereno 1998). These include posterodorsally retracted external nares; the loss of the anterior process of the prefrontal; the extension of the infratemporal fenestra beneath the orbit of a point well anterior of the mid length of the orbit; loss of the suborbital bar of the jugal; a ventrally pinched orbit; a short, deep temporal bar; and a transversely orientated slot-shaped supratemporal fenestra. In contrast, the skull of *Melanorosaurus* is as elongate as that of *Plateosaurus*. The second suite of characters is, at least potentially, linked to increased bite size, strengthening of the jaws and teeth, and oral processing of food before swallowing (Upchurch and Barrett 2000). These include strong dorsoventral expansion of the anterior dentary, precise tooth-on-tooth occlusion, broad U-shaped jaws, lateral plates bracing the premaxillary, maxillary and dentary teeth, and loss of buccal emargination on the posterior dentary (Wilson and Sereno 1998; Upchurch and Barrett 2000; Barrett *et al.* 2005*b*). There are specializations of the jaws and teeth that occur at earlier points on the lineage leading from *Plateosaurus* to Eusauropoda but these do not appear to have greatly affected food gathering and oral processing, except perhaps the reduction in size of the external mandibular fenestra. According to the phylogeny preferred here the latter character evolved early during massopod phylogeny, and was present in the common ancestor of *Jingshanosaurus xinwaensis* and Anchisauria. It may be indicative of a

reorganization of the adductor musculature but its full significance remains obscure (Upchurch and Barrett 2000). *Melanorosaurus* still maintained narrow, V-shaped jaws, long, shallow mandibular rami with elongate retroarticular processes, and upper teeth that overbite the lower dentition, without tooth-on-tooth occlusion or bracing from lateral plates. Character optimization also suggests that it had a buccal emargination on the dentary, which is indicative of the presence of fleshy cheeks and hence a limited gape (Upchurch and Barrett 2000; Barrett *et al.* 2005*b*). Such a suite of features suggests simple cropping with little oral processing. It is probable that *Melanorosaurus* selectively browsed softer, nutritious food items and may still have indulged in omnivory as has been suggested for other basal sauropodomorphs (Barrett 2000).

All of this suggests that the initial inception of the sauropod body plan involved numerous changes of the postcranial skeleton (quadrupedalism, manus pronation, increased number of sacral vertebrae, increased size, increased emphasis on graviportalism) before the cranium became dramatically transformed or specialized for bulk browsing of vegetation.

Acknowledgements. The specimen described here was excavated by J. Welman, J. Nyaphuli and J. Mahoi (National Museum) and prepared by J. Nyaphuli. Without their hard work this research would not have been possible. For access to the specimen I thank R. Nutall and J. Botha (National Museum). I thank E. Butler (National Museum) for the invaluable assistance she gave during my visit to the National Museum. Most of the photographs were taken by J. Van Heerden (University of Fort Hare) assisted by C. Vasconcelos (University of the Witwatersrand), except for those on Text-figures 4, 11 and 14, which were taken by D. Pol (Museo Palaeontológico Egidio Feruglio). English translations of the articles by Young and Chao (1972), Dong *et al.* (1983), Bonaparte (1986*a*, 1999), Bonaparte and Pumares (1995) and He *et al.* (1998) were provided by the Polyglot Paleontologist (http://ravenel.si.edu/paleo/paleoglot/index.cfm). M. Langer, R. Irmis and O. Rauhut are thanked for providing detailed photographs of *Staurikosaurus*, *Eoraptor*, *Chindesaurus*, *Guaibasaurus*, *Saturnalia*, *Unaysaurus*, *Plateosaurus* and *Riojasaurus*. Lastly, I thank P. Barrett and P. Upchurch for reading the original version of this paper and for their constructive criticism. The work was funded in part by a National Research Foundation, South Africa (NRF) grant to the Bernard Price Institute. Travel to China (used to code taxa in the phylogenetic analysis) was funded by P.A.S.T.

REFERENCES

ALLAIN, R., AQUESBI, N., DEJAX, J., MEYER, C., MONBARON, M., MONTENAT, C., RICHIR, P., ROCHDY, M., RUSSELL, D. and TAQUET, P. 2004. A basal sauropod dinosaur from the Early Jurassic of Morocco. *Comptes Rendus Palevol*, **3**, 199–208.

BARRETT, P. M. 2000. Prosauropod dinosaurs and iguanas: speculations on the diets of extinct reptiles. 42–78. *In* SUES, H.-D. (ed.). *Evolution of herbivory in terrestrial vertebrates: perspectives from the fossil record.* Cambridge University Press, Cambridge, 256 pp.

—— and YATES, A. M. 2006. New information on the palate and lower jaw of *Massospondylus* (Dinosauria: Sauropodomorpha). *Palaeontologia Africana*, **41**, 123–130.

—— UPCHURCH, P. and WANG XIAO-LIN 2005*a*. Cranial osteology of *Lufengosaurus huenei* Young (Dinosauria: Prosauropoda) from the Lower Jurassic of Yunnan, People's Republic of China. *Journal of Vertebrate Paleontology*, **25**, 806–822.

—— —— XU XING and ZHAO XI-JIN 2005*b*. *Chinshakiangosaurus* and the early evolution of herbivory in sauropod dinosaurs. *Journal of Vertebrate Paleontology*, **25** (Supplement to No. 3), 34A.

BENTON, M. J. and WALKER, A. D. 2002. *Erpetosuchus*, a crocodile-like basal archosaur from the Late Triassic of Elgin, Scotland. *Zoological Journal of the Linnean Society*, **136**, 25–47.

—— JUUL, L., STORRS, G. W. and GALTON, P. M. 2000. Anatomy and systematics of the prosauropod dinosaur *Thecodontosaurus antiquus* from the Upper Triassic of southwest England. *Journal of Vertebrate Paleontology*, **20**, 77–108.

BONAPARTE, J. F. 1972. Los tetrápodos del sector superior de la Formación Los Colorados, La Rioja, Argentina (Triásico Superior). I. Parte. *Opera Lilloana*, **22**, 1–183.

—— 1978. *Coloradia brevis* n. g. et n. sp. (Saurischia, Prosauropoda), dinosaurio Plateosauridae superior de la Formación Los Colorados, Triásico Superior de La Rioja, Argentina. *Ameghiniana*, **15**, 327–332.

—— 1986*a*. Les dinosaurs (carnosaures, allosauridés, sauropodes, cétiosauridés) du Jurassique moyen de Cerro Cóndor (Chubut, Argentine). Part 2. *Annales de Paléontologie*, **72**, 325–386.

—— 1986*b*. The early radiation and phylogenetic relationships of the Jurassic sauropod dinosaurs, based on vertebral anatomy. 247–258. *In* PADIAN, K. (ed.). *The beginning of the age of dinosaurs.* Cambridge University Press, Cambridge, 378 pp.

—— 1999. Evolución de las vertebras presacras en Sauropodomorpha. *Ameghiniana*, **36**, 115–187.

—— and PUMARES, J. A. 1995. Notas sobre el primer craneo de *Riojasaurus incertus* (Dinosauria, Prosauropoda, Melanorosauridae) del Triasico Superior de La Rioja, Argentina. *Ameghiniana*, **32**, 341–349.

—— FERIGOLO, J. and RIBEIRO, A. M. 1999. A new early Late Triassic saurischian dinosaur from Rio Grande do Sul State, Brazil. *National Science Museum Monographs*, **15**, 89–109.

BONNAN, M. F. and YATES, A. M. 2007. A new description of the forelimb of the basal sauropodomorph *Melanorosaurus*: implications for the evolution of pronation, manus shape and quadrupedalism in sauropod dinosaurs. 157–168. *In* BARRETT, P. M. and BATTEN, D. J. (eds). *Evolution and palaeobiology of early sauropodomorph dinosaurs.* Special Papers in Palaeontology, **77**, 289 pp.

BUFFETAUT, E., SUTEETHORN, V., CUNY, G., TONG, H., LE LOEUFF, J., KHANSUBHA, S. and JONG-AUTCHARIYAKUL, S. 2000. The earliest known sauropod dinosaur. *Nature*, **407**, 72–74.

BUTLER, R. J. 2005. The 'fabrosaurid' ornithischian dinosaurs of the Upper Elliot Formation (Lower Jurassic) of South Africa and Lesotho. *Zoological Journal of the Linnean Society*, **145**, 175–218.

CALVO, J. O. and SALGADO, L. 1995. *Rebbachisaurus tessonei* sp. nov., a new Sauropoda from the Albian–Cenomanian of Argentina; new evidence on the origin of the Diplodocidae. *Gaia*, **11**, 13–33.

CARRANO, M. T. 2000. Homoplasy and the evolution of dinosaur locomotion. *Paleobiology*, **26**, 489–512.

CHARIG, A. J., ATTRIDGE, J. and CROMPTON, A. W. 1965. On the origin of the sauropods and the classification of the Saurischia. *Proceedings of the Linnean Society*, **176**, 197–221.

CHATTERJEE, S. 1978. A primitive parasuchid (phytosaur) reptile from the Upper Triassic Maleri Formation of India. *Palaeontology*, **21**, 83–128.

—— and ZHENG ZHONG 2002. Cranial anatomy of *Shunosaurus*, a basal sauropod dinosaur from the Middle Jurassic of China. *Zoological Journal of the Linnean Society*, **136**, 145–169.

COLBERT, E. H. 1970. A saurischian dinosaur from the Triassic of Brasil. *American Museum Novitates*, **2405**, 1–39.

COOPER, M. R. 1980. The first record of the prosauropod dinosaur *Euskelosaurus* from Zimbabwe. *Arnoldia*, **9** (3), 1–17.

—— 1981. The prosauropod dinosaur *Massospondylus carinatus* Owen from Zimbabwe: its biology, mode of life and phylogenetic significance. *Occasional Papers, National Museums and Monuments of Rhodesia, Series B*, **6**, 689–840.

—— 1984. A reassessment of *Vulcanodon karibaensis* Raath (Dinosauria: Saurischia) and the origin of the Sauropoda. *Palaeontologia Africana*, **25**, 203–231.

CRUICKSHANK, A. R. I. 1975. Origin of sauropod dinosaurs. *South African Journal of Science*, **71**, 89–90.

CURRIE, P. J. and ZHAO XI-JIN 1993. A new carnosaur (Dinosauria, Theropoda) from the Jurassic of Xinjiang, People's Republic of China. *Canadian Journal of Earth Sciences*, **30**, 2037–2081.

DONG ZHI-MING, ZHOU SHI-WU and ZHANG YI-HONG 1983. Dinosaurs from the Jurassic of Sichuan. *Palaeontologica Sinica, Series C*, **23**, 1–136. [In Chinese, English summary].

DZIK, J. 2003. A beaked herbivorous archosaur with dinosaur affinities from the early Late Triassic of Poland. *Journal of Vertebrate Paleontology*, **23**, 556–574.

EWER, R. F. 1965. The anatomy of the thecodont reptile *Euparkeria capensis* Broom. *Philosophical Transactions of the Royal Society of London, Series B*, **248**, 379–435.

FRASER, N. C., PADIAN, K., WALKDEN, G. M. and DAVIS, A. L. M. 2002. Basal dinosauriform remains from Britain and the diagnosis of the Dinosauria. *Palaeontology*, **45**, 79–96.

GALTON, P. M. 1973. On the anatomy and relationships of *Efraasia diagnostica* (V. Huene) n. gen., a prosauropod dinosaur (Reptilia: Saurischia) from the Upper Triassic of Germany. *Paläontologische Zeitschrift*, **47**, 229–255.

—— 1974. The ornithischian dinosaur *Hypsilophodon* from the Wealden of the Isle of Wight. *Bulletin of the British Museum (Natural History), Geology*, **25**, 1–152.

—— 1976. Prosauropod dinosaurs (Reptilia: Saurischia) of North America. *Postilla*, **169**, 1–98.

—— 1977. On *Staurikosaurus pricei*, an early saurischian dinosaur from the Triassic of Brazil, with notes on the Herrerasauridae and Poposauridae. *Paläontologische Zeitschrift*, **51**, 234–245.

—— 1984a. Cranial anatomy of the prosauropod dinosaur *Plateosaurus* from the Knollenmergel (Middle Keuper, Upper Triassic) of Germany. 1. Two complete skulls from Trossingen/Württ., with comments on the diet. *Geologica et Palaeontologica*, **18**, 139–171.

—— 1984b. An early prosauropod dinosaur from the Upper Triassic of Nordwürttemberg, West Germany. *Stuttgarter Beiträge zur Naturkunde, Series B*, **106**, 1–25.

—— 1985a. Cranial anatomy of the prosauropod dinosaur *Sellosaurus gracilis* from the Middle Stubensandstein (Upper Triassic) of Nordwürttemberg, West Germany. *Stuttgarter Beiträge zur Naturkunde, Series B*, **118**, 1–39.

—— 1985b. Diet of prosauropods from the Late Triassic and Early Jurassic. *Lethaia*, **18**, 105–123.

—— 1985c. Cranial anatomy of the prosauropod dinosaur *Plateosaurus* from the Knollenmergel (Middle Keuper, Upper Triassic) of Germany. II. All the cranial material and details of the soft-part anatomy. *Geologica et Palaeontologica*, **19**, 119–159.

—— 1986. Prosauropod dinosaur *Plateosaurus* (=*Gresslyosaurus*) (Saurischia: Sauropodomorpha) from the Upper Triassic of Switzerland. *Geologica et Palaeontologica*, **20**, 167–183.

—— 1990. Basal Sauropodomorpha–Prosauropoda. 320–344. *In* WEISHAMPEL, D. B., DODSON, P. and OSMÓLSKA, H. (eds). *The Dinosauria*. University of California Press, Berkeley, CA, 733 pp.

—— 1998. Saurischian dinosaurs from the Upper Triassic of England: *Camelotia* (Prosauropoda, Melanorosauridae) and *Avalonianus* (Theropoda, ?Carnosauria). *Palaeontographica Abteilung A*, **250**, 155–172.

—— 2000. Are *Spondylosoma* and *Staurikosaurus* (Santa Maria Formation, Middle–Upper Triassic, Brazil) the oldest saurischian dinosaurs? *Paläontologische Zeitschrift*, **74**, 393–423.

—— 2001. The prosauropod dinosaur *Plateosaurus* Meyer, 1837 (Saurischia: Sauropodomorpha; Upper Triassic). II. Notes on the referred species. *Revue de Paléobiologie, Genève*, **20**, 435–502.

—— and BAKKER, R. T. 1985. The cranial anatomy of the prosauropod dinosaur 'Efraasia diagnostica', a juvenile individual of *Sellosaurus gracilis* from the Upper Triassic of Nordwürttemberg, West Germany. *Stuttgarter Beiträge zur Naturkunde, Series B*, **117**, 1–15.

—— and UPCHURCH, P. 2004. Prosauropoda. 232–258. *In* WEISHAMPEL, D. B., DODSON, P. and OSMÓLSKA, H. (eds). *The Dinosauria*. Second edition. University of California Press, Berkeley, CA, 861 pp.

—— and VAN HEERDEN, J. 1998. Anatomy of the prosauropod dinosaur *Blikanasaurus cromptoni* (Upper Triassic, South Africa), with notes on other tetrapods from the lower Elliot Formation. *Paläontologische Zeitschrift*, **72**, 163–177.

—— —— and YATES, A. M. 2005. Postcranial anatomy of referred specimens of *Melanorosaurus* from the Upper Triassic of South Africa. 1–37. *In* TIDWELL, V. and CARPENTER, K. (eds). *Thunder-lizards: the sauropodomorph dinosaurs*. Indiana University Press, Bloomington, IN, 495 pp.

GAUFFRE, F.-X. 1993. The most recent Melanorosauridae (Saurischia, Prosauropoda), Lower Jurassic of Lesotho, with remarks on the prosauropod phylogeny. *Neues Jahrbuch für Geologie und Paläontologie, Monatschefte*, **1993**, 648–654.

GAUTHIER, J. 1986. Saurischian monophyly and the origin of birds. *Memoirs of the Californian Academy of Sciences*, **8**, 1–55.

GILMORE, C. W. 1920. Osteology of the carnivorous Dinosauria in the United States National Museum, with special reference to the genera *Antrodemus* (*Allosaurus*) and *Ceratosaurus*. *Bulletin of the United States National Museum*, **110**, 1–154.

—— 1936. Osteology of *Apatosaurus* with special reference to specimens in the Carnegie Museum. *Memoirs of the Carnegie Museum*, **11**, 175–300.

GOW, C. E. 1990. Morphology and growth of the *Massospondylus* braincase (Dinosauria, Prosauropoda). *Palaeontologia Africana*, **27**, 59–75.

HATCHER, J. B. 1903. Osteology of *Haplocanthosaurus*, with description of a new species, and remarks on the probable habits of these Sauropoda and the age and origin of the *Atlantosaurus* beds. *Memoirs of the Carnegie Museum*, **2**, 1–75.

HAUGHTON, S. H. 1924. The fauna and stratigraphy of the Stormberg Series. *Annals of the South African Museum*, **12**, 323–497.

HE XIN-LU, LI KUI and CAI KAI-JI 1988. *The Middle Jurassic dinosaur fauna from Dashanpu, Zigong, Sichuan. IV. Sauropod dinosaurs (2) Omeisaurus tianfuensis*. Sichuan Publishing House of Science and Technology, Chengdu, 143 pp. [In Chinese, English summary].

—— WANG CHANG-SHENG, LIU SHANG-ZHONG, ZHOU FENG-YUN, LIU TU-QIANG, CAI KAI-JI and DAI BING 1998. A new species of sauropod from the Early Jurassic of Gongxian County, Sichuan. *Acta Geologica Sichuan*, **18**, 1–6. [In Chinese, English abstract].

HOLTZ, T. R. Jr 1994. The phylogenetic position of Tyrannosauridae: implications for theropod systematics. *Journal of Paleontology*, **68**, 1100–1117.

HUENE, F. von 1908. Die dinosaurier der europäischen Triasformation mit Berücksichtigung der aussereuropäischen Vorkommnisse. *Geologische und Palaeontologische Abhandlungen, Supplement*, **1**, 1–419.

—— 1915. Beiträge zur Kenntnis einiger Saurischier der schwäbischen Trias. *Neuen Jahrbuch für Mineralogy, Geologie und Paläontologie*, **1915**, 1–27.

—— 1926. Vollständige Osteologie eines Plateosauriden aus der schwäbischen Trias. *Geologische und Palaeontologische Abhandlungen*, **15**, 129–179.

—— 1934. Ein neuer Coelurosaurier in der thüringischen Trias. *Palaeontologische Zeitschrift*, **16**, 145–170.

HUXLEY, T. H. 1866. On the remains of large dinosaurian reptiles from the Stormberg Mountains, South Africa. *Geological Magazine*, **3**, 563.

JAIN, S. L., KUTTY, T. S., ROYCHOWDHURY, T. and CHATTERJEE, S. 1979. Some characteristics of *Barapasaurus tagorei*, a sauropod dinosaur from the Lower Jurassic of Deccan, India. 204–216. *In* LASKAR, B. and RAJA RAO, C. S. (eds). *Proceedings of the IV International Gondwana Symposium, Calcutta*. Volume 1. Hindusthan Publishing Corporation, Dehli, 384 pp.

JANENSCH, W. 1935–36. Die Schädel der Sauropoden *Brachiosaurus, Barosaurus* und *Dicraeosaurus* aus den Tendaguruschichten Deutsch-Ostafrikas. *Palaeontographica (Supplement 7)*, **2**, 147–298.

KERMACK, D. 1984. New prosauropod material from South Wales. *Zoological Journal of the Linnean Society*, **92**, 67–104.

LANGER, M. C. 2003. The pelvic and hindlimb anatomy of the stem-sauropodomorph *Saturnalia tupiniquim* (Late Triassic, Brazil). *PaleoBios*, **23**, 1–40.

—— 2004. Basal Saurischia. 25–46. *In* WEISHAMPEL, D. B., DODSON, P. and OSMÓLSKA, H. (eds). *The Dinosauria*. Second edition. University of California Press, Berkeley, CA, 861 pp.

—— ABDALA, F., RICHTER, M. and BENTON, M. J. 1999. A sauropodomorph dinosaur from the Upper Triassic (Carnian) of southern Brazil. *Comptes Rendus de l'Académie des Sciences de Paris, Série 2, Sciences de la Terre et des Planètes*, **329**, 511–517.

LEAL, L. A., AZEVEDO, S. A. K., KELLNER, A. W. A. and DA ROSA, Á. A. S. 2004. A new early dinosaur (Sauropodomorpha) from the Caturrita Formation (Late Triassic), Paraná Basin, Brazil. *Zootaxa*, **690**, 1–24.

LONG, R. A. and MURRY, P. A. 1995. Late Triassic (Carnian and Norian) tetrapods from the south-western United States. *Bulletin of the New Mexico Museum of Natural History and Science*, **4**, 1–254.

LUCAS, S. G. and HANCOX, P. J. 2001. Tetrapod-based correlation of the nonmarine Upper Triassic of southern Africa. *Albertiana*, **25**, 5–9.

MADSEN, J. H. 1976. *Allosaurus fragilis*: a revised osteology. *Utah Geological Survey, Bulletin*, **109**, 1–163.

—— and WELLES, S. P. 2000. *Ceratosaurus* (Dinosauria, Theropoda) – a revised osteology. *Utah Geological Survey, Miscellaneous Publications*, **00-2**, 1–80.

—— McINTOSH, J. S. and BERMAN, D. S. 1995. Skull and atlas-axis complex of the Upper Jurassic sauropod *Camarasaurus* Cope (Reptilia: Saurischia). *Bulletin of the Carnegie Museum of Natural History*, **31**, 1–115.

MOLNAR, R. E., KURZANOV, S. M. and DONG ZHI-MING 1990. Carnosauria. 169–209. *In* WEISHAMPEL, D. B., DODSON, P. and OSMÓLSKA, H. (eds). *The Dinosauria*. University of California Press, Berkeley, CA, 733 pp.

MOSER, M. 2003. *Plateosaurus engelhardti* Meyer, 1837 (Dinosauria: Sauropodomorpha) aus dem Feuerletten (Mittelkeuper; Obertrias) von Bavaria. *Zitteliana*, **24**, 3–186.

NORMAN, D. B., SUES, H.-D., WITMER, L. M. and CORIA, R. A. 2004. Basal Ornithopoda. 393–412. *In* WEISHAMPEL, D. B., DODSON, P. and OSMÓLSKA, H. (eds). *The Dinosauria*. Second edition. University of California Press, Berkeley, CA, 861 pp.

NOVAS, F. E. 1989. The tibia and tarsus in Herrerasauridae (Dinosauria, *incertae sedis*) and the origin and evolution of the dinosaurian tarsus. *Journal of Paleontology*, **63**, 677–690.

—— 1992. Phylogenetic relationships of the basal dinosaurs, the Herrerasauridae. *Palaeontology*, **35**, 51–62.

—— 1993. New information on the systematics and postcranial skeleton of *Herrerasaurus ischigualastensis* (Theropoda: Herrerasauridae) from the Ischigualasto Formation (Upper Triassic) of Argentina. *Journal of Vertebrate Paleontology*, **13**, 400–423.

—— 1996. Dinosaur monophyly. *Journal of Vertebrate Paleontology*, **16**, 723–741.

OSBORN, H. F. and MOOK, C. C. 1921. *Camarasaurus, Amphicoelias*, and other sauropods of Cope. *Memoirs of the American Museum of Natural History, New Series*, **3**, 249–386.

OUYANG HUI and YE YONG 2002. *The first mamenchisaurian skeleton with complete skull*, Mamenchisaurus youngi. Sichuan Science and Technology Press, Chengdu, 111 pp. [In Chinese, English summary].

PÉREZ-MORENO, B. P., SANZ, J. L., BUSCALIONI, A. D., MORATALLA, J. J., ORTEGA, F. and RASSKIN-GUTMAN, D. 1994. A unique multi-toothed ornithomimosaur dinosaur from the Lower Cretaceous of Spain. *Nature*, **370**, 363–367.

POL, D. and POWELL, J. E. 2007. New information on *Lessemsaurus sauropoides* (Dinosauria: Sauropodomorpha) from the Upper Triassic of Argentina. 223–243. *In* BARRETT, P. M. and BATTEN, D. J. (eds). *Evolution and palaeobiology of early sauropodomorph dinosaurs*. Special Papers in Palaeontology, **77**, 289 pp.

RAATH, M. A. 1969. A new coelurosaurian dinosaur from the Forest Sandstone of Rhodesia. *Arnoldia*, **4** (28), 1–25.

—— 1972. Fossil vertebrate studies in Rhodesia: a new dinosaur (Reptilia: Saurischia) from near the Trias-Jurassic boundary. *Arnoldia*, **5** (30), 1–37.

RAUHUT, O. W. M. 2003*a*. The interrelationships and evolution of basal theropod dinosaurs. *Special Papers in Palaeontology*, **69**, 1–213.

—— 2003*b*. A dentary of *Patagosaurus* (Sauropoda) from the Middle Jurassic of Patagonia. *Ameghiniana*, **40**, 425–432.

ROMER, A. S. 1956. *Osteology of the reptiles*. University of Chicago Press, Chicago, IL, 772 pp.

ROWE, T. 1989. A new species of the theropod dinosaur *Syntarsus* from the Early Jurassic Kayenta Formation of Arizona. *Journal of Vertebrate Paleontology*, **9**, 125–136.

RUSSELL, D. A. and ZHENG ZHONG 1993. A large mamenchisaurid from the Junggar Basin, Xinjiang, People's Republic of China. *Canadian Journal of Earth Sciences*, **30**, 2082–2095.

SALGADO, L., CALVO, J. O. and CORIA, R. A. 1997. Evolution of the titanosaurid sauropods. 1. Phylogenetic analysis based on postcranial evidence. *Ameghiniana*, **34**, 3–32.

SANTA LUCA, A. P. 1980. The postcranial skeleton of *Heterodontosaurus tucki* from the Stormberg of South Africa. *Annals of the South African Museum*, **79**, 159–211.

SERENO, P. C. 1991. *Lesothosaurus*, 'fabrosaurids', and the early evolution of Ornithischia. *Journal of Vertebrate Paleontology*, **11**, 168–197.

—— 1993. The pectoral girdle and forelimb of the basal theropod *Herrerasaurus ischigualastensis*. *Journal of Vertebrate Paleontology*, **13**, 425–450.

—— 1998. A rationale for phylogenetic definitions, with application to the higher-level phylogeny of Dinosauria. *Neues Jahrbuch für Geologie und Paläontologie, Abhandlungen*, **210**, 41–83.

—— 1999. The evolution of dinosaurs. *Science*, **284**, 2137–2147.

—— and NOVAS, F. E. 1993. The skull and neck of the basal theropod *Herrerasaurus ischigualastensis*. *Journal of Vertebrate Paleontology*, **13**, 451–476.

—— DUTHEIL, D. B., IAROCHENE, M., LARSSON, H. C. E., LYON, G. H., MAGWENE, P. M., SIDOR, C. A., VARRICCHIO, D. J. and WILSON, J. A. 1996. Predatory dinosaurs from the Sahara and Late Cretaceous faunal differentiation. *Science*, **272**, 986–991.

—— FORSTER, C. A., ROGERS, R. R. and MONETTA, A. M. 1993. Primitive dinosaur skeleton from Argentina and the early evolution of Dinosauria. *Nature*, **361**, 64–66.

SUES, H.-D., REISZ, R. R., HINIC, S. and RAATH, M. A. 2004. On the skull of *Massospondylus carinatus* Owen, 1854 (Dinosauria: Sauropodomorpha) from the Elliot and Clarens formations (Lower Jurassic) of South Africa. *Annals of the Carnegie Museum*, **73**, 239–257.

SWOFFORD, D. L. 2002. PAUP* Phylogenetic Analysis Using Parsimony (*and Other Methods). Version 4. Sinauer Associates, Sunderland, MA.

TANG FENG, JIN XING-SHENG, KANG XI-MIN and ZHANG GUO-JUN 2001. Omeisaurus maoianus. *A complete Sauropoda from Jingyan, Sichuan*. China Ocean Press, Beijing, 128 pp. [In Chinese, English summary].

THULBORN, R. A. 1972. The post-cranial skeleton of the Triassic ornithischian dinosaur *Fabrosaurus australis*. *Palaeontology*, **15**, 29–60.

UPCHURCH, P. 1995. The evolutionary history of sauropod dinosaurs. *Philosophical Transactions of the Royal Society of London, Series B*, **349**, 365–390.

—— 1998. The phylogenetic relationships of sauropod dinosaurs. *Zoological Journal of the Linnean Society*, **124**, 43–103.

—— and BARRETT, P. M. 2000. The evolution of sauropod feeding mechanisms. 79–122. *In* SUES, H.-D. (ed.). *Evolution of herbivory in terrestrial vertebrates: perspectives from the fossil record*. Cambridge University Press, Cambridge, 256 pp.

—— —— and DODSON, P. 2004. Sauropoda. 259–322. *In* WEISHAMPEL, D. B., DODSON, P. and OSMÓLSKA, H. (eds). *The Dinosauria*. Second edition. University of California Press, Berkeley, CA, 861 pp.

—— and MARTIN, J. 2002. The Rutland *Cetiosaurus*: the anatomy and relationships of a Middle Jurassic British sauropod dinosaur. *Palaeontology*, **45**, 1049–1074.

—— —— 2003. The anatomy and taxonomy of *Cetiosaurus* (Saurischia, Sauropoda) from the Middle Jurassic of England. *Journal of Vertebrate Paleontology*, **23**, 208–231.

VAN HEERDEN, J. 1979. The morphology and taxonomy of *Euskelosaurus* (Reptilia: Saurischia; Late Triassic) from South Africa. *Navorsinge van die Nasionale Museum*, **4**, 21–84.

—— and GALTON, P. M. 1997. The affinities of *Melanorosaurus* – a Late Triassic prosauropod dinosaur from South Africa.

Neues Jahrbuch für Geologie und Paläontologie, Monatschefte, **1997**, 39–55.

VAN HOEPEN, E. C. N. 1920. Contributions to the knowledge of the reptiles of the Karoo Formation. 6. Further dinosaurian material in the Transvaal Museum. *Annals of the Transvaal Museum*, **7**, 93–140.

WALKER, A. D. 1961. Triassic reptiles from the Elgin area: *Stagonolepis*, *Dasygnathus* and their allies. *Philosophical Transactions of the Royal Society of London, Series B*, **244**, 103–204.

WELLES, S. P. 1984. *Dilophosaurus wetherilli* (Dinosauria, Theropoda). Osteology and comparisons. *Palaeontographica A*, **185**, 85–180.

WELMAN, J. 1999. The basicranium of a basal prosauropod from the *Euskelosaurus* range zone and thoughts on the origin of dinosaurs. *Journal of African Earth Sciences*, **29**, 227–232.

WILSON, J. A. 1999. A nomenclature for vertebral laminae in sauropods and other saurischian dinosaurs. *Journal of Vertebrate Paleontology*, **19**, 639–653.

—— 2002. Sauropod dinosaur phylogeny: critique and cladistic analysis. *Zoological Journal of the Linnean Society*, **136**, 217–276.

—— and SERENO, P. C. 1998. Early evolution and higher-level phylogeny of sauropod dinosaurs. *Memoir of the Society of Vertebrate Paleontology*, **5**, 1–68.

WITMER, L. M. 1997. The evolution of the antorbital cavity of archosaurs: a study in soft-tissue reconstruction in the fossil record with an analysis of the function of pneumaticity. *Memoir of the Society of Vertebrate Palaeontology*, **3**, 1–73.

YATES, A. M. 2003a. A definite prosauropod dinosaur from the lower Elliot Formation (Norian: Upper Triassic) of South Africa. *Palaeontologia Africana*, **39**, 63–68.

—— 2003b. A new species of the primitive dinosaur, *Thecodontosaurus* (Saurischia: Sauropodomorpha) and its implications for the systematics of early dinosaurs. *Journal of Systematic Palaeontology*, **1**, 1–42.

—— 2003c. The species taxonomy of the sauropodomorph dinosaurs from the Löwenstein Formation (Norian, Late Triassic) of Germany. *Palaeontology*, **46**, 317–337.

—— 2004. *Anchisaurus polyzelus* (Hitchcock): the smallest known sauropod dinosaur and the evolution of gigantism amongst sauropodomorph dinosaurs. *Postilla*, **230**, 1–58.

—— 2006. Solving a dinosaurian puzzle: the identity of *Aliwalia rex* Galton. *Historical Biology*, First article, DOI:10.1080/08912960600866953.

—— and KITCHING, J. W. 2003. The earliest known sauropod dinosaur and the first steps towards sauropod locomotion. *Proceedings of the Royal Society of London, Series B*, **270**, 1753–1758.

YE YONG, OUYANG HUI and FU QUIAN-MING 2001. New material of *Mamenchisaurus hochuanensis* from Zigong, Sichuan. *Vertebrata Palasiatica*, **39**, 266–271. [In Chinese, English summary].

YOUNG CHUNG-CHIEN 1941. A complete osteology of *Lufengosaurus huenei* Young (gen. et sp. nov.). *Palaeontologica Sinica, Series C*, **7**, 1–53.

—— 1942. *Yunnanosaurus huangi* Young (gen. et sp. nov.), a new Prosauropoda from the Red Beds at Lufeng, Yunnan. *Bulletin of the Geological Society of China*, **22**, 63–104.

—— 1951. The Lufeng saurischian fauna in China. *Palaeontologica Sinica, Series C*, **13**, 1–96.

—— and CHAO XI-JIN 1972. *Mamenchisaurus* sp. nov. *Monograph of the Institute of Vertebrate Paleontology and Paleoanthropology, Series I*, **8**, 1–32. [In Chinese].

ZHANG YI-HONG 1988. *The Middle Jurassic dinosaur fauna from Dashanpu, Zigong, Sichuan: sauropod dinosaurs (1).*

Shunosaurus. Sichuan Publishing House of Science and Technology, Chengdu, 89 pp. [In Chinese, English summary].

—— and YANG ZHAO-LONG 1994. *A new complete osteology of Prosauropoda in Lufeng Basin, Yunnan, China:* Jingshanosaurus. Yunnan Publishing House of Science and Technology, Kunming, 100 pp. [In Chinese, English summary].

APPENDIX

Character list

1. Skull to femur ratio: greater than (0), or less than (1), 0·6 (modified from Gauthier 1986).

2. Lateral plates appressed to the labial side of the premaxillary, maxillary and dentary teeth: absent (0) or present (1) (Upchurch 1995).

3. Relative height of the rostrum at the posterior margin of the naris: more than (0), or less than (1), 0·6 of the height of the skull at the middle of the orbit (Langer 2004).

4. Foramen on the lateral surface of the premaxillary body: absent (0) or present (1).

5. Distal end of the dorsal premaxillary process: tapered (0) or transversely expanded (1) (Sereno 1999).

6. Profile of premaxilla: convex (0) or with an inflection at the base of the dorsal process (1) (Upchurch 1995).

7. Size and position of the posterolateral process of premaxilla: large and lateral to the anterior process of the maxilla (0) or small and medial to the anterior process of the maxilla (1).

8. Relationship between posterolateral process of the premaxilla and the anteroventral process of the nasal: broad sutured contact (0), point contact (1), or separated by maxilla (2) (modified from Gauthier 1986). Ordered.

9. Posteromedial process of the premaxilla: absent (0) or present (1) (Rauhut 2003a).

10. Shape of the anteromedial process of the maxilla: narrow, elongated and projecting anterior to lateral premaxilla-maxilla suture (0) or short, broad and level with lateral premaxilla-maxilla suture (1).

11. Development of external narial fossa: absent to weak (0) or well developed with sharp posterior and anteroventral rims (1).

12. Development of narial fossa on the anterior ramus of the maxilla: weak and orientated laterally to dorsolaterally (0) or well developed and forming a horizontal shelf (1) (modified from Upchurch 1995).

13. Size and position of subnarial foramen: absent (0), small (no larger than adjacent maxillary neurovascular foramina) and positioned outside of narial fossa (1), or large and on the rim of, or inside, the narial fossa (2) (modified from Sereno *et al.* 1993). Ordered.

14. Shape of subnarial foramen: rounded (0) or slot-shaped (1).

15. Maxillary contribution to the margin of the narial fossa: absent (0) or present (1).

16. Diameter of external naris: less than (0), or greater than (1), 0·5 of the orbital diameter (Wilson and Sereno 1998).

17. Shape of the external naris (in adults): rounded (0) or subtriangular with an acute posteroventral corner (1) (Galton and Upchurch 2004).

18. Level of the anterior margin of the external naris: anterior to (0) or posterior to (1) the midlength of the premaxillary body (Rauhut 2003a).

19. Level of the posterior margin of external naris: anterior to, or level with the premaxilla-maxilla suture (0), posterior to the first maxillary alveolus (1), or posterior to the midlength of the maxillary tooth row and the anterior margin of the antorbital fenestra (2) (modified from Wilson and Sereno 1998). Ordered.

20. Dorsal profile of the snout: straight to gently convex (0) or with a depression behind the naris (1).

21. Elongate median nasal depression: absent (0) or present (1) (Sereno 1999).

22. Width of anteroventral process of nasal at its base: less than (0) or greater than (1) width of anterodorsal process at its base (modified from Sereno 1999).

23. Nasal relationship with dorsal margin of antorbital fossa: not contributing to the margin of the antorbital fossa (0), lateral margin overhangs the antorbital fossa and forms its dorsal margin (1), overhang extensive, obscuring the dorsal lachrymal-maxilla contact in lateral view (2) (modified from Sereno 1999).

24. Pointed caudolateral process of the nasal overlapping the lachrymal: absent (0) or present (1) (Sereno 1999).

25. Anterior profile of the maxilla: slopes continuously towards the rostral tip (0) or with a strong inflection at the base of the ascending ramus, creating a rostral ramus with parallel dorsal and ventral margins (1) (Sereno *et al.* 1996).

26. Length of rostral ramus of the maxilla: less than (0), or greater than (1), its dorsoventral depth (Sereno *et al.* 1996).

27. Shape of the main body of the maxilla: tapering posteriorly (0) or dorsal and ventral margins parallel for most of their length (1).

28. Shape of the ascending ramus of the maxilla in lateral view: tapering dorsally (0) or with an anteroposterior expansion at the dorsal end (1).

29. Rostrocaudal length of the antorbital fossa: greater (0), or less (1), than that of the orbit (Yates 2003b).

30. Posteroventral extent of medial wall of antorbital fossa: reaching (0), or terminating anterior to (1), the anterior tip of the jugal (modified from Galton and Upchurch 2004).

31. Development of the antorbital fossa on the ascending ramus of the maxilla: deeply impressed and delimited by a

sharp, scarp-like rim (0) or weakly impressed and delimited by a rounded rim or a change in slope (1).

32. Shape of the antorbital fossa: crescentic with a strongly concave posterior margin that is roughly parallel to the rostral margin of the antorbital fossa (0), subtriangular with a straight to gently concave posterior margin (1), or antorbital fossa absent (2) (modified from Galton 1985*a*).

33. Size of the neurovascular foramen at the caudal end of the lateral maxillary row: not larger than the others (0) or distinctly larger than the others in the row (1) (Yates 2003*b*).

34. Direction that the neurovascular foramen at the caudal end of the lateral maxillary row opens: caudally (0) or rostrally, ventrally or laterally (1) (modified from Sereno 1999).

35. Arrangement of lateral maxillary neurovascular foramina: linear (0) or irregular (1) (modified from Sereno 1999).

36. Longitudinal ridge on the posterior lateral surface of the maxilla: absent (0) or present (1) (Barrett *et al.* 2005*a*).

37. Dorsal exposure of the lachrymal: present (0) or absent (1) (Gauthier 1986).

38. Shape of the lachrymal: dorsoventrally short and block-shaped (0) or dorsoventrally elongate and shaped like an inverted L (1) (Rauhut 2003*a*).

39. Orientation of the lachrymal orbital margin: strongly sloping anterodorsally (0) or erect and close to vertical (1).

40. Length of the anterior ramus of the lachrymal: greater than (0), or less than (1), half the length of the ventral ramus, or absent altogether (2) (modified from Galton 1990). Ordered.

41. Web of bone spanning junction between anterior and ventral rami of lachrymal: absent and antorbital fossa laterally exposed (0) or present, obscuring posterodorsal corner of antorbital fossa (1).

42. Extension of the antorbital fossa onto the ventral end of the lachrymal: present (0) or absent (1) (modified from Wilson and Sereno 1998).

43. Length of the caudal process of the prefrontal: short (0), or elongated (1), so that total prefrontal length is equal to the rostrocaudal diameter of the orbit (Galton 1985*a*).

44. Ventral process of prefrontal extending down the posteromedial side of the lachrymal: present (0) or absent (1) (Wilson and Sereno 1998).

45. Maximum transverse width of the prefrontal: less than (0), or more than (1), 0·25 of the skull width at that level (modified from Galton 1990).

46. Shape of the orbit: subcircular (0) or ventrally constricted making the orbit subtriangular (1) (Wilson and Sereno 1998).

47. Slender anterior process of the frontal intruding between the prefrontal and the nasal: absent (0) or present (1) (modified from Sereno 1999).

48. Jugal-lachrymal relationship: lachrymal overlapping lateral surface of jugal or abutting it dorsally (0), or jugal overlapping lachrymal laterally (1) (Sereno *et al.* 1993).

49. Shape of the suborbital region of the jugal: an anteroposteriorly elongate bar (0) or an anteroposteriorly shortened plate (1).

50. Jugal contribution to the antorbital fenestra: absent (0) or present (1) (Holtz 1994).

51. Dorsal process of the anterior jugal: present (0) or absent (1) (modified from Rauhut 2003*a*)

52. Ratio of the minimum depth of the jugal below the orbit to the distance between the rostral end of the jugal and the rostroventral corner of the infratemporal fenestra: less than (0), or greater than (1), 0·2 (modified from Galton 1985*a*).

53. Transverse width of the ventral ramus of the postorbital: less than (0), or greater than (1), its rostrocaudal width at mid-shaft (Wilson and Sereno 1998).

54. Shape of the dorsal margin of postorbital in lateral view: straight to gently curved (0) or with a distinct embayment between the anterior and posterior dorsal processes (1).

55. Height of the postorbital rim of the orbit: flush with the posterior lateral process of the postorbital (0) or raised so that it projects laterally to the posterior dorsal process (1).

56. Postfrontal bone: present (0) or absent (1) (Sereno *et al.* 1993).

57. Position of the rostral margin of the infratemporal fenestra: behind the orbit (0), extends under the rear half of the orbit (1), or extends as far forward as the midlength of the orbit (2) (modified from Upchurch 1995). Ordered.

58. Frontal contribution to the supratemporal fenestra: present (0) or absent (1) (modified from Gauthier 1986).

59. Orientation of the long axis of the supratemporal fenestra: longitudinal (0) or transverse (1) (Wilson and Sereno 1998).

60. Medial margin of supratemporal fossa: simple smooth curve (0) or with a projection at the frontal/postorbital-parietal suture producing a scalloped margin (1) (Leal *et al.* 2004).

61. Length of the quadratojugal ramus of the squamosal relative to the width at its base: less than (0), or greater than (1), four times its width (Sereno 1999).

62. Proportion of infratemporal fenestra bordered by squamosal: more than (0), or less than (1), 0·5 of the depth of the infratemporal fenestra.

63. Squamosal-quadratojugal contact: present (0) or absent (1) (Gauthier 1986).

64. Angle of divergence between jugal and squamosal rami of quadratojugal: close to 90 degrees (0) or close to parallel (1).

65. Length of jugal ramus of quadratojugal: no longer than (0), or longer than (1), the squamosal ramus (Wilson and Sereno 1998).

66. Shape of the rostral end of the jugal ramus of the quadratojugal: tapered (0) or dorsoventrally expanded (1) (Wilson and Sereno 1998).

67. Relationship of quadratojugal to jugal: jugal overlaps the lateral surface of the quadratojugal (0), quadratojugal overlaps the lateral surface of the jugal (1), or quadratojugal sutures along the ventrolateral margin of the jugal (2). Unordered.

68. Position of the quadrate foramen: on the quadrate-quadratojugal suture (0), deeply incised into, and partly encircled by, the quadrate (1), or on the quadrate-squamosal suture, just below the quadrate head (2) (modified from Rauhut 2003*a*). Unordered.

69. Shape of posterolateral margin of quadrate: sloping anterolaterally from posteromedial ridge (0), everted posteriorly creating a posteriorly facing fossa (1), posterior fossa deeply excavated, invading quadrate body (2) (Wilson and Sereno 1998). Ordered.

70. Exposure of the lateral surface of the quadrate head: absent, covered by lateral sheet of the squamosal (0) or present (1) (Sereno *et al.* 1993).

71. Proportion of the length of the quadrate that is occupied by the pterygoid wing: at least 70 per cent (0) or greater than 70 per cent (1) (Yates 2003b).

72. Depth of the occipital wing of the parietal: less than (0), or more than (1), 1·5 times the depth of the foramen magnum (Wilson and Sereno 1998).

73. Position of foramina for mid-cerebral vein on occiput: between supraoccipital and parietal (0) or on the supraoccipital (1) (modified from Yates 2003b).

74. Postparietal fenestra between supraoccipital and parietals: absent (0) or present (1).

75. Shape of the supraoccipital: diamond-shaped, at least as high as wide (0), or semilunate and wider than high (1) (Yates 2003c).

76. Orientation of the supraoccipital plate: erect to gently sloping (0) or strongly sloping forward so that the dorsal tip lies level with the basipterygoid processes (1) (Galton and Upchurch 2004).

77. Orientation of the paroccipital processes in occipital view: slightly dorsolaterally directed to horizontal (0) or ventrolaterally directed (1) (Rauhut 2003a).

78. Orientation of the paroccipital processes in dorsal view: posterolateral forming a V-shaped occiput (0) or lateral forming a flat occiput (1) (Wilson 2002).

79. Size of the post-temporal fenestra: large fenestra (0) or a small hole that is much less than half the depth of the paroccipital process (1).

80. Exit of the mid-cerebral vein: through trigeminal foramen (0) or through a separate foramen anterodorsal to trigeminal foramen (1) (Rauhut 2003a).

81. Shape of the floor of the braincase in lateral view: relatively straight with the basal tuberae, basipterygoid processes and parasphenoid rostrum roughly aligned (0), bent with the basipterygoid processes and the parasphenoid rostrum below the level of the basioccipital condyle and the basal tuberae (1), or bent with the basal tuberae lowered below the level of the basioccipital and the parasphenoid rostrum raised above it (2) (modified from Galton 1990). Unordered.

82. Shape of basal tuberae: knob-like, with basisphenoidal component rostral to basioccipital component (0), or forming a transverse ridge with the basisphenoidal component lateral to the basioccipital component (1).

83. Length of the basipterygoid processes (from the top of the parasphenoid to the tip of the process): less than (0), or greater than (1), the height of the braincase (from the top of the parasphenoid to the top of the supraoccipital) (Benton et al. 2000).

84. Ridge formed along the junction of the parabasisphenoid and the basioccipital, between the basal tuberae: present with a smooth rostral face (0), present with a median fossa on the rostral face (1), or absent with the basal tuberae being separated by a deep caudally opening U-shaped fossa (2). Unordered.

85. Deep septum spanning the interbasipterygoid space: absent (0) or present (1) (Galton 1990).

86. Dorsoventral depth of the parasphenoid rostrum: much less than (0), or about equal to (1), the transverse width (Yates 2003b).

87. Shape of jugal process of ectopterygoid: gently curved (0) or strongly recurved and hook-like (1) (Yates 2003b).

88. Pneumatic fossa on the ventral surface of the ectopterygoid: present (0) or absent (1) (Sereno et al. 1996).

89. Relationship of the ectopterygoid to the pterygoid: ectopterygoid overlapping the ventral (0), or dorsal (1), surface of the pterygoid (Sereno et al. 1993).

90. Position of the maxillary articular surface of the palatine: along the lateral margin of the bone (0) or at the end of a narrow anterolateral process due to the absence of the posterolateral process (1) (Wilson and Sereno 1998).

91. Centrally located tubercle on the ventral surface of palatine: absent (0) or present (1).

92. Medial process of the pterygoid forming a hook around the basipterygoid process: absent (0), flat and blunt-ended (1), or bent upward and pointed (2) (modified from Wilson and Sereno 1998). Ordered.

93. Length of the vomers: less than (0), or more than (1), 0·25 of the total skull length.

94. Position of jaw joint: no lower than the level of the dorsal margin of the dentary (0) or depressed well below this level (1) (Sereno 1999).

95. Shape of upper jaws in ventral view: narrow with an acute rostral apex (0) or broad and U-shaped (1) (Wilson and Sereno 1998).

96. Length of the external mandibular fenestra: more than (0), or less than (1), 0·1 of the length of the mandible (modified from Upchurch 1995).

97. Caudal end of dentary tooth row medially inset with a thick lateral ridge on the dentary forming a buccal emargination: absent (0) or present (1) (Gauthier 1986).

98. Height : length ratio of the dentary: less than (0), or greater than (1), 0·2 (modified from Benton et al. 2000).

99. Orientation of the symphyseal end of the dentary: in line with the long axis of the dentary (0) or strongly curved ventrally (1) (Sereno 1999).

100. Position of first dentary tooth: adjacent to symphysis (0) or inset one tooth's width from the symphysis (1) (Sereno 1999).

101. Dorsoventral expansion at the symphyseal end of the dentary: absent (0) or present (1) (Wilson and Sereno 1998).

102. Splenial foramen: absent (0), present and enclosed (1), or present and open anteriorly (2) (Rauhut 2003a). Ordered.

103. Splenial-angular joint: flattened sutured contact (0), synovial joint surface between tongue-like process of angular fitting in groove of the splenial (1) (Sereno et al. 1993).

104. A stout, triangular, medial process of the articular, behind the glenoid: present (0) or absent (1) (Yates 2003b).

105. Length of the retroarticular process: less than (0), or greater than (1), than the depth of the mandible below the glenoid (Yates 2003b).

106. Strong medial embayment behind glenoid of the articular in dorsal view: absent (0), or present (1) (Yates and Kitching 2003).

107. Number of premaxillary teeth: four (0) or more than four (1) (Galton 1990).

108. Number of dentary teeth (in adults): less than 18 (0) or 18 or more (1) (modified from Wilson and Sereno 1998).

109. Arrangement of teeth within the jaws: linearly placed, crowns not overlapping (0) or imbricated with distal side of tooth overlapping mesial side of the succeeding tooth (1).

110. Orientation of the maxillary tooth crowns: erect (0) or procumbent (1) (modified from Gauthier 1986).

111. Orientation of the dentary tooth crowns: erect (0) or procumbent (1) (modified from Gauthier 1986).

112. Teeth with basally constricted crowns: absent (0) or present (1) (Gauthier 1986).

113. Tooth–tooth occlusal wear facets: absent (0) or present (1) (Wilson and Sereno 1998).

114. Mesial and distal serrations of the teeth: fine and set at right angles to the margin of the tooth (0) or coarse and angled upwards at an angle of 45 degrees to the margin of the tooth (1) (Benton *et al.* 2000).

115. Distribution of serrations on the maxillary and dentary teeth: present on both the mesial and distal carinae (0), absent on the posterior carinae (1), or absent on both carinae (2) (Wilson 2002). Unordered.

116. Long axis of the tooth crowns distally recurved: present (0) or absent (1) (Gauthier 1986).

117. Texture of the enamel surface: entirely smooth (0), finely wrinkled in some patches (1), or extensively and coarsely wrinkled (2) (modified from Wilson and Sereno 1998).

118. Lingual concavities of the teeth: absent (0) or present (1) (Upchurch 1995).

119. Longitudinal labial grooves on the teeth: absent (0) or present (1) (Upchurch 1998).

120. Distribution of the serrations along the mesial and distal carinae of the tooth: extend along most of the length of the crown (0) or are restricted to the upper half of the crown (1) (Yates 2003*b*).

121. Number of cervical vertebrae: eight or fewer (0), 9–10 (1), 12–13 (2), or more than 13 (3) (modified from Wilson and Sereno 1998). Ordered.

122. Shallow, dorsally facing fossa on the atlantal neurapophysis bordered by a dorsally everted lateral margin: absent (0) or present (1) (Yates and Kitching 2003).

123. Width of axial intercentrum: less than (0), or greater than (1), width of axial centrum (Sereno 1999).

124. Position of axial prezygapophyses: on the anterolateral surface of the neural arch (0) or mounted on anteriorly projecting pedicels (1).

125. Posterior margin of the axial postzygapophyses: overhang the axial centrum (0) or are flush with the caudal face of the axial centrum (1) (Sereno 1999).

126. Length of the axial centrum: less than (0), or at least (1), three times the height of the centrum.

127. Length of the anterior cervical centra (cervicals 3–5): no more than (0), or greater than (1), the length of the axial centrum.

128. Length of middle to posterior cervical centra (cervicals 6–8): no more than (0), or greater than (1), the length of the axial centrum.

129. Dorsal excavation of the cervical parapophyses: absent (0) or present (1) (Upchurch 1998).

130. Lateral compression of the anterior cervical vertebrae: centra are no higher than they are wide (0) or are approximately 1·25 times higher than wide (1) (Upchurch 1998).

131. Relative elongation of the anterior cervical centra (cervicals 3–5): lengths of the centra are less than 2·5 times the height of their anterior faces (0), lengths are 2·5–4 times the height of their anterior faces (1) or the length of at least cervical 4 or 5 exceeds 4 times the anterior centrum height (2) (modified from Sereno 1999). Ordered.

132. Ventral keels on cranial cervical centra: present (0) or absent (1) (modified from Upchurch 1998).

133. Height of the mid cervical neural arches: no more than (0), or greater than (1), height of the posterior centrum face.

134. Cervical epipophyses on the dorsal surface of the postzygapophyses: absent (0), or present on at least some cervical vertebrae (1).

135. Caudal ends of cranial, postaxial epipophyses: with a free pointed tip (0) or joined to the postzygapophysis along their entire length (1).

136. Shape of the epipophyses: tall ridges (0) or flattened, horizontal plates (1) (Yates 2003*b*).

137. Epipophyses overhanging the rear margin of the postzygapophyses: absent (0), or present in at least some postaxial cervical vertebrae (1) (Sereno *et al.* 1993).

138. Anterior spur-like projections on mid-cervical neural spines: absent (0) or present (1).

139. Shape of mid-cervical neural spines: less than (0), or at least (1), twice as long as high.

140. Shape of cervical rib shafts: short and posteroventrally directed (0) or longer than the length of their centra and extending parallel to cervical column (1) (Sereno 1999).

141. Position of the base of the cervical rib shaft: level with, or higher than the ventral margin of the cervical centrum (0) or located below the ventral margin due to a ventrally extended parapophysis (1) (Wilson and Sereno 1998).

142. Postzygodiapophyseal lamina in cervical neural arches 4–8: present (0) or absent (1) (Yates 2003*b*).

143. Laminae of the cervical neural arches 4–8: well-developed tall laminae (0) or weakly developed low ridges (1) (Wilson and Sereno 1998).

144. Shape of anterior centrum face in cervical centra: concave (0), flat (1), or convex (2) (modified from Gauthier 1986). Ordered.

145. Ventral surface of the centra in the cervicodorsal transition: transversely rounded (0) or with longitudinal keels (1) (Rauhut 2003*a*).

146. Number of vertebrae between cervicodorsal transition and primordial sacral vertebrae: 15–16 (0) or no more than 14 (1) (modified from Wilson and Sereno 1998).

147. Lateral surfaces of the dorsal centra: with at most vague, shallow depressions (0), with deep fossae that approach the midline (1), or with invasive, sharp-rimmed pleurocoels (2) (Gauthier 1986). Ordered.

148. Oblique ridge dividing pleural fossa of cervical vertebrae: absent (0) or present (1) (Wilson and Sereno 1998).

149. Laterally expanded tables at the midlength of the dorsal surface of the neural spines: absent in all vertebrae (0), present on the pectoral vertebrae (1) or present on the pectoral and cervical vertebrae (2) (Yates and Kitching 2003). Ordered.

150. Dorsal centra: entirely amphicoelous to amphiplatyan (0), first two dorsals are opisthocoelous (1), or cranial half of dorsal column is opisthocoelous (2) (Wilson and Sereno 1998). Ordered.

151. Shape of the posterior dorsal centra: relatively elongated for their size (0) strongly axially compressed for their size (1) (modified from Novas 1993).

Various authors have noted that the posterior dorsal vertebrae of some dinosaur taxa, notably *Herrerasaurus*, are much shorter than they are tall. This is contrast to the condition in basal dinosauromorphs (e.g. *Marasuchus lilloensis*) and most basal dinosaurs (e.g. *Lesothosaurus diagnosticus*, *Dilophosaurus wetherilli*, *Thecodontosaurus antiquus*). Thus, the shortening of the posterior dorsal vertebral centra, so that the length : height ratio is less than 1·0, has been used as a character in several cladistic analyses. Conversely some analyses take a roughly equant centrum to be the primitive condition and describe the elongation of the posterior dorsal centra as derived. Regardless of the polarity of this character, these analyses fail to take into account the allometric changes that occur in the posterior dorsal centra with increasing body size. If the lengths of the posterior dorsal centra of various-sized basal sauropodomorphs are plotted against their elongation index (EI = centrum length/posterior centrum height) it can be seen that they plot close to a line with a negative slope. The very largest basal sauropodomorphs have short posterior dorsal vertebrae with EIs of less than 1 whereas small ones approach 1·5. So it would seem that EI is dependent upon size and that simple variation in EI should not be used to discriminate different evolutionary states. However, it is clear that there is variation among the EI of dinosauromorphs that cannot be accounted for by size alone. When a range of dinosauromorphs is added to the plot, most form a linear cloud that is centred upon the regression line for the basal sauropodomorph data. Early neotheropods such as *Liliensternus* and *Dilophosaurus* tend to plot above the line, indicating that they have slightly elongated vertebrae, for their size, compared with other early dinosauromorphs. However, this variation is not pronounced, as they still fall within the bounds of the main cloud and either reflects a derived condition within Neotheropoda or a synapomorphy of Neotheropoda as a whole. In either case this variation is not parsimony informative for the present analysis and is ignored. On the other hand, *Herrerasaurus* and *Staurikosaurus* plot well below the main dinosauromorph cloud and are separated by a distinct gap in morphospace. It is clear that, for their size, the latter taxa have posterodorsal centra that are markedly compressed in the anteroposterior dimension and they are coded as having a different character state from other dinosauromorphs.

152. Laminae bounding triangular infradiapophyseal fossae (chonae) on dorsal neural arches: absent (0) or present (1) (Wilson 1999).

153. Location of parapophysis in first two dorsals: at the anterior end of the centrum (0), or located at the mid-length of the centrum, within the middle chonos (1).

154. Parapophyses of the dorsal column completely shift from the centrum to the neural arch: anterior (0), or posterior (1), to the thirteenth presacral vertebra (Langer 2004).

155. Orientation of the transverse processes of the dorsal vertebrae: most horizontally directed (0) or all upwardly directed (1) (Upchurch 1998).

156. Contribution of the paradiapophyseal lamina to the margin of the anterior chonos in mid-dorsal vertebrae: present (0) or prevented by high placement of parapophysis (1).

157. Hyposphenes in the dorsal vertebrae: absent (0), present but less than the height of the neural canal (1), or present and equal to the height of the neural canal (2) (modified from Gauthier 1986). Ordered.

158. Prezygodiapophyseal lamina and associated anterior triangular fossa (chonos): present on all dorsals (0) or absent in mid-dorsals (1) (Yates 2003*b*).

159. Anterior centroparapophyseal lamina in dorsal vertebrae: absent (0) or present (1) (Wilson 2002).

160. Prezygoparapophyseal lamina in dorsal vertebrae: absent (0) or present (1).

161. Accessory lamina dividing posterior chonos from postzygapophysis: absent (0) or present (1).

162. Lateral pneumatic fenestra in middle chonos of middle and posterior dorsal vertebrae opening into neural cavity: absent (0) or present (1) (Wilson and Sereno 1998).

163. Separation of lateral surfaces of anterior dorsal neural arches under transverse processes: widely spaced (0) or only separated by a thin midline septum (1) (Upchurch *et al.* 2004).

164. Height of dorsal neural arches, from neurocentral suture to level of zygapophyseal facets: much less than (0), or subequal to or greater than (1), height of centrum.

165. Form of anterior surface of neural arch: simple centroprezygopophyseal ridge (0) or broad anteriorly facing surface bounded laterally by centroprezygopophyseal lamina (1) (Bonaparte 1999).

166. Shape of posterior dorsal neural canal: subcircular (0) or slit-shaped (1) (Wilson and Sereno 1998).

167. Height of middle dorsal neural spines: less than the length of the base (0), higher than the length of the base but less than 1·5 times the length of the base (1) or greater than 1·5 times the length of the base (2) (modified from Bonaparte 1986*b*). Ordered.

168. Shape of anterior dorsal neural spines: lateral margins parallel in anterior view (0) or transversely expanding towards dorsal end (1).

169. Cross-sectional shape of dorsal neural spines: transversely compressed (0), broad and triangular (1), or square-shaped in posterior vertebrae (2) (modified from Bonaparte 1986*b*).

170. Spinodiapophyseal lamina on dorsal vertebrae: absent (0), present and separated from spinopostzygapophyseal lamina (1) or present and joining spinopostzygapophyseal lamina to create a composite posterolateral spinal lamina (Wilson and Sereno 1998).

171. Well-developed, sheet-like suprapostzygapophyseal laminae: absent (0), present on at least the caudal dorsal vertebrae (1) (Bonaparte 1986*b*).

172. Shape of the spinopostzygapophyseal lamina in middle and posterior dorsal vertebrae: singular (0) or bifurcated at its distal end (1) (Wilson 2002).

173. Shape of posterior margin of middle dorsal neural spines in lateral view: approximately straight (0) or concave with a projecting posterodorsal corner (1) (Yates 2003*c*).

174. Transversely expanded plate-like summits of posterior dorsal neural spines: absent (0) or present (1) (Novas 1993).

175. Last presacral rib: free (0) or fused to vertebra (1).

176. Sacral rib much narrower than the transverse process of the first primordial sacral vertebra (and dorsosacral if present) in dorsal view: absent (0) or present (1) (Yates and Kitching 2003).

177. Number of dorsosacral vertebrae: none (0), one (1), or two (2) (modified from Gauthier 1986). Ordered.

178. Caudosacral vertebra: absent (0) or present (1) (Galton and Upchurch 2004).

179. Shape of the iliac articular facets of the first primordial sacral rib: singular (0) or divided into dorsal and ventral facets separated by a non-articulating gap (1).

180. Depth of the iliac articular surface of the primordial sacrals: less than (0), or greater than (1), 0·75 of the depth of the ilium (modified from Novas 1992).

181. Sacral ribs contributing to the rim of the acetabulum: absent (0) or present (1) (Wilson 2002).

182. Posterior and anterior expansion of the transverse processes of the first and second primordial sacral vertebrae, respectively, partly roofing the intercostal space: absent (0) or present (1) (Langer 2004).

183. Length of first caudal centrum: greater than (0), or less than (1), its height (Yates 2003*b*).

184. Length of base of the proximal caudal neural spines: less than (0), or greater than (1), half the length of the neural arch (Gauthier 1986).

185. Position of postzygapophyses in proximal caudal vertebrae: protruding with an interpostzygapophyseal notch visible in dorsal view (0) or placed on either side of the caudal end of the base of the neural spine without any interpostzygapophyseal notch (1) (Yates 2003*b*).

186. A hyposphenal ridge on caudal vertebrae: absent (0) or present (1) (Upchurch 1995).

187. Depth of the bases of the proximal caudal transverse processes: shallow, restricted to the neural arches (0), deep, extending from the centrum to the neural arch (1) (Upchurch 1998).

188. Position of last caudal vertebra with a protruding transverse process: distal (0), or proximal (1), to caudal 16 (Wilson 2002).

189. Orientation of posterior margin of proximal caudal neural spines: sloping posterodorsally (0) or vertical (1) (Novas 1992).

190. Longitudinal ventral sulcus on proximal and middle caudal vertebrae: present (0) or absent (1) (modified from Upchurch 1995).

191. Length of midcaudal centra: greater than (0), or less than (1), twice the height of their anterior faces (Yates 2003*b*).

192. Cross-sectional shape of the distal caudal centra: oval with rounded lateral and ventral sides (0) or square-shaped with flattened lateral and ventral sides (1).

193. Length of distal caudal prezygapophyses: short, not overlapping the preceding centrum by more than a quarter (0) or long and overlapping the preceding the centrum by more than a quarter (Gauthier 1986).

194. Shape of the terminal caudal vertebrae: unfused, size decreasing toward tip (0) or expanded and fused to form a club-shaped tail (1) (Upchurch 1995).

195. Length of the longest chevron: is less than (0), or greater than (1), twice the length of the preceding centrum (modified from Yates 2003*b*).

196. Anteroventral process on distal chevrons: absent (0) or present (1) (Upchurch 1995).

197. Mid-caudal chevrons with a ventral slit: absent (0) or present (1) (Upchurch 1995).

198. Longitudinal ridge on the dorsal surface of the sternal plate: absent (0) or present (1) (Upchurch 1998).

199. Craniocaudal length of the acromion process of the scapula: less than (0), or greater than (1), 1·5 times the minimum width of the scapula blade (Wilson and Sereno 1998).

200. Minimum width of the scapula: less than (0), or greater than (1), 20 per cent of its length (Gauthier 1986).

201. Caudal margin of the acromion process of the scapula: rises from the blade at angle that is less than (0), or greater than (1), 65 degrees from the long axis of the scapula, at its steepest point (modified from Novas 1992).

202. Width of dorsal expansion of the scapula: less than (0), or equal to (1), the width of the ventral end of the scapula (Pol and Powell 2007).

203. Flat caudoventrally facing surface on the coracoid between glenoid and coracoid tubercle: absent (0) or present (1) (Yates and Kitching 2003).

204. Coracoid tubercle: present (0) or absent (1) (modified from Pérez-Moreno *et al.* 1994).

205. Length of the humerus: less than 55 per cent (0), 55–65 per cent (1), 65–70 per cent (2), or more than 70 per cent (3), of the length of the femur (modified from Gauthier 1986). Ordered.

206. Shape of the deltopectoral crest: subtriangular (0) or subrectangular (1) (Gauthier 1986).

207. Length of the deltopectoral crest of the humerus: less than 30 per cent (0), 30–50 per cent (1), or greater than 50 per cent (2), of the length of the humerus (modified from Sereno *et al.* 1993). Ordered.

208. Shape of the anterolateral margin of the deltopectoral crest of the humerus: straight (0) or strongly sinuous (1) (Yates 2003*b*).

209. Rugose pit centrally located on the lateral surface of the deltopectoral crest: absent (0) or present (1).

210. Well-defined fossa on the distal flexor surface of the humerus: present (0) or absent (1) (Yates and Kitching 2003).

211. Transverse width of the distal humerus: is less than (0), or greater than (1), 33 per cent of the length of the humerus (Langer 2004).

212. Shape of the entepicondyle of the distal humerus: rounded process (0) or with a flat distomedially facing surface bounded by a sharp proximal margin (1).

213. Length of the radius: greater than (0), or less than (1), 80 per cent of the humerus (Langer 2004).

214. Deep radial fossa, bounded by an anterolateral process, on proximal ulna: absent (0) or present (1) (Wilson and Sereno 1998).

215. Olecranon process on proximal ulna: present (0) or absent (1) (Wilson and Sereno 1998).

216. Maximum linear dimensions of the ulnare and radiale: exceed that of at least one of the first three distal carpals (0) or are less than any of the distal carpals (1) (Yates 2003*b*).

217. Transverse width of the first distal carpal: less than (0), or greater than (1), 120 per cent of the transverse width of the second distal carpal (Sereno 1999).

218. Sulcus across the medial end of the first distal carpal: absent (0) or present (1).

219. Lateral end of first distal carpal: abuts (0), or overlaps (1), second distal carpal (Yates 2003*b*).

220. Second distal carpal: does (0), or does not (1), completely cover the proximal end of the second metacarpal (Yates and Kitching 2003).

221. Ossification of the fifth distal carpal: present (0) or absent (1).

222. Length of the manus: less than 38 per cent (0), 38–45 per cent (1), or greater than 45 per cent (2), of the humerus + radius (modified from Sereno *et al.* 1993). Ordered.

223. Shape of metacarpus: flattened to gently curved and spreading (0) or a colonnade of subparallel metacarpals tightly curved into a U-shape (1) (Wilson and Sereno 1998).

224. Proximal width of first metacarpal: less than (0), or greater than (1), the proximal width of the second metacarpal (modified from Gauthier 1986).

225. Minimum transverse shaft width of first metacarpal: less than (0), or greater than (1), twice the minimum transverse shaft width of second metacarpal.

226. Proximal end of first metacarpal: flush with other metacarpals (0) or inset into the carpus (1) (Sereno 1999).

227. Shape of the first metacarpal: proximal width less than 65 per cent (0), 65–80 per cent (1), 80–100 per cent (2), or greater than 100 per cent (3), of its length (modified from Sereno 1999). Ordered.

228. Strong asymmetry in the lateral and medial distal condyles of the first metacarpal: absent (0) or present (1) (Gauthier 1986).

229. Deep distal extensor pits on the second and third metacarpals: absent (0) or present (1) (Novas 1993).

230. Shape of the distal ends of second and third metacarpals: subrectangular in distal view (0) or trapezoidal with flexor rims of distal collateral ligament pits flaring beyond extensor rims (1).

231. Shape of the fifth metacarpal: longer than wide at the proximal end with a flat proximal surface (0) or close to as wide as it is long with a strongly convex proximal articulation surface (1) (Yates 2003*b*).

232. Length of the fifth metacarpal: less than (0), or greater than (1), 75 per cent of the length of the third metacarpal (Upchurch 1998).

233. Length of manual digit one: less than (0), or greater than (1), the length of manual digit two (Yates 2003*b*).

234. Ventrolateral twisting of the transverse axis of the distal end of the first phalanx of manual digit one relative to its proximal end: absent (0), present but much less than 60 degrees (1), or 60 degrees (2) (Sereno 1999). Ordered.

235. Length of the first phalanx of manual digit one: less than (0), or greater than (1), the length of the first metacarpal (Gauthier 1986).

236. Shape of the proximal articular surface of the first phalanx of manual digit one: rounded (0) or with an embayment on the medial side (1) (modified from Sereno 1999).

237. Shape of the first phalanx of manual digit one: elongate and subcylindrical (0) or strongly proximodistally compressed and wedge-shaped (1) (Wilson 2002).

238. Length of the penultimate phalanx of manual digit two: less than (0), or greater than (1), the length of the second metacarpal (Rauhut 2003*a*).

239. Length of the penultimate phalanx of manual digit three: less than (0), or greater than (1), the length of the third metacarpal (Rauhut 2003*a*).

240. Shape of non-terminal phalanges of manual digits two and three: longer than wide (0) or as long as wide (1) (Yates 2003*b*).

241. Shape of the unguals of manual digits two and three: straight (0), or strongly curved with tips projecting well below flexor margin of proximal articular surface (1) (Sereno *et al.* 1993).

242. Length of the ungual of manual digit two: greater than the length of the ungual of manual digit one (0), 75–100 per cent of the ungual of manual digit one (1), less than 75 per cent of the ungual of manual digit one (2), or the ungual of manual digit two is absent (3) (modified from Gauthier 1986). Ordered.

243. Phalangeal formula of manual digits two and three: three and four, respectively (0), or with at least one phalanx missing from each digit (1) (modified from Wilson and Sereno 1998).

244. Phalangeal formula of manual digits four and five: greater than (0), or less than (1), 2–0, respectively (Gauthier 1986).

245. Strongly convex dorsal margin of the ilium: absent (0) or present (1) (Gauthier 1986).

246. Cranial extent of preacetabular process of ilium: does not (0), or does (1), project further forward than cranial end of the pubic peduncle (Yates 2003*b*).

247. Shape of the preacetabular process: blunt and rectangular (0) or with a pointed, projecting cranioventral corner and a rounded dorsum (1) (modified from Sereno 1999).

248. Depth of the preacetabular process of the ilium: much less than (0), or subequal to (1), the depth of the ilium above the acetabulum (modified from Gauthier 1986).

249. Length of preacetabular process of the ilium: less than (0), or greater than (1), twice its depth.

250. Buttress between preacetabular process and the supraacetabular crest of the ilium: present (0) or absent (1) (Gauthier 1986).

251. Medial wall of acetabulum: fully closing acetabulum with a triangular ventral process between the pubic and ischial peduncles (0), partially open acetabulum with a straight ventral margin between the peduncles (1), partially open acetabulum with a concave ventral margin between the peduncles (2), or fully open acetabulum with medial ventral margin closely approximating lateral rim of acetabulum (3) (modified from Gauthier 1986). Ordered.

252. Length of the pubic peduncle of the ilium: less than (0), or greater than (1), twice the craniocaudal width of its distal end (Sereno 1999).

253. Caudally projecting 'heel' at the distal end of the ischial peduncle: absent (0) or present (1) (Yates 2003*c*).

254. Length of the ischial peduncle of the ilium: similar to pubic peduncle (0), much shorter than pubic peduncle (1), or virtually absent so that the chord connecting the distal end of the pubic peduncle with the ischial articular surface contacts the postacetabular process (2) (Upchurch *et al.* 2004). Ordered.

255. Length of the postacetabular process of the ilium: between 40 and 100 per cent of the distance between the pubic and ischial peduncles (0), less than 40 per cent of this distance (1), or more than 100 per cent of this distance (2). Unordered.

256. Well developed brevis fossa with sharp margins on the ventral surface of the postacetabular process of the ilium: absent (0) or present (1) (Gauthier 1986).

257. Anterior end of ventrolateral ridge bounding brevis fossa: not connected to (0), or joining (1) supracetabular crest.

258. Shape of the caudal margin of the postacetabular process of the ilium: rounded to bluntly pointed (0), square ended (1), or with a pointed ventral corner and a rounded caudodorsal margin (2) (Yates 2003c). Unordered.

259. Width of the conjoined pubes: less than (0), or greater than (1), 75 per cent of their length (Cooper 1984).

260. Pubic tubercle on the lateral surface of the proximal pubis: present (0) or absent (1) (Yates 2003b).

261. Proximal anterior profile of pubis: anterior margin of pubic apron smoothly confluent with anterior margin of iliac pedicel (0) or iliac pedicel set anterior to the pubic apron creating a prominent inflection in the proximal anterior profile of the pubis (1).

262. Minimum transverse width of the pubic apron: much more than (0), or less than (1), 40 per cent of the width across the iliac peduncles of the ilium.

263. Position of the obturator foramen of the pubis: at least partially occluded by the iliac pedicel (0), or completely visible (1), in anterior view (Galton and Upchurch 2004).

264. Lateral margins of the pubic apron in anterior view: straight (0) or concave (1) (Yates and Kitching 2003).

265. Orientation of distal third of the blades of the pubic apron: confluent with the proximal part of the pubic apron (0) or twisted posterolaterally relative to proximal section so that the anterior surface turns to face laterally (1) (Langer 2004).

266. Orientation of the entire blades of the pubic apron: transverse (0) or twisted posteromedially (1) (Wilson and Sereno 1998).

267. Craniocaudal expansion of the distal pubis: absent (0), less than 15 per cent (1), or greater than 15 per cent (2), of the length of the pubis (modified from Gauthier 1986). Ordered.

268. Notch separating posteroventral end of the ischial obturator plate from the ischial shaft: present (0) or absent (1) (Rauhut 2003a).

269. Elongate interischial fenestra: absent (0) or present (1) (Yates 2003c).

270. Longitudinal dorsolateral sulcus on proximal ischium: absent (0) or present (1) (Yates 2003b).

271. Shape of distal ischium: broad and plate-like, not distinct from obturator region (0) or with a discrete rod-like distal shaft (1).

272. Length of ischium: less than (0) or greater than (1) that of the pubis (Salgado et al. 1997).

273. Ischial component of acetabular rim: larger than (0), or equal to (1), the pubic component (Galton and Upchurch 2004).

274. Shape of the transverse section of the ischial shaft: ovoid to subrectangular (0) or triangular (1) (Sereno 1999).

275. Orientation of the long axes of the transverse section of the distal ischia: meet at an angle (0) or are coplanar (1) (Wilson and Sereno 1998).

276. Depth of the transverse section of the ischial shaft: much less than (0) at least as great as (1), the transverse width of the section (Wilson and Sereno 1998).

277. Distal ischial expansion: absent (0) or present (1) (Holtz 1994).

278. Transverse width of the conjoined distal ischial expansions: greater than (0), or less than (1), their sagittal depth (Yates 2003b).

279. Length of the hindlimb: greater than (0), or less than (1), the length of the trunk (Gauthier 1986).

280. Longitudinal axis of the femur in lateral view: strongly bent with an offset between the proximal and distal axes greater than 15 degrees (0), weakly bent with an offset of less than 10 degrees (1), or straight (2) (Cooper 1984). Ordered.

281. Shape of the cross-section of the mid-shaft of the femur: subcircular (0) or strongly elliptical with the long axis orientated mediolaterally (1) (Wilson and Sereno 1998).

282. Angle between the long axis of the femoral head and the transverse axis of the distal femur: about 30 degrees (0) or close to 0 degrees (1) (Carrano 2000).

283. Shape of femoral head: roughly rectangular in profile with a sharp medial distal corner (0) or roughly hemispherical with no sharp medial distal corner (1).

This character only applies to taxa with a medially, or anteromedially protruding femoral head. It does not apply to outgroup taxa (*Euparkeria* or Crurotarsi) with proximally directed femoral heads and is coded as unknown in these taxa.

284. Posterior proximal tubercle on femur: well developed (0) or indistinct to absent (1) (Novas 1996).

285. Shape of the lesser trochanter: small rounded tubercle (0), proximodistally orientated, elongate ridge (1), or absent (2) (modified from Gauthier 1986). Unordered.

286. Position of proximal tip of lesser trochanter: level with (0), or distal to (1), the femoral head (Galton and Upchurch 2004).

287. Projection of the lesser trochanter: just a scar upon the femoral surface (0) or a raised process (1).

288. Transverse ridge extending laterally from the lesser trochanter: absent (0) or present (1) (Rowe 1989).

289. Height of the lesser trochanter in cross section: less than (0), or at least as high as (1), basal width (modified from Galton 1990).

290. Position of the lesser trochanter: near the centre of the anterior face (0), or close to the lateral margin (1), of the femoral shaft in anterior view.

291. Visibility of the lesser trochanter in posterior view: not visible (0) or visible (1) (Galton and Upchurch 2004).

292. Height of the fourth trochanter: tall crest (0) or a low rugose ridge (1) (Gauthier 1986).

293. Position of the fourth trochanter along the length of the femur: in the proximal half (0) or straddling the midpoint (1) (Galton 1990).

294. Symmetry of the profile of the fourth trochanter of the femur: subsymmetrical without a sharp distal corner (0) or asymmetrical with a steeper distal slope than the proximal slope and a distinct distal corner (1) (Langer 2004).

295. Shape of the profile of the fourth trochanter of the femur: rounded (0) or subrectangular (1).

296. Position of fourth trochanter along the mediolateral axis of the femur: centrally located (0) on the medial margin (1) (Galton 1990).

297. Extensor depression on anterior surface of the distal end of the femur: absent (0) or present (1) (Molnar et al. 1990).

298. Size of the medial condyle of the distal femur: subequal to (0), or larger than (1), the fibular + lateral condyles (modified from Wilson 2002).

299. Tibia : femur length ratio: greater than 1·0 (0), between 0·6 and 1·0 (1) or less than 0·6 (2) (modified from Gauthier 1986). Ordered.

300. Orientation of cnemial crest: projects anteriorly to anterolaterally (0) or projecting laterally (1) (Wilson and Sereno 1998).

301. Paramarginal ridge on lateral surface of cnemial crest: absent (0) or present (1).

302. Position of the tallest point of the cnemial crest: close to the proximal end of the crest (0) or about half-way along the length of the crest, creating an anterodorsally sloping proximal margin of the crest (1).

303. Proximal end of tibia with a flange of bone that contacts the fibula: absent (0) or present (1) (Gauthier 1986).

304. Position of the posterior end of the fibular condyle on the proximal articular surface tibia: anterior to (0) or level with (1), the posterior margin of proximal articular surface.

305. Shape of the proximal articular surface of the tibia: ovoid, anteroposteriorly longer than transversely wide (0) or subcircular and as wide transversely as anteroposteriorly long (1) (Wilson and Sereno 1998).

306. Transverse width of the distal tibia: subequal to (0), or greater than (1), its craniocaudal length (Gauthier 1986).

307. Anteroposterior width of the lateral side of the distal articular surface of the tibia: as wide (0), or narrower than (1), the anteroposterior width of the medial side.

308. Relationship of the posterolateral process of the distal end of the tibia with the fibula: not flaring laterally and not making significant contact with the fibula (0) or flaring laterally and backing the fibula (1).

309. Shape of the distal articular end of the tibia in distal view: ovoid (0) or subrectangular (1).

310. Shape of the anteromedial corner of the distal articular surface of the tibia: forming a right angle (0) or forming an acute angle (1) (Langer 2004).

311. Position of the lateral margin of descending caudoventral process of the distal end of the tibia: protrudes laterally at least as far as (0), or set well back from (1), the craniolateral corner of the distal tibia (Wilson and Sereno 1998).

312. A triangular rugose area on the medial side of the fibula: absent (0) or present (1) (Wilson and Sereno 1998).

313. Transverse width of the midshaft of the fibula: greater than 0·75 (0), between 0·5 and 0·75 (1), or less than 0·5 (2), of the transverse width of the midshaft of the tibia (Langer 2004). Ordered.

314. Position of fibula trochanter: on anterior surface of fibula (0), laterally facing (1), or anteriorly facing but with strong lateral bulge (2) (modified from Wilson and Sereno 1998).

315. Depth of the medial end of the astragalar body in cranial view: roughly equal to the lateral end (0) or much shallower creating a wedge-shaped astragalar body (1) (Wilson and Sereno 1998).

316. Shape of the posteromedial margin of the astragalus in dorsal view: forming a moderately sharp corner of a subrectangular astragalus (0) or evenly rounded without formation of a caudomedial corner (1) (Wilson and Sereno 1998).

317. Dorsally facing horizontal shelf forming part of the fibular facet of the astragalus: present (0) or absent with a largely vertical fibular facet (1) (Sereno 1999).

318. Pyramidal dorsal process on the posteromedial corner of the astragalus: absent (0) or present (1).

319. Shape of the ascending process of the astragalus: anteroposteriorly deeper than transversely wide (0) or transversely wider than anteroposteriorly deep (1).

320. Posterior extent of ascending process of the astragalus: well anterior to (0), or close to the posterior margin of (1), the astragalus (Wilson and Sereno 1998).

321. Sharp medial margin around the depression posterior to the ascending process of the astragalus: absent (0) or present (1) (Novas 1996).

322. Buttress dividing posterior fossa of astragalus and supporting ascending process: absent (0) or present (1) (Wilson and Sereno 1998).

323. Vascular foramina set in a fossa at the base of the ascending process of the astragalus: present (0) or absent (1) (Wilson and Sereno 1998).

324. Transverse width of the calcaneum: greater than (0), or less than (1), 30 per cent of the transverse width of the astragalus (Yates and Kitching 2003).

325. Lateral surface of calcaneum: simple (0) or with a fossa (1).

326. Medial peg of calcaneum fitting into astragalus: present, even if rudimentary (0) or absent (1) (Sereno *et al.* 1993).

327. Calcaneal tuber: large and well developed (0) or highly reduced to absent (1).

328. Shape of posteromedial heel of distal tarsal four (lateral distal tarsal): proximodistally deepest part of the bone (0) or no deeper than the rest of the bone (1) (Sereno *et al.* 1993).

329. Shape of posteromedial process of distal tarsal four in proximal view: rounded (0) or pointed (1) (Langer 2004).

330. Ossified distal tarsals: present (0) or absent (1) (Gauthier 1986).

331. Proximal width of the first metatarsal: is less than (0), or at least as great as (1), the proximal width of the second metatarsal (modified from Wilson and Sereno 1998).

332. Orientation of proximal articular surface of metatarsal one: horizontal (0) or sloping proximolaterally relative to the long axis of the bone (1) (Wilson 2002).

333. Orientation of the transverse axis of the distal end of metatarsal one: horizontal (0) or angled proximomedially (1) (Wilson 2002).

334. Shape of the medial margin of the proximal surface of the second metatarsal: straight (0) or concave (1) (modified from Sereno 1999).

335. Shape of the lateral margin of the proximal surface of the second metatarsal: straight (0) or concave (1) (modified from Sereno 1999).

336. Length of the third metatarsal: greater than (0), or less than (1), 40 per cent of the length of the tibia (Gauthier 1986).

337. Minimum transverse shaft diameters of third and fourth metatarsals: greater than (0), or less than (1), 60 per cent of the minimum tansverse shaft diameter of the second metatarsal (Wilson and Sereno 1998).

338. Transverse width of the proximal end of the fourth metatarsal: less than (0), or at least (1), twice the anteroposterior depth of the proximal end (modified from Sereno 1999).

339. Transverse width of the proximal end of the fifth metatarsal: less than 25 per cent (0), between 30 and 49 per cent (1), or greater than 50 per cent (2), of the length of the fifth metatarsal (modified from Sereno 1999). Ordered.

340. Transverse width of distal articular surface of metatarsal four in distal view: greater (0), or less than (1), anteroposterior depth (Sereno 1999).

341. Pedal digit five: reduced, non-weight bearing (0) or large (fifth metatarsal at least 70 per cent of fourth metatarsal), robust and weight bearing (1) (Wilson and Sereno 1998).

342. Length of non-terminal pedal phalanges: all longer than wide (0), proximalmost phalanges longer than wide while more distal phalanges are as wide as long (1), or all non-terminal phalanges are as wide, if not wider, than long (2) (modified from Wilson and Sereno 1998). Ordered.

343. Length of the first phalanx of pedal digit one: greater than (0), or less than (1), the length of the ungual of pedal digit one (Yates and Kitching 2003).

344. Length of the ungual of pedal digit one: less than at least some non-terminal phalanges (0) or longer than all non-terminal phalanges (1).

345. Shape of the ungual of pedal digit one: shallow, pointed, with convex sides and a broad ventral surface (0) or deep, abruptly tapering, with flattened sides and a narrow ventral surface (1) (Wilson and Sereno 1998).

346. Shape of proximal articular surface of pedal unguals: proximally facing, visible on medial and lateral sides (0) or proximomedially facing and visible only in medial view, causing medial deflection of pedal unguals in articulation (1) (Wilson and Sereno 1998).

347. Penultimate phalanges of pedal digits two and three: well-developed (0) or reduced disc-shaped elements if they are ossified at all (1) (Wilson and Sereno 1998).

348. Shape of the unguals of pedal digits two and three: dorsoventrally deep with a proximal articulating surface that is at least as deep as it is wide (0) or dorsoventrally flattened with a proximal articulating surface that is wider than deep (1) (Wilson and Sereno 1998).

349. Length of the ungual of pedal digit two: greater than (0), between 90 and 100 per cent of (1), or less than 90 per cent of (2), the length of the ungual of pedal digit one (modified from Gauthier 1986). Ordered.

350. Size of the ungual of pedal digit three: greater than (0), or less than (1), 85 per cent of the ungual of pedal digit two in all linear dimensions (Yates 2003*b*).

351. Number of phalanges in pedal digit four: four (0) or fewer than four (1) (Gauthier 1986).

352. Phalanges of pedal digit five: present (0) or absent (1) (Gauthier 1986).

353. Femoral length: less than 200 mm (0), between 200 and 399 mm (1), between 400 and 599 mm (2), between 600 and 799 mm (3), between 800 and 1000 mm (4), or greater than 1000 mm (modified from Yates 2004). Ordered.

Character-taxon matrix used in the phylogenetic analysis

Euparkeria
00000 000?0 000?0 00000 0?000
?1001 00000 00000 00000 00?00
00000 00000 10000 00000 00000
00?00 00000 ?00?0 00?00 00000
00?10 00000 00000 00000 00?00
00000 0000? ??000 0??00 00000
00?00 ?0?00 ??000 00000 00000
00000 00000 ?0?0? ????? ???00
0?002 00000 00000 ????? ?0000
00000 01?0? 0?000 0???0 0?000
0?000 0?001 00000 00000 01000
00000 00?02 ?000? ?0100 0001?
000?1 00000 ?0000 000?? 0?000
10??0 000?? 0001? 00000 0000?
000

Crurotarsi
00000 00000 000?0 (01)0000 00000
?0000 00000 0(01)000 00000 00000
00000 00000 00000 00000 00000
00?00 00000 00000 00000 00000
00000 00000 00000 00000 00000
00000 0000? ??000 0??00 00020
00?00 ?0?00 ??000 02000 00000
00000 00000 00000 0000(01) 00?0(01)
0?002 00000 00000 00000 00000

00000 00000 00000 0(01)000 00000
0?000 0?0(01)0 00000 00000 00000
00000 00?00 ?0000 00000 0001?
00000 00000 ?0000 000?? 0?000
00000 00000 00000 00000 00000
00(01)

Marasuchus
00??? ????0 ?0??? ????? ????0
?00?? ????? 0???? ????? ??0??
????? ????? ????? ????? ?????
00??? 00000 ????? ????? ?????
????? ???00 ?0000 0000? 1?100
00000 0010? ??00? 0??00 ?0000
00?00 ???00 ??000 ?0000 0000?
00000 0000? ?0101 00000 00?00
0?012 1000? 0000? ????? ?????
????? ????? ????? ????0 00001
00000 0?000 00100 00000 00000
00010 00000 01100 01000 00000
00000 00010 00100 10000 0?000
00000 00000 0000? 10?0? ?00??
?10

Agnosphitys
?0??? ????0 ?0??? ????? ????1
1?0?0 00100 0???? ????? ?????
????? ????? ????? ????? ?????
????? ????? ????? ????? ?????

```
?????   ???00   ?0000   00000   ?????
?????   ?????   ?????   ?????   ?????
?????   ?????   ?????   ?????   ?????
?000?   ?????   ?????   ?????   ?????
?????   1100?   00???   ?????   ?????
?????   ?????   ?????   ????0   00001
20000   100??   ?????   ?????   ?????
?????   ?????   ?????   ?????   ?????
?????   ?10??   ????0   11000   100??
0????   ?????   ?????   ?????   ?????
???
```

Anchisaurus

```
10???      00??0   102?1   ??11?      ????1
11011      ?010?   00101   01100      01100
11110      1200?   10???   ??201      ?0011
01?1?      20020   10???   ???10      ?10?0
0??10      0?000   0101?   1(12)?01   11?00
11???      10010   01111   011?1      0000?
010?0      01100   ?000?   ?0000      000??
0100?      0??0?   01?00   ?10??      ???11
(01)0??2   11001   01100   ??0??      10010
11100      10020   10000   02000      (01)1010
31000      00?11   11100   01001      1?001
01000      01011   01000   010(01)1   11010
00000      11011   1?1?0   1?010      ???10
111?0      ?01??   0011?   10100      00010
011
```

Antetonitrus

```
?1???   ?????     ?????   ?????      ?????
?????   ?????     ?????   ?????      ?????
?????   ?????     ?????   ?????      ?????
?????   ?????     ?????   ?????      ?????
?????   ?????     ?1?10   12000      ?????
???0?   (01)1???  ?????   0??00      ?0000
01??0   02100     10010   12100      1000?
?????   ??100     00?01   100??      ?0?00
11??3   11000     10110   ?????      ?0010
?3100   ???10     10???   ?????      ?????
?????   ???11     10100   01???      ?????
????1   11011     11001   11111      11010
00000   11011     101??   ?????      ?????
?????   10111     10???   ?(12)1?0   0?0??
??3
```

Barapasaurus

```
?????     ?????   ?????   ?????   ?????
?????     ?????   ?????   ?????   ?????
?????     ?????   ?????   ?????   ?????
?????     ?????   ?1?11   1211?   ?????
?????     1?1??   ??00?   ?002?   ?1001
01??0     02011   01111   12012   1100?
0(12)100  1?10?   11??1   ?????   ???10
1?00?     1????   ??11    ?????   ?????
?????     ?????   ?????   ????1   11100
31011     00001   11100   11???   11110
010??     ?????   ?????   ?????   ??1?1
?1???     ?????   ?1?1?   ?????   ?1???
```

```
?????   ?0???   ?????   ???1?      1????
??5
```

Blikanasaurus

```
?????   ?????   ?????   ?????   ?????
?????   ?????   ?????   ?????   ?????
?????   ?????   ?????   ?????   ?????
?????   ?????   ?????   ?????   ?????
?????   ?????   ?????   ?????   ?????
?????   ?????   ?????   ?????   ?????
?????   ?????   ?????   ?????   ?????
?????   ?????   ?????   ?????   ?????
?????   ?????   ?????   ?????   ?????
?????   ?????   ?????   ?????   ?????
?????   ?????   ?????   ?????   ????0
000?0   11011   1?1?0   00010   11010
11100   10111   11120   12110   00010
0?2
```

Camelotia

```
?????   ?????   ?????   ?????       ?????
?????   ?????   ?????   ?????       ?????
?????   ?????   ?????   ?????       ?????
?????   ?????   ?????   ?????       ?????
?????   ?????   ?????   ?????       ?????
?????   ?1???   ?????   ???0?       ?0??0
01??0   02100   00000   0??00       ?0???
?????   ??1??   ?0??0   11???       ?????
?????   ?????   ?????   ?????       ?????
?????   ?????   ?????   ?????       ?????
?????   ????1   ?????   ?????       1??00
111?1   11011   11011   1?1??       110?0
00000   ?????   ?????   ?????       ?????
?????   ?????   ?????   ?(12)???    000??
??5
```

Cetiosaurus

```
1????      ?????   ?????   ?????   ?????
?????      ?????   ?????   ?????   ?????
?????      ?????   ??01?   ?????   ?????
?????      ?????   ?????   ?????   ?????
?????      ?????   ?????   ?????   2??00
11110      11111   01001   00020   ?2001
?10?1      020??   10111   02010   100??
?(12)???   ??100   1??00   100??   ?0111
10103      11001   00011   ?????   ??1??
?????      ?????   ?????   ????1   11100
3?021      0?001   0?1?0   11??1   101?0
0???2      11112   ??0??   ?010?   11121
01000      10010   111??   ?????   ?????
?????      ?????   ?????   ?????   ?????
??5
```

Chindesaurus

```
?????   ?????   ?????   ?????   ?????
?????   ?????   ?????   ?????   ?????
?????   ?????   ?????   ?????   ?????
?????   ?????   ?????   ?????   ?????
??100   ?????   ?????   ???01   ?0??0
```

?1??0 ?1??? ???00 0??00 ?????
????? ?1??0 ????1 ?00?? ?????
????? ????? ????? ????? ?????
?0??? 101?1 ????0 ????? ?????
????0 0?010 11100 00000 100??
00?10 11110 0???0 11110 100??
1???? ????? ????? ????? ?????
??1

Coloradisaurus
?00?1 002?0 102?1 11111 ?1??1
110?1 11100 0?10? ?00?0 01?01
11010 1?100 10010 0?101 10110
11??? 11011 1???0 ?1?10 01011
01011 10210 01010 10000 ?1?00
?1100 20010 0011? 0110? ?00?0
01?10 01100 00000 00000 00101
110?? 01001 00?0? ????? ???01
10111 121?0 11??? ??0?? ?????
????? ????? ????? ????0 01000
31100 ?0?11 10110 01001 11110
111?0 01011 01000 01011 11010
000?0 11011 0???0 11110 ?0001
01??? 00111 0012? 101?0 00021
002

Efraasia
100?1 001?0 10??1 ?111? 112?1
110?1 00??? 00100 10000 0??0?
10??0 1?1?? 1???? ??10? 000?1
????? 01101 00??? ???10 01001
0??11 ?0210 01010 10000 1???1
?1100 11010 01011 01101 00000
01??0 01?00 00000 00000 00000
11000 01010 00?01 10000 ?0?01
00112 12000 11100 11010 12010
?1100 00011 10100 01000 01000
31000 10110 00100 00001 10010
11010 00011 11000 01011 00010
00000 1101? 001?0 11010 ?0?01
?1110 001?? 0011? 10100 00010
002

Eoraptor
0010? 000?0 1010? 00110 00001
01001 001?0 00110 10000 0??01
10010 10000 10000 01001 ???0?
?1??? ????? ????? ??00 0000?
0?0?0 ?0?00 ?1000 00000 ?????
?1000 01010 000?1 0???? 000??
00?0? ???00 ?000? ?0000 0001?
?10?? ?100? ????? ????? ???01
?0??0 11??? 0?0?? ????? ?20?0
00?1? 000?0 ?0100 00011 0100?
2?001 110?? 0???0 ?0??? 10?00
10?10 ????0 ?100? ?0011 ??000
?0?10 ?0110 ??1?? ??0?? ?????
????0 ?0??? 00?00 10000 000??
000

Eucnemosaurus
????? ????? ????? ????? ?????
????? ????? ????? ????? ?????
????? ????? ????? ????? ?????
????? ????? ????? ????? ?????
????? ????? ????? ????? ?0???
01??0 01100 10000 0??00 ?0???
????? ??100 00?01 100?? ?????
??11? ????? ????? ????? ?????
????? ????? ????? ????? ?????
????? ????1 10??? 0???? ?????
????0 01101 ?1010 ?1010 110??
00000 11011 0??? ????? ?????
????? ????? ????? ????? ?????
?? (34)

Gongxianosaurus
1???? ?0??? ????? ??12? ?????
????? ????? ????? ????? ?????
????? ????? ????? ????? ?????
????? ?00?? ?1??2 121?? ?????
????? 0???? ????? ???1? ?0??0
01??0 ????? ?0?0? ?0000 0????
????? ???0? ?0?0? 1?0?0 00?01
1???2 110?? ??1?1 ????? ?????
????? ????? ????? ????0 ?100?
???0? ????? ????? ????? ?????
???? (12) 111?2 ????? ?0??? 1101?
????? 1?0?? 1???0 ?1?1? ???1?
11??0 10??? 10?2? 01110 00020
005

Plateosaurus (=Gresslyosaurus) ingens
????? ????? ????? ????? ?????
????? ????? ????? ????? ?????
????? ????? ????? ????? ?????
????? ????? ????? ????? ?????
????? ????? ????? ????? ?????
????? ????? ????? ????? ?????
?01?? 01??? ????? 10??? ?????
????? ????? ????? ????? ?????
????? ????? ????? ????? ?????
????? ????? ????? ????? ????0
0?010 ????? ????? ????? ?????
????? ????? ????? ? (01)??? 0?0??
??4

Guaibasaurus
????? ????? ????? ????? ?????
????? ????? ????? ????? ?????
????? ????? ????? ????? ?????
????? ????? ????? ????? ?????
????? ????? ????? ????? ?????
????? ????? ????? ????? ?0???
01??0 11001 00000 0??00 000??
????? 0?011 00??? ????1 ????1

? 0??? ????? ????? ????? ?????
????? ????? ????? ????0 ?????
1?002 10110 01000 00000 10?10
110?0 00010 01000 010?1 ?1010
000?0 01111 001?0 ?01?? ??001
01??0 000?? 10?00 10000 00000
011

Herrerasaurus
00000 000?0 10100 00000 00000
?0001 00000 00000 10000 00100
01010 10100 01000 01101 00000
0101? 00000 00?10 00?00 00000
0?100 00000 00000 00000 10100
01000 00010 0000? 00001 00000
11??0 010(01)1 00000 02000 00010
00001 01100 00011 00101 00?11
10??0 11000 00000 00000 12000
00110 00001 10110 10010 00001
20000 00010 00001 0???0 1?010
10010 00010 11100 01011 00010
00010 00010 00100 10010 10000
00110 00100 00100 10000 00000
?02

Isanosaurus
????? ????? ????? ????? ?????
????? ????? ????? ????? ?????
????? ????? ????? ????? ?????
????? ????? ????? ????? ?????
????? ????? ????? ????? ?????
???0? ?0??? ????? ???2? ?10??
0???? ????? ????? ?2?01 1000?
????? ????? ????? ????? ??101
?0??? ????? ????? ????? ?????
????? ????? ????? ????? ?????
????2 11112 ??0?? ?001? 1?0??
????? ????? ????? ????? ?????
????? ????? ????? ????? ?????
??3

Jingshanosaurus
1001? 002?? 10211 11110 0?0?1
11011 10110 0?102 001?? 1?101
11011 10100 1000? ???01 10000
0101? 21100 1000? ?1?10 11001
0??11 10210 01012 11001 ????1
11??0 10??? ???1? ??101 00??0
01110 0?1?? ?0000 00000 0000?
?10?? 01000 00000 10000 00?00
10??1 12010 11100 ??00? ?0011
?3100 100?0 10001 02000 01000
31000 00211 10010 02??1 10010
110?0 10011 01000 01011 11010
10000 11011 0?120 11010 ?00??
????0 10111 10120 10100 00001
0?4

Lessemsaurus
????? ????? ????? ????? ?????

????? ????? ????? ????? ?????
????? ????? ????? ????? ?????
????? ????? ????? ????? ?????
????? ????? ????? ????? ?????
???0? ?1111 0100? ?1101 ?0000
01110 02100 10010 12100 0000?
????? ????? ????? ????? ???00
01??? 11100 10110 ????? ??0?1
?310? ????1 ?0??1 ????0 01000
31000 00?11 ?1100 02??? 1??10
110?1 11011 ?10?? ?1111 1?0?0
00000 10010 1???0 11011 001??
????? ???11 ??121 11??0 0??0?
??3

Lufengosaurus
100?? ?02?? ?0211 ?1?11 ?1111
11111 10100 10101 10100 00101
11010 11100 10?1? ???01 10??0
110?? 01010 100?0 11010 010??
??011 1?210 11010 10001 1?000
11100 20011 0111? 01101 00020
01110 01100 00000 00000 00100
11010 01000 0000(01) 10000 00000
10111 12(01)10 11100 11011 11011
13100 10120 10001 02000 01000
31100 00211 10110 02101 10010
11000 01011 01000 01021 11010
10000 11011 00100 11110 ?0001
?1110 10111 00120 10110 000(12)(01)
004

Mamenchisaurus
11000 113?1 11201 10120 00000
01011 12??0 01112 01010 10111
11110 12010 00001 11221 01000
0111? 201?0 0???? ?2?11 (01)0000
10010 00111 11111 12111 3?000
11101 21010 00011 0002? 11101
01010 02011 00011 02012 10000
01101 11111 11101 10010 11111
10003 11001 00110 11001 10110
00100 011?0 ?10?1 ?31?1 11100
31011 ?0001 11100 11000 11010
01002 11112 ??0?? ?010? 01121
010?1 ??0?? ????1 01011 111??
????1 101?? 21?2? 02111 1102?
?05

Massospondylus
10011 00210 10211 11111 11211
01111 10100 00101 10100 01100
11011 11100 10001 02?01 ?0110
01011 01010 1??10 10010 01011
01010 10210 (01)1010 11001 1100?
11100 20010 01111 01101 00020
01110 01100 00000 00000 00101
11010 01(01)00 00000 00000 00?01
(01)0111 12010 11100 11011 11011
12100 10121 10000 02000 01000

31000 10011 10110 01001 10010
11100 01011 01000 01021 01010
00000 11011 00100 11110 10001
11110 00111 00120 10100 00021
002

Melanorosaurus
1001? 10310 10211 11110 1?111
01011 11??? 10101 01101 01100
11010 11?00 00001 02201 10110
010?? 20020 10011 01110 1?001
11?10 00210 ?1002 11001 10?01
01100 11011 01001 0111? 00000
01??0 02100 00000 01000 0000?
02010 01100 10?00 1100? ???00
00??2 12000 00110 ????? ?00?0
?2100 101(12)0?0001 02000 01000
31000 00111 10100 01??? 1?010
110?1 11011 11001 11001 11010
00000 11011 00102 11010 10000
11100 10111 001?1 11110 00011
0?3

Neosauropoda
11000 11311 11201 10120 00001
01011 1(02)111 01112 01010 1011(01)
11110 12010 0(01)101 12221 11001
01111 20(01)20 00001 02011 10000
12010 00(02)11 1111(02) 12(01)11 2(01)000
01110 01111 01001 10020 12101
0101(01) 02011 10011 02012 11000
02100 11100 1111(01) 10000 10011
10003 11001 (01)0011 11000 10100
00000 01100 01??1 ?3001 11100
31021 00001 11100 11001 110(01)1
0(01)002 11112 ??0?? ?010? 11121
010?1 10000 11111 01011 11110
11??1 11100 21020 02111 11021
105

Neotheropoda
00(01)0(01) 002(01)0 10201 00110 0001(01)
00000 00000 00110 00000 0010(01)
100(01)0 10100 01(01)00 01(012)01 00000
01011 00000 01110 00100 00000
01?00 00000 00000 00000 10110
01100 00010 000(01)1 00001 02000
01010 11001 00000 01000 00000
0(12)100 01100 00000 00101 00001
00010 11000 00100 01000 12000
00010 ?0001 00110 1(12)010 10100
30002 11111 01000 01111 10000
11010 0001? 01100 01000 00000
(01)0110 11111 00200 10110 00011
11100 0?0?0 00001 10000 00000
012

Omeisaurus
11000 11311 11201 10120 00001
01011 ?2010 01112 01010 10111
11010 12010 00001 12221 11001

0111? ??0?0 0???1 ???11 10000
1?0?0 00211 01111 12111 3??00
11101 20011 01011 1002? 12101
01011 02011 00011 02012 1100?
021?? 1?100 11100 100?0 11111
10003 11001 00011 ????? ?0100
00100 01100 01001 ?3101 11100
31011 00001 11100 11000 11110
01002 11012 ??0?? ?010? 01021
01001 1?0?? 111?0 11010 ?11??
1???? 11110 21020 02111 1102?
105

Ornithischia
00100 00000 000?0 00000 0?000
?1011 00010 0100? 10000 00100
10000 10100 00(01)00 01001 00000
00010 00000 00010 010(01)0 1(01)000
00010 01110 01110 10000 10000
00000 01010 0(01)000 0??01 00000
00?00 ?0?00 ?0000 00000 00000
0(12)100 00000 00001 00001 00001
00111 11000 00100 00000 00000
00000 00000 00000 00000 11010
21000 10011 ??000 ?0001 10000
00010 0001? 01011 11020 00000
000(01)0 11110 002?0 1?(01)10 00001
11100 00000 0000(01) 10000 00000
010

Patagosaurus
11??? ????? ????? ????? ?????
????? ????? ????? ????? ?????
????? ????? ????? ????? ?????
????? ????? ????? ????1 ?0000
1???? ??0?? 01110 12111 ?????
?1100 10110 00001 10020 ?2001
010?0 0201? 01111 12012 1?00?
02100 11100 ?1?0? 100?? ?0?11
1?00? 11001 00?11 ????? ?????
????? ????? ????? ????1 11100
31010 00001 11100 11001 11110
010?2 11112 ??0?? ?0011 111?1
01000 ??0?? ????? ????? ?????
????? ????? ????? ????? ?????
??5

Plateosaurus engelhardti
10011 00110 10201 1(01)111 11211
11001 01100 00100 10101 01100
11001 10101 10010 01101 10110
11011 11011 00010 11110 01011
01011 11210 01010 10000 11001
11100 11010 0(01)011 01101 00020
01110 01100 00000 00000 00101
10100 01000 00000 10000 00001
00111 12000 11100 11011 11010
11100 10020 10000 01000 01000
31100 10210 10100 01111 10010
11110 00011 11000 01011 00010

00010	11111	00100	11010	10001
01110	00110	00111	10100	00010
004				

Plateosaurus gracilis

?00??	001?0	102?1	10111	?1??1
110?1	01??0	0?10?	1010?	0??0?
10001	101??	10???	?????	?????
?????	1????	?????	????0	??011
0????	?1210	01010	10?00	1????
?1100	??01?	??011	01101	?00?0
01?10	01100	00000	00000	0010?
10100	01010	00000	1000?	00???
?????	1200?	11100	?1011	?1010
10100	00???	10???	??0?0	01000
31100	10210	10100	01111	10010
11??0	00011	11000	01011	0?010
000?0	??011	0????	?????	?????
?????	?0???	?????	?0?0?	?????
?? (23)				

Plateosauravus

?????	?????	?????	?????	?????
?????	?????	?????	?????	?????
?????	?????	?????	?????	?????
?????	?????	?????	?????	?????
?????	?????	?????	?????	?????
??10?	??010	0101?	01101	?00(12)0
011?0	01100	00000	00100	0000?
?10??	011??	?0??1	10???	?????
0???(23)	11100	11100	?????	?????
??1??	?????	?????	????0	01000
31100	101?0	1????	????1	1??10
110?0	00011	11000	01011	00010
00000	11011	0????	??0??	?????
?????	???11	?0???	?0???	?????
??4				

Riojasaurus

1001?	00??0	10201	1?110	??001
11011	01100	0?100	00100	0?100
10?00	11100	10000	01101	?0?10
0101?	01011	?????	?1?00	1000?
1??10	1?200	01010	10000	10??1
11100	11010	01011	01101	00010
01110	01100	00000	00000	00101
11010	01000	00001	1000?	?0?01
00112	12101	11100	11111	1?011
11101	10010	10001	020?0	01000
31100	10211	10100	01101	10010
11001	01101	11010	11020	11010
00000	11011	00100	11010	100??
011?0	00111	00121	11100	0001?
?13				

Ruehleia

?????	?????	?????	?????	?????
?????	?????	?????	?????	?????
?????	?????	?????	?????	?????
?????	?????	?????	?????	?????
?????	?????	?????	?????	?????

??10?	??0??	??01?	??101	?0020
01??0	01100	10000	01100	0000?
11000	01?00	?0?01	10???	???01
00111	12000	11100	?0100	??010
11101	000??	?0??0	0???0	01000
31100	0021?	?1100	01101	10?10
110?1	01011	01000	01021	00010
?00?0	11011	001?0	00010	100??
?????	?????	?????	?????	?????
??3				

Saturnalia

10???	?????	?????	?????	?????
?1??1	????0	0?100	?0???	?????
?????	?????	10???	??1??	00???
0?0??	0?000	0????	?????	?0100
0????	??000	01000	10000	?????
???00	00010	010??	01101	?0000
01000	01100	10000	00000	00000
00000	0101?	?????	?????	???11
10111	11000	10100	?????	?????
?????	?????	?????	????0	??0?1
11002	10210	00000	01001	10010
01000	00010	01100	01011	00010
00000	01010	00100	10010	10000
00010	00000	00101	1000?	000?0
010				

Shunosaurus

11000	113?1	10201	10120	00000
01011	?2010	01112	01010	10100
11010	12100	01101	11210	10001
011?1	20020	??11	02001	00000
12010	00111	11112	12111	2??01
011?1	001?1	?1001	0012?	11000
01011	021?0	?001?	02012	10000
?11?0	0?100	11101	1001?	111?1
?0002	11001	100?1	1000?	10110
01?00	01100	00??1	?3101	11100
31011	00001	111?0	1100?	11110
01002	11112	??0??	?000?	01121
010?0	1?0??	??110	?101?	??110
11??1	11110	2002?	02111	?1021
105				

Silesaurus

00?0?	?00?0	000?0	01000	00??0
?00?0	?0101	0????	?????	0010?
?0???	?01??	?????	??00?	0?0?0
000??	00000	0????	?1?00	?0?01
0???0	?0000	01000	10000	10?00
00000	0000?	??001	00000	00000
01010	00011	00000	00000	0000?
?0000	01000	00?0?	0????	??11
1?111	00001	00000	?????	?????
?????	?????	?????	????0	00001
00000	00010	00000	000?0	10100
?0?10	00011	01100	01011	00010
00110	00000	000?0	00000	0??11
?1??0	0??00	0000?	10???	000?0

011

Staurikosaurus

```
00???   ?????   ?????   ?????   ?????
?????   ?????   ?????   ?????   ?????
?????   ?????   ?????   ?????   ?????
?????   ?????   ?????   ???00   00000
00100   0?00?   00000   00000   ?????
???00   010??   ??00?   01001   ?0000
11?00   0?000   00000   02000   0001?
?0101   01100   00011   0010?   ???1?
1????   ?????   ?????   ?????   ?????
?????   ?????   ?????   ????0   00001
20011   00011   10?01   0?000   10?00
00010   0001?   ?000?   0101?   00000
00010   00000   001??   ??01?   ?????
?????   ?????   ?????   ?????   ?????
??1
```

Tazoudasaurus

```
11???   ?????   ?????   ?????   ?????
?????   ?????   ?????   ?????   ?????
??010   ?????   ?????   ??00?   1????
?????   ?????   ?????   ???00   00000
1???0   ??20?   11110   1210?   ?????
?????   ?????   ?????   ??02?   ?1??0
01??1   02011   ?001?   ?2011   1?00?
?????   ??00?   ???11   0????   11???
?????   ?????   ?????   ?????   ?????
?????   ??11    10100   01???   ?????
????(12) 1???1   ?1001   ?0???   ????0
????0   ?????   ????1   0101?   111??
?????   ?????   ?????   ?1??1   0?1??
??5
```

Thecodontosaurus antiquus

```
?0???   ?????   ?????   ?????   ?????
?1??1   ??100   0????   ??0?0   ?0?0?
??000   ?????   ?????   ??10?   0?0?1
010??   01100   0????   ???00   ?1101
0????   ??210   01010   10000   ?1???
???00   01010   10?1?   01101   ?0000
01??0   01100   00000   00000   0000?
1000?   01011   00?01   000??   00?01
00112   11000   10100   11000   12010
00110   00010   10000   01010   00000
21000   001??   ?0?00   0??01   1??01
010?0   00011   11000   01011   00010
00000   01011   00??0   11010   100??
??110   00111   0?1?0   10??0   0?0??
??1
```

Thecodontosaurus caducus

```
?0???   0????   ?????   ???(01)0  0?0??
?1??1   ????0   00100   10000   0010?
10000   1?10?   ?????   ??00?   000?1
?10??   00100   011?0   00?00   ?0101
0??00   00?10   01010   10000   1001?
?1100   ?1010   10011   01101   ?00??
?1???   ?1???   ?????   ?????   ?????
```

```
?????   0??11   ?0001   00001   00???
??11?   1?00?   ?????   ?????   ?????
?????   ?????   ?????   ????0   01000
21100   001??   ?????   ??001   1??10
110?0   0????   ?????   ?1?1?   000?0
00000   0?01?   001??   ?????   ?????
?????   00111   0011?   10000   00010
01?
```

Unaysaurus

```
10011   00110   10??1   ?111?   ?1??1
010?1   01100   0????   ?????   ???0?
?00??   ??101   ???00   0??0?   10??0
?10??   11010   0????   ??010   ?1011
0????   ?0210   01010   10000   ???0?
1????   ?0?1?   0????   ?????   ?0??0
?11?0   0?100   00?0?   ?0000   0????
?????   ????0   ????0   1000?   00?01
0011?   12000   01100   ?????   ?1???
?110?   ???21   1???0   ??0??   ?????
?????   ?????   ?????   ?????   ?????
?????   ?????   ?????   ?????   ?????
?????   11011   0??0    11010   ?00??
?????   0?110   ???0?   ?0???   0????
??1
```

Vulcanodon

```
?????   ?????   ?????   ?????   ?????
?????   ?????   ?????   ?????   ?????
?????   ?????   ?????   ?????   ?????
?????   ?????   ?????   ?????   ?????
?????   ?????   ?????   ?????   ?????
?????   ?(12)??? ?????   ?????   ?????
??1??   ????0   111?0   ?00?0   ?0??0
1???2   110?1   0?011   ?????   ?????
?????   ?????   ?????   ?????   ?????
310(12)? 00?11   11100   01001   11110
010?1   11011   11001   0011?   1??20
?1000   1?0??   ?11?1   01011   01110
11??1   1?010   10120   01111   0?121
?05
```

Yunnanosaurus

```
110??   002?0   10?11   11110   ????1
01?11   ?0101   0?10(12) ?????   0??0?
??011   1?00?   ?0001   0??01   ?0??0
010??   ?????   ?????   ???10   ?0001
0??0    ?0110   01012   12111   1???1
11100   11011   0100?   01101   00000
01100   01100   00000   00100   0000?
01011   01000   00?00   ?0??0   ??001
?0??1   1200?   11100   ??01?   ?0011
?2101   10020   1000(01) 020?0   01000
31000   00011   10100   010?1   10010
111?0   01011   01000   01011   01010
10000   10011   001?0   11010   10001
11100   10111   001??   ?11?0   00011
0?2
```

[Special Papers in Palaeontology, 77, 2007, pp. 57–90]

A PHYLOGENETIC ANALYSIS OF BASAL SAUROPODOMORPH RELATIONSHIPS: IMPLICATIONS FOR THE ORIGIN OF SAUROPOD DINOSAURS

by PAUL UPCHURCH*, PAUL M. BARRETT† *and* PETER M. GALTON‡

*Department of Earth Sciences, University College London, Gower Street, London WC1E 6BT, UK; e-mail: P.Upchurch@ucl.ac.uk
†Department of Palaeontology, The Natural History Museum, Cromwell Road, London SW7 5BD, UK; e-mail: P.Barrett@nhm.ac.uk
‡Professor Emeritus, University of Bridgeport, Bridgeport, CT, USA; current address: 315 Southern Hills Drive, Rio Vista, CA 94571, USA; e-mail: pgalton@juno.com

Typescript received 1 February 2006; accepted in revised form 29 July 2006

Abstract: New discoveries, revision of existing taxa and the application of cladistic analysis have all shed light on the relationships of basal sauropodomorphs. Nevertheless, the interrelationships proposed in recent studies have varied widely, with some authors advocating the view that Prosauropoda and Sauropoda are monophyletic sister-taxa, whereas others favour an extreme form of prosauropod paraphyly with respect to sauropods. A data set comprising 292 characters for seven outgroups and 27 ingroup sauropodomorph taxa is presented and analysed. The most parsimonious trees suggest that *Efraasia*, *Mussaurus*, *Thecodontosaurus* and *Saturnalia* are increasingly more distant sister-taxa to the remaining sauropodomorphs. The latter are divided into two monophyletic sister-groups: a plateosaurian clade containing *Plateosaurus*, *Lufengosaurus*, *Massospondylus*, *Coloradisaurus* and others, and a sauropod clade, which includes melanorosaurs (near its base), *Antetonitrus*, *Chinshakiangosaurus*, *Vulcanodon*, *Barapasaurus* and eusauropods. Bootstrap values and constrained analyses with Templeton's tests indicate that support for many of the proposed relationships is relatively weak. This results from the inclusion of poorly known taxa, such as *Blikanasaurus*, and from considerable levels of character conflict. Character mapping indicates several apomorphic features that support the monophyly of a plateosaurian clade or subgroups within it. In addition, it appears that approximately 20 apomorphies are acquired early in basal sauropodomorph evolution, but are reversed to the plesiomorphic state in basal sauropods and eusauropods. Aside from their impact on phylogenetic uncertainty, these reversals may reflect important aspects of early sauropod evolution that relate to shifts in the ecological niches occupied by these taxa.

Key words: cladistic analysis, Dinosauria, phylogeny, Prosauropoda, Sauropoda, Sauropodomorpha.

SAUROPODOMORPHS represent a major radiation of dinosaurian herbivores. Members of this group first appeared in the late Ladinian or Carnian in South America and achieved a virtually global distribution by the end of the Triassic (Langer 2004). Sauropodomorphs range in size from 1 to 30 m, and are typically characterized by small heads and relatively long necks. The latter feature probably reflects their status as the first specialist 'high-browsers' among vertebrates (Galton 1985*a*, 1986), although the details of their feeding strategies remain complex and controversial (Upchurch and Barrett 2000; Upchurch *et al.* 2004; Parrish 2005), and it should be remembered that neck elongation commenced in relatively small bipeds (such as *Saturnalia*) that are unlikely to have exploited such a niche.

Recent reviews of the Prosauropoda and Sauropoda indicate that there are currently at least 130 valid or potentially valid sauropodomorph genera (Galton and Upchurch 2004; Upchurch *et al.* 2004). Less than half of these taxa have been incorporated into cladistic analyses and it is only in the past 5 years that this group has started to receive intensive scrutiny. As a result, the phylogenetic relationships of sauropodomorphs remain confused and controversial (see below), and much further work is required. In particular, the extent to which the various prosauropod taxa represent the monophyletic sister-group to Sauropoda, or are paraphyletic with respect to the latter, has developed into a major issue (Benton *et al.* 2000; Yates 2003*a*, 2004*a*, 2005; Yates and Kitching 2003; Galton and Upchurch 2004; Leal *et al.* 2004; Pol and Powell 2005; Sereno 2005; Upchurch *et al.* 2005). Therefore, the aims of the current paper are to: (1) present a detailed phylogenetic analysis of the relationships of basal sauropodomorphs; (2) test some of the most important competing phylogenetic hypotheses using the large dataset and constrained topologies; and (3) briefly consider the implications of this work for our understanding of the origin of sauropods.

Institutional abbreviations. AMNH, American Museum of Natural History, New York, USA; IVPP, Institute of Vertebrate Palaeontology and Palaeoanthropology, Beijing, China; NGMJ, Nanjing Geology Museum, Nanjing, China; QG, Queen Victoria Museum, Salisbury, Zimbabwe; SMNS, Staatliches Museum für Naturkunde, Stuttgart, Germany; UCR, University College of Rhodesia (now University of Zimbabwe).

PREVIOUS WORK

Pre-cladistic studies

By the end of the 19th century, collections of dinosaur material (especially in the USA) contained representatives of most major groups (e.g. Ornithopoda, Theropoda). O. C. Marsh, E. D. Cope, H. G. Seeley and others created classifications of the Dinosauria that included well-known groups such as the Theropoda, Sauropoda and Saurischia. Despite these systems of classification, the picture of sauropodomorph relationships remained very confused. Prosauropod material had been studied by Marsh and assigned to the Anchisauridae; however, he placed this family within the Theropoda (Marsh 1881, 1896). Von Huene (1914*a*, p. 154; 1914*b*, p. 36) divided the Saurischia into two groups: the Coelurosauria (small theropods) and the 'Pachypodosauria' (carnosaurs and the prosauropod-sauropod lineage). Matthew (1915) and Abel (1919) split the Saurischia into four groups, the Coelurosauria, Carnosauria, 'Pachypodosauria' and Sauropoda. It is implicit, however, that two subsets of the pachypodosaurs, the 'Zanclodontidae' and Plateosauridae, gave rise to the Carnosauria and Sauropoda, respectively. This is perhaps the first instance of the suggestion that 'prosauropods', or at least a subset of that group, were ancestral to the sauropods. Von Huene coined the terms 'Prosauropoda' and 'Sauropodomorpha' in 1920 and 1932, respectively. The latter was meant to contain the sauropods and prosauropods, but does not seem to have become widely accepted until the work of Charig *et al.* (1965), which provided more detailed evidence supporting a close relationship between these two groups.

The first explicit statement of the view that prosauropods gave rise to sauropods was given by Romer (1956), who divided the prosauropods into three families: Thecodontosauridae, Plateosauridae and Melanorosauridae. These three groups, if taken as a transformation series, show a trend towards increased body size and quadrupedality. Thus, the large melanorosaurids were described as

'... perhaps transitional to the sauropods' (Romer 1956, p. 618). Charig *et al.* (1965) proposed a novel variation of this 'traditional' view. They accepted the three-family division of Prosauropoda proposed by Romer, and also agreed that melanorosaurids were most closely related to sauropods. They suggested, however, that the ancestral sauropodomorph was probably quadrupedal, and that this stance had been retained by melanorosaurids and sauropods (Charig *et al.* 1965). The bipedality of thecodontosaurids and plateosaurids was therefore a specialization. The main evidence for this came from the stratigraphic distribution of the various prosauropods, including the discovery of South African melanorosaurid material in rocks older than those producing the bipedal forms.

Cruickshank (1975) appears to be the first researcher to have proposed an alternative view of prosauropod-sauropod relationships. As well as questioning the prosauropod affinities of *Vulcanodon* Raath, 1972 (Lower Jurassic, Zimbabwe), he suggested that sauropods were not descended from prosauropods. He noted that sauropods have a well-developed fifth metatarsal, whereas in prosauropods it is somewhat reduced and therefore more derived. Cooper (1981, 1984), however, proposed a version of the more traditional position, arguing that prosauropods represent the ancestral stock from which all dinosaurs descended. He also revised the description of *Vulcanodon*: while agreeing with Cruickshank's (1975) reinterpretation of this genus as a basal sauropod, Cooper (1984) also noted several prosauropod features. He therefore argued that *Vulcanodon* was a transitional form between the Prosauropoda and Sauropoda and suggested that the former were ancestral to the latter. Furthermore, Bonaparte (1986) produced an impressive list of derived characters that united some of the prosauropods (notably *Riojasaurus* and melanorosaurids) with the sauropods, again proposing that prosauropods are paraphyletic. These studies are important because they represent the first attempts to list a large series of characters relevant to the question of prosauropod-sauropod relationships. The majority of these pre-cladistic studies typically regarded prosauropods as ancestral to sauropods, with the former giving rise to the latter during the Late Triassic through an increase in body size and a concomitant shift from bipedality to quadrupedality.

Cladistic studies

Gauthier (1986) presented the first cladistic analysis of Sauropodomorpha. The focus of this publication was the relationship between theropods and birds: as a result, the details of the sauropodomorph part of the analysis are brief. Nevertheless, Gauthier's description of his results makes it clear that prosauropods were found to be paraphyletic with respect to sauropods. Furthermore, Gauthier's results suggested that basal sauropodomorphs included small bipedal forms such as *Efraasia* and *Thecodontosaurus*, with larger quadrupedal forms (e.g. *Riojasaurus*) being more closely related to true sauropods (Text-fig. 1A). The segnosaurs (= therizinosaurs) were also found to be basal sauropodomorphs, but the relationships among the *Plateosaurus*-like prosauropods were not discussed in detail.

Sereno (1989) seems to have been the first worker to suggest that prosauropods are monophyletic on the basis of a cladistic analysis (but see also Paul 1984). The details of this analysis were not published, although lists of synapomorphies for various clades were provided. Sereno also proposed that Sauropodomorpha included the segnosaurs, as the sister-taxon to Prosauropoda. Galton (1990) reviewed the anatomy, taxonomy and phylogeny of prosauropods and presented his conclusions in the form of a cladogram (Text-fig. 1B). This topology, however, was not based on a formal cladistic analysis, although synapomorphies for some of the nodes were listed. This tree depicts prosauropods as a monophyletic group that contains two major clades: the 'Plateosauridae' and 'Melanorosauridae'.

Small bipedal taxa, such as *Thecodontosaurus* and *Anchisaurus*, were considered to lie basal to these two clades.

Gauffre (1996) analysed a data matrix comprising 18 ingroup sauropodomorphs and 44 characters, thus providing the first detailed analysis in which the data matrix is available for examination (see also Gauffre 1993). Gauffre's most parsimonious topology (Text-fig. 1C) depicts sauropods and prosauropods as monophyletic sister-groups, with the latter group divided into plateosaur and melanorosaur clades, although the exact contents of these groups differ in detail from those proposed by Galton (1990). Sereno (1997, 1999*a*) presented the results of a cladistic analysis of 32 characters for nine ingroup prosauropods and two outgroups (Theropoda and Sauropoda). Several of the ingroups were unstable in the most parsimonious trees and were deleted *a posteriori*: thus Sereno effectively presented the reduced consensus tree. This topology (Text-fig. 1D) placed *Riojasaurus* at the base of Prosauropoda, and found two derived clades, the Massospondylidae and Plateosauridae. Although this might appear to conflict with the results of Galton (1990) and Gauffre (1996), it is not possible to confirm this because these analyses have only one potential 'melanorosaurid' (*Riojasaurus*) in common,

TEXT-FIG. 1. Cladograms representing the results of previous phylogenetic analyses of basal sauropodomorph relationships, A, Gauthier (1986); B, Galton (1990); C, Gauffre (1996); D, Sereno (1999*a*).

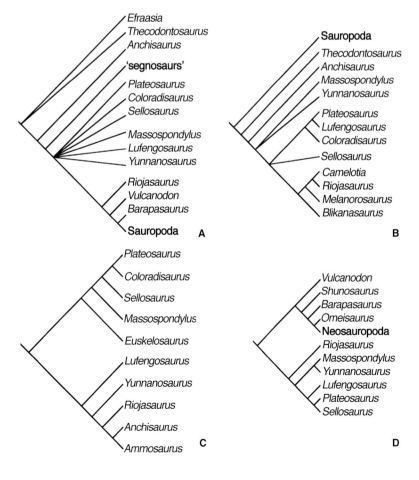

and Sereno's analyses do not provide definitive positions for basal forms such as *Thecodontosaurus*.

Benton *et al.* (2000) reviewed the anatomy of *Thecodontosaurus* and presented a cladistic analysis based on 49 characters for 13 sauropodomorphs and two outgroups. The most parsimonious trees indicated that prosauropods and sauropods are monophyletic sister-taxa and placed *Coloradisaurus* and *Thecodontosaurus* at the base of the Prosauropoda (Text-fig. 2A). The more derived prosauropods form two clades, again corresponding approximately to the 'plateosaurid' and 'melanorosaurid' clades found by Galton (1990) and Gauffre (1996).

It is interesting to note that most of the analyses that have recovered a monophyletic Prosauropoda have not presented a full character list and data matrix, have been based on relatively small data sets (< 50 characters), or have assumed that sauropods represent an outgroup rather than including them as true ingroup taxa. Despite these potential drawbacks, a monophyletic Prosauropoda rapidly became the orthodox view, with several cladistic analyses of sauropod interrelationships employing Prosauropoda as the nearest outgroup (Calvo and Salgado 1995; Salgado *et al.* 1997; Upchurch 1998; Wilson and Sereno 1998; Wilson 2002; Upchurch *et al.* 2004). This standpoint created a gap in our understanding of the origins of sauropods, especially as no members of this clade were recognized from the Late Triassic. In the past 5 years, however, new discoveries, re-examination of existing specimens, and further application of cladistic analysis have revealed a more detailed and complex picture of basal sauropodomorph relationships and the origin of sauropods. For example, *Isanosaurus* from Thailand (Buffetaut *et al.* 2000) and *Blikanasaurus* (Galton and Van Heerden 1985; Galton and Upchurch 2004; Yates 2004*a*) and *Antetonitrus* (Yates and Kitching 2003) from South Africa represent very basal sauropods from the Late Triassic. Furthermore, a number of recent cladistic analyses (Yates 2003*a*, 2004*a*, 2005; Yates and Kitching 2003; Pol and Powell 2005) have cast doubt on prosauropod monophyly and have indicated that some taxa (such as *Anchisaurus* and *Melanorosaurus*) might be regarded as basal sauropods. For example, on the basis of a cladistic analysis of 164 characters for 16 ingroup and five outgroup taxa, Yates (2003*a*) found that prosauropods were almost completely paraphyletic with respect to sauropods (Text-fig. 2B). Later, larger analyses (Yates and Kitching 2003; Yates 2004*a*) found most parsimonious topologies in which some prosauropods (*Plateosaurus* and similar taxa) formed a monophyletic group, but continued to place small bipedal forms (*Saturnalia*, *Thecodontosaurus* and *Efraasia*) at the base of the sauropodomorph tree and 'melanorosaurids' within basal Sauropoda (Text-fig. 2C). It appears, however, that the most recent analyses (Pol and Powell 2005; Yates 2005) favour the extreme form of prosauropod paraphyly (N.B. the results of these two studies

had not been published in full by the time this contribution went to press so they cannot be discussed in detail herein).

Despite the potential 'paradigm shift' towards a partially or even fully paraphyletic prosauropod assemblage, one recent cladistic analysis (Galton and Upchurch 2004) supported the view that most basal sauropodomorphs from the Late Triassic and Early Jurassic belong to a monophyletic Prosauropoda (Text-fig. 2D). This study even placed *Saturnalia* as a basal member of Prosauropoda, in a position slightly more derived than *Thecodontosaurus*. The remaining prosauropods form two clades, one comprising *Anchisaurus* and 'melanorosaurids' and the other containing *Plateosaurus*-like taxa (Text-fig. 2D).

The brief review presented above clearly indicates that the phylogenetic relationships of basal sauropodomorphs are currently in a state of flux. Not only is there disagreement concerning the relationships of major groups of taxa such as 'plateosaurids', 'melanorosaurids' and sauropods, but also there is no clear consensus regarding the relationships within most of these clades (compare the positions of the various plateosaurs in Text-figs 1–2). For example, although several studies have recovered monophyletic Prosauropoda, Plateosauridae and Melanorosauridae, others have separated these taxa and identified them as a paraphyletic array leading to Sauropoda. Work on basal sauropodomorph phylogeny is still in its infancy, and it will be interesting to observe whether future studies converge on a stable and robust consensus.

PHYLOGENETIC ANALYSIS

The dataset

The data matrix used for the cladistic analyses described below consists of 292 osteological characters that were coded for 34 taxa. The operational taxonomic units (OTUs) include seven outgroups and 27 ingroup sauropodomorphs. The outgroups were chosen on the basis of recent phylogenetic analyses of basal dinosaurian relationships (Langer 2004) and include *Coelophysis* and *Guaibasaurus* (basal theropods), *Herrerasaurus* and *Eoraptor* (basal saurischians), *Lesothosaurus* and *Heterodontosaurus* (basal ornithischians), and *Marasuchus* (the nearest dinosauromorph to true Dinosauria). The sources of data for these outgroups are listed in the Appendix. The 27 ingroup taxa include *Saturnalia*, a large number of traditional 'prosauropods' and several representative sauropods. Information on the anatomy of these ingroups has been obtained from the literature and supplemented with numerous personal observations (see Appendix).

The 292 characters represent a synthesis of the character lists provided by Yates and Kitching (2003) and

TEXT-FIG. 2. Cladograms representing the results of previous phylogenetic analyses of basal sauropodomorph relationships, A, Benton *et al.* (2000); B, Yates (2003*a*); C, Yates and Kitching (2003); D, Galton and Upchurch (2004).

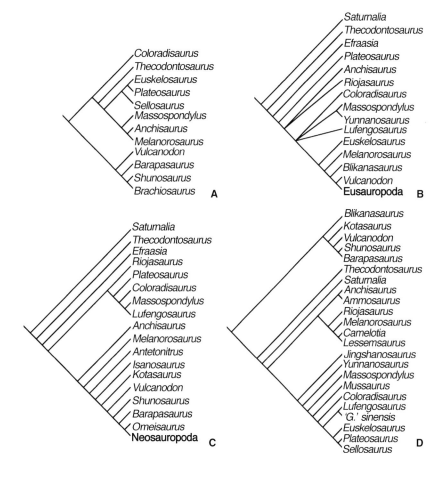

Galton and Upchurch (2004). Many of these characters have been modified in terms of their definitions, state boundaries, ordering and codings. For example, most of the ordered multistate characters from Yates and Kitching (2003) have been re-formatted using additive binary coding in the current dataset. Full details of the character list and modifications are provided in the Appendix. The data matrix is also presented in the Appendix. To avoid copying errors, an electronic version of the original Nexus file (which also contains the backbone constraints described below) can be obtained from the first author on request, or downloaded from the Palaeontological Association website (http://www.palass.org).

Analyses and results

In all analyses, the following assumptions were made: (1) polymorphisms were treated as uncertainty; (2) character 144 was treated as an ordered multistate character (all other characters are binary); (3) the topology of the basal part of the tree (i.e. the non-sauropodomorph taxa) was constrained to conform to that found by Langer (2004; Text-figs 3–4) because the data matrix contains no

characters relevant to outgroup relationships; and (4) all analyses using the heuristic searches in PAUP employed the default settings.

The data matrix was analysed using the heuristic search in PAUP 4.0b10 (Swofford 2002). This analysis found 36 most parsimonious trees (MPTs) with tree lengths (TL) of 724 steps (CI = 0·40, RI = 0·66, RCI = 0·27; Text-fig. 3). The data matrix was also analysed using PAUPRat, a parsimony ratchet program (Sikes and Lewis 2001) in order to increase the probability of finding the globally most parsimonious solutions. The PAUPRat analyses employed 200 replicates with the character perturbation rate set at 15 per cent. This analysis produced MPTs with TL = 724, many of which were identical to each other. Inspection of these PAUPRat MPTs indicated that there were no new most parsimonious topologies that had been missed by the original heuristic search. We conclude therefore that the 36 MPTs recovered originally represent the most parsimonious solutions that can be obtained at present.

The 36 MPTs were subjected to reduced consensus techniques (Wilkinson 1994), which revealed that the *a posteriori* deletion of *Blikanasaurus* had the greatest impact on topological resolution, reducing the number of separate tree topologies from 36 to six. A strict cladistic

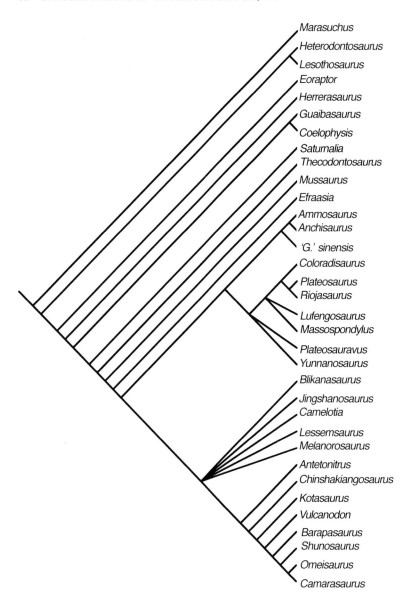

Marasuchus
Heterodontosaurus
Lesothosaurus
Eoraptor
Herrerasaurus
Guaibasaurus
Coelophysis
Saturnalia
Thecodontosaurus
Mussaurus
Efraasia
Ammosaurus
Anchisaurus
'G.' sinensis
Coloradisaurus
Plateosaurus
Riojasaurus
Lufengosaurus
Massospondylus
Plateosauravus
Yunnanosaurus
Blikanasaurus
Jingshanosaurus
Camelotia
Lessemsaurus
Melanorosaurus
Antetonitrus
Chinshakiangosaurus
Kotasaurus
Vulcanodon
Barapasaurus
Shunosaurus
Omeisaurus
Camarasaurus

TEXT-FIG. 3. Strict consensus cladogram (SCC) of the 36 MPTs found by the heuristic search in PAUP 4.0b10 (Swofford 2002) when applied to the data matrix in the Appendix. Tree statistics for the original MPTs are: TL = 724 steps; CI = 0·40, RI = 0·66, RCI = 0·27. See text for further details.

consensus (SCC) of these six reduced topologies is shown in Text-figure 4. This tree indicates support for the monophyly of Sauropodomorpha, a plateosaurian clade and a sauropod clade that includes *Jingshanosaurus* and 'melanorosaurids' at its base (see 'Discussion').

The data matrix was subjected to a Permutation Tail-Probability (PTP) test using PAUP 4.0b10 (Swofford 2002). This test employed 10,000 random replicates (ingroup taxa only) and resulted in a *P*-value of <0·0001. This result indicates that there is non-random structure to the distribution of character states and suggests that the matrix contains a strong phylogenetic signal. Bootstrap analyses were carried out using 10,000 replicates and the heuristic search in PAUP. The results are summarized in Text-figure 4. It is clear that the majority of nodes are weakly supported with bootstrap values of less than 50 per cent. This weak support occurs at many

of the nodes of greatest interest, such as at the base of the monophyletic plateosaur clade and the nodes linking *Jingshanosaurus* and 'melanorosaurids' to Sauropoda. It is clear from these results that the relationships within Sauropoda represent the most robust part of the tree, with several nodes achieving values of 90 per cent or higher. The lack of robustness for most of the remaining relationships may reflect both missing data and widespread character conflict within the data. In particular, there is considerable homoplasy among the more basal taxa, which is apparently caused by numerous character state reversals: this issue will be explored further below. In addition, it is common for large cladistic analyses to perform poorly on bootstrap tests because of the presence of a small number of highly unstable taxa (Wilkinson 1996). For example, the reduced consensus analysis identified *Blikanasaurus* as the most unstable taxon. This is not surprising, given that this OTU can be coded for only 12 per cent of the characters because it is known solely on the basis of a lower hindlimb and pes. However, *a priori* deletion of *Blikanasaurus* results in only a marginal increase in bootstrap values for the nodes at, and below, the base of Sauropoda.

Constrained analyses

In order to explore the relative strengths of several competing phylogenetic hypotheses, a number of topological constraints were created. Each constraint was enforced in turn: the resulting MPTs were collected and then compared with the original (unconstrained) MPTs using Templeton's tests in PAUP. The results of these tests are summarized in Table 1. Four separate competing hypotheses have been considered: (1) 'melanorosaurids' and plateosaurs form a monophyletic group; (2) plateosaurs are paraphyletic with respect to the 'melanorosaurid' + sauropod clade; (3) *Anchisaurus* + *Ammosaurus* is a basal sauropod; and (4) all prosauropods form a monophyletic clade (N.B. *Saturnalia* and *Blikanasaurus* were not considered to be prosauropods in this constraint). Table 1 also provides information on the taxa included/excluded in particular groups, and cites some recent studies that have supported the hypotheses concerned. It should be noted that the second hypothesis listed above (i.e. that plateosaurs are paraphyletic with respect to 'melanorosaurids' + Sauropoda) cannot be truly represented by writing a backbone constraint in PAUP because it is not possible to force

a group to be paraphyletic in a generalized way. Here, therefore, the constraint simply forces *Plateosaurus* (as well as taxa such as *Saturnalia*, *Thecodontosaurus*, *Efraasia* and *Mussaurus*) to lie outside of a clade containing all remaining ingroup taxa. This constraint did result in complete prosauropod paraphyly, although there is no guarantee that the resulting topologies represent the most parsimonious trees in which such paraphyly occurs. A more general method for exploring paraphyly would be to build constraints in which each of the taxa in the monophyletic group termed Prosauropoda in Text-figure 4 were excluded in turn in the manner just described for *Plateosaurus*. For each such analysis, the shortest trees could be collected, and a review of all such trees would reveal which constraint yielded the shortest MPTs in which total prosauropod paraphyly occurs. Alternatively, when the detailed structure of the phylogenies proposed by Pol and Powell (2005) and Yates (2005) become available, then a precise constraint could be constructed which would test their particular versions of prosauropod paraphyly.

The results of the Templeton's tests (Table 1) indicate that none of these alternative arrangements represents a statistically worse explanation of the data than the 36 MPTs found when no constraint was enforced. In the case of the 'grand prosauropod monophyly' constraint, the resulting constrained topologies are 22 steps longer than the unconstrained MPTs, and the *P*-values straddle the 0·05 cut-off for statistical significance. Thus, the current dataset is nearly strong enough to reject decisively the 'grand prosauropod monophyly' concept, but caution is still required.

TABLE 1. Summary of the results of constrained phylogenetic analyses and Templeton's tests (see text for details). Abbreviations: No. MPTs, number of MPTs generated when the constraint was enforced; TL, tree length of the constrained topologies; + steps, the number of extra character steps required by the constrained topology relative to the original unconstrained MPTs; *P*, the *P*-values found by the Templeton's tests. The precise constraints are presented in the electronic version of the data matrix. In order to facilitate replication of this study, however, the taxon clusters listed in the table are defined as follows: 'melanorosaurids' = *Camelotia*, *Melanorosaurus* and *Lessemsaurus* (i.e. Node O in Text-fig. 4); 'plateosaurs' = *Coloradisaurus*, *Plateosauravus*, *Lufengosaurus*, *Massospondylus*, *Plateosaurus*, *Riojasaurus* and *Yunnanosaurus* (i.e. Node I in Text-fig. 4); Sauropoda = *Blikanasaurus* plus the taxa united at Node Q in Text-figure 4.

Constraint/hypothesis	No. MPTs	TL	+ steps	*P*
'Melanorosaurids' are placed within Prosauropoda (Galton and Upchurch 2004)	6	733	9	0·14–0·20
Anchisaurus and *Ammosaurus* placed within Sauropoda (Yates and Kitching 2003; Yates 2004*a*)	45	728	4	0·54–0·58
Plateosaurs are paraphyletic with respect to 'melanorosaurids' + sauropods (Yates 2003*a*, 2005; Pol and Powell 2005)	30	732	8	0·38
'Grand prosauropod monophyly'	5	746	22	0·042–0·053

DISCUSSION

Classification and nomenclature

The current analysis has produced a set of cladogram topologies that have implications for basal sauropodomorph classification and nomenclature. Therefore, before discussing the impact of these results on our understanding of basal sauropodomorph evolution, it is vital that we clarify the names applied to various clades. The phylogenetic definitions for the major sauropodomorph clades are briefly summarized below, with comments and revisions where appropriate.

Sauropodomorpha von Huene, 1932. This clade was defined as a node-based taxon by Salgado *et al.* (1997, p. 6), such that it contains '… the most recent common ancestor of Prosauropoda and Sauropoda and all of its descendants' (see also Langer 2002, 2003). It has also been defined as a stem-based taxon that includes all taxa more closely related to *Saltasaurus* than to Theropoda (Sereno 1999*b*; Galton and Upchurch 2004). Here, we prefer the stem-based definition because it is more stable. The node-based taxon of Salgado *et al.* uses 'Prosauropoda' as one of its 'specifiers': this is problematic because the content of Prosauropoda is unstable, and it is even possible that such a clade may eventually be reduced to a single genus (*Plateosaurus*) if complete prosauropod paraphyly becomes the consensus view. This instability means that the contents of Sauropodomorpha also vary when the node-based definition is applied to the various competing cladograms.

Anchisauria Galton and Upchurch, 2004. This clade has been defined as a node-based taxon that includes the most recent common ancestor of *Anchisaurus* and *Melanorosaurus*, and all of that ancestor's descendants (Galton and Upchurch 2004). When originally defined, the Anchisauria comprised a subset of Prosauropoda: *Anchisaurus*, *Ammosaurus* and the Melanorosauridae. However, the changes in tree topology proposed by the current analysis mean that this taxon now applies to a more inclusive clade of sauropodomorphs (essentially all taxa except the most basal forms: *Saturnalia*, *Thecodontosaurus*, *Mussaurus* and *Efraasia*). Consequently, Anchisauria becomes the name applied to the clade that contains Prosauropoda and Sauropoda as currently defined (Text-fig. 4). If subsequent analyses support the view that *Anchisaurus* is more closely related to Sauropoda (Yates 2004*a*), or that prosauropods form a completely paraphyletic assemblage at the base of Sauropoda (Pol and Powell 2005; Yates 2005), the contents of Anchisauria will inevitably change to include fewer (or perhaps none) of the *Plateosaurus*-like forms.

Prosauropoda von Huene, 1920. This clade has been defined as a stem-based taxon that includes all taxa more closely related to *Plateosaurus* than to either *Saltasaurus* (Sereno 1999*b*) or Sauropoda (Galton and Upchurch 2004). According to the current cladistic analysis, the Prosauropoda persists as a monophyletic clade that contains the Plateosauria (see below) and a clade comprising *Anchisaurus*, *Ammosaurus* and '*Gyposaurus*' (Text-fig. 4).

Plateosauria Tornier, 1913. This clade has been defined as a node-based taxon containing the most recent common ancestor of Plateosauridae and Massospondylidae, and all of that ancestor's descendants (Sereno 1998, 1999*b*). Galton and Upchurch (2004) also defined Plateosauria as a node-based taxon, but employed *Jingshanosaurus* instead of Massospondylidae as one of the reference taxa. Unfortunately, *Jingshanosaurus* has now been placed as a basal sauropod (Text-fig. 4), so Galton and Upchurch's definition would now include all sauropodomorphs except for the most basal forms, and would therefore have the same content as Anchisauria (see above). We therefore apply Sereno's (1998, 1999*b*) definition, despite the potential instability of *Massospondylus*. This means that Plateosauria currently includes Plateosauridae, *Massospondylus* and possibly *Lufengosaurus*, but would exclude *Yunnanosaurus* and *Plateosauravus*.

Anchisauridae Marsh, 1885. This clade has been defined as a stem-based taxon that includes all taxa more closely related to *Anchisaurus* than to *Melanorosaurus* (Galton and Upchurch 2004). When originally defined, the Anchisauridae included just *Anchisaurus* and *Ammosaurus*. However, changes to tree topology mean that this clade would now include the same taxa as Prosauropoda. To avoid confusion therefore we suggest that use of the name Anchisauridae should be suspended in favour of Prosauropoda, pending further study of basal sauropodomorph phylogeny.

Massospondylidae von Huene, 1914b. This clade has been defined as a stem-based taxon that includes all taxa more closely related to *Massospondylus* than to *Plateosaurus* (Sereno 1999*b*). Under this definition, the Massospondylidae included either *Massospondylus* and *Yunnanosaurus* (Sereno 1999*a*), or *Massospondylus*, *Coloradisaurus* and *Lufengosaurus* (Yates and Kitching 2003). However, the topologies proposed here, and also by Yates (2003*a*, 2005) and Pol and Powell (2005), imply that the Massospondylidae contains only *Massospondylus*.

Plateosauridae Marsh, 1895. This clade has been defined as a stem-based taxon that includes all taxa more closely related to *Plateosaurus* than to either Massospondylidae

TEXT-FIG. 4. Strict consensus cladogram (SCC) depicting the phylogenetic relationships of basal sauropodomorphs found by the current study. The analysis found 36 MPTs that were reduced to six topologies via *a posteriori* deletion of *Blikanasaurus*. These six trees then formed the basis for the SCC shown here. Outgroup taxa are shown using white branches; the solid black branches depict the ingroup relationships. Upper case letters (A–W) are used to designate individual nodes. Percentages at nodes are bootstrap values (those nodes without a value can be assumed to have scored less than 50 per cent in the bootstrap analyses). Stem-based names are marked by *; node-based names lack the *.

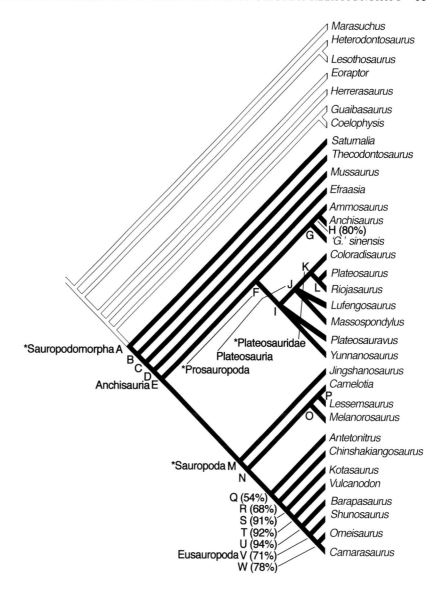

(Sereno 1999*b*) or *Yunnanosaurus* and *Massospondylus* (Galton and Upchurch 2004). The selection of both *Yunnanosaurus* and *Massospondylus* as reference taxa by Galton and Upchurch (2004) reflected uncertainty in the tree topologies recovered by that study. The relationships of these taxa remain unstable in the current analysis, although *Massospondylus* appears to be more closely related to plateosaurids than is *Yunnanosaurus*. Thus, as currently defined, Plateosauridae includes *Plateosaurus*, *Riojasaurus* and *Coloradisaurus*, whereas the inclusion of *Lufengosaurus* in this family is equivocal (Text-fig. 4).

Sauropoda Marsh, 1878. This clade has been defined as a node-based taxon by Salgado *et al.* (1997, p. 6), such that it contains '… the most recent common ancestor of *Vulcanodon karibaensis* and Eusauropoda and all of its descendants'. A stem-based definition (Sauropoda includes

sauropodomorphs more closely related to *Saltasaurus* than to *Plateosaurus*: Wilson and Sereno 1998; Sereno 1999*b*; Upchurch *et al.* 2004) has also been introduced. There is little to choose between these two definitions, except that the node-based version depends on the definition of Eusauropoda (which itself is disputed) and that the stem-based definition seems to have been adopted by the majority of recent studies (although debate continues). Here, we prefer the stem-based definition which, when combined with the tree topology in Text-figure 4, means that Sauropoda includes Eusauropoda, *Barapasaurus*, *Vulcanodon*, *Kotasaurus*, *Chinshakiangosaurus*, *Antetonitrus*, *Blikanasaurus*, 'melanorosaurids' and *Jingshanosaurus*.

Melanorosauridae von Huene, 1929. This clade has been defined as a stem-based taxon that includes 'all taxa more closely related to *Melanorosaurus* than to *Anchisaurus*'

(Galton and Upchurch 2004, p. 251). Originally, this definition meant that the 'Melanorosauridae' included a small number of taxa such as *Melanorosaurus*, *Camelotia* and *Lessemsaurus*. However, changes to tree topology mean that such a clade would now be virtually synonymous with Anchisauria (see above). The current analysis recovered a clade containing *Melanorosaurus*, *Camelotia* and *Lessemsaurus*, and it would be possible to redefine 'Melanorosauridae' so that it corresponds with this clade. However, the relationships of these taxa to each other and to other basal sauropods remain uncertain (Pol and Powell 2005; Yates 2005) and any new definition may create confusion in the future. For the present therefore we suggest that the name 'Melanorosauridae' be suspended until it can be defined using a more stable and widely accepted tree topology, and we use the informal term 'melanorosaurids' to refer to the clade containing *Melanorosaurus*, *Camelotia* and *Lessemsaurus*.

Eusauropoda Upchurch, 1995. Although no phylogenetically based definition was provided, Upchurch (1995) stated that Eusauropoda referred to a clade containing *Shunosaurus* and more derived sauropods, but excluding *Barapasaurus* and *Vulcanodon*, effectively defining this taxon as a node-based clade. Salgado *et al.* (1997, p. 7) also defined Eusauropoda as a node-based taxon, but suggested that it should include '... the most recent common ancestor of *Barapasaurus tagorei* and Neosauropoda and all of its descendants'. Wilson and Sereno (1998) defined Eusauropoda as a stem-based taxon containing all sauropods more closely related to *Saltasaurus* than to *Vulcanodon*. Upchurch *et al.* (2004) redefined Eusauropoda as a node-based name that is more consistent with Upchurch's original intention. This definition therefore is that the Eusauropoda includes the most recent common ancestor of *Shunosaurus* and *Saltasaurus* and all of that ancestor's descendants (Upchurch *et al.* 2004). The results of the current cladistic analysis do not create major problems for either the stem- or node-based definitions of Eusauropoda; however, it is worth noting that the current analysis has placed *Barapasaurus* further from the neosauropods than *Shunosaurus*, as originally proposed by Upchurch (1995, 1998). It would be interesting to investigate the extent to which this situation has been created by new character data and/or the exclusion of taxa such as *Cetiosaurus*: however, this represents an aspect of the relationships of more derived sauropods, and will not be addressed here. In short, despite varying definitions and topologies, the contents of Eusauropoda are relatively stable, with the main problems relating to whether *Barapasaurus* or *Shunosaurus* are included or excluded.

The phylogenetically based classification outlined above, and also shown in Text-figure 4, illustrates the difficulties generated when this approach to systematics is applied before stable and widely accepted phylogenies are available. While definitions remain stable (an inherent property of phylogenetic classification), the content of several of the named groups varies widely, depending on which cladogram topology is employed. Furthermore, the changes enforced by the discovery of new topologies can bring two or more clade names into conflict or synonymy, as has occurred here with names such as Anchisauria and 'Melanorosauridae'. Here, we have attempted to clarify the classification, retaining original definitions and clade contents whenever possible.

One key issue concerns the contents of the Sauropoda. The extreme prosauropod paraphyly proposed by Pol and Powell (2005) and Yates (2003*a*, 2005), combined with the phylogenetic definition of Sauropoda, would mean that this clade would contain all sauropodomorphs except for *Plateosaurus* and more basal forms (*Saturnalia*, *Thecodontosaurus* and *Efraasia*). This would seem to stretch the contents of Sauropoda far beyond that intended when this group name is normally cited (e.g. Romer 1956; McIntosh 1990; Calvo and Salgado 1995; Upchurch 1995, 1998; Wilson and Sereno 1998; Wilson 2002; Upchurch *et al.* 2004). Yates (2005) therefore has suggested that the definition of Sauropoda might need adjustment, or that a new name for sauropods plus more derived prosauropods could be created. We suggest, however, that revised definitions and the creation of new clade names would be premature at this stage and is likely to generate even greater confusion in the future. This is because: (1) current cladograms for basal sauropodomorphs are relatively unstable and no clear consensus is available; and (2) once a consensus on basal sauropodomorph relationships is achieved, existing names (such as Anchisauria, Sauropoda and Eusauropoda) may provide a satisfactory means of classifying taxa into manageable groups. Indeed, given the current status of basal sauropodomorph relationships, the classification of this group would benefit greatly from a general moratorium on the creation of new clade names, pending future phylogenetic work.

Basal sauropodomorphs

The SCCs in Text-figures 3–4 indicate that four taxa (*Saturnalia*, *Thecodontosaurus*, *Mussaurus* and *Efraasia*) lie at the base of the sauropodomorph radiation, outside of a clade containing prosauropods, 'melanorosaurids' and sauropods. This is in general agreement with several previous analyses, including those of Yates (2003*a*, 2004*a*, 2005) and Yates and Kitching (2003) (see Text-fig. 2). One key aspect of the disintegration of prosauropod monophyly therefore is the discovery that several small bipedal 'prosauropods' are actually basal sauropodomorphs. The alternative view, that these small forms are members of a monophyletic Prosauropoda, is virtually rejected by the constrained analyses described above (Table 1).

The position of *Mussaurus* as a basal sauropodomorph is interesting because this taxon has been incorporated into only one previous published analysis (Galton and Upchurch 2004; Text-fig. 2D), where it was recovered as a more derived prosauropod. The relationships of *Mussaurus* should be treated with caution, however, as this taxon is based on juvenile specimens: ontogenetic factors may have affected character state codings and distorted the evidence for its phylogenetic position. It is conceivable that additional character data will alter the placement of *Mussaurus*, potentially to a more derived position closer to sauropods.

The skull of *Mussaurus* displays some similarities with those of sauropods, including teeth that become larger towards the rostral end of the snout and an infratemporal opening that extends beneath the orbit (Bonaparte and Vince 1979). Such features have been observed in the skulls of several juvenile prosauropods (e.g. *Massospondylus*: Gow *et al.* 1990) and have led to suggestions that sauropods may have arisen from prosauropods as a result of heterochronic changes (Bonaparte and Vince 1979; see also Reisz *et al.* 2005). Our knowledge of growth series within basal sauropodomorphs is still very limited, however, and a strong case for the heterochronic origin of sauropods from prosauropods has yet to be made. Nevertheless, the increased paraphyly of prosauropods with respect to sauropods, found both here and in other studies, makes the 'heterochronic origin' hypothesis more feasible, because it was effectively ruled out previously by prosauropod monophyly.

Prosauropod/plateosaurian monophyly

The phylogenetic analyses of Yates (2003*a*, 2005) and Pol and Powell (2005) recovered tree topologies in which traditional prosauropod taxa form a completely paraphyletic group with respect to Sauropoda. Although the details differ (see 'Previous work', above), these arrangements even posit that plateosaurian taxa form a completely paraphyletic group, in which *Plateosaurus* itself occupies a more basal position than similar forms such as *Coloradisaurus* and *Lufengosaurus*. The current analysis, however, has recovered a monophyletic Prosauropoda containing *Anchisaurus*, '*Gyposaurus*', *Plateosaurus*, *Coloradisaurus*, *Lufengosaurus*, *Massospondylus* and others (Text-figs 3–4). Although bootstrap values are low for this clade, and the constrained analyses do not decisively reject complete prosauropod paraphyly (Table 1), it does require an additional eight steps to decompose completely the Prosauropoda and its constituent clades into a pectinate paraphyletic array of taxa basal to Sauropoda.

The Delayed Transformation Optimization (DELTRAN) option in PAUP 4.0b10 (Swofford 2002) and MacClade 4.07 (Maddison and Maddison 2004) were used to map character states onto the SCC tree in Text-figures 3–4, in order to investigate the apomorphies that potentially support the monophyly of Prosauropoda and its subgroups. A combination of homoplasy and missing data means that there are few unequivocal synapomorphies, but a number of apomorphic features do appear to be unique to these taxa (see Appendix). The monophyly of the Prosauropoda is supported by at least nine potential synapomorphies, including: modifications to the neurovascular foramina on the lateral surface of the maxilla (C16, C17); the presence of a postparietal opening between the supraoccipital and parietal (C57); metacarpal V has an expanded proximal end that is divided into two articular surfaces (C192); the first manual phalanx has proximal and distal articular axes twisted by at least 60 degrees to each other (rather than 45 degrees or less in other basal sauropodomorphs: C196); and the proximal end of metatarsal IV is strongly compressed dorsoventrally and expanded transversely (C275) (Text-fig. 5, Node F). Monophyly of the Plateosauria is supported by six potential synapomorphies, including: a subtriangular external naris (C4); the distal end of the premaxillary ascending process is widened transversely (possibly also present in *Yunnanosaurus*: C9); the supratemporal fenestra is obscured in lateral view by the upper temporal bar (C48); and the second distal carpal does not completely cover the proximal end of metacarpal II (C183) (Text-fig. 5, Node J). Finally, the monophyly of the Plateosauridae is supported by at least four potential synapomorphies, including: a deep transverse wall of bone between the basipterygoid processes (C61); the deltopectoral crest has a sigmoid outline in cranial view (N.B. the distribution of this state is somewhat equivocal because it is also seen in '*Gyposaurus*': C167); and a 'heel'-like projection of the caudal margin of the ischial articulation of the ilium (C215) (Text-fig. 5, Node K). Thus, there is considerable evidence that at least some of the taxa currently regarded as 'prosauropods' represent a monophyletic taxon that can be characterized by a suite of derived characters. Abandonment of the name Prosauropoda, based on the view that its constituent taxa form a completely paraphyletic array at the base of Sauropoda, may therefore be premature.

Sauropod origins

The proposal that Prosauropoda represented a monophyletic clade encompassing virtually all early basal sauropodomorphs (Sereno 1989, 1999*b*; Galton 1990; Gauffre 1993, 1996; Galton and Upchurch 2004) meant that little could

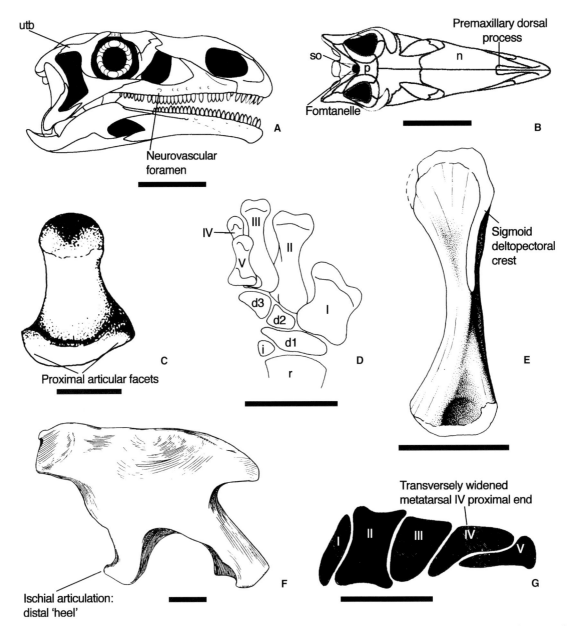

TEXT-FIG. 5. Examples of potential synapomorphic character states that unite the Prosauropoda, Plateosauria or Plateosauridae (see text for details). A, skull of *Plateosaurus* in lateral view (AMNH 6810). B, skull of *Plateosaurus* in dorsal view (AMNH 6810). C, right metacarpal V of *Massospondylus* in medial view (UCR 9558). D, right carpus and metacarpus of *Massospondylus* in palmar view (UCR 9558). E, left humerus of *Massospondylus* in anterior view (QG 1396). F, right ilium of *Plateosaurus* in lateral view (SMNS 13200). G, metatarsus of *Massospondylus* in proximal view (QG 1159). A and B after Galton (1985c); C–E and G after Cooper (1981); F after von Huene (1926). Abbreviations: d1–3, distal carpals 1–3; i, intermedium; n, nasal; p, parietal; r, radius; so, supraoccipital; utb, upper temporal bar. Scale bars represent 50 mm (A–B, D–G) and 10 mm (C).

be said about sauropod origins. No Late Triassic sauropods could be identified, resulting in a stratigraphical and morphological 'gap' at the base of Sauropoda. The most that could be inferred about sauropod origins was that the clade must have diverged from the prosauropod lineage by the Carnian, and that they were probably descended from small bipedal forms (Upchurch 1995; Sereno 1997). During the past decade, however, a series of new discoveries and revised cladistic analyses have greatly enhanced our knowledge of sauropod origins. Table 2 summarizes those taxa from the Late Triassic and Early Jurassic that have been identified (usually via cladistic analysis) as being genuine sauropods that are probably more basal than *Vulcanodon*. Several of these taxa are well preserved and include cranial material. For example, the most basal sauropod identified by the current analysis,

TABLE 2. A summary of the taxa that have been identified as basal sauropods (i.e. forms that are more closely related to *Saltasaurus* than to *Plateosaurus*, but which are more basal than *Vulcanodon*).

Taxon	Country	Age	Reference
Antetonitrus	South Africa	Norian	Yates and Kitching (2003)
Blikanasaurus	South Africa	Norian	Galton and Van Heerden (1985)
Camelotia	UK	Rhaetian	Galton (1985*b*, 1998)
Chinshakiangosaurus	China	Early Jurassic	Dong (1992), Upchurch *et al.* (in press)
Gongxianosaurus	China	Early Jurassic	He *et al.* (1998)
Isanosaurus	Thailand	Norian/Rhaetian	Buffetaut *et al.* (2000)
Jingshanosaurus	China	Early Jurassic	Zhang and Yang (1994)
Kotasaurus	India	Early Jurassic	Yadagiri (1988, 2001)
Lessemsaurus	Argentina	Norian	Bonaparte (1999), Pol and Powell (2005)
Melanorosaurus	South Africa	Norian	Haughton (1924), Yates (2005, 2007)
New Indian taxon	India	Early Jurassic	Kutty *et al.* (in press)

Jingshanosaurus, is known from a virtually complete skull and postcranial skeleton (Zhang and Yang 1994). Cranial material is also now known in *Melanorosaurus* (Yates 2005, 2007), *Chinshakiangosaurus* (Upchurch *et al.* in press) and a new Indian taxon (Kutty *et al.* in press), and substantial portions of the postcranial skeleton are available for these forms and also for *Antetonitrus* (Yates and Kitching 2003) and *Kotasaurus* (Yadagiri 2001). This wealth of new data has implications for our understanding of sauropod origins, a topic that is explored below and in Upchurch *et al.* (in press) and Barrett and Upchurch (2007).

Examination of character state distributions using MacClade 4.07 (Maddison and Maddison 2004) reveals an unexpected phenomenon. A large number of derived character states (approximately 20) are acquired at several basal nodes in Sauropodomorpha, but are then reversed to the plesiomorphic state within basal Sauropoda (Table 3). It should be noted that the total number of such reversals would increase further if, as proposed by Pol and Powell (2005) and Yates (2003*a*, 2005), prosauropods form a completely paraphyletic assemblage at the base of Sauropoda. In either case, such a major set of character state reversals must have contributed considerably to the causes of controversy and disagreement regarding basal sauropodomorph relationships: aside from the normal difficulties associated with cladistic analysis (e.g. subjective character coding, missing data, inconsistent taxon and character sampling strategies), there might also exist a coherent homoplastic signal that runs counter to the primary phylogenetic signal. It is interesting to note, for example, that approximately 12 of the 20 reversals involve characters that, in their derived state, have previously been cited as synapomorphies of the Prosauropoda (Table 3). Thus, the discoveries of taxa such as *Antetonitrus* and the new Indian form, as well as skull material of *Melanorosaurus* and *Chinshakiangosaurus*, have had a profound impact on basal sauropodomorph phylogeny because they demonstrate that many supposed prosauropod synapomorphies are actually also present in basal sauropods. For example, the 45 degree rotation of the articular axes of the first manual phalanx was initially believed to be a clear prosauropod synapomorphy (Galton 1990; Sereno 1999*a*), but it is now also known to occur in *Antetonitrus* and the new Indian taxon (Yates and Kitching 2003; Kutty *et al.* in press). Thus, these new specimens have simultaneously undermined support for prosauropod monophyly and produced the large number of reversals listed in Table 3.

It is not clear why these reversals should have occurred in such a coherent fashion. We might speculate that such a reversal pattern could be consistent with the 'heterochronic origin' hypothesis discussed above. Juvenile prosauropods may have retained plesiomorphic states, while the adult forms developed apomorphies. If sauropods then arose from prosauropods via some form of paedomorphosis, then adult sauropods would acquire character states present in juvenile prosauropods (i.e. plesiomorphic states). However, a closer examination of the relevant characters indicates that the 'heterochronic origin' hypothesis is not sufficient (at least by itself) to explain the observed reversals. For example, the 'twisted' first manual phalanx mentioned above is present not only in adult prosauropods, but also small basal forms (e.g. *Efraasia*) and juveniles (e.g. '*Gyposaurus*': PU and PMB, pers. obs. 2002). It seems more plausible therefore that the character reversals reflect the new demands placed on basal sauropods by their invasion into new ecological niches. For example, increased body size (perhaps associated with a shift towards bulk processing of large quantities of poor quality vegetation) would have necessitated greater reliance on quadrupedal locomotion with concomitant modifications to the forelimb. Thus, disappearance of the 'twisted' first manual phalanx might represent the 'overwriting' of a pre-existing derived state by a new version of the plesiomorphic condition, because of biomechanical

TABLE 3. Derived character states that originate in basal sauropodomorphs and then reverse at nodes in basal Sauropoda; *, character states that have previously been cited as synapomorphies of the Prosauropoda (see text for details).

1.		Rostral process of the maxilla is longer than high.
*2.		5–6 large neurovascular foramina on the lateral surface of the maxilla.
*3.		Lateral lamina along the ventral margin of the antorbital fossa.
4.		Rostral process of the jugal is sharply pointed.
*5.		Ridge on the lateral surface of the caudal part of the dentary.
*6.		Inset first dentary tooth.
*7.		Axial postzygapophyses are flush with the caudal end of the centrum.
*8.		Axial centrum has a length : height ratio of 3·0 or more.
9.		Strong constriction between the sacral rib and transverse process of the first primordial sacral.
10.		Length of the base of the proximal caudal neural spines is greater than half the length of the neural arch.
11.		Flat caudoventrally facing surface on the coracoid between the glenoid and the tubercle.
*12.		Deltopectoral crest is perpendicular to the transverse axis of the distal condyles.
13.		Ratio of the transverse width of the distal end to total humerus length > 0·33.
14.		Lateral end of first distal carpal overlaps the second distal carpal.
*15.		Proximal width of metacarpal I divided by metacarpal length > 1·0.
*16.		Proximal 'heel' on first phalanx of manual digit I.
*17.		Phalanx I of manual digit I has proximal and distal surfaces with their axes twisted at approximately 45 degrees.
18.		Transverse width of the distal tibia is greater than its craniocaudal width.
*19.		Lateral margin of the proximal end of metatarsal II is concave.
*20.		Metatarsal IV proximal end transverse width : dorsoventral height ratio > 2·0.

demands (Yates 2001, 2003a). Similarly, Upchurch et al. (in press) have argued that the lateral ridge on the dentary, potentially associated with a fleshy cheek in prosauropods, was probably present in basal sauropods but was lost in more derived forms because of changes to the feeding system.

In short, new information on basal sauropod anatomy and phylogenetic relationships is creating important opportunities for examining the origins of the group. The basal taxa listed in Table 2 have the potential to modify greatly our current understanding of character state distributions, which in turn could have ramifications for the relationships among more derived sauropods (by, for example, affecting outgroup comparisons) and will also affect hypotheses concerning the ecological and biomechanical evolution of sauropods.

CONCLUSION

The datasets available for investigating basal sauropodomorph evolution are steadily increasing in size as new taxa and characters are added, and as existing species are re-studied. It is possible therefore to examine basal sauropodomorph interrelationships using a large amount of relevant data: c. 300 characters for 30 or more taxa. A comparison between recent cladistic analyses, and that presented here, indicates that several elements of a consensus are beginning to emerge. For example, most recent studies agree that Saturnalia, Thecodontosaurus and Efraasia represent the most basal sauropodomorphs and lie outside of any 'Prosauropoda + Sauropoda' clade. In

addition, it appears that 'melanorosaurids' are probably basal sauropods, and that we now have access to information on at least ten basal sauropod taxa that lie on the stem of this group below the position of Vulcanodon. Other aspects of basal sauropodomorph relationships, however, are much more controversial. In particular, it is not clear whether the various Plateosaurus-like taxa form a monophyletic clade or are paraphyletic with respect to sauropods. Even if Prosauropoda, Plateosauria and/or Plateosauridae are monophyletic, the relationships within these clades require considerable further study. Various measures of tree robustness indicate that many of the proposed relationships are poorly supported, except perhaps those among several sauropod taxa. This result probably reflects a combination of factors, such as large amounts of missing data (e.g. Blikanasaurus) and considerable character conflict. Disagreement between the phylogenies proposed by different authors may reflect the usual difficulties associated with, for example, the formulation of characters and character states, incorporation of different OTUs, and disagreement over character state codings. However, it also seems likely that many problems may have stemmed from the unusual pattern of character state reversal in basal sauropods. The significance of this pattern is currently unclear, although the various reversals may perhaps indicate the selective pressures imposed by functional/ecological shifts during the transition from small bipedal omnivores, through medium-sized herbivores, to gigantic graviportal bulk-feeding herbivores. Heterochrony, in the form of paedomorphosis, may have played a role in the origin of sauropods via the retention into the adult stage of features present in juvenile prosauropods. However, there is currently little evidence to suggest that the 20 character reversals observed during the origin of sauropods pertain to those character states that have been identified as potentially paedomorphic in sauropods. Finally, the influx of new data on basal sauropodomorphs has important implications for the future study of their evolution. Scenarios regarding the early diversity and biogeography of sauropodomorphs will need to be reconsidered in the light of new spatial and temporal distributions.

Acknowledgements. We are grateful to the Royal Society of London, the Jurassic Foundation, the Natural Environment Research Council (PU's research fellowship No. GT59906ES), the Cambridge Philosophical Society and the Palaeontological Research Fund (Natural History Museum) for funding this research. We thank the following for access to specimens in their care and for their kindness and hospitality when visiting their institutions: S. Kaal and R. Smith (SAM); M. Raath (BPI); Xu Xing, Zhou Zhonghe, Wang Yuan and Wang Xiao-Lin (IVPP); Zhou Xiao-Dan (NGMJ); M. Norell and C. Mehling (AMNH); A. Arcucci (ULR); W.-D. Heinrich (MB); C. Chandler (YPM); S. Chapman and A. Milner (BMNH). He *et al.* (1998) was translated by W. Downs and obtained courtesy of the Polyglot Paleontologist website (http://ravenel.si.edu/paleo/paleoglot/index.cfm). Reviews by A. Yates and M. Langer greatly improved an earlier version of this paper.

REFERENCES

ABEL, O. 1919. *Die stämme der Wirbeltiere.* De Gruyter, Berlin and Leipzig, 914 pp.

ATTRIDGE, J., CROMPTON, A. W. and JENKINS, F. A. Jr 1985. The southern African Liassic prosauropod *Massospondylus* discovered in North America. *Journal of Vertebrate Paleontology*, **5**, 128–132.

BARRETT, P. M. and UPCHURCH, P. 2007. The evolution of feeding mechanisms in early sauropodomorph dinosaurs. 91–112. *In* BARRETT, P. M. and BATTEN, D. J. (eds). *Evolution and palaeobiology of early sauropodomorph dinosaurs.* Special Papers in Palaeontology, **77**, 289 pp.

—— —— and WANG XIAO-LIN 2005. Cranial osteology of *Lufengosaurus huenei* Young (Dinosauria: Prosauropoda) from the Lower Jurassic of Yunnan, People's Republic of China. *Journal of Vertebrate Paleontology*, **25**, 806–822.

—— —— ZHOU XIAO-DAN and WANG XIAO-LIN in press. The skull of *Yunnanosaurus huangi* Young (Dinosauria: Prosauropoda) from the Lower Lufeng Formation (Lower Jurassic) of Yunnan, China. *Zoological Journal of the Linnean Society*.

BENTON, M. J., JUUL, L., STORRS, G. W. and GALTON, P. M. 2000. Anatomy and systematics of the prosauropod dinosaur *Thecodontosaurus antiquus* from the Upper Triassic of southwest England. *Journal of Vertebrate Paleontology*, **20**, 77–108.

BERMAN, D. S and McINTOSH, J. S. 1978. Skull and relationships of the Upper Jurassic sauropod *Apatosaurus* (Reptilia, Saurischia). *Bulletin of the Carnegie Museum of Natural History*, **8**, 1–35.

BONAPARTE, J. F. 1969. Los nuevas 'faunas' de reptiles Triásicos de Argentina. I. *Gondwana Symposium, Mar del Plata, Ciencias Tierra*, **2**, 283–306.

—— 1972. Los tetrapodos del sector superior de la formacion Los Colorados, La Rioja, Argentina, (Triásico Superior). *Opera Lilloana*, **22**, 1–183.

—— 1978. *Coloradia brevis* n. g. et n. sp. (Saurischia, Prosauropoda), dinosaurio Plateosauridae de la Formacion Los Colorados, Triásico superior de La Rioja, Argentina. *Ameghiniana*, **15**, 327–332.

—— 1986. The early radiation and phylogenetic relationships of Jurassic sauropod dinosaurs, based on vertebral anatomy. 247–258. *In* PADIAN, K. (ed.). *The beginning of the age of dinosaurs.* Cambridge University Press, Cambridge, 378 pp.

—— 1999. Evolución de las vértebras presacras en Sauropodomorpha. *Ameghiniana*, **36**, 115–187.

—— and PUMARES, J. A. 1995. Notas sobre el primer craneo de *Riojasaurus incertus* (Dinosauria, Prosauropoda, Melanorosauridae) del Triásico Superior de La Rioja, Argentina. *Ameghiniana*, **32**, 341–349.

—— and VINCE, M. 1979. El hallazgo del primer nido de dinosaurios Triásicos, (Saurischia, Prosauropoda), Triásico Superior de Patagonia, Argentina. *Ameghiniana*, **16**, 173–182.

—— FERIGOLO, J. and RIBEIRO, A. M. 1999. A new Early Triassic saurischian dinosaur from Rio Grande do Sul State, Brazil. 89–109. *In* TOMIDA, Y., RICH, T. H. and VICKERS-RICH, P. (eds). *Proceedings of the Second Gondwana Dinosaur Symposium.* National Science Museum Monographs, **15**, 296 pp.

BUFFETAUT, E., SUTEETHORN, V., CUNY, G., TONG, H., LE LOEUFF, J., KHANSUBHA, S. and JONGAUTCHARIYAKUL, S. 2000. The earliest known sauropod dinosaur. *Nature*, **407**, 72–74.

CALVO, J. O. and SALGADO, L. 1995. *Rebbachisaurus tessonei* sp. nov. A new Sauropoda from the Albian–Cenomanian of Argentina; new evidence on the origin of the Diplodocidae. *Gaia*, **11**, 13–33.

CHARIG, A. J., ATTRIDGE, J. and CROMPTON, A. W. 1965. On the origin of the sauropods and the classification of the Saurischia. *Proceedings of the Linnean Society of London*, **176**, 197–221.

COLBERT, E. H. 1989. The Triassic dinosaur *Coelophysis*. *Bulletin of the Museum of Northern Arizona*, **57**, 1–160.

COOPER, M. R. 1981. The prosauropod dinosaur *Massospondylus carinatus* Owen from Zimbabwe: its biology, mode of life and phylogenetic significance. *Occasional Papers of the National Museums and Monuments of Rhodesia, Series B, Natural Sciences*, **6**, 689–840.

—— 1984. A reassessment of *Vulcanodon karibaensis* Raath (Dinosauria: Saurischia) and the origin of the Sauropoda. *Palaeontologia Africana*, **25**, 203–231.

CRUICKSHANK, A. R. 1975. The origin of sauropod dinosaurs. *South African Journal of Science*, **71**, 89–90.

DONG ZHI-MING 1992. *The dinosaurian faunas of China.* Springer-Verlag, Berlin, 188 pp.

GALTON, P. M. 1971. The prosauropod dinosaur *Ammosaurus*, the crocodile *Protosuchus*, and their bearing on the age of the Navajo Sandstone of northeastern Arizona. *Journal of Paleontology*, **45**, 781–795.

—— 1976. Prosauropod dinosaurs (Reptilia: Saurischia) of North America. *Postilla*, **169**, 1–98.

—— 1984. Cranial anatomy of the prosauropod dinosaur *Plateosaurus* from the Knollenmergel (Middle Keuper, Upper Triassic) of Germany. I. Two complete skulls from Trossingen/Württ. with comments on the diet. *Geologica et Palaeontologica*, **18**, 139–171.

—— 1985*a*. Diet of prosauropods from the Late Triassic and Early Jurassic. *Lethaia*, **18**, 105–123.

—— 1985*b*. Notes on the Melanorosauridae, a family of large prosauropod dinosaurs (Saurischia; Sauropodomorpha). *Geobios*, **18**, 671–676.

—— 1985*c*. Cranial anatomy of the prosauropod dinosaur *Plateosaurus* from the Knollenmergel (Middle Keuper, Upper Triassic) of Germany. II. All the cranial material and details of the soft part anatomy. *Geologica et Palaeontologica*, **19**, 119–159.

—— 1985*d*. Cranial anatomy of the prosauropod dinosaur *Sellosaurus gracilis* from the Middle Stubensandstein (Upper Triassic) of Nordwürttemberg, West Germany. *Stuttgarter Beiträge zur Naturkunde, Series B*, **118**, 1–39.

—— 1986. Herbivorous adaptations of Late Triassic and Early Jurassic dinosaurs. 203–221. *In* PADIAN, K. (ed.). *The beginning of the age of dinosaurs*. Cambridge University Press, Cambridge, 378 pp.

—— 1990. Basal Sauropodomorpha–Prosauropoda. 320–344. *In* WEISHAMPEL, D. B., DODSON, P. and OSMÓLSKA, H. (eds). *The Dinosauria*. University of California Press, Berkeley, CA, 733 pp.

—— 1998. Saurischian dinosaurs from the Upper Triassic of England: *Camelotia* (Prosauropoda, Melanorosauridae) and *Avalonianus* (Theropoda, ?Carnosauria). *Palaeontographica Abteilung A*, **250**, 155–172.

—— 1999. Sex, sacra and *Sellosaurus gracilis* (Saurischia, Sauropodomorpha, Upper Triassic, Germany) – or why the character 'two sacral vertebrae' is plesiomorphic for Dinosauria. *Neues Jahrbuch für Geologie und Paläontologie, Abhandlungen*, **213**, 19–55.

—— 2000. The prosauropod dinosaur *Plateosaurus* Meyer 1837 (Saurischia: Sauropodomorpha). I. The syntypes of *Plateosaurus engelhardti* Meyer 1837 (Upper Triassic, Germany) with notes on other European prosauropods with 'distally straight' femora. *Neues Jahrbuch für Geologie und Paläontologie, Abhandlungen*, **216**, 233–275.

—— 2001*a*. The prosauropod dinosaur *Plateosaurus* Meyer 1837 (Saurischia: Sauropodomorpha: Upper Triassic). II. Notes on the referred species. *Revue de Paléobiologie*, **20**, 435–502.

—— 2001*b*. Prosauropod dinosaur *Sellosaurus gracilis* (Upper Triassic, Germany): third sacral as either a dorsosacral or a caudosacral. *Neues Jahrbuch für Geologie und Paläontologie, Monatschefte*, **2001**, 688–704.

—— 2005. Basal sauropodomorph dinosaur taxa *Thecodontosaurus* Riley & Stutchbury, 1836, *T. antiquus* Morris, 1843 and *T. caducus* Yates, 2003: their status re humeral morphs from the 1834 fissure fill (Upper Triassic) in Clifton, Bristol, UK. *Journal of Vertebrate Paleontology*, **25** (Supplement to No. 3), 61A.

—— and BAKKER, R. T. 1985. The cranial anatomy of the prosauropod dinosaur '*Efraasia diagnostica*', a juvenile individ

ual of *Sellosaurus gracilis* from the Upper Triassic of Nordwürttemberg, West Germany. *Stuttgarter Beiträge zur Naturkunde, Series B*, **117**, 1–15.

—— and CLUVER, M. A. 1976. *Anchisaurus capensis* (Broom) and a revision of the Anchisauridae (Reptilia, Saurischia). *Annals of the South African Museum*, **69**, 121–159.

—— and UPCHURCH, P. 2000. Prosauropod dinosaurs: homeotic transformations ('frame shifts') with third sacral as a caudosacral or a dorsosacral. *Journal of Vertebrate Paleontology*, **20** (Supplement to No. 3), 43A.

—— —— 2004. Prosauropoda. 232–258. *In* WEISHAMPEL, D. B., DODSON, P. and OSMÓLSKA, H. (eds). *The Dinosauria*. Second edition. University of California Press, Berkeley, CA, 861 pp.

—— and VAN HEERDEN, J. 1985. Partial hindlimb of *Blikanasaurus cromptoni* n. gen. and n. sp., representing a new family of prosauropod dinosaurs from the Upper Triassic of South Africa. *Geobios*, **18**, 509–516.

—— —— 1998. Anatomy of the prosauropod dinosaur *Blikanasaurus cromptoni* (Upper Triassic, South Africa), with notes on the other tetrapods from the lower Elliot Formation. *Paläontologische Zeitschrift*, **72**, 163–177.

—— —— and YATES, A. M. 2005. The postcranial anatomy of referred specimens of the sauropodomorph dinosaur *Melanorosaurus* from the Upper Triassic of South Africa. 1–37. *In* TIDWELL, V. and CARPENTER, K. (eds). *Thunderlizards. The sauropodomorph dinosaurs*. Indiana University Press, Bloomington, IN, 495 pp.

GAUFFRE, F.-X. 1993. The prosauropod dinosaur *Azendohsaurus laaroussii* from the Upper Triassic of Morocco. *Palaeontology*, **36**, 897–908.

—— 1996. Phylogénie des dinosaures prosauropodes et étude d'un prosauropode du Trias supérieur d'Afrique australe. Unpublished PhD dissertation, Muséum National d'Histoire Naturelle, Paris, 156 pp.

GAUTHIER, J. 1986. Saurischian monophyly and the origin of birds. *Memoirs of the California Academy of Sciences*, **8**, 1–55.

GILMORE, C. W. 1925. A nearly complete articulated skeleton of *Camarasaurus*, a saurischian dinosaur from The Dinosaur National Monument. *Memoirs of the Carnegie Museum of Natural History*, **10**, 347–384.

GOW, C. E. 1990. Morphology and growth of the *Massospondylus* braincase (Dinosauria, Prosauropoda). *Palaeontologia Africana*, **27**, 59–75.

—— KITCHING, J. W. and RAATH, M. A. 1990. Skulls of the prosauropod dinosaur *Massospondylus carinatus* Owen in the collections of the Bernard Price Institute for Palaeontological Research. *Palaeontologia Africana*, **27**, 45–58.

HAUGHTON, S. H. 1924. The fauna and stratigraphy of the Stormberg Series. *Annals of the South African Museum*, **12**, 323–497.

HE XIN-LU, LI KUI and CAI KAI-JI 1988. *The Middle Jurassic dinosaur fauna from Dashanpu, Zigong, Sichuan: Sauropod Dinosaurs (2). Omeisaurus tianfuensis*. Sichuan Publishing House of Science and Technology, Chengdu, 143 pp., 20 pls [In Chinese, English summary].

—— —— —— and GAO YU-HUI 1984. *Omeisaurus tianfuensis* – a new species of *Omeisaurus* from Dashanpu, Zigong,

Sichuan. *Journal of the Chengdu College of Geology, Supplement*, **2**, 13–32. [In Chinese, English summary].

—— WANG CHANG-SHENG, LIU SHANG-ZHONG, ZHOU FENG-YUN, LIU TU-QIANG, CAI KAI-JI and DAI BING 1998. A new species of sauropod from the Early Jurassic of Gongxian County, Sichuan. *Acta Geologica Sichuan*, **18**, 1–6. [In Chinese, English abstract].

HUENE, F. von 1907–08. Die Dinosaurier der europäischen Triasformation mit Berücksichtigung der aussereuropäischen Vorkommnisse. *Geologie und Paläontologie Abhandlungen, Supplement*, **1**, 1–419.

—— 1914a. Nachtrage zu meinen fruheren Beschreibungen triassischer Saurischia. *Geologie und Paläontologie Abhandlungen*, **12**, 69–82.

—— 1914b. Beiträge zur Geschichte der Archosaurier. *Geologie und Paläontologie Abhandlungen*, **13**, 1–53.

—— 1920. Bemerkungen zur systematik und stammesgeschichte einiger reptilien. *Zeitschrift für Induktive Abstammungs- und Vererbungslehre*, **24**, 162–166.

—— 1926. Vollstandige Osteologie eines Plateosauriden aus dem Schwabischen Trias. *Geologie und Paläontologie Abhandlungen*, **15**, 129–179.

—— 1929. Los Saurisquios y Ornithisquios de Cretacéo Argentino. *Annales de Museo de la Plata, Series 2*, **3**, 1–196.

—— 1932. Die fossile Reptile-Ordnung Saurischia ihre Entwicklung und Geschichte. *Monographie für Geologie und Paläontologie*, **1932**, 1–361.

JAIN, S. L. 1980. The continental Lower Jurassic fauna from the Kota Formation, India. 99–123. *In* JACOBS, L. L. (ed.). *Aspects of vertebrate history*. Museum of Northern Arizona Press, Flagstaff, AZ, 407 pp.

—— KUTTY, T. S., ROY-CHOWDHURY, T. and CHATTERJEE, S. 1975. The sauropod dinosaur from the Lower Jurassic Kota Formation of India. *Proceedings of the Royal Society of London, Series B*, **188**, 221–228.

—— —— —— —— 1979. Some characteristics of *Barapasaurus tagorei*, a sauropod dinosaur from the Lower Jurassic of Deccan, India. 204–216. *In* LASKAR, B. and RAO, C. S. R., (eds). *IV International Gondwana Symposium, Calcutta*. Volume I. Hindustan Publishing Corporation, India, 384 pp.

JENSEN, J. A. 1988. A fourth new sauropod dinosaur from the Upper Jurassic of the Colorado Plateau and sauropod bipedalism. *Great Basin Naturalist*, **48**, 121–145.

KERMACK, D. 1984. New prosauropod material from South Wales. *Zoological Journal of the Linnean Society of London*, **82**, 101–117.

KUTTY, T. S., CHATTERJEE, S., GALTON, P. M. and UPCHURCH, P. in press. Three basal sauropodomorphs (Dinosauria: Saurischia) from the Lower Jurassic of India: their anatomy and relationships. *Journal of Paleontology*.

LAMBERT, D. 1983. *A field guide to dinosaurs*. Avon Books, New York, NY, 256 pp.

LANGER, M. C. 2002. Is *Saturnalia tupiniquim* really a sauropodomorph? *Boletim de Resumenes del I Congresso Latino-Americo de Paleontologia de Vertebrados*, **1**, 38–39.

—— 2003. The pelvic and hindlimb anatomy of the stem-sauropodomorph *Saturnalia tupiniquim* (Late Triassic, Brazil). *Paleobios*, **23**, 1–40.

—— 2004. Basal Saurischia. 25–46. *In* WEISHAMPEL, D. B., DODSON, P. and OSMÓLSKA, H. (eds). *The Dinosauria*. Second edition. University of California Press, Berkeley, CA, 861 pp.

—— ABDALA, F. and RICHTER, M. 1999a. New record of Dinosauria from the Santa Maria Formation (upper Carnian of the Paraná Basin – Brazil). *Ameghiniana*, **36**, 103.

—— —— —— and BENTON, M. J. 1999b. A sauropodomorph dinosaur from the Upper Triassic (Carnian) of southern Brazil. *Comptes Rendus de l'Academie des Sciences de Paris. Série 2 (Sciences de la Terre et des Planètes)*, **329**, 511–517.

LEAL, L. A., AZEVEDO, S. A. K., KELLNER, A. W. A. and DA ROSA, Á. A. S. 2004. A new early dinosaur (Sauropodomorpha) from the Caturrita Formation (Late Triassic), Paraná Basin, Brazil. *Zootaxa*, **690**, 1–24.

MADDISON, W. P. and MADDISON, D. R. 2004. MacClade: Analysis of Phylogeny and Character Evolution, Version 4.07. Sinauer Associates, Sunderland, MA.

MADSEN, J. H., McINTOSH, J. S. and BERMAN, D. S. 1995. Skull and atlas-axis complex of the Upper Jurassic sauropod *Camarasaurus* Cope (Reptilia: Saurischia). *Bulletin of the Carnegie Museum of Natural History*, **31**, 1–115.

MARSH, O. C. 1878. Principal characters of American Jurassic dinosaurs. Part I. *American Journal of Science, Series 3*, **16**, 411–416.

—— 1881. Principal characters of American Jurassic dinosaurs. Part V. *American Journal of Science, Series 3*, **21**, 417–423.

—— 1885. Names of extinct reptiles. *American Journal of Science, Series 3*, **29**, 169.

—— 1895. On the affinities and classification of dinosaurian reptiles. *American Journal of Science, Series 3*, **50**, 483–498.

—— 1896. The dinosaurs of North America. *United States Geological Survey, 16th Annual Report*, **1894–1895**, 133–244.

MATTHEW, W. D. 1915. Dinosaurs. *American Museum Handbook*, **5**, 73–74.

McINTOSH, J. S. 1990. Sauropoda. 345–401. *In* WEISHAMPEL, D. B., DODSON, P. and OSMÓLSKA, H. (eds). *The Dinosauria*. University of California Press, Berkeley, CA, 733 pp.

—— MILLER, W. E., STADTMAN, K. L. and GILLETTE, D. D. 1996. The osteology of *Camarasaurus lewisi* (Jensen, 1988). *Brigham Young University Geological Studies*, **41**, 73–115.

NORMAN, D. B., SUES, H.-D., WITMER, L. M. and CORIA, R. A. 2004a. Basal Ornithopoda. 393–412. *In* WEISHAMPEL, D. B., DODSON, P. and OSMÓLSKA, H. (eds). *The Dinosauria*. Second edition. University of California Press, Berkeley, CA, 861 pp.

—— WITMER, L. M. and WEISHAMPEL, D. B. 2004b. Basal Ornithischia. 325–334. *In* WEISHAMPEL, D. B., DODSON, P. and OSMÓLSKA, H. (eds). *The Dinosauria*. Second edition. University of California Press, Berkeley, CA, 861 pp.

NOVAS, F. E. 1994. New information on the systematics and postcranial skeleton of *Herrerasaurus ischigualastensis* (Theropoda: Herrerasauridae) from the Ischigualasto Formation (Upper Triassic) of Argentina. *Journal of Vertebrate Paleontology*, **13**, 400–423.

—— 1996. Dinosaur monophyly. *Journal of Vertebrate Paleontology*, **16**, 723–741.

PARRISH, J. M. 2005. Evolutionary and paleoecological aspects of the Triassic–Early Jurassic sauropodomorph radiation. *Journal of Vertebrate Paleontology*, **25** (Supplement to No. 3), 99A.

PAUL, G. S. 1984. The segnosaurian dinosaurs: relics of the prosauropod–ornithischian transition? *Journal of Vertebrate Palaeontology*, **4**, 507–515.

POL, D. and POWELL, J. 2005. New information on *Lessemsaurus sauropoides* (Dinosauria, Sauropodomorpha) from the Late Triassic of Argentina. *Journal of Vertebrate Paleontology*, **25** (Supplement to No. 3), 100A.

RAATH, M. A. 1972. Fossil vertebrate studies in Rhodesia: a new dinosaur (Reptilia, Saurischia) from near the Triassic–Jurassic boundary. *Arnoldia*, **5** (30), 1–37.

REISZ, R., SCOTT, D., SUES, H.-D., EVANS, D. C. and RAATH, M. A. 2005. Embryos of an Early Jurassic prosauropod dinosaur and their evolutionary significance. *Science*, **309**, 761–764.

ROMER, A. S. 1956. *Osteology of the reptiles*. University of Chicago Press, Chicago, IL, 772 pp.

SALGADO, L., CORIA, R. A. and CALVO, J. O. 1997. Evolution of titanosaurid sauropods. I: phylogenetic analysis based on the postcranial evidence. *Ameghiniana*, **34**, 3–32.

SANTA LUCA, A. P. 1980. The postcranial skeleton of *Heterodontosaurus tucki* (Reptilia, Ornithischia) from the Stormberg of South Africa. *Annals of the South African Museum*, **79**, 159–211.

SERENO, P. C. 1989. Prosauropod monophyly and basal sauropodomorph phylogeny. *Journal of Vertebrate Paleontology*, **9** (Supplement to No. 3), 38A.

—— 1991. *Lesothosaurus*, 'fabrosaurids', and the early evolution of Ornithischia. *Journal of Vertebrate Paleontology*, **11**, 168–197.

—— 1994. The pectoral girdle and forelimb of the basal theropod *Herrerasaurus ischigualastensis*. *Journal of Vertebrate Paleontology*, **13**, 425–450.

—— 1997. The origin and evolution of dinosaurs. *Annual Reviews of Earth and Planetary Sciences*, **25**, 435–489.

—— 1998. A rationale for phylogenetic definitions, with application to the higher-level taxonomy of Dinosauria. *Neues Jahrbuch für Geologie und Paläontologie, Abhandlungen*, **210**, 41–83.

—— 1999a. The evolution of dinosaurs. *Science*, **284**, 2137–2147.

—— 1999b. A rationale for dinosaurian taxonomy. *Journal of Vertebrate Paleontology*, **19**, 788–790.

—— 2005. Basal sauropodomorph phylogeny: a comparative analysis. *Journal of Vertebrate Paleontology*, **25** (Supplement to No. 3), 114A.

—— and ARCUCCI, A. B. 1994. Dinosaurian precursors from the Middle Triassic of Argentina: *Marasuchus lilloensis*, gen. nov. *Journal of Vertebrate Paleontology*, **14**, 53–73.

—— and NOVAS, F. E. 1994. The skull and neck of the basal theropod *Herrerasaurus ischigualastensis*. *Journal of Vertebrate Paleontology*, **13**, 451–476.

—— FORSTER, C. A., ROGERS, R. R. and MONETTA, A. M. 1993. Primitive dinosaur skeleton from Argentina and the early evolution of the Dinosauria. *Nature*, **361**, 64–66.

SIKES, D. S. and LEWIS, P. O. 2001. PAUPRat. Download site: http://www.ucalgary.ca/~dsikes/software2.htm.

STORRS, G. W. 1994. Fossil vertebrate faunas of the British Rhaetian (latest Triassic). *Zoological Journal of the Linnean Society*, **112**, 217–259.

SUES, H.-D., REISZ, R. R., HINIC, S. and RAATH, M. A. 2004. On the skull of *Massospondylus carinatus* Owen, 1854 (Dinosauria: Sauropodomorpha) from the Elliot and Clarens formations (Lower Jurassic) of South Africa. *Annals of Carnegie Museum*, **73**, 239–257.

SWOFFORD, D. L. 2002. PAUP: Phylogenetic Analysis Using Parsimony, Version 4.0b10. Macmillan Publishers, London.

TORNIER, G. 1913. Reptilia (Paläontologie). *Handwörterbuch Naturwissenschaften*, **8**, 337–376.

UPCHURCH, P. 1993. The anatomy, phylogeny and systematics of the sauropod dinosaurs. Unpublished PhD dissertation, University of Cambridge, 489 pp.

—— 1994. Sauropod phylogeny and palaeoecology. *Gaia*, **10**, 249–260.

—— 1995. The evolutionary history of sauropod dinosaurs. *Philosophical Transactions of the Royal Society of London, Series B*, **349**, 365–390.

—— 1998. The phylogenetic relationships of sauropod dinosaurs. *Zoological Journal of the Linnean Society*, **124**, 43–103.

—— 1999. The phylogenetic relationships of the Nemegtosauridae (Saurischia, Sauropoda). *Journal of Vertebrate Paleontology*, **19**, 106–125.

—— and BARRETT, P. M. 2000. The evolution of sauropod feeding mechanisms. 79–122. *In* SUES, H.-D. (ed.). *The evolution of herbivory in terrestrial vertebrates: perspectives from the fossil record*. Cambridge University Press, Cambridge, 256 pp.

—— —— and DODSON, P. 2004. Sauropoda. 259–322. *In* WEISHAMPEL, D. B., DODSON, P. and OSMÓLSKA, H. (eds). *The Dinosauria*. Second edition. University of California Press, Berkeley, CA, 861 pp.

—— —— and GALTON, P. M. 2005. The phylogenetic relationships of basal sauropodomorphs: implications for the origins of sauropods. *Journal of Vertebrate Paleontology*, **25** (Supplement to No. 3), 126A.

—— —— ZHAO XI-JIN and XU XING in press. A re-evaluation of *Chinshakiangosaurus chunghoensis* Ye *vide* Dong 1992 (Dinosauria, Sauropodomorpha): implications for cranial evolution in basal sauropod dinosaurs. *Geological Magazine*.

—— and MARTIN, J. 2003. The anatomy and taxonomy of *Cetiosaurus* (Saurischia, Sauropoda) from the Middle Jurassic of England. *Journal of Vertebrate Paleontology*, **23**, 208–231.

VAN HEERDEN, J. 1979. The morphology and taxonomy of *Euskelosaurus* (Reptilia: Saurischia: Late Triassic) from South Africa. *Navorsinge Van die Nasionale Museum, Bloemfontein*, **4**, 21–84.

—— and GALTON, P. M. 1997. The affinities of *Melanorosaurus*, a Late Triassic prosauropod dinosaur from South Africa. *Neues Jahrbuch für Geologie und Paläontologie, Abhandlungen*, **1997**, 39–55.

WILKINSON, M. 1994. Common cladistic information and its consensus representation: reduced Adams and cladistic consensus trees and profiles. *Systematic Biology*, **43**, 343–368.

—— 1996. Majority-rule reduced consensus trees and their use in bootstrapping. *Molecular Biology and Evolution*, **13**, 437–444.

WILSON, J. A. 1999. A nomenclature for vertebral laminae in sauropods and other saurischian dinosaurs. *Journal of Vertebrate Paleontology*, **19**, 639–653.

—— 2002. Sauropod dinosaur phylogeny: critique and cladistic analysis. *Zoological Journal of the Linnean Society*, **136**, 215–277.

—— and SERENO, P. C. 1998. Early evolution and higher-level phylogeny of sauropod dinosaurs. *Memoirs of the Society of Vertebrate Paleontology*, **5**, 1–68.

WITMER, L. M. 1997. The evolution of the antorbital cavity of archosaurs: a study of soft-tissue reconstruction in the fossil record with an analysis of the function of pneumaticity. *Memoirs of the Society of Vertebrate Paleontology*, **4**, 1–73.

YADAGIRI, P. 1988. A new sauropod *Kotasaurus yamanpalliensis* from Lower Jurassic Kota Formation of India. *Records of the Geological Survey of India*, **11**, 102–127.

—— 2001. The osteology of *Kotasaurus yamanpalliensis*, a sauropod dinosaur from the Early Jurassic Kota Formation of India. *Journal of Vertebrate Paleontology*, **21**, 242–252.

YATES, A. M. 2001. A new look at *Thecodontosaurus* and the origin of sauropod dinosaurs. *Journal of Vertebrate Paleontology*, **22** (Supplement to No. 3), 116A.

—— 2003a. A new species of the primitive dinosaur *Thecodontosaurus* (Saurischia: Sauropodomorpha) and its implications for the systematics of early dinosaurs. *Journal of Systematic Palaeontology*, **1**, 1–42.

—— 2003b. The species taxonomy of the sauropodomorph dinosaurs from the Löwenstein Formation (Norian, Late Triassic) of Germany. *Palaeontology*, **46**, 317–337.

—— 2004a. *Anchisaurus polyzelus* (Hitchcock): the smallest known sauropod dinosaur and the evolution of gigantism among sauropodomorph dinosaurs. *Postilla*, **230**, 1–58.

—— 2004b. A definite prosauropod dinosaur from the Lower Elliot Formation (Norian: Upper Triassic) of South Africa. *Palaeontologia Africana*, **39**, 63–68.

—— 2005. The skull of the Triassic sauropodomorph, *Melanorosaurus readi*, from South Africa and the definition of

Sauropoda. *Journal of Vertebrate Paleontology*, **25** (Supplement to No. 3), 132A.

—— 2007. The first complete skull of the Triassic dinosaur *Melanorosaurus* Haughton (Sauropodomorpha: Anchisauria). 9–55. *In* BARRETT, P. M. and BATTEN, D. J. (eds). *Evolution and palaeobiology of early sauropodomorph dinosaurs*. Special Papers in Palaeontology, **77**, 289 pp.

—— and KITCHING, J. 2003. The earliest known sauropod dinosaur and the first steps towards sauropod locomotion. *Proceedings of the Royal Society of London, Series B*, **270**, 1753–1758.

YOUNG CHUNG-CHIEN 1941a. A complete osteology of *Lufengosaurus huenei* Young (gen. et sp. nov.) from Lufeng, Yunnan, China. *Palaeontologica Sinica, Series C*, **7**, 1–53, pls 4–5.

—— 1941b. *Gyposaurus sinensis*, Young (sp. nov.) a new Prosauropoda from the Upper Triassic Beds at Lufeng, Yunnan. *Bulletin of the Geological Society of China*, **21**, 207–253, pls 1–9.

—— 1942. *Yunnanosaurus huangi* (gen. et sp. nov.) a new Prosauropoda from the Red Beds of Lufeng, Yunnan. *Bulletin of the Geological Society of China*, **22**, 63–104.

YU CHAO 1990. Sauropod phylogeny: a preliminary cladistic analysis. *Journal of Vertebrate Paleontology*, **10** (Supplement to No. 3), 51A.

ZHANG YI-HONG 1988. *The Middle Jurassic dinosaur fauna from Dashanpu, Zigong, Sichuan: sauropod dinosaurs (1). Shunosaurus.* Sichuan Publishing House of Science and Technology, Chengdu, 89 pp., 15 pls. [In Chinese, English summary].

—— and YANG ZHAO-LONG 1994. *A complete osteology of Prosauropoda in the Lufeng Basin, Yunnan, China. Jingshanosaurus.* Yunnan Science and Technology Publishing House, Kunming, 100 pp., 24 pls. [In Chinese, English summary].

—— YANG DAI-HUAN and PENG GUANG-CHAO 1984. New material of *Shunosaurus* from the Middle Jurassic of Dashanpu, Zigong, Sichuan. *Journal of the Chengdu College of Geology, Supplement*, **2**, 1–12. [In Chinese, English summary].

ZHENG ZHONG 1991. Morphology of the braincase of *Shunosaurus*. *Vertebrata Palasiatica*, **4**, 108–118. [In Chinese, English summary].

APPENDIX

List of outgroup taxa and sources of information on their anatomy

Marasuchus lilloensis: Sereno and Arcucci (1994). This taxon was previously known as 'Lagosuchus' lilloensis.

Heterodontosaurus tucki: Santa Luca (1980); Norman *et al.* (2004a).

Lesothosaurus diagnosticus: Sereno (1991); Norman *et al.* (2004b).

Eoraptor lunensis: Sereno *et al.* (1993); Langer (2004).

Herrerasaurus ischigualastensis: Novas (1994); Sereno (1994); Sereno and Novas (1994); Yates and Kitching (2003); Langer (2004).

Coelophysis bauri: Colbert (1989).

Guaibasaurus candelariensis: Bonaparte *et al.* (1999); Langer (2004).

List of ingroup sauropodomorphs and sources of information on their anatomy

Ammosaurus major: Galton (1976); PMG pers. obs. This taxon is frequently considered to be a junior synonym of *Anchisaurus*. Here, we treat *Ammosaurus* and *Anchisaurus* as separate taxa in order to test whether they cluster as sister-taxa.

Anchisaurus polyselus: Galton (1976); Yates (2004a); PMG pers. obs.

Antetonitrus ingenipes: Yates and Kitching (2003).

Barapasaurus tagorei: Jain *et al.* (1975, 1979); Wilson and Sereno (1998).

Blikanasaurus cromptoni: Galton and Van Heerden (1985, 1998).

Camarasaurus spp.: Gilmore (1925); Jensen (1988); Madsen *et al.* (1995); McIntosh *et al.* (1996); PMB and PU pers. obs.

This genus contains at least three species, but is believed to be monophyletic (Wilson and Sereno 1998).

Camelotia borealis: Galton (1985*b*, 1998); Storrs (1994); PMG pers. obs.

Chinshakiangosaurus chunghoensis: Dong (1992); Upchurch *et al.* in press); PMB and PU pers. obs. This taxon is apparently known from at least one partial skeleton. However, PMB and PU were only able to examine the left dentary.

Coloradisaurus brevis: Bonaparte (1978). Originally known as '*Coloradia*' *brevis* (see Lambert 1983).

Efraasia minor: Galton and Bakker (1985); PMG pers. obs. Some of this material was mixed with *Plateosaurus gracilis* to form '*Sellosaurus*'. This taxonomic problem has been resolved by Yates (2003*b*). Note that this taxon is also known as *E. diagnostica* (the latter being a junior synonym).

'*Gyposaurus*' *sinensis*: Young (1941*a*); PMB and PU pers. obs. This taxon has often been referred to *Lufengosaurus* (Galton 1990), but we believe that the partial skeleton on display at the Nanjing Geological Museum is a separate taxon (see also Sereno 1999*a*).

Jingshanosaurus xinwaensis: Zhang and Yang (1994).

Kotasaurus yamanpalliensis: Yadagiri (1988, 2001).

Lessemsaurus sauropoides: Bonaparte (1999).

Lufengosaurus huenei: Young (1941*a*); Barrett *et al.* (2005); PMB and PU pers. obs. This OTU has been coded using only the holotype (IVPP V15).

Massospondylus carinatus: Cooper (1981); Gow (1990); Gow *et al.* (1990); Sues *et al.* (2004); PMB and PU pers. obs. The taxonomy and anatomy of this taxon require revision. Here, we have restricted the OTU to specimens from South Africa.

Melanorosaurus readi: Van Heerden and Galton (1997); Yates (2005); Galton *et al.* (2005). We have restricted this OTU to the type species.

Mussaurus patagonicus: Bonaparte and Vince (1979). This OTU is based on juvenile material.

Omeisaurus spp.: He *et al.* (1984, 1988); PMB and PU pers. obs. This OTU is mainly based on the well-preserved holotype and referred material of *O. tianfuensis*.

Plateosauravus cullingworthi: Van Heerden (1979). This specimen was originally referred to *Euskelosaurus* (Van Heerden 1979), but is now regarded as *Plateosauravus* following Yates (2004*b*).

Plateosaurus spp.: von Huene (1926); Galton (1984, 1985*c*, 2000, 2001*a*); PMB, PMG and PU pers. obs. This genus contains several species. Here, the OTU is based on *P. engelhardti* and *P. longiceps*.

Riojasaurus incertus: Bonaparte (1969, 1972); Bonaparte and Pumares (1995).

Saturnalia tupiniquim: Langer (2003, 2004); Langer *et al.* (1999*a*, *b*); Yates and Kitching (2003).

Shunosaurus lii: Zhang *et al.* (1984); Zhang (1988); Zheng (1991); PMB and PU pers. obs.

Thecodontosaurus: Kermack (1984); Benton *et al.* (2000); Yates (2003*a*); PMG pers. obs. This genus contains two species, *T. antiquus* and *T. caducus*, which have been combined in this OTU. However, recent work by Galton (2005) indicates that future analyses may need to separate these species into two distinct OTUs.

Vulcanodon karibaensis: Raath (1972); Cooper (1984).

Yunnanosaurus huangi: Young (1942); Barrett *et al.* (in press); PMB and PU pers. obs. We have coded this OTU solely on the basis of the holotype specimen (NGMJ 004546).

Character description and discussion
Skull roof

C1. Skull length: >50 per cent of femur length (0); <50 per cent of femur length (1) (Benton *et al.* 2000).

C2. Caudal rim of external naris: lies rostral to the rostral margin of the antorbital fenestra (0); lies caudal to the rostral margin of the antorbital fenestra (1) (Upchurch 1998).

C3. External naris is: <50 per cent of maximum diameter of orbit (0); >50 per cent of orbit diameter (1) (Wilson and Sereno 1998).

C4. External narial shape: oval/elliptical (0); subtriangular with right-angle at caudoventral corner (1). Juvenile *Massospondylus* specimens have state '0' whereas the adults have state '1' (see Galton and Upchurch 2004, figs 3L–M, 4E, G). This character was originally defined as a 'fully individualised' ascending process of the maxilla (Gauffre 1993): however, this original definition is vague and this variation is therefore redefined here in terms of external narial shape at its caudoventral corner.

C5. External narial margin formed by the premaxilla and nasal alone, with a broad sutured contact between these processes (0); there is a point contact, or a gap between these elements filled by the maxilla (1) (Gauthier 1986; Yates and Kitching 2003; Galton and Upchurch 2004). *Plateosaurus* displays some variation in this character because of post-mortem distortion (Galton 1985*c*, *d*; Yates and Kitching 2003).

C6. Contact between premaxilla and nasal at the caudoventral margin of the external nares: is a broad sutured contact or reduced to a point contact (0); is lost and replaced by a portion of the maxilla contributing to the margin (1). This represents a modified version of 'C5' in Galton and Upchurch (2004) and 'C6' in Yates and Kitching (2003).

C7. Internarial bar: wide (transverse width equals or exceeds rostrocaudal width) (0); laterally compressed (transverse width is less than rostrocaudal width) (1) (Gauthier 1986; modified).

C8. Dorsal (ascending) process of premaxilla: curves dorsocaudally throughout its length (0); becomes horizontal in its caudal half (1) (Sereno 1989).

C9. Distal end of premaxillary dorsal (ascending) process: tapers to a slender point (0); maintains or even increases its transverse width (1) (Sereno 1999*a*).

C10. Caudolateral process of the premaxilla: present (0); absent (1) (Upchurch 1998; Yates and Kitching 2003).

C11. Profile of the rostrodorsal margin of the premaxilla in lateral view: convex (0); has an inflexion at the base of the dorsal process (1) (Yates and Kitching 2003).

C12. Lateral plate on premaxilla, maxilla and dentary (supporting the bases of the tooth crowns labially): absent (0); present (1) (Upchurch 1995).

C13. Profile of the rostral end of the maxilla in lateral view: slopes continuously towards the rostral tip (0); has an inflexion at the base of the ascending process which creates a rostral process with parallel dorsal and ventral margins (1) (Yates and Kitching 2003). This character may be correlated with 'C4' above. However, we treat 'shape of the external naris' and 'shape of the rostral process of the maxilla' as separate characters because there is some evidence for their independence. For example, although *Camarasaurus* possesses the derived state for C13, its external nares are elliptical in outline rather than subtriangular.

C14. Length of the rostral process of the maxilla: less than its dorsoventral height (0); greater than its dorsoventral height (1) (Yates and Kitching 2003).

C15. Neurovascular foramina on the lateral surface of the maxilla: numerous (0); reduced to 5–6 large foramina (1) (Sereno 1999*a*).

C16. Size of neurovascular foramen at the caudal end of the lateral maxillary row: not larger than the others (0); distinctly larger than the others (1) (Yates and Kitching 2003).

C17. Direction of the neurovascular foramen at the caudal end of the lateral maxillary row: opens ventrally, laterally and rostrally (0); opens caudally (1) (Yates and Kitching 2003).

C18. Arrangement of lateral maxillary neurovascular foramina: linear (0); irregular (1) (Yates and Kitching 2003).

C19. Lamina from the back edge of the maxillary ascending process creates an antorbital fossa: present (0); absent (1) (Upchurch 1998).

C20. Shape of the rostral margin of the antorbital fenestra in lateral view: strongly concave, creating a narrow antorbital fossa (0); straight or slightly concave, creating a broad subtriangular antorbital fossa (1) (Yates and Kitching 2003).

C21. Lateral lamina (extending caudally from the ascending process of the maxilla, along the ventral rim of the antorbital fenestra): absent (0); present (1) (Galton 1990; Witmer 1997).

C22. Maxillary lateral lamina: short (its rostrocaudal length is less than twice its height) (0); long (rostrocaudal length is more than twice its height) (1) (Galton 1985*c*; Galton and Upchurch 2004).

C23. Dorsally open neurovascular canal on the floor of the antorbital fossa: absent (0); present (1) (Yates and Kitching 2003).

C24. Maxilla–lachrymal contact above antorbital fenestra: visible in lateral view (0); obscured in lateral view by an overhanging portion of the nasal (1). Sereno (1999*a*) only gave the derived state to *Plateosaurus* and *Sellosaurus*, but this character seems to be polymorphic in *Plateosaurus* (Galton 1985*d*). In addition, Yates (2003*b*) has shown that '*Sellosaurus gracilis*' is actually referable to *Plateosaurus*. Benton *et al*. (2000) proposed a different distribution for the character states among prosauropods: *Thecodontosaurus* and *Anchisaurus* were given state '1', while *Massospondylus* was given state '0'. In the case of *Anchisaurus* the correct state is clearly '?' because of damage to the relevant area of the skull (Galton 1976). The Arizona '*Massospondylus*' described by Attridge *et al*. (1985) does have state '0', but we doubt that this specimen is congeneric with the South African *Massospondylus* material. Furthermore, dorsoventral crushing of the Arizona specimen may have artificially produced the plesiomorphic state. The South African *Massospondylus* skulls clearly possess the derived state (Gow *et al*. 1990; Sues *et al*. 2004). See also C18 in Yates and Kitching (2003).

C25. Rostrocaudal length of the antorbital fossa: greater than that of the orbit (0); less than that of the orbit (1) (Yates and Kitching 2003).

C26. 'Shelf'-like area lateral to the external naris, extending onto the rostral end of the maxilla: absent (0); present (1) (McIntosh 1990).

C27. Rostrolateral process of the nasal: has a basal width equal to that of the rostromedial process (0); is 50 per cent wider than the rostromedial process (1) (Sereno 1999*a*).

C28. Median nasal depression: absent (0); present (1) (Sereno 1999*a*; Yates and Kitching 2003; Galton and Upchurch 2004).

C29. Pointed caudolateral process of the nasal overlapping the lachrymal: absent (0); present (1) (Yates and Kitching 2003).

C30. Dorsal exposure of the lachrymal: present (0); absent (1) (Yates and Kitching 2003).

C31. Length of the dorsal process of the lachrymal compared to its ventral process: > 0·5 (0); < 0·5 (1) (Yates and Kitching 2003).

C32. Extension of the antorbital fossa onto the ventral end of the lachrymal: present (0); absent (1) (Yates and Kitching 2003).

C33. Prefrontal ventral process: short, lies on the caudal surface of the lachrymal (0); long and extends down the medial side of the lachrymal (1) (Galton and Upchurch 2004).

C34. Prefrontal : frontal length ratio: < 0·75 (0); > 0·75 (1) (Galton 1990; Yates and Kitching 2003; Galton and Upchurch 2004). Benton *et al*. (2000) gave *Anchisaurus*, *Sellosaurus* and *Thecodontosaurus* state '0' (this may reflect their use of a combined character involving prefrontal length and width).

C35. Jugal contribution to the antorbital fenestra: absent (0); present (1) (Yates and Kitching 2003).

C36. Profile of the rostral end of the jugal in lateral view: blunt (0); sharply pointed (1) (Yates and Kitching 2003).

C37. Ratio of the minimum depth of the jugal beneath the orbit to the length of the jugal from its rostral tip to the rostroventral corner of the infratemporal fenestra: < 0·2 (0); > 0·2 (1) (Yates and Kitching 2003).

C38. Transverse width of the ventral process of the postorbital: is less than its rostrocaudal width at mid-shaft (0); greater than its rostrocaudal width at mid-shaft (1) (Wilson and Sereno 1998; Yates and Kitching 2003).

C39. Angle between the rostral and dorsal rami of the quadratojugal: 90 degrees or more (0); *c*. 60 degrees or less (1) (Galton 1990).

C40. Length of the rostral process of the quadratojugal: less than or equal to that of the dorsal process (0); longer than the dorsal process (1) (Upchurch 1998).

C41. Distal end of the quadratojugal rostral process: tapers to a point (0); is dorsoventrally expanded (1) (Upchurch 1998).

C42. Rounded 'heel'-like caudoventral process of the quadratojugal: present (0); absent (1) (Yates and Kitching 2003).

C43. Rostral margin of infratemporal fenestra: lies caudal to the orbit (0); lies below the orbit (1) (Gauthier 1986; Upchurch 1995). Benton *et al*. (2000) gave *Thecodontosaurus* and *Anchisaurus* state '0', and *Sellosaurus* state '1'.

C44. Rostral margin of infratemporal fenestra: lies below the midpoint of the orbit or more caudally (0); lies level with the rostral margin of the orbit (1) (Upchurch 1998).

C45. Frontal contribution to dorsal margin of the orbit: substantial (0); small or absent (as a result of the prefrontal and postfrontal approaching close to each other) (1) (Galton 1985*c*, 1990).

C46. Frontal exposure between prefrontal and nasal: absent (0); present (1) (Sereno 1999*a*). Sereno (1999*a*) gave *Coloradisaurus* state '?'.

C47. Frontal contribution to supratemporal fossa: present (0); absent (excluded by a parietal-postorbital contact) (1) (Wilson and Sereno 1998; Yates and Kitching 2003). Wilson and Sereno (1998) stated that prosauropods possess the plesiomorphic state, whereas the derived condition occurs in most sauropods. However, prosauropod taxa vary in terms of which state they possess, and the outgroup comparisons reverse the polarity of this character relative to that claimed by Wilson and Sereno (1998).

C48. Supratemporal fenestra: obscured laterally by the postorbital bar (0); visible in lateral view because the postorbital bar lies substantially below the dorsal rim of the orbit (1) (Wilson and Sereno 1998). The derived state may occur in juvenile prosauropods, e.g. *Massospondylus*, and is replaced by the plesiomorphic state in adults (Gow *et al.* 1990; Sues *et al.* 2004).

C49. Supratemporal fenestra: longer rostrocaudally than wide transversely (0); wider transversely than rostrocaudally (1) (Wilson and Sereno 1998).

C50. Parietals: paired, sutured on the midline (0); fused on the midline (1). *Massospondylus* (Gow *et al.* 1990) and *Plateosaurus* (Galton 1984, 1985*c*) are polymorphic, *Efraasia* and *Thecodontosaurus* are only known from juvenile skulls and are provisionally assigned state '?' because parietals may only fuse in adults.

C51. Ventral process of the squamosal: is tab-like (0); is strap-like (1) (Sereno 1999*a*). The definitions of the two states need clarification. In sauropods and theropods the ventral process seems to be formed by a gradually narrowing extension of the main body of the squamosal, whereas in prosauropods the area posterior to the ventral process is more sharply distinguished from the main body, giving the bone a distorted T-shape in lateral view.

C52. Quadratojugal process of the squamosal is: less than four times as long as its basal width (0); more than four times as long as its basal width (1) (Yates and Kitching 2003).

C53. Squamosal-quadratojugal contact: present (0); absent (1) (Upchurch 1995).

C54. Position of the quadrate foramen: deeply incised into, and partly encircled by, the quadrate (0); lies on the quadrate-quadratojugal suture (1) (Yates and Kitching 2003). The polarity of this character is effectively reversed in the current analysis. Yates and Kitching (2003) assigned state '0' to *Herrerasaurus*, whereas this taxon appears to have state '1' as defined above. Furthermore, *Eoraptor* and *Lesothosaurus* also possess state '1' (Sereno 1991; Sereno *et al.* 1993).

C55. Proportion of the length of the quadrate that is occupied by the pterygoid wing: 0·7 or more (0); <0·7 (1) (Yates and Kitching 2003).

Braincase

C56. Location of the foramen for the vena capitis media: between the parietal, supraoccipital and exoccipital-opisthotic complex (0); fully enclosed by the supraoccipital (1) (Yates and Kitching 2003, modified). N.B. Yates and Kitching (2003) use the 'post-temporal fenestra' rather than the foramen for the vena capitis media, but this has been corrected as a result of discussion with A. M. Yates (pers. comm. 2006).

C57. Postparietal fenestra between the parietal and supraoccipital: absent (0); present (1) (Upchurch 1995; Yates and Kitching 2003). N.B. Yates and Kitching (2003) used the term 'fontanelle', but A. M. Yates (pers. comm. 2006) has suggested that the wording used here is more appropriate.

C58. Supraoccipital: is inclined at 75 degrees to the vertical so its rostral tip lies caudal to the basipterygoid processes (0); is inclined at 45 degrees so that its rostral tip lies above the basipterygoid processes (1) (Galton 1990).

C59. Shape of the supraoccipital in caudal view: diamond-shaped, at least as high as wide (0); semilunate and wider than high (1) (Yates and Kitching 2003).

C60. Notch in proötic, above the opening for cranial nerve V, for a separate exit of the vena cerebralis medialis: absent (0); present (1) (Galton 1976). The derived state is present in adult *Massospondylus* but absent in juveniles (Gow 1990).

C61. Deep transverse wall of bone between basipterygoid processes: absent (0); present (1) (Galton 1985*c*, 1990; Galton and Bakker 1985; Yates and Kitching 2003; Galton and Upchurch 2004).

C62. Shape of the floor of the braincase in lateral view: relatively straight with the basal tubera, basipterygoid processes and parasphenoid rostrum roughly aligned (0); bent with the basipterygoid processes and parasphenoid rostrum below the level of the basioccipital condyle and the basal tubera (1); bent with the basal tubera below the basioccipital condyle and the parasphenoid rostrum above it (2) (Galton 1985*c*, 1990; Yates and Kitching 2003). Sereno (1999*a*) gave state '0' to *Lufengosaurus*, whereas it actually has state '1' (Barrett *et al.* 2005). This multistate character is treated here as unordered.

C63. Ridge formed along the junction of the parabasisphenoid and the basioccipital, between the basal tubera: present with a smooth rostral face (0); present with a median fossa on the rostral face (1); absent, with the basal tubera separated by a deep fossa that opens caudally into a U-shaped fossa (2) (Yates and Kitching 2003). This multistate character is treated here as unordered.

C64. Ossification of the extremity of the basal tubera: complete, so that the basioccipital and parabasisphenoid form a single rugose tuber (0); unossified, with the basioccipital forming a ventrally facing platform of unfinished bone that abuts a similarly unfinished caudally facing wall of the parabasisphenoid (1) (Yates and Kitching 2003).

C65. Shape of the basal tubera: knob-like, with the basisphenoidal component rostral to the basioccipital component (0); a transverse ridge with the basisphenoidal component lying lateral to the basioccipital component (1) (Yates and Kitching 2003).

C66. Dorsoventral depth of the parasphenoid rostrum: much less than its transverse width (0); approximately equal to its transverse width (1) (Yates and Kitching 2003).

C67. Length of the basipterygoid processes (from the top of the parabasisphenoid to the tip of the process): less than the height of the braincase (from the top of the parabasisphenoid to the top of the supraoccipital) (0); greater than the height of the braincase (1) (Yates and Kitching 2003).

Palate

C68. Shape of the jugal process of the ectopterygoid: gently curved (0); strongly recurved and 'hook-like' (1) (Yates and Kitching 2003).

C69. Pneumatic fossa on the ventral surface of the ectopterygoid: present (0); absent (1) (Yates and Kitching 2003).

C70. Position of the maxillary articulating surface of the palatine: on the lateral margin of the bone (0); at the end of a narrow rostrolateral process (1) (Yates and Kitching 2003).

C71. Medial process of the pterygoid forming a 'hook' around the basipterygoid process: absent (0); flat and blunt ended (1); pointed and bent upward (2) (Upchurch 1998; Wilson and Sereno 1998; Yates and Kitching 2003). N.B. Yates and Kitching treated this as an ordered character, but here it is unordered.

Mandible

C72. Rostral end of dentary: is narrower dorsoventrally than the caudal portion (0); is wider dorsoventrally, and more robust, than the caudal portion (1) (Upchurch and Barrett 2000; Yates and Kitching 2003).

C73. Dentary in lateral view: is essentially straight or curves slightly upwards towards its rostral tip (0); curves ventrally towards its rostral tip (1) (Sereno 1989; Gauffre 1996; Yates and Kitching 2003). Gauffre (1996) used the opposite polarity to that employed here. Sereno (1999*a*) restricted the derived state to *Coloradisaurus*, *Plateosaurus* and 'Sellosaurus', whereas here we regard it as being more widespread among prosauropods.

C74. Ridge on lateral surface of the dentary (possibly associated with a fleshy cheek in life): absent (0); present (1) (Paul 1984; Sereno 1989; Galton 1990; Yates and Kitching 2003; Galton and Upchurch 2004).

C75. Outline of lower jaw in dorsal view: the rami meet each other at an acute angle (0); the rostral ends of the dentaries curve toward each other, creating a U-shaped outline (1) (Upchurch 1998; Yates and Kitching 2003).

C76. Height : length ratio of the dentary: < 0·2 (0); > 0·2 (1) (Yates and Kitching 2003).

C77. Long diameter of external mandibular fenestra: is 10–15 per cent of mandible length (0); is 5 per cent of mandible length, or less (1) (Upchurch 1994, 1995).

C78. External mandibular fenestra: present (albeit in a reduced form) (0); closed (1) (Upchurch 1998).

C79. Jaw articulation: lies above the dorsal margin of the dentary (0); lies well below the dorsal margin of the dentary (at or close to the level of the ventral margin of the dentary) (1) (Galton 1976, 1990; Gauthier 1986; Gauffre 1996; Galton and Upchurch 2004). Gauffre (1996) used the opposite polarity to that employed here.

C80. Retroarticular process length divided by its height at its base: < 1·0 (0); > 1·0 (approaching 2·0) (1) (Gauffre 1996, modified; Yates and Kitching 2003; Galton and Upchurch 2004. The polarity used by Yates and Kitching (2003) is effectively reversed in the current study: these authors assigned state '0' to their outgroups (*Herrerasaurus* and Neotheropoda), but *Heterodontosaurus*, *Lesothosaurus*, *Eoraptor*, *Herrerasaurus* and *Coelophysis* all have retroarticular processes that are longer than high (state '1').

C81. Stout triangular medial process of the articular behind the glenoid: present (0); absent (1) (Yates and Kitching 2003).

C82. Strong medial embayment behind the glenoid of the articular in dorsal view: absent (0); present (1) (Yates and Kitching 2003).

Dentition

C83. Number of teeth in the premaxilla: 4 (0); 5 or more (1) (Sereno 1999*a*; Galton and Upchurch 2004). Sereno (1999*a*) restricted the derived state to *Plateosaurus* and 'Sellosaurus', and gave '*Gyposaurus*' *sinensis* state '0'.

C84. Adjacent tooth crowns: are aligned so they do not overlap in lateral view (0); are angled relative to the long axis of the jaw so tooth crowns appear to overlap in lateral view (each tooth has its mesial margin lying lingual to the distal margin of the crown immediately in front) (1) (Galton 1985*c*, 1990). Wilson and Sereno (1998) used a combined character in which crowns are overlapping and in contact: here we separate these features into two characters (see below).

C85. Adjacent tooth crowns: not in contact (0); in contact (1) (Wilson and Sereno 1998, modified).

C86. Tooth crown serrations: project approximately perpendicular to the long axis of the crown (0); project at approximately 45 degrees to the long axis of the crown (1); serrations absent (2) (Galton 1984, 1985*c*, 1990; Yates and Kitching 2003). Benton *et al.* (2000) gave *Barapasaurus* state '0'; however, the only published information at that tine was based on unfigured teeth that are reported to be coarsely serrated (Jain 1980). At least one tooth has denticles with a 45 degree orientation (state '1') according to Kutty *et al.* (in press). Unordered.

C87. First dentary tooth: lies at the extreme rostral end of the dentary (0); is inset a short distance from the rostral tip of the dentary (1) (Sereno 1989; Yates and Kitching 2003).

C88. Lingual surfaces of tooth crowns: are convex or flat mesiodistally (0); have a concave area (either mildly concave or strongly concave) (1) (Upchurch 1995; Galton and Upchurch 2004).

C89. Lingual surfaces of tooth crowns: are convex, nearly flat, or slightly concave mesiodistally (0); are deeply concave mesiodistally (1) (Upchurch 1994, 1995).

C90. Prominent grooves near the distal margin of the labial surface of each tooth crown: absent (0); present (1) (Upchurch 1994).

C91. Prominent grooves near the mesial margin of the labial surface of each tooth crown: absent (0); present (1) (Upchurch 1994).

C92. Tooth crowns: are all recurved (0); are lanceolate in at least the middle and caudal part of the tooth row (1) (Yates and Kitching 2003).

C93. Number of dentary teeth: 18 or more (0); 17 or fewer (1) (Wilson and Sereno 1998; Yates and Kitching 2003).

C94. Orientation of the dentary tooth crowns: erect (0); procumbent (1) (Yates and Kitching 2003).

C95. Orientation of maxillary tooth crowns: erect (0); procumbent (1) (Yates and Kitching 2003).

C96. Teeth with basally constricted crowns: absent (0); present (1).

C97. Tooth-tooth occlusion: absent (0); present (1) (Wilson and Sereno 1998).

C98. Tooth crown enamel: smooth (0); wrinkled (1) (Wilson and Sereno 1998).

C99. Tooth crown serrations: distributed along the mesial and distal margins of the crown (0); restricted to the apical half of the crown (1) (Yates and Kitching 2003).

Cervical vertebrae

C100. Number of cervical vertebrae: nine or fewer (0); ten or more (1) (Gauthier 1986; Benton *et al.* 2000).

C101. Number of cervical vertebrae: ten or fewer (0); 12 or more (1) (Upchurch 1994, 1995, 1998).

C102. Shallow, dorsally facing fossa on the atlantal neurapophysis: absent (0); present (1) (Yates and Kitching 2003).

C103. Axial postzygapophyses: project caudally beyond end of centrum (0); are flush with end of centrum (1) (Sereno 1999*a*). Yates and Kitching (2003) gave *Anchisaurus* the plesiomorphic state: this is used here also, but this differs from Galton and Upchurch (2004).

C104. Length : height ratio of axis centrum: < 3·0 (0); 3·0 or more (1) (Upchurch 1993; Sereno 1999*a*).

C105. Length : height ratio of longest postaxial cervical centrum: < 3·0 (0); 3·0 or more (1) (Upchurch 1998). Yates and Kitching (2003) used the elongation of the third cervical and the mid-cervicals.

C106. Articulations between cervical centra: are amphicoelous/amphiplatyan (0); are opisthocoelous (i.e. a cranial hemispherical convexity articulates with a corresponding concavity on the caudal surface of the preceding cervical) (1) (Gauthier 1986; Upchurch 1995).

C107. Dorsal excavation of the cervical parapophyses: absent (0); present (1) (Upchurch 1998).

C108. Strong lateral compression of cranial cervical vertebrae: absent (0); present (1) (Upchurch 1998; Yates and Kitching 2003).

C109. Ventral keels on caudal cervical centra: present (0); absent (1) (Upchurch 1998). Yates and Kitching (2003) coded *Vulcanodon* as having the keel, but here it is coded as '?'. The one partial cervical known in this taxon cannot be positively identified as a caudal cervical and, in any case, it is too incomplete to be sure if a keel is present or absent.

C110. Height of neural arches of mid-cervicals: is less than centrum diameter (0); is equal to or greater than centrum diameter (1) (Galton and Upchurch 2004).

C111. Height of mid-cervical neural arches: is equal to, or less than, centrum diameter (0); is greater than centrum diameter (1) (Wilson and Sereno 1998).

C112. Centrodiapophyseal lamina system: is restricted to the dorsal vertebrae and caudal cervicals (0); is found on all presacral vertebrae (1) (Galton 1990, modified; Upchurch 1995). Character 82 in Yates and Kitching (2003) is the same, but the polarity is reversed.

C113. Short cranially projected pedicels bearing axial prezygapophyses: absent (0); present (1) (Yates and Kitching 2003).

C114. Epipophyses overhanging the rear margin of the postzygapophyses: present (0); absent (1) (Yates and Kitching 2003).

C115. Caudal ends of cranial postaxial epipophyses: with a free pointed tip (0); joined to the postzygapophyses along their entire length (1) (Yates and Kitching 2003).

Dorsal vertebrae

C116. Length : height ratios for caudal dorsal centra: < 1·0 (0); > 1·0 (1) (Galton and Upchurch 2004).

C117. Lateral surfaces of dorsal centra: with at most a shallow depression (0); strongly excavated (either deep fossae or true pleurocoels) (1) (Upchurch 1998; Yates and Kitching 2003). This has been modified from Yates and Kitching (2003) who used a single ordered character (No. 90) whereas here it is split into two characters via additive binary coding.

C118. Lateral surfaces of dorsal centra: have a shallow or deep depression (0); have a deep pleurocoel that is sharp-edged and ramifies within the centrum (1) (Upchurch 1998).

C119. Height of dorsal neural arches (i.e. from top of centrum to the level of the zygapophyses): low (i.e. less than that of the centrum) (0); high (i.e. subequal to, or greater than, the height of the centrum) (1) (Bonaparte 1986).

C120. Cranial face of dorsal neural arch: is flat or shallowly excavated (0); is deeply excavated, forming a large cavity above the neural canal (1) (Bonaparte 1986; Yates and Kitching 2003).

C121. Cranial dorsal transverse processes are directed: laterally or slightly upwards (0); strongly dorsolaterally (1) (Galton and Upchurch 2004).

C122. Laminae (prezygodiapophyseal) linking the prezygapophyses to the transverse processes on caudal dorsal vertebrae: present (0); absent (1) (Bonaparte 1986, modified; Yates and Kitching 2003, modified; Galton and Upchurch 2004).

C123. Prezygodiapophyseal lamina on cranial dorsals: present (0); absent (1) (Galton and Upchurch 2004).

C124. Spinodiapophyseal lamina on middle and caudal dorsal vertebrae: absent (0); present (1) (Upchurch 1998; Wilson and Sereno 1998; Wilson 1999; Yates and Kitching 2003).

C125. Laterally expanded tables at the mid-length of the distal surface of the neural spines: absent in all vertebrae (0); present on the cervical vertebrae (1) (Yates and Kitching 2003, modified).

C126. Laterally expanded tables at mid-length on the distal surface of the neural spines: absent, or present on cranial dorsal vertebrae alone (0); present on the cervical and cranial dorsal vertebrae (1) (Yates and Kitching 2003, modified).

C127. Dorsoventral height of the hyposphene: much less than the dorsoventral height of the neural canal (0); equal to the dorsoventral height of the neural canal (1) (Yates and Kitching 2003).

C128. Ratio of the height of the neural spine to its craniocaudal basal width: > 1·5 (0); < 1·5 (1) (Yates and Kitching 2003).

C129. Cross-sectional shape of dorsal neural spines: narrow and elliptical (0); broad and triangular (1) (Upchurch 1998; Yates and Kitching 2003).

C130. Composite lateral laminae on dorsal neural spines: absent (0); present (1) (Yates and Kitching 2003).

C131. Spinoprezygapophyseal laminae: absent (0); present on caudal or all dorsal vertebrae (1) (Yates and Kitching 2003).

C132. Spinoprezygapophyseal laminae: absent, or present as low ridges on caudal dorsal vertebrae only (0); present on all dorsals as thin laminae (1).

C133. Well-developed spinopostzygapophyseal laminae: absent (0); present on at least the caudal dorsal vertebrae (1) (Yates and Kitching 2003, modified; Galton and Upchurch 2004, modified).

C134. Well-developed spinopostzygapophyseal laminae: absent or restricted to the caudal dorsal vertebrae (0); present on all dorsals (1).

C135. Accessory infrapostzygapophyseal laminae on dorsal vertebrae: present (0); absent (1) (Upchurch 1998; Yates and

Kitching 2003). Here, the polarity follows that employed by Yates and Kitching (2003) and is reversed with respect to that used by Upchurch (1998).

C136. Hindlimb : trunk length ratio: is 1·0 or lower (0); > 1·0 (1) (Galton 1971, 1990; Benton *et al.* 2000; Yates and Kitching 2003). Benton *et al.* (2000) used the opposite polarity and coded the states with a different distribution.

Thoracic ribs

C137. Last presacral rib: free (0); fused to vertebra (1) (Yates and Kitching 2003).

Sacrum

C138. Sacral number is: two (0); three (via the addition of a caudosacral) (1) (Charig *et al.* 1965; Galton 1999, 2001*b*; Galton and Upchurch 2000). '*Sellosaurus*' has two or three sacrals (Galton 1999), partly because this taxon represents a mixture of *Efraasia* and *Plateosaurus* specimens (Yates 2003*b*). Yates and Kitching (2003) gave *Anchisaurus* a dorsosacral rather than a caudosacral as proposed by Galton and Upchurch (2004) and here.

C139. Sacral number is: two (0); three (via the addition of a dorsosacral) (1) (Novas 1996; Galton 1999).

C140. Sacral number: four or fewer (0); five or more (1) (Upchurch 1998).

C141. Sacricostal yoke (distal ends of sacral ribs fuse together): absent (0); present (1) (Wilson and Sereno 1998, modified).

C142. Sacricostal yoke: does not contribute to the dorsal rim of the acetabulum (0); contributes to the dorsal margin of the acetabulum (1) (Wilson and Sereno 1998, modified).

C143. Strong constriction between the sacral rib and transverse process of the first primordial sacral (and dorsosacral if present) in dorsal view: absent (0); present (1) (Yates and Kitching 2003).

Caudal vertebrae

C144. Centrum length : height ratio: > 1·0 (0); 1·0–0·7 (1); < 0·7 (2) (Gauthier 1986, modified; Upchurch 1998, modified; Yates and Kitching 2003). 'Centrum length' excludes the length of the articular 'ball' in procoelous and opisthocoelous caudals. *Massospondylus* displays polymorphism. Ordered.

C145. Length of mid-caudal centra compared with height of the cranial articular face: > 2·0 (0); < 2·0 (1) (Yates and Kitching 2003).

C146. Longitudinal sulcus on the ventral surface of caudal centra: absent (0); present (1) (Upchurch and Martin 2003; Yates and Kitching 2003).

C147. Caudal 'hyposphenal' ridge: absent (0); present (1) (Upchurch 1998; Yates and Kitching 2003). In many sauropods, the cranial caudals possess a 'hyposphenal' ridge that extends from the ventral midline junction of the postzygapophyses to the top of the neural canal.

C148. Length of base of the proximal caudal neural spines: greater than (0); or less than (1) half the length of the neural arch (Yates and Kitching 2003).

C149. Position of postzygapophyses in proximal caudal vertebrae: protruding with an interpostzygapophyseal notch visible in dorsal view (0); placed on either side of the base of the neural spine, without an interpostzygapophyseal notch (1) (Yates and Kitching 2003).

C150. Disappearance of caudal ribs occurs: on caudal 20 or more distally (0); on caudals 14–16 or more cranially (1) (Upchurch 1998).

C151. 'Forked' or 'skid'-like middle and distal chevrons: absent (0); present (1) (Berman and McIntosh 1978; Upchurch 1995). 'Forked' chevrons are defined as possessing a prominent cranial process, resulting in the craniocaudal length of the chevron greatly exceeding its height.

C152. Mid-caudal chevrons with ventral midline slit: absent (0); present (1) (Upchurch 1998).

C153. Length of the longest chevron divided by the length of the centrum preceding it: < 1·0 (0); > 1·0 (1) (Yates and Kitching 2003). The polarity of this character is effectively reversed in the current study. Yates and Kitching (2003) coded *Herrerasaurus* and Neotheropoda as possessing state '0', whereas in fact *Marasuchus*, basal Ornithischia, *Eoraptor*, *Guaibasaurus* and *Coelophysis* possess state '1'.

Pectoral girdle

C154. Longitudinal ridge along the dorsal surface of the sternal plate: absent (0); present (1) (Upchurch 1998).

C155. Craniocaudal length of the acromion process of the scapula: < 1·5 (0) or > 1·5 (1) times the minimum width of the scapula blade (Yates and Kitching 2003).

C156. Minimum width of the scapula divided by scapular length: < 0·2 (0); > 0·2 (1) (Yates and Kitching 2003).

C157. Scapular blade in lateral view: with strap-shaped midsection that has straight, subparallel margins (0); waisted with curved margins (1) (Yates and Kitching 2003).

C158. Caudal margin of the acromion process of the scapula rises at an angle to the blade that, at its steepest point is: < 65 degrees (0); > 65 degrees (1) (Yates and Kitching 2003).

C159. Flat caudoventrally facing surface on the coracoid between the glenoid and the coracoid tubercle: absent (0); present (1) (Yates and Kitching 2003).

C160. Coracoid tubercle: present (0); absent (1) (Yates and Kitching 2003).

Forelimb

C161. Forelimb : hindlimb length ratio is: < 0·60 (0): 0·60 or more (1) (McIntosh 1990; Upchurch 1994).

C162. Forelimb : hindlimb length ratio is: < 0·75 (0); 0·75 or more (1) (Gauthier 1986; McIntosh 1990; Upchurch 1994).

C163. Deltopectoral crest: slants at 45–60 degrees to the transverse axis of the distal condyles (0); is perpendicular to the transverse axis of the distal condyles (1) (Sereno 1999*a*).

C164. Deltopectoral crest: terminates less than 50 per cent of humerus length from its proximal end (0); terminates at least 50 per cent of humerus length from the proximal end (1) (Sereno 1999*a*; Benton *et al.* 2000; Yates and Kitching 2003; Galton and Upchurch 2004). *Plateosauravus* (Van Heerden 1979) and *Riojasaurus* (Bonaparte 1972) have the plesiomorphic state in smaller individuals and the derived state in larger ones, and are therefore treated as being polymorphic here. This contrasts with Benton *et al.* (2000) who gave *Plateosauravus* the plesiomorphic state. Sereno (1999*a*) gave '*Gyposaurus*' *sinensis* the derived state. Young's (1941*b*) figures of '*G.*' *sinensis* suggest that the deltopectoral crest terminates *c.* 44 per cent of humerus length from the proximal end and he stated that the crest ends some way above the mid-point of the bone. Yates and Kitching (2003) gave *Saturnalia* the plesiomorphic state, whereas in Galton and Upchurch (2004) and here it is given the derived state.

C165. Deltopectoral crest is: visible in caudal view (because the crest projects laterally beyond the rest of the shaft) (0); not visible in caudal view (1) (Gauffre 1996).

C166. Deltopectoral crest: prominent (0); reduced to a low ridge (1) (Wilson and Sereno 1998). Yates and Kitching (2003) gave *Kotasaurus* the plesiomorphic state, whereas in Galton and Upchurch (2004) and here it is given the derived state.

C167. Craniolateral margin of the deltopectoral crest in cranial view: straight (0); sigmoid (1) (Yates and Kitching 2003).

C168. Humerus : femur length ratio: <0.55 (0); >0.55 (1) (Yates and Kitching 2003).

C169. Humerus : femur length ratio: <0.65 (0); >0.65 (1) (Yates and Kitching 2003).

C170. Humerus : femur length ratio: <0.8 (0); >0.8 (1) (Yates and Kitching 2003).

C171. Well-defined semicircular fossa on the distal flexor surface of the humerus: present (0); absent (1) (Yates and Kitching 2003).

C172. Ratio of the transverse width of the distal end to total humerus length: <0.33 (0); >0.33 (1) (Yates and Kitching 2003).

C173. Proximal end of ulna is: subtriangular in outline and lacks a groove for the radius (0); triradiate because of a deep groove for reception of the radius (1) (Wilson and Sereno 1998; Yates and Kitching 2003).

C174. Olecranon: present as a prominent projection (0); almost completely absent (1) (Upchurch 1995, 1998, modified; Wilson and Sereno 1998; Yates and Kitching 2003).

C175. Ratio of the lengths of the craniomedial and craniolateral processes of the proximal end of the ulna: *c.* 1.0 (0); >1.0 (1) (Yates and Kitching 2003, modified).

C176. Radius : humerus length ratio: <0.80 (0); 0.80 or more (1) (Yates and Kitching 2003; Galton and Upchurch 2004).

C177. Distal condyle of radius: is subcircular or oval in outline (0); is subrectangular with a flattened caudal margin for articulation with the ulna (1) (Wilson and Sereno 1998).

C178. Proximal carpals: present as ossifications (0); absent or fail to ossify (1) (Gauthier 1986).

C179. Maximum linear dimensions of ulnare and radiale: exceed those of at least one of the first three distal carpals (0); are less than any of the distal carpals (1) (Yates and Kitching 2003).

C180. First distal carpal: is narrower transversely than metacarpal I (0); is subequal to, or greater than, the transverse width of metacarpal I (1) (Sereno 1989, modified). Benton *et al.* (2000) gave *Thecodontosaurus* state '?' and *Anchisaurus* the plesiomorphic state. See also character 131 in Yates and Kitching (2003) where the width of the first distal carpal is measured relative to the second distal carpal.

C181. Lateral end of first distal carpal: abuts the second distal carpal (0); overlaps the second distal carpal (1) (Yates and Kitching 2003).

C182. Proximal end of first metacarpal: flush with the proximal ends of other metacarpals (0); inset into the wrist (1) (Yates and Kitching 2003).

C183. Second distal carpal: completely covers the proximal end of the second metacarpal (0); does not cover this surface completely (1) (Yates and Kitching 2003).

C184. Length of the manus divided by humerus + radius length: >0.45 (0); <0.45 (1) (Yates and Kitching 2003).

C185. Length of manus divided by humerus + radius length: >0.40 (0); <0.40 (1) (Yates and Kitching 2003).

C186. Proximal width of metacarpal I divided by proximal width of metacarpal II: <1.0 (0); >1.0 (1) (Yates and Kitching 2003).

C187. Proximal width of metacarpal I divided by metacarpal length: <0.65 (0); >0.65 (1) (Yates and Kitching 2003).

C188. Proximal width of metacarpal I divided by metacarpal length: <0.8 (0); >0.8 (1) (Yates and Kitching 2003).

C189. Proximal width of metacarpal I divided by metacarpal length: <1.0 (0); >1.0 (1) (Galton and Cluver 1976; Benton *et al.* 2000, modified).

C190. Strong asymmetry in the lateral and medial condyles of the first metacarpal: absent (0); present (1) (Galton 1971, 1990; Yates and Kitching 2003; Galton and Upchurch 2004).

C191. Deep distal extensor pits on the distal end of metacarpals II and III: present (0); absent (1) (Yates and Kitching 2003).

C192. Shape of metacarpal V: longer than wide at the proximal end, with a flat proximal surface (0); nearly as wide as long with a strongly convex proximal surface (1) (Yates and Kitching 2003).

C193. Metacarpal V: is reduced or absent (0): is large, robust and approximately 90 per cent of the length of the longest metacarpal (1) (Upchurch 1998).

C194. Proximal 'heel' on first phalanx of manual digit I: absent (0); present (1) (Sereno 1999*a*).

C195. First phalanx of manual digit I: has its proximal and distal articular surfaces with their axes in the same plane (0); has proximal and distal articular surfaces with their axes twisted so that they are at approximately 45 degrees to each other (1) (Galton 1971, 1990, 2001*a*; Benton *et al.* 2000; Yates and Kitching 2003; Galton and Upchurch 2004).

C196. First phalanx of manual digit I: has its proximal and distal articular surfaces with their axes in the same plane or twisted by no more than 45 degrees (0); has proximal and distal articular surfaces with their axes twisted so that they are at approximately 60 degrees to each other (1) (Yates and Kitching 2003, modified).

C197. Length of manual digit I divided by length of manual digit II: <1.0 (0); >1.0 (1) (Yates and Kitching 2003).

C198. Length of ungual on manual digit II divided by length of ungual on manual digit I: >1.0 (0); <1.0 (1) (Yates and Kitching 2003).

C199. Length of ungual on manual digit II divided by length of ungual on manual digit I: >0.75 (0); <0.75 (1) (Yates and Kitching 2003).

C200. Length of ungual on manual digit II divided by length of ungual on manual digit I: >0.75 (0); ungual on manual digit II absent (1) (Yates and Kitching 2003).

C201. Shape of non-terminal manual phalanges: longer than wide (0); as wide as long (1) (Yates and Kitching 2003).

C202. Phalangeal formula of manual digits IV and V: less than 2–0 (0); at least equal to or greater than 2–0, respectively (1) (Yates and Kitching 2003).

Pelvic girdle

C203. Cranial process of the ilium: lacks a scar (0); scar present (1) (Sereno 1999*a*).

C204. Cranial process of the ilium: terminates behind the level of the distal end of the pubic process (0); projects further cranially than the distal end of the pubic process (1) (Galton 1976, 1990; Gauffre 1996).

C205. Cranial process of the ilium: is long and slender or very short (0); is relatively large and has a broad triangular outline in lateral view (1) (Upchurch 1998; Yates and Kitching 2003).

C206. Depth of the cranial process of the ilium: much less than the depth of the ilium above the acetabulum (0); is approximately the same depth as the ilium immediately above the acetabulum (1) (Yates and Kitching 2003).

C207. Length of cranial process of the ilium divided by its maximum depth: < 2·0 (0); > 2·0 (1) (Yates and Kitching 2003).

C208. Area between the cranial process of the ilium and the pubic peduncle: is gently curved in lateral view (0); is acute in lateral view (1) (Gauffre 1996). Gauffre (1996) gave *Ammosaurus* state '?'.

C209. Iliac portion of acetabulum: partially backed by a wall of bone (0); almost completely open (1) (Gauthier 1986; Yates and Kitching 2003). Benton *et al.* (2000) gave *Thecodontosaurus* the derived state but this is contradicted by von Huene 1907–08: pl. 24).

C210. Concave area on the lateral surface of the ilium: extends ventrally to a point close to the acetabular margin (0); is restricted to the dorsal half of the blade (1) (Gauffre 1996). Gauffre (1996) gave several prosauropods state '0' instead of '?' and gave *Plateosaurus* the plesiomorphic state although von Huene (1926) clearly shows the derived condition in this taxon. This is possibly the same character as no. 149 in Yates and Kitching (2003).

C211. Dorsal margin of ilium is smoothly convex (mildly or strongly) in lateral view (0) has a 'step-like' sigmoid profile in lateral view (1) (Gauffre 1996; Van Heerden and Galton 1997). Gauffre (1996) gave the derived state to other prosauropods, e.g. *Massospondylus*. There may be two derived states, 'irregularity' and 'stepped'. Van Heerden and Galton (1997) cited the stepped margin as a character diagnosing the Melanorosauridae.

C212. Lateral profile of the dorsal margin of the ilium: is straight or sinusoidal (0); is strongly convex (1) (Gauthier 1986).

C213. Length of the pubic peduncle of the ilium divided by the craniocaudal width of the peduncle: < 2·0 (0); > 2·0 (1) (Yates and Kitching 2003).

C214. Ischial peduncle of the ilium: is subequal in length to the pubic peduncle, giving the long axis of the iliac blade a nearly horizontal orientation (0); is reduced so that the long axis of the iliac blade slopes strongly craniodorsally in lateral view (1) (Salgado *et al.* 1997, modified; Yates and Kitching 2003).

C215. Caudally projecting 'heel' at the distal end of the ischial peduncle of the ilium: absent (0); present (1) (Yates and Kitching 2003).

C216. Length of the postacetabular process of the ilium divided by total length of the ilium: > 0·30 (0); < 0·30 (1) (Yates and Kitching 2003).

C217. Well-developed brevis fossa with sharp margins on the ventral surface of the postacetabular process of the ilium: absent (0); present (1) (Gauthier 1986; Yates and Kitching 2003). Yates and Kitching (2003) reversed the polarity used by Gauthier (1986).

C218. Shape of the caudal margin of the postacetabular process of the ilium: rounded and bluntly pointed (0); square ended (1); with a pointed ventral corner and a rounded caudodorsal margin (2) (Yates and Kitching 2003). Unordered.

C219. Pubic acetabular margin: is approximately subequal in length to the ischial acetabular margin (0); is approximately half the length of the ischial acetabular margin (1) (Galton and Upchurch 2004).

C220. Pubis in cranial view: lateral margin of the 'apron' is straight or bows laterally (0); has a concave profile (1) (Gauffre 1996, modified; Yates and Kitching 2003).

C221. Ischium : pubis length ratio: < 0·90 (0); 0·90 or more (1) (Yu 1990, modified; Calvo and Salgado 1995, modified; Yates and Kitching 2003).

C222. Pubic obturator foramen: is absent or very small (0); is large, at least 50 per cent of acetabulum diameter (1) (Gauffre 1996, modified; Benton *et al.* 2000).

C223. Pubic obturator foramen: partially obscured in cranial view of the pubis (0); completely visible in cranial view of the pubis (1) (Gauffre 1996).

C224. Middle and distal portions of the pubis: form a transverse sheet of bone that is twisted with respect to the proximal end (0); lies in approximately the same plane as the proximal end (1) (Cooper 1984; Upchurch 1995, modified; Yates and Kitching 2003).

C225. Pubic tubercle on the lateral surface of the proximal pubis: present (0); absent (1) (Yates and Kitching 2003).

C226. Width of the conjoined pubes divided by their length: < 0·7 (0); > 0·7 (1) (Yates and Kitching 2003).

C227. Minimum transverse width of the pubic 'apron' divided by the width across the pubic peduncles of the ilium: > 0·4 (0); < 0·4 (1) (Yates and Kitching 2003).

C228. Craniocaudal length of the distal pubic expansion divided by pubis length: < 0·15 (0); > 0·15 (1) (Yates and Kitching 2003).

C229. Notch separating caudoventral end of the ischial obturator plate from the ischial shaft: present (0); absent (1) (Yates and Kitching 2003).

C230. Elongate interischial fenestra: present (0); absent (1) (Yates and Kitching 2003). The polarity of this character is effectively reversed in this analysis because most of the outgroup taxa possess state '1'. This suggests that the presence of the elongate interischial fenestra within certain sauropodomorphs is actually the derived state.

C231. Long-axes of the distal ends of the ischia: are set at an angle to each other (0); are co-planar (1) (Wilson and Sereno 1998).

C232. Distal end of ischium: is only slightly expanded relative to the rest of the shaft (0); is strongly expanded dorsoventrally (so that the thickness of the shaft appears to have doubled at the distal end in lateral view) (1) (Berman and McIntosh 1978, modified; McIntosh 1990, modified; Upchurch 1995, 1998, modified).

C233. Distal end of ischium: maximum thickness is less than three times the minimum thickness (0); maximum thickness is at least three times minimum thickness (1) (Wilson and Sereno 1998; Yates and Kitching 2003).

C234. Outline of the distal end of ischium: rounded or flattened (0); subtriangular (1) (Sereno 1989; Wilson and Sereno 1998; Yates and Kitching 2003).

Hindlimb

C235. Proximal end of the lesser trochanter: terminates below the femoral head (0); terminates level with the femoral head (1) (Gauffre 1996, modified).

C236. Lesser trochanter: all of lateral edge lies medial to the lateral edge of femur (0); projects beyond the lateral edge of the femur so that it is visible in caudal view (1) (Van Heerden and Galton 1997; Yates and Kitching 2003).

C237. Lesser trochanter on the femur: is well developed (0); is absent or greatly reduced (1) (McIntosh 1990; Gauffre 1996; Upchurch 1998; Yates and Kitching 2003).

C238. Lesser trochanter: is well developed or reduced to a still visible ridge (0); is completely absent (1) (Upchurch 1998; Yates and Kitching 2003).

C239. Lesser trochanter: is a ridge-like structure or reduced (0); is developed into a prominent sheet-like structure (1) (Galton 1990; Benton *et al.* 2000; Yates and Kitching 2003). Benton *et al.* (2000) gave the derived state to *Anchisaurus* and *Massospondylus*, whereas here they are coded with state '0'.

C240. Fourth trochanter: is a prominent plate-like structure (0); is reduced to a low ridge (1) (Yates and Kitching 2003).

C241. Fourth trochanter on the femoral shaft: lies in the proximal half (0); lies over the mid-point (1) (Yates and Kitching 2003; Galton and Upchurch 2004).

C242. Profile of the fourth trochanter in lateral/medial view: rounded and symmetrical (0); asymmetrical with the distal margin steeper than the rounded dorsal margin (1) (Yates and Kitching 2003).

C243. Fourth trochanter of the femur: lies centrally on the caudal surface (0); lies near or on the caudomedial margin (1) (Upchurch 1998; Yates and Kitching 2003).

C244. Femoral head: projects medially or ventromedially (0); projects dorsomedially (1) (Galton and Upchurch 2004).

C245. Proximal end of femur in cranial or caudal view: merges smoothly with the lateral margin of the shaft (0); meets the lateral margin at an abrupt angle (approximately 90 degrees) (1) (Van Heerden and Galton 1997, modified).

C246. Femoral shaft: has a sigmoid curve (0); is straight (in cranial or caudal view) (1) (Galton 1985d; Gauthier 1986). Gauffre (1996) treated the sigmoid shaft as the derived state, but we believe the outgroups demonstrate that the straight femoral shaft is in fact derived.

C247. The cranial face of the femur in lateral view: is convex (0); is straight (1) (Gauffre 1996, modified; Yates and Kitching 2003).

C248. Horizontal cross-section through the femoral shaft: is subcircular (0); is elliptical or subrectangular (with the transverse diameter wider than the craniocaudal diameter) (1) (Gauffre 1993; Upchurch 1995, 1998; Yates and Kitching 2003).

C249. Angle between the long-axis of the femoral head and the transverse axis of the distal end: close to 30 degrees (0); close to 0 degrees (1) (Yates and Kitching 2003).

C250. Tibia : femur length ratio: 1·0 or more (0); < 1·0 (1) (Yates and Kitching 2003).

C251. Tibia : femur length ratio: approximately 0·65 or higher (0); < 0·65 (1) (McIntosh 1990; Yates and Kitching 2003).

C252. Extensor depression on the distal end of the femur: absent (0); present (1) (Yates and Kitching 2003).

C253. Cnemial crest on the tibia: is directed cranially (0); is directed laterally (1) (Wilson and Sereno 1998).

C254. Medial malleolus of the tibia: extends caudoventrally to cover the astragalus in caudal view (0); is reduced, exposing the posterior fossa of the astragalus in caudal view (1) (Wilson and Sereno 1998, modified; Yates and Kitching 2003).

C255. Transverse width of the distal tibia: subequal to its craniocaudal width (0); greater than its craniocaudal width (1) (Salgado *et al.* 1997; Upchurch 1999; Yates and Kitching 2003).

C256. Trigonal striated articular crest on medial surface of the proximal end of the fibula: absent (0); present (1) (Wilson and Sereno 1998, modified; Yates and Kitching 2003). Wilson and Sereno (1998) gave *Vulcanodon* state '0'.

C257. Muscle scar/trochanter on the lateral surface of fibula (at mid-length): absent (0); present (1) (Wilson and Sereno 1998, modified; Yates and Kitching 2003).

C258. Ascending process of the astragalus: extends dorsally in front of the distal end of the tibia (0); keys into the distal end surface of the tibia (1) (Charig *et al.* 1965; Gauthier 1986; Galton 1990; Benton *et al.* 2000).

C259. Depression and vascular foramina in front of the base of the astragalar ascending process: present (0); absent (1) (Wilson and Sereno 1998; Yates and Kitching 2003). Wilson and Sereno (1998) gave *Vulcanodon* the derived state.

C260. Caudal fossa of the astragalus: is undivided (0); is divided into lateral and medial portions by the presence of a ridge or crest that descends ventromedially from the apex of the ascending process (1) (Wilson and Sereno 1998, modified). In basal sauropods like *Kotasaurus* and *Vulcanodon* there is a single ridge, whereas in more advanced forms there are two. Wilson and Sereno (1998) gave *Vulcanodon* state '0'.

C261. Depth of the medial end of the astragalus in cranial view: equal to the depth of the lateral end (0); much less than the lateral end, making the astragalus wedge-shaped (1) (Upchurch 1998, modified; Yates and Kitching 2003).

C262. Shape of the caudomedial margin of the astragalus in dorsal view: forms a moderately sharp corner of a subrectangular outline (0); smoothly rounded without a caudomedial corner (1) (Yates and Kitching 2003).

C263. Dorsally facing horizontal shelf forming part of the fibular facet of the astragalus: present (0); absent, so that the fibular facet faces laterally and is vertical (1) (Yates and Kitching 2003).

C264. A lateral horizontal groove on the calcaneum: absent (0); present (1) (Yates and Kitching 2003).

C265. Transverse width of the calcaneum divided by the transverse width of the astragalus: > 0·3 (0); < 0·3 (1) (Yates and Kitching 2003).

C266. Number of ossified distal tarsals: two or more (0); none (1) (Gauthier 1986; Benton *et al.* 2000; Yates and Kitching 2003). Benton *et al.* (2000) gave most prosauropods except *Riojasaurus* the derived state.

C267. Metatarsal III length : tibia length ratio: 0·4 or higher (0); <0·4 (1) (McIntosh 1990; Wilson and Sereno 1998; Yates and Kitching 2003).

C268. Metatarsal III length : tibia length: >0·3 (0); <0·3 (1) (Yates and Kitching 2003).

C269. Metatarsal I length : width ratio: >1·5 (0); <1·5 (1) (Yates and Kitching 2003, modified).

C270. Proximal ends of metatarsals I and V: are smaller in area than those of metatarsals II and IV (0); have areas equal to or larger than metatarsals II and IV (1) (Wilson and Sereno 1998).

C271. Medial margin of the proximal end of metatarsal II: straight or convex (0); concave (1) (Sereno 1999*a*; Yates and Kitching 2003).

C272. Lateral margin of the proximal end of metatarsal II: straight or convex (0); concave (1) (Yates and Kitching 2003).

C273. Metatarsals II and III have proximal width : length ratios of: <0·25 (0); 0·25 or higher (1) (Benton *et al.* 2000, modified). Benton *et al.* (2000) used a radically different distribution of character states, which, when considered with the adjacent column in their data matrix, suggests that characters 48 and 49 may have been transposed by mistake.

C274. Metatarsal IV proximal end transverse width : dorsoventral height ratio: <2·0 (0); >2·0 (1) (Sereno 1999*a*, modified).

C275. Metatarsal IV proximal end transverse width : dorsoventral height ratio: <3·0 (0); approximately 3·0 (1) (Sereno 1999*a*; Yates and Kitching 2003).

C276. Minimum shaft widths of metatarsals III and IV divided by minimum shaft width of metatarsal II: <0·6 (0); >0·6 (1) (Wilson and Sereno 1998).

C277. Metatarsal V in dorsal (cranial) view: has proximal and distal ends approximately subequal in width (0); has a transversely widened proximal end and narrowed distal end so that the metatarsal is 'funnel' or 'paddle'-shaped (1) (Upchurch 1995, 1998; Wilson and Sereno 1998).

C278. Metatarsal V length: <50 per cent of metatarsal III length (0); >50 per cent of metatarsal III length (1) (Upchurch 1993; Wilson and Sereno 1998, modified).

C279. Proximal width : length ratio of metatarsal V: <0·25 (0); >0·25 (1) (Yates and Kitching 2003).

C280. Proximal width : length ratio of metatarsal V: <0·5 (0); >0·5 (1) (Yates and Kitching 2003).

C281. At least some pedal phalanges, apart from unguals: are longer proximodistally than their transverse widths (0); are wider transversely than their proximodistal lengths (1) (Wilson and Sereno 1998, modified).

C282. Number of phalanges on pedal digit IV: four (0); fewer than four (1) (Upchurch 1995; Yates and Kitching 2003).

C283. Phalanges of pedal digit V: absent (0); present (1) (Yates and Kitching 2003).

C284. Ungual on pedal digit I: is shorter than other pedal unguals (0); is longer than other pedal unguals (1) (Galton 1976, 1990; Wilson and Sereno 1998).

C285. Ungual on pedal digit I: is shorter than other pedal phalanges (0); is subequal to, or longer than, all other pedal phalanges (1) (Benton *et al.* 2000; Yates and Kitching 2003). Benton *et al.* (2000) gave the derived state to all sauropodomorphs, whereas here it has a more restricted distribution. This

discrepancy may reflect the need to redefine the character states. Many prosauropods have a first digit ungual that is larger than some phalanges and smaller than others. Wilson and Sereno (1998) also used an enlarged ungual on pedal digit I as a sauropodomorph synapomorphy.

C286. Ungual on pedal digit I: shorter than metatarsal I (0); longer than metatarsal I (1) (Yates and Kitching 2003).

C287. Shape of ungual on pedal digit I: shallow, pointed, with convex sides and broad ventral surface (0); deep, abruptly tapering, with flat sides and a narrow ventral surface (1) (Yates and Kitching 2003).

C288. Proximal ends of unguals on pedal digits II and III: have subequal transverse and dorsoventral diameters (or are dorsoventrally compressed) (0); are transversely compressed (1) (Wilson and Sereno 1998).

C289. Length of ungual on pedal digit II divided by length of ungual on pedal digit I: >1·0 (0); <1·0 (1) (Yates and Kitching 2003, modified).

C290. Length of ungual on pedal digit II divided by length of ungual on pedal digit I: >0·9 (0); <0·9 (1) (Yates and Kitching 2003. modified).

C291. Length of ungual on pedal digit III divided by length of ungual on pedal digit II: >0·85 (0); <0·85 (1) (Yates and Kitching 2003, modified).

C292. Pedal unguals: directed forwards (0); deflected forwards and laterally, so that the articular surfaces of the unguals are only visible in lateral view (1) (Wilson and Sereno 1998; Yates and Kitching 2003).

Summary of synapomorphies supporting the nodes labelled in Text-figure 4.

The following character distributions are based on the Delayed Transformation Optimization in PAUP 4.0b10 (Swofford 2002). Typically only characters with individual consistency indices of 1·0 are listed. Those characters marked by an asterisk (*) have equivocal distributions (i.e. transform at different nodes under Accelerated Transformation Optimization).

Node A (Sauropodomorpha)

1. Skull length is less than 50 per cent of femur length (C1) (reversed in *Mussaurus*).

2. Flat caudoventrally facing area on the coracoid between the glenoid and the coracoid tubercle (C159) (reversed in Sauropoda).

3. Pubic peduncle of the ilium is at least twice as long as its craniocaudal width (C213).

Node B

*1. Ulnare and radiale are smaller than any of the distal carpals (C179).

*2. First distal carpal is equal to, or wider than, the transverse width of metacarpal I (C180).

*3. Ungual on manual digit II is shorter than that on manual digit I (C198).

4. Fibular facet of the astragalus is vertical and faces laterally (C263).

5. Lateral margin of the proximal end of metatarsal II is concave (C272) (reversed in *Barapasaurus* + Eusauropoda).

6. Proximal width:length ratio of metatarsal V is greater than 0·25 (C279).

Node C

*1. 'Funnel'-shaped metatarsal V (C277).

2. Ungual on pedal digit I is longer than any other pedal ungual (C284) (reversed in *Anchisaurus*).

Node D

1. Maxilla has a distinct rostral process that is set off from the ascending process by a distinct inflexion (C13) (reversed in *Shunosaurus*).

*2. Articular lacks the stout triangular process behind the glenoid (C81).

*3. Metacarpals II and III lack extensor pits at their distal ends (C191).

Node E (Anchisauria)

*1. Transversely compressed internarial bar (C7).

2. Antorbital fossa does not extend onto the ventral end of the lachrymal (C32) (reversed in all plateosaurians except *Yunnanosaurus*).

3. The pterygoid wing of the quadrate occupies less than 0·7 of the length of the quadrate (C55).

4. Basal tubera separated on the midline by a caudally opening U-shaped fossa (C63).

5. Dental serrations restricted to the apical half of the crown (C99) (reversed in Plateosauridae).

6. Manus length divided by humerus + radius length is less than 0·45 (C184).

7. Long-axes of the proximal and distal ends of the femur are nearly parallel (C249) (reversed in *Plateosaurus*).

8. Extensor depression on the distal end of the femur (C252).

Node F (Prosauropoda)

1. Neurovascular foramen at the end of the maxillary row is at least twice as large as the other foramina (C16) (convergently acquired in *Camarasaurus*).

2. Neurovascular foramen at the end of the maxillary row faces caudally (C17) (reversed in *Lufengosaurus*).

3. Postparietal opening between parietal and supraoccipital (C57).

4. Notch in the proötic above the opening for cranial nerve V, for a separate exit of the vena cerebralis medialis (C60).

5. Dorsoventral depth of the parasphenoid rostrum is subequal to its transverse width (C66) (reversed in *Plateosaurus*).

6. Shallow dorsally facing fossa on the atlantal neurapophysis (C102) (reversed in *Riojasaurus*).

7. Metacarpal V is nearly as wide at its proximal end as it is long, with a convex proximal articular surface (C192).

8. The axes through the proximal and distal ends of the first phalanx of manual digit I are twisted at approximately 60 degrees to each other (C196) (reversed in *Riojasaurus*).

9. Transverse width : dorsoventral height ratio of the proximal end of metatarsal IV is approximately 3·0 (C275).

Node G

This is a weakly supported node. Only three synapomorphies support this node, and their individual c_i values are 0·333 or lower.

Node H (*Ammosaurus* + *Anchisaurus*)

This is a weakly supported node. Only five synapomorphies support this node, and their individual c_i values are 0·333 or lower. Support for this node could be increased by including characters that are currently considered to be autapomorphies of *Anchisaurus* (in studies where *Ammosaurus* is viewed as a junior synonym of this genus).

Node I

1. Premaxillary ascending process with transversely expanded distal end (C9) (convergently acquired in *Efraasia*).

*2. Rostrolateral process of the nasal is 50 per cent wider than the rostromedial process (C27).

Node J (Plateosauria)

1. External naris is subtriangular in outline (C4).

2. 'Shelf'-like area lateral to the external naris (C26) (convergently acquired in Eusauropoda).

3. Prefrontal ventral process is long and extends down the medial surface of the lachrymal (C33) (convergently acquired in *Mussaurus*).

4. Supratemporal fenestra is obscured in lateral view by the upper temporal bar (C48) (this is a reversal).

*5. Laterally expanded 'tables' at the mid-length of the distal surface of cervical neural spines (C125).

*6. Second distal carpal does not completely cover the proximal surface of metacarpal II (C183).

Node K (Plateosauridae)

1. Ascending process of the premaxilla becomes horizontal towards its distal end (C8).

2. Deep transverse wall of bone between the basipterygoid processes (C61) (convergently acquired in *Efraasia*).

3. Dental serrations present on both mesial and distal margins of tooth crowns (C99) (this is a reversal).

4. Caudally projecting 'heel' at the distal end of the ischial articulation of the ilium (C215).

Node L

1. Prefrontal : frontal length ratio is greater than 0·75 (C34) (convergently acquired in *Anchisaurus*).

2. Notch separating the obturator proximal plate of the ischium from the distal shaft (C229) (this is a reversal).

Node M (Sauropoda)

1. Caudal margin of the external naris lies caudal to the rostral margin of the antorbital fenestra (C2).

2. External mandibular fenestra is reduced to less than 5 per cent of mandible length (C77).

3. Cranial postaxial epipophyses are fused with the postzygapophyses (C115).

4. Caudal 'hyposphenal' ridge on proximal caudal vertebrae (C147) (reversed in *Antetonitrus*).

5. Loss of the flat caudoventrally facing area between the glenoid and coracoid tubercle (C159) (this is a reversal).

6. Coracoid tubercle absent (C160).

7. Non-terminal manual phalanges are as wide as long (C201) (convergently acquired in *Lufengosaurus*).

8. Ungual on pedal digit I is longer than metatarsal I (C286).

Node N

1. Premaxilla has an inflexion at the base of the ascending process (C11).

2. Dorsal neural spines have height:basal width ratios of more than 1·5 (C128) (this is a reversal).

*3. Proximal end of the ulna is triradiate, with a deep groove for reception of the radius (C173).

4. Medial malleolus of the tibia is reduced so that the caudal fossa of the astragalus is visible in caudal view (C254).

Node O ('melanorosaurids')

1. At least some pedal phalanges (except for the terminal ones) are wider than long (C281) (convergently acquired in Eusauropoda).

Node P

This is a weakly supported node. There is only one synapomorphy (C116), and this character has an individual ci value of 0·2.

Node Q

1. Dorsal neural arches are subequal to or higher than their centra (C119).

2. Spinoprezygapophyseal laminae present on caudal dorsal vertebrae (C131).

3. Well-developed spinopostzygapophyseal lamina present on at least the caudal dorsal vertebrae (C133).

4. Humerus : femur length ratio is 0·8 or higher (C170).

*5. Craniomedial proximal process of the ulna is longer than the craniolateral process (C175).

Node R

*1. Lateral plate on premaxilla, maxilla and dentary (C12).

*2. Concavity on the lingual surfaces of tooth crowns (C88).

3. Cranial face of the femur is straight in lateral view (C247).

Node S

1. Lingual surfaces of tooth crowns are deeply concave (C89).

2. Prominent groove on the labial surface of each tooth crown, near the mesial margin (C91).

*3. 12 or more cervical vertebrae (C101).

*4. Centrodiapophyseal laminae system is present on cervical vertebrae (C112).

*5. Dorsal centra have deep lateral depressions (C117).

*6. Spinoprezygapophyseal lamina present on all dorsal vertebrae (C132).

*7. Greatly reduced or absent lesser trochanter on the femur (C237).

8. Fourth trochanter reduced to a low ridge (C240).

*9. Medial surface of the proximal end of the fibula bears a trigonal striated crest (C256).

*10. Caudal surface of the astragalus is divided into lateral and medial portions by a ridge (C260).

Node T

1. Caudal ribs disappear on caudal 16 or more cranially (C150).

2. Deltopectoral crest of the humerus is not visible in caudal view (C165).

*3. Olecranon absent (C174).

*4. Distal end of the radius is subrectangular with a flattened caudal margin (C177).

5. Ischial peduncle of the ilium is strongly reduced (C214).

*6. Maximum thickness of the distal end of the ischium is at least three times the minimum thickness (C233).

7. Tibia : femur length ratio is less than 0·65 (C251).

8. Muscle scar at mid-length on the lateral surface of the fibula (C257) (convergently acquired in *Massospondylus*).

*9. Distal tarsals absent or unossified (C266).

10. Metatarsal III : tibia length ratio is less than 0·3 (C268).

*11. Proximal ends of metatarsals II and IV are larger than those of metatarsals I and V (C270).

Node U

*1. Tooth-tooth occlusion present (C97).

*2. Opisthocoelous cervical centra (C106).

*3. Height of mid-cervical neural arches is greater than centrum height (C111).

*4. Spinodiapophyseal lamina present on middle and caudal dorsal vertebrae (C124).

*5. Well-developed spinopostzygapophyseal laminae present on all dorsal vertebrae (C134).

*6. 'Forked' middle and distal chevrons (C151).

*7. Cranial process of the ilium is approximately the same depth as the main body of the ilium above the pubic process (C206).

*8. Dorsal margin of the ilium is strongly convex (C212).

9. Middle and distal portions of the pubis lie in approximately the same plane as the proximal end (C224).

10. Width of the conjoined pubes divided by their length is greater than 0·7 (C226).

11. Lesser trochanter of the femur is completely absent (C238).

12. Cnemial crest of the tibia is directed laterally (C253).

13. Depression and vascular foramina at the base of the ascending process of the astragalus are absent (C259).

14. Transversely compressed proximal ends of the ungual of pedal digits II and III (C288).

Node V (Eusauropoda)

*1. Loss of the caudolateral process of the premaxilla (C10).

*2. Rostral process of the quadratojugal is greater than that of the dorsal process (C40).

*3. Distal end of the rostral process of the quadratojugal is expanded (C41).

*4. Loss of contact between the squamosal and quadratojugal (C53).

*5. Palatine has a rostral process, at the end of which is an articular area for the maxilla (C70).

*6. Rostral end of dentary is dorsoventrally expanded (C72).

Node W

1. Neurovascular foramina on the lateral surface of the maxilla are arranged irregularly (C18).

2. External mandibular fenestra is absent (C78).

3. Cervical parapophyses have excavated dorsal surfaces (C107).

4. Dorsal centra have deep well-developed pleurocoels (C118).

5. Minimal shaft widths of metatarsals III and IV divided by minimum shaft width of metatarsal II equals 0·6 or less (C276).

6. Pedal unguals directed craniolaterally (C292).

Data matrix

```
                      1         10        20        30        40        50        60
Marasuchus        ? ? ? ? ? ? ? ? ? ? ? ? ? ? ? ? ? ? ? ? ? ? ? ? ? ? ? ? ? ? ? ? ? ? ? ? ? ? ? ? ? ? ? ? ? ? ? ? ? ? ? ? ? ? ? ? ? ? ? 0
Heterodontosaurus 0 0 0 0 0 0 ? ? 0 0 0 0 0 ? ? ? ? 0 0 ? ? ? 0 1 0 ? 1 ? ? 0 0 ? ? ? 0 0 0 ? 0 0 0 0 0 ? ? ? 0 ? ? 0 0 0 ? ? ? 0 0 ?
Lesothosaurus     0 0 0 0 0 0 0 0 0 0 0 0 0 0 0 0 0 0 0 0 0 0 0 1 0 0 0 0 1 0 0 0 0 1 0 0 0 0 0 0 0 0 0 0 0 0 0 0 1 1 1 0 1 0 ? 0 0 0 0
Eoraptor          0 0 0 1 0 0 ? ? 0 0 ? 0 0 0 0 ? 0 0 0 ? ? ? 0 0 0 ? 1 1 ? 0 0 ? ? 0 0 0 ? 1 0 0 0 0 0 0 ? ? 0 ? ? 1 1 1 0 1 0 ? ? ? ?
Herrerasaurus     0 0 0 0 0 0 0 0 0 0 0 0 1 0 0 0 1 0 0 ? ? 0 0 0 0 0 0 0 0 0 0 0 0 0 0 0 1 0 0 0 0 0 0 0 0 0 0 0 0 0 0 0 0 1 0 0 0 0 0 ?
Guaibasaurus      ? ? ? ? ? ? ? ? ? ? ? ? ? ? ? ? ? ? ? ? ? ? ? ? ? ? ? ? ? ? ? ? ? ? ? ? ? ? ? ? ? ? ? ? ? ? ? ? ? ? ? ? ? ? ? ? ? ? ? ?
Coelophysis       0 0 0 0 1 1 0 0 0 0 0 0 0 0 0 0 0 0 0 ? 0 0 0 0 0 0 0 0 0 0 0 0 1 1 0 0 0 0 0 0 1 0 0 0 0 0 0 1 0 0 ? 0 0 0 0 0 ?
Ammosaurus        1 ? ? ? ? ? ? ? ? ? ? ? ? ? ? ? ? ? ? ? ? ? ? ? ? ? ? ? ? ? ? ? ? ? ? ? ? ? ? ? ? ? ? ? ? ? ? ? ? ? ? ? ? ? ? ? ? ? ? ?
Anchisaurus       1 0 0 ? ? ? ? ? ? 0 ? 0 1 1 1 1 1 ? 0 0 1 0 1 ? 1 ? 0 ? ? ? ? 1 1 0 1 0 1 1 1 0 ? ? ? 1 1 0 0 1 1 0 0 1 1 ? 1 ? 1 0 1 0 1 1
Antetonitrus      ? ? ? ? ? ? ? ? ? ? ? ? ? ? ? ? ? ? ? ? ? ? ? ? ? ? ? ? ? ? ? ? ? ? ? ? ? ? ? ? ? ? ? ? ? ? ? ? ? ? ? ? ? ? ? ? ? ? ? ?
Barapasaurus      ? ? ? ? ? ? ? ? ? ? ? ? ? ? ? ? ? ? ? ? ? ? ? ? ? ? ? ? ? ? ? ? ? ? ? ? ? ? ? ? ? ? ? ? ? ? ? ? ? ? ? ? ? ? ? ? ? ? ? ?
Blikanasaurus     ? ? ? ? ? ? ? ? ? ? ? ? ? ? ? ? ? ? ? ? ? ? ? ? ? ? ? ? ? ? ? ? ? ? ? ? ? ? ? ? ? ? ? ? ? ? ? ? ? ? ? ? ? ? ? ? ? ? ? ?
Camarasaurus      1 1 1 0 1 1 1 0 0 1 1 1 0 0 1 0 1 1 ? 0 0 0 0 1 1 0 0 0 1 1 1 0 0 ? 0 1 1 0 1 1 1 1 1 0 ? ? 1 1 0 0 0 1 1 1 0 0 0 1 0
Camelotia         ? ? ? ? ? ? ? ? ? ? ? ? ? ? ? ? ? ? ? ? ? ? ? ? ? ? ? ? ? ? ? ? ? ? ? ? ? ? ? ? ? ? ? ? ? ? ? ? ? ? ? ? ? ? ? ? ? ? ? ?
Chinshakiangosaurus ? ? ? ? ? ? ? ? ? ? 1 ? ? ? ? ? ? ? ? ? ? ? ? ? ? ? ? ? ? ? ? ? ? ? ? ? ? ? ? ? ? ? ? ? ? ? ? ? ? ? ? ? ? ? ? ? ? ? ? ?
Coloradisaurus    ? 0 0 1 1 1 1 1 0 0 0 1 1 1 1 1 0 0 1 1 1 1 0 ? 1 1 ? ? ? ? 0 ? 0 1 1 1 0 1 0 0 ? 0 0 0 1 1 0 0 1 1 1 0 0 1 1 1 1 0 1
Efraasia          1 0 ? ? 1 0 ? ? 1 0 0 0 1 1 ? ? ? ? 0 0 ? ? ? ? ? ? ? 0 0 0 0 ? 0 ? 1 0 ? ? ? ? ? ? ? ? 0 ? ? ? 1 1 ? 0 0 0 ? ? 1 ?
Gyposaurus        ? ? ? ? ? ? ? ? ? ? ? 0 ? ? ? ? ? ? ? ? ? ? ? ? ? ? ? ? ? ? ? ? ? ? ? ? ? ? ? ? ? ? ? ? ? ? ? ? ? ? ? ? ? ? ? ? ? ? ? ?
Jingshanosaurus   1 1 1 0 1 1 1 0 ? 0 0 0 1 1 1 ? ? ? 0 0 1 0 ? ? 1 0 0 0 ? 0 0 ? ? 0 0 1 0 0 0 0 0 0 1 0 ? ? ? 1 0 ? ? ? ? ? ? 0 0 0 0 ?
Kotasaurus        ? ? ? ? ? ? ? ? ? ? ? ? ? ? ? ? ? ? ? ? ? ? ? ? ? ? ? ? ? ? ? ? ? ? ? ? ? ? ? ? ? ? ? ? ? ? ? ? ? ? ? ? ? ? ? ? ? ? ? ?
Lessemsaurus      ? ? ? ? ? ? ? ? ? ? ? ? ? ? ? ? ? ? ? ? ? ? ? ? ? ? ? ? ? ? ? ? ? ? ? ? ? ? ? ? ? ? ? ? ? ? ? ? ? ? ? ? ? ? ? ? ? ? ? ?
Lufengosaurus     1 0 ? 1 1 1 ? ? ? 0 ? 0 1 1 1 1 0 0 0 1 1 ? ? ? 1 ? ? ? 0 0 1 0 1 ? 0 1 1 0 1 0 0 0 1 0 1 0 0 0 0 ? 1 1 ? ? 1 0 ? 1 0 ?
Massospondylus    1 0 1 / 1 1 1 0 1 0 0 0 0 1 0 1 1 1 0 0 0 1 1 ? 1 1 1 1 1 1 0 1 0 1 0 0 1 1 0 1 0 0 0 1 0 0 1 1 / 0 / 1 1 0 ? ? 1 1 0 0 /
Melanorosaurus    ? ? ? ? ? ? ? ? ? 1 ? ? ? ? ? ? ? ? ? ? ? ? ? ? ? ? ? 1 ? ? ? 1 ? ? ? ? ? ? ? ? ? ? 1 ? ? ? ? ? 1 1 ? ? ? ? ? ? ? ? ? ?
Mussaurus         0 0 ? ? 1 1 ? ? ? 0 ? ? 0 ? ? 0 1 ? ? ? ? ? ? 1 1 ? 0 0 ? ? ? ? ? ? 1 0 ? ? ? ? ? 1 ? ? 1 ? 1 ? ? ? ? ? ? ? ? 0 ? ?
Omeisaurus        1 1 1 0 1 1 1 0 0 1 1 1 0 0 0 0 1 1 ? 0 0 0 0 1 1 0 0 0 1 1 1 0 0 1 0 1 ? 0 1 1 1 1 0 0 ? ? ? 1 0 0 0 1 1 1 0 0 0 1 0
Plateosauravus    ? ? ? ? ? ? ? ? ? 0 ? ? ? ? ? ? ? ? ? ? ? ? ? ? ? ? ? ? ? ? ? ? ? ? ? ? ? ? ? ? ? ? ? ? ? ? ? ? ? ? ? ? ? ? ? ? ? ? ? ?
Plateosaurus      1 0 1 1 1 0 1 1 1 0 0 0 1 1 1 1 1 0 0 1 1 1 1 1 0 1 1 1 1 0 0 0 1 1 0 1 1 0 1 0 0 0 0 0 1 1 0 0 0 / 1 1 0 0 1 1 1 1 0 1
Riojasaurus       1 0 1 1 ? ? 1 ? ? 0 0 0 1 1 1 1 1 0 0 1 1 1 ? 0 1 1 ? ? 0 1 0 1 0 ? 0 0 0 0 0 1 0 0 0 0 1 0 0 1 1 0 0 ? 1 0 0 ? 1 0 0 ?
Saturnalia        1 ? ? ? ? ? ? ? ? ? 0 ? ? 0 0 0 0 ? ? ? ? ? ? ? ? 0 0 ? ? ? ? ? ? ? ? ? 0 0 ? ? ? ? ? ? ? ? ? ? ? ? ? ? 1 0 0 0 ? ? ? ?
Shunosaurus       1 1 1 0 1 1 1 0 0 1 1 1 0 0 0 0 0 0 ? 1 ? 0 0 0 0 1 1 0 0 0 1 1 1 0 ? 0 0 1 0 0 1 0 0 1 1 1 1 0 0 0 0 1 0 0 0 0 1 1 1 0 0 0 1 ?
Thecodontosaurus  1 0 ? ? ? ? ? ? ? ? 0 0 ? ? 1 / / 0 ? ? 1 0 1 0 ? ? ? ? 0 0 0 0 ? ? ? 1 0 0 ? ? ? 0 1 0 0 ? 0 1 0 ? ? 0 1 0 ? 1 ? ? / 0 0 ? 0 1 0
Vulcanodon        ? ? ? ? ? ? ? ? ? ? ? ? ? ? ? ? ? ? ? ? ? ? ? ? ? ? ? ? ? ? ? ? ? ? ? ? ? ? ? ? ? ? ? ? ? ? ? ? ? ? ? ? ? ? ? ? ? ? ? ?
Yunnanosaurus     1 0 0 0 1 1 1 0 1 0 0 0 1 ? ? ? ? ? 1 ? 0 ? ? ? 1 0 1 0 0 0 0 1 1 0 0 0 1 1 0 1 0 0 1 1 0 0 1 1 0 0 ? 1 1 0 1 1 1 0 ? ? 0 ? 0 0 ?
```

```
                      61        70        80        90        100       110       120
Marasuchus        0 ? ? 0 0 ? 0 ? ? ? ? ? ? ? ? ? ? ? ? ? ? ? ? ? ? ? ? ? ? ? ? ? ? ? ? ? ? ? ? ? ? ? ? 0 0 0 ? ? 0 0 0 ? 1 ? 1 0 0 0 ?
Heterodontosaurus ? ? ? ? ? ? ? ? ? ? 0 0 1 0 1 1 0 1 1 ? ? 0 ? ? ? 1 0 0 0 0 1 1 0 0 1 ? 0 0 0 0 ? 1 0 0 0 0 0 0 0 0 0 ? 0 0 1 0 0 0 ?
Lesothosaurus     0 0 ? 0 1 0 0 0 ? 0 0 0 0 0 1 0 0 1 1 0 0 1 1 1 0 0 0 0 0 0 ? 0 0 1 1 0 0 0 1 0 0 0 ? ? 0 1 0 0 0 0 ? 0 1 0 0 1 0 0 0 ?
Eoraptor          ? ? ? ? ? ? ? ? 1 ? ? 0 0 0 0 0 0 1 1 ? 0 ? 0 1 1 ? 0 ? ? 0 0 0 0 0 1 0 ? 0 / 0 0 0 ? ? ? ? 0 0 ? ? ? 0 ? 0 1 ? ? 0 0 ? ?
Herrerasaurus     0 0 1 0 0 0 0 0 ? 0 0 0 0 0 0 0 0 0 0 1 0 0 0 1 1 0 0 0 0 0 0 1 0 0 0 0 0 0 0 ? ? 0 0 ? ? ? 0 0 0 ? 0 ? 0 ? 0 0 0 0 0 0 0
Guaibasaurus      ? ? ? ? ? ? ? ? ? ? ? ? ? ? ? ? ? ? ? ? ? ? ? ? ? ? ? ? ? ? ? ? ? ? ? ? ? ? ? ? ? ? ? ? ? ? ? ? ? ? ? ? ? ? ? ? 0 0 1 0
Coelophysis       0 0 ? 0 0 0 0 ? 0 0 0 0 0 0 0 0 0 0 0 0 0 1 0 0 0 0 ? 0 0 0 0 0 0 0 0 0 0 0 0 0 0 1 0 0 0 0 1 0 0 0 1 0 0 0 1 1 ? 1 0 0 0 0
Ammosaurus        ? ? ? ? ? ? ? ? ? ? ? ? ? ? ? ? ? ? ? ? ? ? ? ? ? ? ? ? ? ? ? ? ? ? ? ? ? 1 0 ? ? ? ? ? ? ? ? ? ? ? 0 ? ?
Anchisaurus       0 2 2 1 0 1 0 0 ? ? 0 1 1 0 0 ? ? 0 0 1 0 1 1 ? 1 1 ? ? ? ? 1 1 ? 1 1 0 1 1 ? ? 1 0 1 1 0 ? ? 0 ? ? 0 0 1 0 1 0 1 0 0
Antetonitrus      ? ? ? ? ? ? ? ? ? ? ? ? ? ? ? ? ? ? ? ? ? ? ? ? ? ? ? ? ? ? ? ? ? ? ? ? ? ? ? 0 0 0 1 ? ? ? ? ? ? 1 0 0 1 0
Barapasaurus      ? ? ? ? ? ? ? ? ? ? ? ? ? ? ? ? ? ? ? ? ? ? ? ? ? 1 1 1 1 1 ? ? ? ? 1 1 ? ? ? ? ? 1 1 ? 0 ? 1 1 1 ? ? ? 0 1 0 1 1
Blikanasaurus     ? ? ? ? ? ? ? ? ? ? ? ? ? ? ? ? ? ? ? ? ? ? ? ? ? ? ? ? ? ? ? ? ? ? ? ? ? ? ? ? ? ? ? ? ? ? ? ? ? ? ? ? ? ? ? ? ? ? ? ?
Camarasaurus      0 2 2 1 0 0 0 0 1 1 2 1 1 0 1 0 1 1 1 0 0 1 ? 1 2 0 1 1 1 1 1 1 1 1 1 1 1 1 1 0 0 0 0 1 1 0 1 1 1 1 1 0 1 1 ? 1 1 1 1
Camelotia         ? ? ? ? ? ? ? ? ? ? ? ? ? ? ? ? ? ? ? ? ? ? ? ? ? ? ? ? ? ? ? ? ? ? ? ? ? ? ? ? ? ? 0 ? ? ? ? ? ? ? 0 0 0 0 0 0
Chinshakiangosaurus ? ? ? ? ? ? ? ? ? ? 0 0 1 1 1 ? ? ? ? ? ? ? ? 1 0 1 0 1 0 1 0 0 ? ? 1 1 ? ? ? ? ? 0 ? ? ? ? ? ? ? ? ? ? ? ? ? ? ? ? ?
Coloradisaurus    1 1 1 0 1 1 0 ? ? 0 1 0 1 1 0 0 0 0 1 1 1 0 ? ? 1 1 0 0 0 0 ? 0 0 0 1 0 0 0 ? ? 1 0 1 1 0 0 0 0 0 0 0 0 0 0 ? 0 0 ? ?
Efraasia          1 0 0 0 1 0 1 0 ? ? ? 0 ? 1 0 0 ? ? 1 1 1 ? ? ? ? ? 1 0 0 0 0 0 1 0 0 0 1 0 ? 1 ? ? 0 0 0 1 ? ? 0 ? ? ? 1 0 0 ? ?
Gyposaurus        ? ? ? ? ? ? ? ? ? ? ? ? ? ? ? ? ? ? ? ? 1 1 ? ? 0 0 0 0 1 ? ? ? ? ? ? 1 0 1 ? 0 1 ? 1 ? 0 0 0 0 0 ? ? ? 0 0 0 0 ?
Jingshanosaurus   0 1 ? ? ? 0 ? ? ? 0 1 1 1 1 0 0 1 ? 0 ? ? 1 1 ? ? 0 0 1 0 0 0 1 ? ? ? 1 0 ? 1 1 1 0 0 0 ? 0 0 0 0 1 1 0 0 0 0 0
Kotasaurus        ? ? ? ? ? ? ? ? ? ? ? ? ? ? ? ? ? ? ? ? ? ? 1 1 1 1 1 ? ? ? 1 ? 1 ? 1 1 ? 0 ? 0 0 ? 0 0 1 0 1 0 ? ? 1 1 0 1 1
Lessemsaurus      ? ? ? ? ? ? ? ? ? ? ? ? ? ? ? ? ? ? ? ? ? ? ? ? ? ? ? ? ? ? ? ? ? ? ? ? ? ? ? ? ? ? ? ? ? ? ? ? ? ? ? 0 ? ? 0 ?
Lufengosaurus     0 1 2 0 0 ? 0 1 1 0 1 ? ? 1 ? ? ? 0 1 1 1 0 ? 1 1 1 ? 0 0 0 0 1 ? 0 1 0 0 ? 1 0 1 ? 1 1 0 0 0 0 0 0 ? ? ? 0 0 0 0
Massospondylus    0 0 1 0 1 1 0 ? ? 0 0 0 1 1 0 0 0 0 0 0 1 1 0 1 ? 1 1 1 0 0 0 0 1 0 1 0 1 0 0 1 1 0 1 ? 1 0 0 0 0 0 0 1 0 1 0 0 0 0
Melanorosaurus    ? 2 ? ? ? ? ? ? ? ? ? / 1 1 0 ? 0 0 / 1 ? ? 0 ? ? ? 1 ? ? ? ? ? 0 1 ? 1 ? ? 1 ? ? 0 0 0 1 0 0 0 ? ? 1 0 0 0 0
Mussaurus         ? ? ? ? ? ? ? ? ? ? / 1 1 0 ? 0 0 / 1 ? ? 0 ? ? ? 1 0 0 ? ? 1 ? ? ? ? ? ? 0 0 0 0 ? ? ? ? ? ? ? ? ? ? ? ?
Omeisaurus        0 2 ? 1 0 0 ? ? 1 ? 1 1 0 1 0 1 1 1 0 ? 0 0 1 ? 1 0 1 1 1 1 1 1 1 1 1 1 1 0 0 1 1 1 1 0 1 1 1 0 1 1 0 1 1 ? 1 1 1 1
Plateosauravus    ? ? ? ? ? ? ? ? ? ? ? ? ? ? ? ? ? ? ? ? ? ? ? ? ? ? ? ? ? ? ? ? ? ? ? ? ? ? ? 1 0 ? ? 0 0 0 ? ? ? 0 ? ? 0 0
Plateosaurus      1 1 1 0 1 0 0 0 1 0 1 0 1 1 0 0 0 0 1 1 1 1 1 ? 1 1 1 0 0 0 1 0 0 0 1 0 0 0 1 0 1 1 1 1 0 0 0 0 / 0 1 0 1 0 0 0
Riojasaurus       1 0 1 0 1 ? 0 ? ? 1 0 1 1 0 0 0 0 0 1 1 1 0 ? ? 0 0 0 0 1 0 0 0 1 0 0 0 1 0 0 0 1 1 0 0 0 1 1 0 0 1 0 0 0 0
Saturnalia        0 0 0 0 ? 0 0 ? ? 0 1 0 ? 1 0 0 ? ? ? ? ? ? 0 0 0 0 0 1 1 0 0 1 0 0 0 ? ? ? ? 0 0 0 0 0 0 0 0 ? ? 1 0 0 0 ?
Shunosaurus       0 2 2 1 ? ? 0 ? ? 1 2 1 1 0 1 0 1 0 0 0 1 0 0 1 ? 1 0 1 1 1 1 1 0 1 1 1 1 1 1 1 0 1 0 1 1 0 1 0 1 1 1 0 1 1 0 1 0 1 0
Thecodontosaurus  0 0 0 0 1 0 1 1 0 0 0 0 1 / 0 0 0 0 0 0 0 0 1 ? 1 1 1 0 0 0 0 0 1 0 0 0 1 0 0 0 1 0 / ? ? 0 0 0 0 1 0 0 0 1 0 0 1 0 0 0 0
Vulcanodon        ? ? ? ? ? ? ? ? ? ? ? ? ? ? ? ? ? ? ? ? ? ? ? ? ? ? ? ? ? ? ? ? ? ? ? ? ? 0 ? ? ? ? ? ? ? ? ? ? 0 ? ? ? ? ? ? ? ?
Yunnanosaurus     ? ? ? ? ? ? ? ? ? ? 0 0 1 0 0 0 0 0 1 ? ? ? 0 0 2 ? ? ? 1 0 1 ? ? 0 1 0 0 ? ? ? ? 1 1 1 0 0 0 ? ? ? ? ? ? 1 0 0 0 0
```

```
                      121        130        140        150        160        170        180
Marasuchus            ? 1 1 0 0 0 ? 1 0 0 0 0 0 0 1 ? 0 0 0 0 0 ? 0 0 ? 0 ? 0 0 0 0 1 ? 0 ? ? ? ? ? 0 0 ? ? ? ? ? 1 1 0 ? 0 ? ? ? 1 ? ? ?
Heterodontosaurus     0 1 1 0 0 0 ? 0 0 0 0 0 0 0 0 0 1 1 1 ? ? ? 0 0 0 0 0 0 0 0 1 0 0 0 0 0 0 0 0 1 0 0 0 1 1 0 0 0 0 0 0 ? 0 0 0
Lesothosaurus         0 1 1 0 0 0 ? 0 0 0 0 0 0 0 ? ? 1 1 1 ? ? ? 0 ? 0 0 ? 1 0 ? ? ? ? 0 0 0 0 ? ? 0 0 0 0 0 0 0 1 0 0 0 1 ? ? ? 0 ? ? ?
Eoraptor              ? ? ? ? ? ? ? 0 ? ? ? ? ? ? 1 ? 0 1 0 ? ? ? 0 ? ? ? ? ? 0 0 1 ? 0 ? 0 0 ? ? 0 0 ? 0 ? 0 ? 1 0 0 ? ? ? ? 0 ? ? ? ?
Herrerasaurus         0 0 0 0 0 0 0 0 0 0 0 0 0 ? 0 0 0 1 0 0 1 0 0 0 0 0 0 0 0 ? ? 0 0 0 1 ? ? 0 0 ? 0 ? 0 0 0 0 0 0 0 0 0 0 1 0 0 0 0
Guaibasaurus          1 0 0 0 ? ? 0 ? 0 0 0 ? ? 0 1 ? 1 0 0 ? ? 0 ? 0 0 0 ? ? ? 1 ? ? 1 ? 0 ? 1 ? 0 ? ? ? ? ? ? ? ? ? ? ? ? ? ? ? ? ? ? ?
Coelophysis           0 0 0 0 0 0 1 0 0 0 0 0 0 0 1 0 1 1 1 0 ? 0 0 0 0 0 0 0 0 0 0 1 0 0 0 0 0 0 0 0 0 0 0 0 0 0 0 0 0 0 0 0 0 0 0 0 0 0
Ammosaurus            1 ? ? 0 ? ? ? ? ? ? ? ? ? ? 1 0 0 1 0 0 ? ? ? ? ? ? ? ? ? ? ? ? ? ? ? ? ? ? ? 0 0 0 ? ? ? ? ? ? ? ? ? ? ? ? ? ? ?
Anchisaurus           ? 1 1 0 ? ? 0 1 0 0 0 0 0 ? 0 ? ? ? ? ? 0 ? ? ? ? ? ? ? ? ? ? ? ? ? 1 0 0 / ? ? 1 0 1 0 0 0 0 1 1 0 1 0 0 0 ? 0 0 1 ? 1
Antetonitrus          ? 0 0 0 0 0 0 1 0 1 0 1 0 1 0 1 ? ? ? ? ? ? ? ? ? ? ? 1 0 0 1 0 ? ? ? ? 0 1 0 0 ? ? 1 1 1 0 0 0 0 1 1 1 1 0 1 0 1 0 0 ? ? ?
Barapasaurus          0 0 0 1 0 0 1 0 1 1 1 1 1 1 ? ? ? 1 1 0 1 1 0 ? ? ? 0 ? 0 ? ? 1 ? ? ? 0 1 0 1 0 1 ? ? ? ? 1 ? ? ? ? ? ? 1 1 1 ? 1 ? ? ?
Blikanasaurus         ? ? ? ? ? ? ? ? ? ? ? ? ? ? ? ? ? ? ? ? ? ? ? ? ? ? ? ? ? ? ? ? ? ? ? ? ? ? ? ? ? ? ? ? ? ? ? ? ? ? ? ? ? ? ? ? ? ? ?
Camarasaurus          0 ? ? 1 0 0 1 0 1 1 1 1 1 1 ? 0 1 1 1 ? ? 0 2 1 1 0 0 1 1 0 1 0 1 0 0 1 0 1 1 1 0 0 1 1 0 1 1 1 1 1 1 1 1 1 1 ? ?
Camelotia             ? ? ? 0 ? ? 0 ? ? ? ? ? 0 ? ? ? ? ? ? ? ? ? ? 2 ? 1 ? ? ? 0 ? ? ? ? ? ? ? ? ? ? ? ? ? ? ? ? ? ? ? ? ? ? ? ? ? ? ? ? ?
Chinshakiangosaurus   ? ? ? ? ? ? ? ? ? ? ? ? ? ? ? ? ? ? ? ? ? ? ? ? ? ? ? ? ? ? ? ? ? ? ? ? ? ? ? ? ? ? ? ? ? ? ? ? ? ? ? ? ? ? ? ? ? ? ?
Coloradisaurus        ? ? ? ? ? 0 1 0 0 0 0 0 1 ? 1 0 1 0 ? 1 ? ? 0 0 1 ? ? ? ? 0 0 0 1 1 0 ? ? ? 1 ? 0 1 1 0 0 0 1 ? ? ? ? ? ? ? ? ?
Efraasia              ? ? ? 0 0 0 1 0 0 0 0 0 0 1 ? 0 0 1 0 0 0 1 0 1 0 0 1 0 0 0 ? 0 0 1 0 1 0 0 0 ? 1 ? 0 0 1 1 0 0 1 0 0 0 0 0 0 1 1
Gyposaurus            1 ? 0 0 ? ? ? ? ? ? ? ? ? 0 ? 0 1 0 0 0 ? ? ? ? ? ? ? ? ? ? ? ? ? ? ? 0 0 1 0 0 0 1 ? ? ? 0 ? 0 0 0 0 0 0 1 ? 1
Jingshanosaurus       0 1 0 0 0 0 ? 1 0 0 0 0 0 0 1 0 ? 0 1 0 1 0 ? 1 1 ? 1 0 0 0 0 0 1 0 1 0 0 0 0 1 0 0 0 0 0 0 0 0 0 0 ? 1 ? 0 ? 0 ? ? ? ?
Kotasaurus            ? ? ? 0 0 0 1 0 0 0 1 1 ? ? ? 1 ? 1 1 1 0 1 1 ? 2 1 ? 1 0 ? 0 ? ? ? ? 0 0 0 0 ? ? 1 1 0 0 0 1 0 ? ? ? 1 0 ? ? 1 1 ? ? ?
Lessemsaurus          0 1 1 0 ? ? ? ? ? ? ? ? ? ? ? ? ? ? ? ? ? ? ? ? ? ? ? ? ? ? ? ? ? ? ? ? ? ? ? ? ? ? ? ? ? ? ? ? ? ? ? ? ? ? ? ? ?
Lufengosaurus         1 1 0 0 1 1 0 1 0 0 0 0 0 0 ? 0 0 0 1 0 0 0 1 2 1 1 0 0 0 0 0 1 0 0 0 1 1 1 0 0 0 1 1 0 0 1 1 0 0 0 1 0 0 0 0 0 0 1 1 1
Massospondylus        0 1 1 0 1 0 0 1 0 0 0 0 0 0 1 ? 1 0 1 0 0 0 1 / 0 1 0 0 0 0 0 0 1 ? 0 0 0 / 1 0 0 0 1 1 0 0 0 1 0 0 0 1 0 0 0 0 0 0 0 1 1
Melanorosaurus        0 1 1 0 0 0 1 0 0 0 0 0 0 0 1 ? ? 1 1 0 0 0 0 2 1 0 1 0 0 0 ? ? ? ? 0 1 0 0 ? ? 1 0 0 1 0 0 0 1 1 0 0 1 1 0 ? 0 ? ? ? ?
Mussaurus             ? ? ? ? ? ? ? ? ? ? ? ? 0 ? 0 1 0 ? ? ? ? ? ? ? ? ? ? ? ? ? ? ? ? 1 0 ? 0 0 ? ? ? 1 0 ? 0 0 0 ? ? ? 0 ? 0 ? ? ? 1
Omeisaurus            0 ? ? 1 0 0 1 0 1 1 1 1 1 1 ? ? ? 1 1 1 ? ? 0 2 1 1 1 0 0 1 1 1 1 1 0 1 1 0 1 1 0 0 1 1 0 1 1 1 1 0 1 1 1 0 1 1 1 ? ?
Plateosauravus        0 1 0 0 ? ? ? ? ? ? ? ? ? ? ? 1 0 0 0 ? ? ? ? ? ? ? 0 ? ? ? ? ? ? ? 0 ? ? ? ? ? ? ? 1 ? 0 0 / 0 0 ? ? ? ? 0 0 ? 0 0 ? ? ?
Plateosaurus          0 1 0 0 1 1 0 1 0 0 0 0 0 0 1 0 1 1 0 0 1 0 1 0 1 1 0 0 0 0 0 0 1 0 0 0 1 0 1 0 0 0 1 1 0 0 0 1 0 0 0 1 0 0 0 0 0 0 1 1
Riojasaurus           0 1 0 0 1 0 0 1 0 0 0 0 0 0 1 ? 1 0 1 0 0 0 1 ? 1 0 0 0 0 ? ? ? ? 0 0 1 0 1 0 1 1 1 / 0 0 1 1 1 0 1 1 0 0 0 0 0 1 1 1
Saturnalia            ? ? ? 0 0 0 1 0 0 0 0 0 0 1 0 0 0 0 0 0 ? 0 0 ? ? 0 0 ? 1 ? ? ? ? ? 0 1 1 1 1 1 0 ? 0 1 ? 0 0 1 0 0 0 0 0 0 0 0 ? ? ?
Shunosaurus           1 0 0 1 0 0 ? 0 1 0 1 1 1 1 ? 1 0 1 1 0 0 0 0 ? 1 1 1 0 0 ? 1 1 1 ? 0 1 ? 0 1 1 0 1 1 0 0 1 1 1 1 1 1 1 1 1 0 1 1 ? ?
Thecodontosaurus      0 1 0 0 0 0 1 0 0 0 0 0 0 0 ? 0 0 0 1 0 1 0 0 0 0 1 1 0 0 0 0 ? 0 0 0 0 1 0 ? ? 1 0 0 0 0 ? ? ? 0 1 0 0 0 0 0 1 1 1
Vulcanodon            ? ? ? ? ? ? ? ? ? ? ? ? ? ? ? ? ? ? 1 1 0 0 0 ? 2 ? 1 1 ? 0 1 ? 0 1 ? ? 1 0 1 ? ? 1 1 ? 0 1 1 0 1 1 0 1 1 1 1 1 0 1 1 1 1 1 ? ? ?
Yunnanosaurus         ? ? ? 0 ? ? ? ? ? ? ? ? ? ? ? ? 0 1 0 0 0 ? ? ? ? ? ? ? ? ? ? ? ? ? ? ? ? ? 0 0 1 1 0 0 ? ? ? ? ? 0 0 ? 0 ? ? ? ?
```

```
                      181        190        200        210        220        230        240
Marasuchus            ? ? ? ? ? ? ? ? ? ? ? ? ? ? ? ? ? ? ? ? 0 0 0 0 0 0 0 0 1 0 1 0 0 0 0 0 0 0 0 0 0 0 1 0 0 0 1 ? 0 ? 0 0 1 0 0 0 0
Heterodontosaurus     0 0 0 0 1 0 0 0 1 0 0 0 1 0 0 0 0 0 0 0 1 0 1 0 0 1 0 1 ? 0 0 1 0 0 1 1 0 1 ? 0 0 0 0 0 ? ? 0 0 0 0 0 0 0 0 1 0 0 0 0 0
Lesothosaurus         ? 0 ? ? 1 0 0 0 1 0 1 0 ? 0 0 0 ? ? 0 1 0 1 0 1 0 0 0 0 0 1 0 0 0 1 0 0 0 ? 0 0 0 0 ? 0 0 ? 0 0 ? 0 0 ? 0 0 ? 0 1 0 1 0
Eoraptor              ? ? ? 0 0 1 ? ? ? ? 0 ? 0 ? ? ? ? ? ? ? 0 0 0 0 0 0 ? 0 0 0 0 0 0 0 ? ? 0 0 ? ? 0 0 ? 0 1 ? 0 0 0 ? 0 0 0 0
Herrerasaurus         0 0 0 0 0 1 0 0 0 1 0 0 0 0 0 0 0 0 0 0 0 0 0 0 0 0 0 0 0 0 0 0 0 0 0 0 0 0 0 0 0 0 0 0 0 0 1 0 1 0 0 0 1 0 1 0 0 0 0
Guaibasaurus          ? ? ? ? ? ? ? ? ? ? ? ? ? ? ? ? ? ? ? ? ? ? ? ? ? ? ? ? 0 0 0 0 0 0 0 ? 1 1 ? 1 1 ? ? 0 0 0 0 0 ? 1 0 1 0 ? 0 0 0 0 0 0
Coelophysis           0 0 0 0 0 0 0 0 1 0 ? 0 0 0 0 0 0 0 0 0 0 0 1 0 0 0 0 0 0 0 0 0 ? ? 0 0 0 0 ? 0 1 0 0 0 1 1 0 0 0 0 0 0 1 0 0 0 1
Ammosaurus            ? ? ? ? ? ? ? ? ? ? ? ? ? ? ? ? ? ? ? ? 0 1 0 0 1 1 1 0 0 0 1 0 ? ? 0 ? ? 1 1 0 ? ? ? ? ? ? ? ? ? ? ? ? ?
Anchisaurus           ? 1 ? 1 1 1 1 0 0 1 1 1 0 1 1 1 0 1 1 0 0 1 0 1 0 0 1 1 1 0 0 0 1 0 0 0 0 ? 0 0 ? 1 1 0 1 0 1 0 1 1 1 0 0 1 0 0 0 0 0 0
Antetonitrus          ? 1 ? 1 1 1 1 1 1 1 1 ? ? ? 1 0 ? ? ? ? ? ? ? ? ? ? ? ? ? ? ? ? ? ? ? ? ? ? 0 ? ? ? 0 1 0 0 0 ? ? ? ? ? 0 1 0 0 1 0
Barapasaurus          ? ? ? ? ? ? ? ? ? ? ? ? ? ? ? ? ? ? ? ? ? ? ? ? 1 0 1 0 1 1 ? ? 1 1 1 1 0 1 0 0 0 1 1 0 0 1 1 1 1 1 ? ? 0 1 1 0 ? ? 1 1 0 1
Blikanasaurus         ? ? ? ? ? ? ? ? ? ? ? ? ? ? ? ? ? ? ? ? ? ? ? ? ? ? ? ? ? ? ? ? ? ? ? ? ? ? ? ? ? ? ? ? ? ? ? ? ? ? ? ? ? ? ? ? ? ? ?
Camarasaurus          0 0 0 1 1 0 0 0 0 0 1 0 1 0 0 0 1 1 1 1 1 0 1 1 1 0 ? 1 0 0 1 1 1 0 1 0 0 0 1 1 ? 1 ? 1 ? ? ? ? ? 1 1 1 ? ? 0 0 1 1 1 0 1
Camelotia             ? ? ? ? ? ? ? ? ? ? ? ? ? ? ? ? ? ? ? ? ? ? ? ? ? ? ? ? ? ? ? ? ? ? ? ? ? ? 1 ? 0 1 ? ? ? ? ? ? 0 0 1 0 0 1 ?
Chinshakiangosaurus   ? ? ? ? ? ? ? ? ? ? ? ? ? ? ? ? ? ? ? ? ? ? ? ? ? ? ? ? ? ? ? ? ? ? ? ? ? ? ? ? ? ? ? ? ? ? ? ? ? ? ? ? ? ? ? ? ? 0
Coloradisaurus        ? ? ? ? ? ? ? ? ? ? ? ? ? ? ? ? ? ? ? ? 0 0 ? ? ? ? 1 ? 1 0 ? ? ? 1 1 ? 0 1 0 0 1 1 1 0 ? 0 1 0 0 1 1 1 0 ? ? ? ? ? ? 0 0
Efraasia              1 ? 0 0 0 1 1 0 0 1 1 0 0 ? 1 0 0 1 0 0 0 1 ? 0 0 0 0 0 0 ? ? 0 0 1 1 1 0 0 0 0 0 0 0 0 0 0 0 0 1 1 0 0 ? 0 0 0 0
Gyposaurus            ? ? ? ? ? 1 1 1 ? ? ? 0 1 1 1 ? ? ? ? ? ? 1 0 0 ? 0 1 ? 0 0 1 0 ? ? ? ? 0 1 1 0 1 0 ? ? ? ? ? ? 1 0 1 0 0 0 0 0 0
Jingshanosaurus       ? ? ? 1 0 1 1 1 1 ? ? 0 ? 1 ? 1 1 1 0 1 1 ? 0 0 0 0 1 0 0 0 1 0 0 0 ? 2 0 0 0 0 0 1 0 0 1 1 1 0 1 0 ? ? ? ? ? ? 0
Kotasaurus            ? ? ? ? ? ? ? ? ? ? ? ? ? ? ? ? ? ? ? 0 ? ? ? ? ? 1 0 0 1 0 1 ? 0 0 1 0 0 0 0 ? ? 1 ? ? ? ? ? ? ? 1 ? 0 1 ? ? ? ? 1 ? 0 1
Lessemsaurus          ? ? ? ? ? ? ? ? ? ? ? ? ? ? ? ? ? ? ? ? ? ? ? ? ? ? ? ? ? ? ? ? ? ? ? ? ? ? ? ? ? ? ? ? ? ? ? ? ? ? ? ? ? ? ? ?
Lufengosaurus         1 1 1 1 0 1 1 1 1 1 1 1 0 1 1 1 1 0 1 1 0 1 0 1 0 0 0 1 1 0 0 1 0 0 0 0 2 1 1 0 0 1 0 1 0 0 1 1 1 1 0 1 0 1 1 ? ? 0 0
Massospondylus        1 1 1 1 0 1 1 1 1 1 1 1 0 1 1 1 1 1 0 0 1 0 0 0 0 1 0 0 0 0 0 1 1 0 0 0 0 1 1 1 0 0 0 0 1 1 1 0 1 0 1 0 1 1 ? 0 0
Melanorosaurus        ? ? ? ? ? ? ? 1 1 ? ? ? ? ? ? ? ? ? ? ? ? ? 0 1 0 0 1 1 0 1 0 1 0 0 0 0 1 ? 0 0 1 1 0 1 0 0 1 ? ? 0 1 0 1 0 1 0 0 1 0
Mussaurus             ? ? ? ? ? ? ? 1 ? ? ? 0 1 1 ? ? ? ? ? ? ? ? ? ? ? ? ? ? ? ? ? ? ? ? ? ? 0 ? ? ? ? ? ? ? ? ? ? 0 ? ? ? ? ? 0 ? ? ? ? ? 0
Omeisaurus            ? 0 ? 1 1 0 0 0 0 1 1 0 1 0 0 0 ? ? ? ? ? 1 1 0 ? 1 0 0 1 1 1 0 1 0 0 0 1 ? ? 1 1 1 1 1 1 1 1 0 ? ? ? ? 1 1 0 1
Plateosauravus        ? ? ? ? ? ? ? 1 ? ? ? 1 ? 1 ? ? ? ? ? ? ? 0 0 ? 0 0 0 0 0 ? 0 ? 0 1 0 ? 0 1 0 ? ? ? ? ? ? ? ? 1 0 1 1 0 0 0 0 0
Plateosaurus          1 1 1 1 0 1 1 1 1 1 1 1 0 1 1 1 0 1 0 0 0 1 ? 0 1 0 0 1 1 1 0 0 1 0 1 0 0 2 1 0 0 / 0 0 0 0 0 0 0 0 0 0 1 0 0 0 0 0 0
Riojasaurus           1 1 1 ? ? 1 1 1 1 1 1 0 1 1 0 0 1 1 1 ? 0 ? 0 1 0 0 1 1 1 1 0 1 0 1 0 1 2 0 0 0 0 1 0 1 0 0 0 0 1 0 1 0 1 0 1 0 0 1 0
Saturnalia            ? ? ? ? ? ? ? ? ? ? ? ? ? ? ? ? ? ? ? ? ? 0 0 0 ? 0 0 ? 0 0 1 0 0 0 0 1 2 0 0 0 1 ? 0 0 0 0 0 1 1 0 1 0 ? ? ? 0 0 0 ?
Shunosaurus           0 0 ? 1 1 0 0 0 0 1 0 1 ? 0 0 1 1 1 1 1 ? 1 1 1 1 0 0 1 1 1 0 1 0 0 0 1 1 0 0 1 1 0 0 1 1 0 0 1 1 1 1 1 1 1 1 0 1 1 0 ? 1 1 0 1
Thecodontosaurus      0 0 0 0 0 1 0 0 0 1 0 0 0 1 0 0 0 0 0 0 0 0 0 0 0 1 0 0 ? 0 0 0 0 0 0 0 1 0 0 ? 1 0 0 ? ? 0 0 1 1 0 0 0 0 0 0 0 0
Vulcanodon            ? ? ? ? ? ? ? ? ? ? ? ? ? 1 ? ? ? ? ? ? ? ? ? ? ? ? ? ? ? ? 1 ? ? ? 1 1 0 ? 0 ? 0 1 1 0 0 0 1 0 1 1 1 1 0 1 1 1 0 0 1 0 0 1
Yunnanosaurus         ? ? ? ? ? ? 1 1 1 ? ? ? 0 1 1 1 ? ? ? ? ? ? 0 0 ? ? 0 1 ? 0 0 1 0 ? ? ? ? 1 0 0 0 1 0 ? ? ? ? ? ? ? 1 0 1 1 0 0 0 0 0
```

```
                        241        250          260          270          280        290 292
Marasuchus          0 0 1 0 0 0 0 0 0 0 0 ? ? 0 0 0 0 ? 0 0 0 0 0 1 0 0 0 0 0 ? ? 0 ? ? 0 ? 0 ? ? 0 0 0 ? ? ? ? ? ? ? ? ?
Heterodontosaurus   0 1 ? ? ? 0 0 0 ? 0 0 0 0 ? ? ? ? ? ? ? ? ? ? 0 0 0 0 ? ? ? 0 ? ? 0 ? ? ? ? 0 0 ? 0 0 0 0 0 0 0 0 0 0
Lesothosaurus       0 1 0 0 0 0 0 0 0 0 0 0 0 0 1 0 0 ? ? ? ? ? ? ? ? ? 0 0 0 0 ? ? 0 ? ? 0 ? ? ? ? 0 ? ? 0 0 0 ? ? 0 0 0 ?
Eoraptor            0 1 ? ? ? 0 0 0 0 0 0 0 0 0 1 ? ? 1 0 ? ? 0 0 ? ? 0 0 0 ? 0 ? ? ? ? ? ? ? 0 ? ? 0 ? ? ? ? ? ? ? ? ? ?
Herrerasaurus       0 1 0 0 0 0 0 0 1 0 0 0 0 0 0 1 0 ? 0 0 0 0 0 0 0 0 0 0 1 0 0 0 0 0 0 1 0 0 0 0 1 0 0 0 0 0 0 0 0 0 0
Guaibasaurus        0 0 0 ? 0 0 0 0 0 0 0 0 1 ? 0 1 0 0 0 0 0 ? 0 0 0 0 0 ? ? 0 ? 0 ? 0 ? ? 0 0 0 1 0 0 0 0 0 0 0 0 0
Coelophysis         0 0 0 0 0 0 0 0 0 0 0 0 0 1 0 0 ? ? ? ? ? ? ? 0 0 0 0 0 ? 0 0 0 0 0 ? 0 ? ? 0 0 0 0 0 0 0 0 0 0 0
Ammosaurus          ? ? ? 0 0 ? 0 ? 1 0 0 ? 0 ? ? ? ? 1 ? ? ? ? ? ? 0 0 0 0 0 1 1 1 1 0 1 0 ? ? ? 0 ? ? 1 ? ? ? 0 0 0 0
Anchisaurus         0 / 1 0 0 1 0 ? 1 0 0 1 0 1 1 ? 0 1 ? ? 0 0 ? 0 1 0 0 0 0 0 1 1 0 1 1 0 1 0 1 0 0 0 0 0 1 0 1 0 ? ? 0 0
Antetonitrus        1 1 1 0 0 0 0 1 1 1 0 1 0 1 1 0 0 ? ? ? ? ? ? ? ? ? 1 0 1 0 1 1 1 ? ? 0 ? ? ? ? ? ? ? ? 1 1 0 ? 1 0 ? 0
Barapasaurus        1 1 ? 1 1 1 1 1 ? ? ? 1 1 1 1 1 1 1 ? ? ? ? ? 1 ? ? ? ? ? ? ? ? ? ? ? ? ? ? ? ? ? ? ? ? ? ? 1 ? ? ? ?
Blikanasaurus       ? ? ? ? ? ? ? ? ? ? 0 1 1 ? 0 1 0 0 0 1 1 ? ? 0 1 0 1 0 1 1 1 1 0 1 1 0 1 1 1 0 ? 1 1 0 0 0 1 0 0 0
Camarasaurus        1 0 1 1 ? 1 1 1 1 1 1 0 1 1 1 1 1 1 1 0 1 1 1 1 1 1 0 0 1 0 0 1 1 1 1 1 1 1 1 1 1 1 1 1 1 1 1 1 1 1
Camelotia           1 ? 1 0 1 1 0 1 1 ? ? 1 0 ? ? ? ? ? ? ? ? ? ? ? ? ? ? ? ? ? ? ? ? ? ? ? 1 ? ? ? ? ? 0 ? ? ? ?
Chinshakiangosaurus ? ? ? ? 1 1 ? ? ? ? ? ? ? ? ? ? ? ? ? ? ? ? ? ? ? ? ? ? ? ? ? ? ? ? ? ? ? ? ? ? ? ? ? ? ? ? ? ? ? ?
Coloradisaurus      0 1 1 ? ? 0 0 1 1 0 1 ? 0 1 ? ? ? 0 ? 0 0 1 0 ? ? 0 0 0 ? 1 1 ? 1 ? 0 ? ? 1 1 ? 0 1 ? 1 0 0 ? 1 0 1 ?
Efraasia            0 1 0 0 0 0 0 0 1 0 0 ? 0 1 0 0 1 ? ? 0 0 1 1 0 0 0 0 0 0 ? ? ? 1 ? 0 1 0 1 0 0 0 1 ? 1 0 0 0 1 1 0 0
Gyposaurus          ? ? ? 0 0 1 0 0 ? 1 0 ? 0 ? ? 0 ? 1 0 ? ? ? ? ? ? 0 0 0 0 1 1 0 1 1 0 1 0 ? ? 0 ? ? 1 1 ? ? 0 ? ? ? ?
Jingshanosaurus     0 0 0 0 0 0 0 ? 1 0 1 0 0 0 ? ? 1 0 0 0 1 1 ? 1 0 1 0 1 ? ? 1 ? ? 0 1 0 1 1 0 0 1 1 1 1 0 0 0 0 1 0
Kotasaurus          0 0 1 1 0 1 1 1 1 1 0 1 0 ? 1 1 0 ? ? 1 1 1 1 ? ? ? 1 0 1 ? ? ? ? ? ? ? ? ? 0 ? ? ? ? ? ? ? ? ? ? ?
Lessemsaurus        ? ? ? ? ? ? ? ? ? ? ? ? ? ? ? ? ? ? ? ? ? ? ? ? ? ? ? ? ? ? ? ? ? ? ? ? ? ? ? ? ? ? ? ? ? ? ? ? ? ?
Lufengosaurus       0 1 0 0 0 1 0 0 1 1 0 1 0 0 0 1 0 ? 0 0 1 1 0 0 0 0 0 ? 1 1 1 1 1 0 1 0 1 1 0 0 1 1 1 ? 0 ? 1 1 1 0
Massospondylus      0 1 0 0 0 0 0 1 1 0 1 0 0 1 0 1 1 0 0 1 1 0 0 0 0 0 1 1 1 1 0 1 0 1 1 0 0 1 1 0 0 0 0 1 1 1 0
Melanorosaurus      0 0 1 0 1 1 0 0 1 1 0 1 0 1 1 0 0 1 0 0 1 0 0 0 1 ? ? ? 0 0 0 1 1 1 1 ? 0 ? ? ? 1 ? ? ? ? ? 0 ? ? ?
Mussaurus           ? ? ? 0 0 0 0 ? ? 0 0 ? ? ? ? ? ? ? ? ? ? ? ? ? ? 0 ? ? 0 ? ? ? / ? ? ? 1 0 ? ? ? ? ? 1 ? ? ? 0 ? ? ?
Omeisaurus          1 0 0 1 ? 1 1 1 1 1 1 1 0 1 1 1 1 0 0 1 ? ? 1 1 1 1 1 0 1 0 0 1 1 1 1 1 0 1 1 1 1 1 1 1 1 1 1
Plateosauravus      ? ? ? 0 0 0 0 0 ? ? ? 0 0 ? 0 0 1 ? 0 ? ? ? ? ? ? ? 0 0 1 1 ? ? ? 0 1 0 ? ? ? ? ? ? ? ? ? 0 ? ? ? ?
Plateosaurus        0 1 0 0 0 0 0 0 1 0 1 0 0 1 0 0 1 0 0 0 0 1 1 0 0 0 0 0 0 1 1 1 1 1 0 1 0 1 0 0 ? 1 1 0 0 0 1 0 0 0
Riojasaurus         0 1 1 0 1 1 0 0 1 1 0 1 0 0 1 0 0 1 0 0 1 ? 0 0 0 1 ? ? 0 0 0 0 0 1 1 1 1 1 0 1 0 1 1 0 0 0 1 0 0 0 ? 1 0 ? ?
Saturnalia          0 1 0 ? ? ? 0 0 0 1 0 0 ? 0 0 0 0 ? 0 ? 0 0 0 0 0 0 0 0 0 ? 0 0 0 1 ? 0 ? 0 0 0 ? ? ? 0 0 0 ? ? ? 0 0
Shunosaurus         0 0 0 1 1 1 1 1 1 1 1 1 ? 1 1 1 ? 0 ? 1 0 1 1 1 1 1 1 0 1 0 0 0 1 1 1 1 1 1 1 1 1 1 1 1 1 1 1 1 0
Thecodontosaurus    0 1 0 0 0 0 0 0 ? ? 0 0 0 0 0 0 1 0 0 0 1 ? ? 0 0 0 0 ? 1 1 0 1 ? 0 ? ? 1 0 ? 0 0 0 0 0 0 0 ? 1 0 0 0
Vulcanodon          1 1 1 0 1 1 ? 1 1 1 1 1 0 ? 1 1 1 1 0 1 1 1 1 0 1 1 1 1 0 1 1 1 0 1 1 0 1 1 0 0 1 1 1 1 0 ? 1 1 1 1 1 0 1 1 1 0
Yunnanosaurus       ? ? ? 0 0 0 0 ? ? 0 ? 0 0 ? 0 0 1 0 0 ? ? ? ? ? ? 0 0 0 ? 1 1 1 1 1 ? ? ? ? ? ? ? ? ? ? ? ? ? ? ? ? ?
```

[Special Papers in Palaeontology 77, 2007, pp. 91–112]

THE EVOLUTION OF FEEDING MECHANISMS IN EARLY SAUROPODOMORPH DINOSAURS

by PAUL M. BARRETT* *and* PAUL UPCHURCH†

*Department of Palaeontology, The Natural History Museum, Cromwell Road, London SW7 5BD, UK; e-mail: P.Barrett@nhm.ac.uk
†Department of Earth Sciences, University College London, Gower Street, London WC1E 6BT, UK; e-mail: P.Upchurch@ucl.ac.uk

Typescript received 10 March 2006; accepted in revised form 22 June 2006

Abstract: Sauropodomorph dinosaurs were the dominant terrestrial herbivores of the Late Triassic to Late Jurassic, but the early evolution of herbivory is poorly documented in this clade. In particular, the transition from the relatively simple feeding mechanisms of basal sauropodomorphs and prosauropods to the more complex feeding apparatus of sauropods has received little attention, owing largely to the paucity of basal sauropod material. Discoveries of Late Triassic and Early Jurassic sauropods and reinterpretation of sauropodomorph phylogeny have alleviated this problem, revealing new information on the sequence of character acquisitions that occurred during the origin of sauropods. The evolution of sauropod herbivory was intimately associated with concurrent trends towards increased body size and quadrupedal locomotion. Recognition of *Jingshanosaurus* and *Melanorosaurus* as basal sauropods closes the morphological gap that existed between more advanced sauropods, such as *Vulcanodon*, and prosauropods.

Key words: Sauropodomorpha, Late Triassic, Early Jurassic, herbivory, evolution.

S AUROPODOMORPHA encompasses the largest terrestrial herbivores of all time, including the titanosaurian sauropod *Argentinosaurus* (body mass of > 70 tonnes: Mazzetta *et al.* 2004) and the diplodocoid sauropod *Seismosaurus* (body length of *c.* 33 m: Lucas *et al.* 2004). Consequently, a detailed understanding of how these dinosaurs were adapted to a diet of plants has the potential to inform not only questions regarding the palaeobiology of individual taxa but also the selection pressures involved in the evolution of large body size, Mesozoic palaeocommunity composition and dynamics, and the evolution of amniote herbivory in general. Although the feeding mechanisms of eusauropod dinosaurs have been closely scrutinized over the past decade (e.g. Barrett and Upchurch 1994, 1995, 2005; Calvo 1994; Upchurch 1994; Fiorillo 1998; Christiansen 1999, 2000; Stevens and Parrish 1999, 2005; Upchurch and Barrett 2000; Sereno and Wilson 2005), little is known of the early evolution of the varied functional complexes found in these animals. In addition, relatively little attention has been paid to the feeding mechanisms of basal sauropodomorphs and prosauropods (see Galton 1984*a*, 1985*a*, *b*, 1986; Crompton and Attridge 1986; reviewed in Galton and Upchurch 2004) or basal sauropods (Upchurch and Barrett 2000). As a result, the transition from the relatively simple feeding mechanisms of basal sauropodomorphs to the more complex systems found in eusauropods is poorly understood.

The main reason for this shortcoming was that, until recently, material of Late Triassic and Early Jurassic saur-

opods was exceptionally rare, with cranial material being especially scarce (Upchurch *et al.* 2004). In addition, there has been no attempt to integrate information on basal sauropodomorph feeding with that from derived sauropods. Consequently, it has been impossible to study the possible sequence of character transformations that occurred in the origin of sauropod feeding. However, new discoveries, reinterpretation of material that has languished undescribed in museum collections, and a re-kindling of interest in early sauropodomorph phylogeny has started to fill this gap. Here, we present a brief review of feeding mechanisms in early sauropodomorph dinosaurs, based on both cranial and postcranial evidence, and posit a scenario for the evolution of sauropodomorph herbivory.

MATERIAL AND METHODS

The interrelationships of basal sauropodomorph and early sauropod dinosaurs are currently controversial. Recent work has recognized that several taxa previously regarded as prosauropods (*Blikanasaurus*, *Jingshanosaurus*, *Camelotia*, *Lessemsaurus*, *Melanorosaurus* and *Chinshakiangosaurus*) are basal (non-eusauropod) sauropods (Yates 2003*a*, 2004*a*; Yates and Kitching 2003; Galton and Upchurch 2004; Upchurch *et al.* 2007). In addition, there is currently little agreement over which taxa, if any, constitute a monophyletic Prosauropoda (Langer *et al.* 1999; Sereno 1999;

Benton *et al.* 2000; Yates 2003*a*, 2004*a*; Yates and Kitching 2003; Galton and Upchurch 2004; Leal *et al.* 2004; Upchurch *et al.* 2007). Although we acknowledge the diversity of published opinions on sauropodomorph phylogeny, we adopt the phylogenetic tree presented by Upchurch *et al.* (2007) as the evolutionary framework for this study (Text-fig. 1), for the following reasons. First, the phylogeny of Upchurch *et al.* (2007) is dependent upon many of the same personal observations of specimens as this review: hence, it forms a framework consistent with the observations provided herein. Secondly, many aspects of the overall cladogram structure are in agreement with those posited by other authors: a monophyletic Prosauropoda is recovered consisting of *Plateosaurus*, *Yunnanosaurus*, *Lufengosaurus* and several other 'traditional' prosauropod taxa; *Saturnalia* and *Thecodontosaurus* form successive sister-taxa to all other sauropodomorphs; and several 'melanorosaurid prosauropods' are reinterpreted as basal sauropods (compare with Yates 2003*a*, 2004*a*; Yates and Kitching 2003). Nevertheless, several areas of disagreement remain, including: the position of *Anchisaurus*, a prosauropod according to Upchurch *et al.* (2007), but a basal sauropod in the phylogeny of Yates (2004*a*); interrelationships within Prosauropoda; and the more recent proposal that almost all non-sauropod sauropodomorphs form a paraphyletic almost wholly pectinate array of taxa with respect to sauropods (Leal *et al.* 2004; Yates 2007). As a consequence, we regard our conclusions as provisional, pending greater consensus between alternative sauropodomorph phylogenies.

A wide range of basal sauropodomorph, prosauropod and basal sauropod material was examined at first hand during the course of this study. These personal observations were supplemented with information from the literature. In order to avoid excessive citation of specimen numbers and references, these sources have been listed in full in Table 1.

Ratios of forelimb length : hindlimb length (FL/HL) and hindlimb length : trunk length (HL/TL) have been used to investigate the habitual stance and locomotor adaptations of various dinosaurs (e.g. Galton 1970; Coombs 1978). This methodology has been adopted here with some modification. Metric data on the proximal segments of the limbs were used to calculate fore- and hindlimb lengths [forelimb length equals the sum of the humeral and radial lengths (or ulna length if the latter is unavailable); hindlimb length equals the sum of femoral and tibial length]. This differs from Galton's (1970) definitions of limb length, which incorporated the longest metapodial in each limb. Many of the specimens used in this study lack metapodials; therefore, they have been excluded from calculation of limb lengths. Neck length : trunk length (NL/TL) ratios also have biological significance as they provide a means for assessing the importance of neck elongation in feeding. Trunk length is defined as the

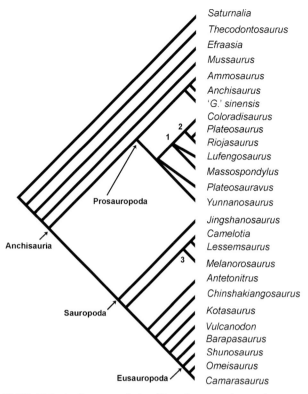

TEXT-FIG. 1. Ingroup relationships of sauropodomorph dinosaur taxa (based on Upchurch *et al.* 2007). Numbers refer to particular clades: 1. Plateosauria; 2. Plateosauridae; 3. 'melanorosaurs'.

length of the dorsal vertebral column; neck length equals the sum of the lengths of the cervical vertebrae. Unfortunately, complete dorsal and cervical series are relatively rare among early sauropodomorph material.

Institutional abbreviations. AMNH, American Museum of Natural History, New York; BMNH, The Natural History Museum, London; BP, Bernard Price Institute for Palaeontological Research, Johannesburg; FMNH, Field Museum of Natural History, Chicago; IVPP, Institute of Vertebrate Paleontology and Paleoanthropology, Beijing; MB, Humboldt Museum für Naturkunde, Berlin; MCZ, Museum of Comparative Zoology, Harvard University, Cambridge (USA); NGMJ, Nanjing Geological Museum, Nanjing; SAM, Iziko: South African Museum, Cape Town; ULR, Museo de Ciencias Naturales, Universidad Nacional de La Rioja, La Rioja; YPM, Peabody Museum of Natural History, Yale University, New Haven.

FEEDING IN BASAL SAUROPODOMORPH DINOSAURS

General comments

The anatomy of basal sauropodomorphs and prosauropods is relatively conservative (Galton and Upchurch

TABLE 1. Sources of comparative anatomical data used in this study. Accession numbers denote specimens examined by the authors at first hand: other data were obtained from the literature. Classification follows the phylogenies presented by Upchurch *et al.* (2007), supplemented with information from Yates and Kitching (2003), Galton and Upchurch (2004) and Upchurch et al. (2004). An asterisk (*) indicates that a taxon has not yet been included as an operational taxonomic unit in a cladistic analysis. It should be noted that *Barapasaurus* is treated as a non-eusauropod sauropod in this review, following Upchurch *et al.* (2007). However, most recent cladistic analyses have placed this taxon within Eusauropoda (e.g. Wilson 2002; Upchurch *et al.* 2004). Note that *Efraasia* is used in the sense of Yates (2003*b*) while in other works (e.g. Galton 1990; Galton and Upchurch 2004) some of this material is referred to as *Sellosaurus* (see Yates 2003*b* for discussion of this issue). In addition, the material referred to *Euskelosaurus* by Van Heerden (1979) is now regarded as *Plateosauravus* (Yates 2004*b*).

Basal sauropodomorphs
    Efraasia minor: Galton (1973*b*, 1985*c*); Galton and Bakker (1985); Yates (2003*b*)
    Mussaurus patagonicus: Bonaparte and Vince (1979)
    Saturnalia tupiniquim: Langer *et al.* (1999); Langer (2003)
    Thecodontosaurus caducus: BMNH RU P24; Kermack (1984); Benton *et al.* (2000); Yates (2003*a*)
Prosauropoda
    Anchisaurus polyzelus: YPM 1883; Galton (1976); Yates (2004*a*)
    Coloradisaurus brevis: Bonaparte (1978)
    '*Gyposaurus sinensis*': NGMJ V0108; Young (1948)
    Lufengosaurus huenei: IVPP V15; Young (1941, 1951); Barrett *et al.* (2005)
    Massospondylus carinatus: BP/1/4376, BP/1/4779; BP/1/4934; BP/1/4952; BP/1/5241; BP/1/5247; SAM-PK-K388;
        SAM-PK-K398; SAM-PK-K1314; SAM-PK-K5135; Cooper (1981); Gow *et al.* (1990); Sues *et al.* (2004); Reisz *et al.* (2005);
        Barrett and Yates (2006)
    '*Massospondylus*' sp.*: MCZ 8893; Attridge *et al.* (1985)
    Unnamed taxon*: SAM-PK-K1325; SAM-PK-K7904; Barrett (2004)
    Plateosauravus cullingworthi: Van Heerden (1979)
    Plateosaurus spp. AMNH 6810; MB XXIV; von Huene (1926); Galton (1984*a*, 1985*a*, 1998*a*); Moser (2003); Yates (2003*b*)
    Riojasaurus incertus: ULR 56; Bonaparte (1972); Bonaparte and Pumares (1995)
    Unaysaurus tolentinoi: Leal *et al.* (2004)
    Yunnanosaurus huangi: NGMJ 004546; Young (1942); Barrett *et al.* (in press)
Basal (non-eusauropod) sauropods
    Antetonitrus ingenipes: BP/1/4952; Yates and Kitching (2003)
    Barapasaurus tagorei: Jain *et al.* (1975, 1979)
    Camelotia borealis: Galton (1998*b*).
    Chinshakiangosaurus chunghoensis: IVPP V14474; Dong (1992); Upchurch *et al.* (in press)
    Gongxianosaurus shibeiensis*: He *et al.* (1998)
    Isanosaurus attavipachi: Buffetaut *et al.* (2000)
    Jingshanosaurus xinwaensis: Zhang and Yang (1994)
    Kotasaurus yamanpalliensis: Yadagiri (1988, 2001*)
    '*Kunmingosaurus wudingensis*'*: IVPP unnumbered; Young (1966); Dong (1992)
    'Lufeng sauropod'*: FMNH CUP 2042; Barrett (1999)
    Melanorosaurus readi: Van Heerden and Galton (1997); Galton *et al.* (2005); Yates (2005)
    Vulcanodon karibaensis: Raath (1972); Cooper (1984)

2004), allowing the feeding mechanisms of these two evolutionary 'grades' to be considered simultaneously. The following account aims to summarize the craniodental and postcranial features present in these taxa (see Table 1): exceptions to the general pattern are noted. Basal sauropodomorphs (considered herein as *Saturnalia*, *Thecodontosaurus*, *Efraasia* and *Mussaurus*) are currently known from late Middle–Late Triassic deposits in Europe and South America (Weishampel *et al.* 2004); prosauropods (including *Plateosaurus*, *Massospondylus*, *Lufengosaurus*, *Yunnanosaurus*, *Riojasaurus*, '*Gyposaurus*' and *Coloradisaurus*) are known from the Norian–late Early

Jurassic (Pliensbachian or Toarcian: Yates *et al.* 2004) and had an almost global distribution (Weishampel *et al.* 2004).

Dietary interpretations of basal sauropodomorphs and prosauropods have been based largely upon tooth morphology, discussions of manus function, the presence/absence of gastroliths and palaeoenvironmental arguments: prosauropods have been variously regarded as predators, scavengers, omnivores and high-fibre herbivores (see reviews in Galton 1985*b* and Barrett 2000). Current evidence suggests that the majority of these taxa were either herbivorous or facultatively omnivorous (Barrett

2000; Galton and Upchurch 2004). The combined presence of an elongate neck and large body size in some prosauropods has led to their recognition as the earliest high-browsing herbivores (Bakker 1978: but see below for an alternative viewpoint).

Craniodental anatomy

Cranial structure and function. The skulls of basal sauropodomorphs and prosauropods are lightweight structures with large external openings (Text-fig. 2). In lateral view, the adult skull is long, but not particularly deep, with an elongated preorbital region. Embryonic and juvenile individuals, such as *Mussaurus* (Bonaparte and Vince 1979) and *Massospondylus* (BP/1/4376; Reisz *et al.* 2005), have relatively larger orbits and shorter preorbital regions. In dorsal view, the skull is relatively narrow and ends in a pointed, V-shaped snout. In some taxa (e.g. *Lufengosaurus*, *Plateosaurus*, *Thecodontosaurus*, *Riojasaurus* and *Unaysaurus*), a shallow fossa surrounds each supratemporal fenestra.

Although many of the articulations within the cranium are simple butt or overlap joints, it appears to have been an akinetic structure (Galton 1985*a*). The skull roof was immobilized by the presence of complex interdigitating articulations between the midline elements, while other joints were probably immobilized by the presence of inelastic connective tissue. Gow *et al.* (1990) suggested the presence of a streptostylic quadrate in *Massospondylus*. However, the elongated ventral process of the squamosal and the dorsal process of the quadratojugal meet to form a continuous bar that contacts the rostral edge of the quadrate, which would have prevented the quadrate from rotating rostrally (Text-fig. 3). A similar situation occurs in all other prosauropods for which appropriate material is known (e.g. *Plateosaurus* and *Lufengosaurus*: see also Galton 1985*a*).

Gow *et al.* (1990) also suggested that the premaxilla of *Massospondylus* was mobile with respect to the maxilla, and that the premaxilla was protracted by the transmission of forces along a longitudinal thrust path through the palate, driven by the contraction of the M. levator pterygoidei [evoking an identical mechanism to that proposed for the coelophysoid theropod *Coelophysis* (= *Syntarsus*) *rhodesiensis*: see Raath 1977]. It was envisaged that the premaxilla rotated dorsally along a sliding contact with the nasals, and around a 'ramp and spur' articulation with the maxilla. This 'ramp and spur' consists of the caudal process of the premaxilla, which fits into a

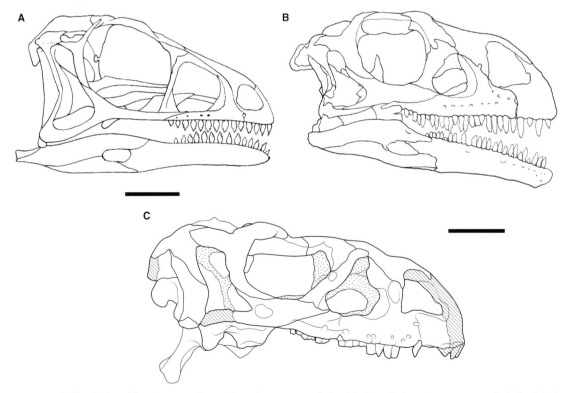

TEXT-FIG. 2. Skulls of selected basal sauropodomorphs and prosauropods in right lateral view. A, reconstructed skull of the basal sauropodomorph *Thecodontosaurus caducus* (after Yates 2003*a*). B, *Massospondylus carinatus* (based on Sues *et al.* 2004). C, *Lufengosaurus huenei* (based on Barrett *et al.* 2005). Cross-hatched areas represent reconstructed areas of the type skull (IVPP V15); stippled areas are obscured by the presence of matrix. Scale bars represent 10 mm in A and 50 mm in B and C.

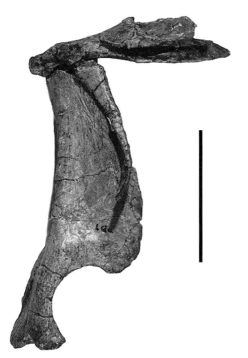

TEXT-FIG. 3. Lateral view of articulated right squamosal and quadrate of *Plateosaurus longiceps* (AMNH 6810: taxonomy after Galton 2001), demonstrating the extensive contact between these two elements. Scale bar represents 50 mm.

as the nasal process of the premaxilla passes beneath and between the nasals, effectively wedging them in place (PMB, pers. obs.). Manipulation of disarticulated jaw material of *Plateosaurus* (AMNH 6810) and *Massospondylus* (SAM-PK-K398) suggests that even the apparently loose 'ramp and spur' contact between the maxilla and premaxilla might not have allowed significant freedom of movement (Text-fig. 4). The maxilla possesses a small premaxillary process that articulates with a shallow groove on the medial surface of the premaxilla, which is bounded ventrally by a thin, acute ridge. The contact between the premaxillary process of the maxilla and the ridge would have prevented dorsal rotation of the premaxilla as would contact between the caudal process of the premaxilla and the premaxillary process of the maxilla. The premaxillary process of the maxilla is orientated slightly laterally, while the caudal process of the premaxilla is orientated slightly medially, so that the two processes overlap rostrally. This overlap would have prevented significant translational or rotational movements along this contact. Nevertheless, a small amount of flexure may have been possible at the premaxilla-maxilla junction as the processes involved are very thin. Although it is unlikely that these movements were involved in active cranial kinesis, they might have functioned as a 'shock-absorber', dampening forces acting on the jaws as they were closed, as has also been proposed for other unfused or non-sutured joints within dinosaur and dicynodont crania [e.g. the quadratojugal/jugal and postorbital/jugal articulations of tyrannosaurids (Rayfield 2005) and the premaxillary/nasal articulation in *Lystrosaurus* (King and Cluver 1993)].

shallow groove on the rostral surface of the ascending process of the maxilla forming a simple overlapping joint. However, the premaxilla-nasal articulation of basal sauropodomorphs and prosauropods (including *Massospondylus*) could not have permitted premaxillary rotation

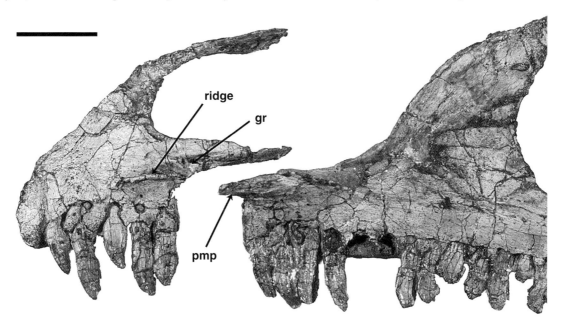

TEXT-FIG. 4. Medial view of the right premaxilla and rostral part of the right maxilla of *Plateosaurus longiceps* (AMNH 6810), showing the articular surfaces between these two elements. Abbreviations: gr, groove for reception of premaxillary process of maxilla; pmp, premaxillary process of maxilla. Scale bar represents 25 mm.

In some genera, the distal portion of the quadrate extended ventral to the level of the maxillary tooth row, resulting in a craniomandibular joint with a pronounced ventral offset (*Plateosaurus*, *Lufengosaurus*, *Coloradisaurus* and *Efraasia*). By contrast, the craniomandibular joints of other taxa are either only slightly offset or lie level with the maxillary tooth row (*Yunnanosaurus*, *Mussaurus*, *Anchisaurus*, *Thecodontosaurus*, *Massospondylus* and *Riojasaurus*).

The mandible is long and slender, terminating in a bluntly rounded symphysis. The symphyseal articulation is rugose, suggesting that the two mandibles were bound firmly to each other by connective tissues. All prosauropods possess an external mandibular fenestra. A well-developed coronoid eminence is present in several genera including *Plateosaurus*, *Lufengosaurus*, *Efraasia* and *Coloradisaurus*. By contrast, the coronoid eminence of *Yunnanosaurus*, *Mussaurus*, *Thecodontosaurus*, *Anchisaurus*, *Massospondylus* and *Riojasaurus* is rather low. The position of the mandibular glenoid also varies between taxa (ventrally offset or on the same level as the dentary tooth row), reflecting the morphology of the quadrate (see above). It is interesting to note that there seems to be at least a partial correlation between the possession of a high coronoid eminence and the presence of a ventrally offset craniomandibular joint. The development of a coronoid eminence is usually viewed as a way of increasing the area of jaw muscle insertion and, more importantly, as a way of increasing the moment arm of the jaw adductors, thereby increasing the bite force (Galton 1974). Ventral offsetting of the craniomandibular joint is considered functionally analogous to the dorsal offsetting of the jaw joint seen in mammals (Greaves 1998): this permits an increase in the angle between the lever arms of the jaw adductors and the tooth row (Crompton and Hiiemäe 1969) and allows the jaws to come together in parallel, rather than closing in a 'scissor-like' fashion. This observation suggests that those taxa possessing ventrally offset craniomandibular joints and high coronoid eminences might have eaten more resistant plant material than taxa lacking these features.

In all basal sauropodomorphs and prosauropods, the mandibular portion of the jaw joint is formed by the articular with a small contribution from the surangular. The glenoid fossa is a cup-shaped depression that is usually divided into two portions, reflecting the shape of the quadrate condyle. The rostral border of the glenoid is formed by the coronoid eminence: a transverse ridge extending across the surface of the articular defines the caudal margin. In all taxa, the glenoid surface is rostrocaudally short. The length of this surface corresponds closely to that of the quadrate condyle, suggesting that translational and rotational movements were not possible while the jaws were articulated, thereby indicating a strictly orthal jaw action.

Dentition. The tooth row of basal sauropodomorphs and prosauropods extends for most of the length of the dentary, terminating just rostral to the coronoid eminence, and for the entire length of the premaxilla and maxilla. The teeth decrease in size caudally, in terms of both crown height and crown width. Dental formulae included 3–6 premaxillary teeth, 11–30 maxillary teeth and 17–28 dentary teeth in adult individuals (Galton and Upchurch 2004). There is some taxonomic variation (particularly in the number of premaxillary teeth: for example, *Coloradisaurus* possesses only three premaxillary teeth, whereas *Plateosaurus* possesses either five or six) and tooth numbers also vary ontogenetically (larger, older individuals possess greater numbers of maxillary and dentary teeth, e.g. Sues *et al.* 2004). In general, maxillary and dentary tooth crowns are lanceolate, mesiodistally expanded, have crowns that are apicobasally longer than they are wide in labial/lingual view, and are symmetrical about their long axes in mesial/distal view (Galton 1985*b*, 1986). The labial and lingual surfaces of the tooth crowns bear no grooves or ridges, but are flat and featureless. Maxillary and dentary teeth are labiolingually compressed, lack recurvature and lack cingula: they can be distinguished from the teeth of contemporary ornithischians that are more triangular in labial/lingual outline (Sereno 1991) and asymmetrical in mesial/distal view (Galton 1984*b*). The teeth are arrayed in an *en echelon* arrangement, with the distal crown margins labially overlapping the mesial margins of successive teeth (Galton 1985*b*, 1986). There is usually a distinct overbite, with the maxillary teeth closing just outside the dentary tooth row. In most taxa, the mesial and distal crown margins are coarsely denticulate, with the serrations projecting outward at an angle of *c.* 45 degrees to the apicobasal axis of the tooth crown (Galton 1985*b*, 1986).

Several taxa have tooth crown morphologies that depart from this general pattern. For example, Bonaparte and Pumares (1995) described the tooth crowns of *Riojasaurus* as 'conical' (with crowns equal in diameter to the roots) rather than 'leaf-shaped' (with mesiodistally expanded tooth crowns), and suggested that this represented the primitive prosauropod condition. However, the teeth in the only described skull of *Riojasaurus* (ULR 56) are too poorly preserved to determine their exact morphology (PMB, pers. obs.) and confirmation of this hypothesis must await the discovery of new material. *Yunnanosaurus* tooth crowns lack marked mesiodistal expansion and are apicobasally elongate, making them appear much more slender than the teeth of other basal sauropodomorphs and prosauropods. In addition, *Yunnanosaurus* teeth lack denticles (Young 1942; Barrett *et al.* in press).

Finally, denticles in *Massospondylus* are confined to the apicalmost region of the tooth crown and are angled at around 10–20 degrees to the long axis of the tooth (Gow *et al.* 1990).

The premaxillary teeth are somewhat different from the maxillary and dentary teeth. The crowns do not exhibit the same degree of labiolingual compression and the crown bases are not mesiodistally expanded, giving the teeth a conical appearance. In many genera these teeth are procumbent, and the crowns may be much longer than those of the maxillary or dentary teeth. The latter is particularly marked in *Massospondylus* (Cooper 1981; Gow *et al.* 1990). Premaxillary teeth may also display some recurvature (e.g. Kayenta '*Massospondylus*', *Anchisaurus*, *Plateosaurus* and *Efraasia*) and sometimes bear fine serrations reminiscent of those on theropod teeth (Kayenta '*Massospondylus*', pers. obs.).

Attridge *et al.* (1985) reported the occurrence of small, conical and slightly recurved palatal teeth in the Kayenta '*Massospondylus*'. This is the only known occurrence of palatal teeth within the Dinosauria. The small size of the teeth (around 1 mm in length) and the lack of dental ankylosis led Attridge *et al.* (1985) to suggest that the teeth were supported by the palatal epithelium.

Small wear facets on isolated teeth of *Massospondylus* (BP/1/4952, BP/1/5247: Gow *et al.* 1990) and *Plateosaurus* (Galton 1998a), and the presence of a lingual wear facet on one of the premaxillary teeth of the Kayenta '*Massospondylus*' (Crompton and Attridge 1986), represent the only evidence for tooth wear in prosauropods. The general absence of wear indicates that these taxa lacked systematic and precise dental occlusion. Several other mechanisms can account for the above-mentioned examples of tooth wear, including tooth-food wear and various dental pathologies (e.g. misalignment of teeth during eruption; breakage during contact with hard ingested items followed by tooth-food wear or taphonomic abrasion).

Galton (1985b, 1986) described large, high-angled mesial and distal wear facets on several isolated teeth that had been referred to *Yunnanosaurus* (Simmons 1965) and inferred the presence of an interlocking dentition in this taxon, which would have been unique among basal sauropodomorphs and prosauropods. However, the identification of these teeth has been questioned. Various features suggest that they may be referable to a basal sauropod (Salgado and Calvo 1997; Wilson and Sereno 1998; Barrett 2000), but the identity of the taxon that possessed these teeth remains unknown (Galton and Upchurch 2004) and they are not considered further herein.

Soft anatomy. There has been some discussion concerning the presence of fleshy non-contractile cheeks in prosauropods. Galton (1973a) initially concluded that prosauro-

pods lacked cheeks, but his views changed gradually in a series of papers dealing with prosauropod cranial anatomy, culminating in the statement 'Unlike lizards, prosauropods probably had cheeks ...' (Galton and Upchurch 2004, p. 256; see also Paul 1984). This conclusion was based on the identification of features proposed as correlates of cheeks (Galton 1973a; Paul 1984): (1) a short ridge lateral to the caudal part of the dentary tooth row; and (2) large nutrient foramina on the lateral surfaces of the maxilla and dentary. These features are present in many basal sauropodomorphs and prosauropods, including *Plateosaurus*, *Massospondylus*, *Sellosaurus*, *Lufengosaurus* and *Riojasaurus*. None of these taxa has an inset maxillary tooth row (Galton 1973a); therefore, although cheeks may have been present they would not have been as well developed as those of ornithischians and would have been limited to the caudal part of the mouth (Paul 1984; Galton 1985a). However, a short ridge extends along the caudal part of the maxilla in *Lufengosaurus*, perhaps indicating that this animal possessed a larger cheek than that seen in other prosauropods (Barrett *et al.* 2005). The well-ossified hyoid bones of *Plateosaurus*, *Massospondylus* and *Anchisaurus* suggest the presence of a large tongue that may have been useful for manipulating food within the mouth.

Several authors have proposed that basal sauropodomorphs and prosauropods possessed keratinous rhamphothecae (Crompton and Attridge 1986; Bonaparte and Pumares 1995; Sereno 1989, 1997). Crompton and Attridge (1986) reconstructed the skull of the Kayenta '*Massospondylus*' with a rhamphotheca at the mandibular symphysis (they found no evidence for a premaxillary beak) and also indicated that a similar structure may have been present in *Massospondylus sensu stricto*. Measurements taken from two specimens (MCZ 8893 and SAM-PK-K1314) appeared to show that the mandibles were not long enough to reach the front of the cranium (Attridge *et al.* 1985; Crompton and Attridge 1986): when placed in articulation with the cranium and with each other, the jaw symphysis was located just rostral to the premaxilla-maxillary junction and the first three premaxillary teeth lay rostral to the tip of the lower jaw (Crompton and Attridge 1986). One of the premaxillary teeth bears a small wear facet on its lingual surface: Crompton and Attridge (1986) suggested that this could have been produced by contact with a mandibular beak. Their proposed reconstruction shows the beak uniting the two mandibular rami and providing additional length to the lower jaw so that it reaches the front of the cranium. However, in extant amniotes that possess a rhamphotheca (chelonians, birds), the beak simply covers the jaw margins: it does not form the primary functional linkage between the rostral ends of the mandibular rami. Furthermore, the rostral ends of the mandibular rami in the

Kayenta 'Massospondylus' are not rugose (although several large foramina are present in this area) and show no evidence for the attachment of a rhamphotheca (PMB, pers. obs.). Gow et al. (1990) suggested that the apparent discrepancy between the length of the cranium and the mandibles seen in the Kayenta 'Massospondylus' was caused by incorrect reconstruction of the occiput: if the occiput had been reconstructed to be too wide it is conceivable that the mandibles would appear to be too short when articulated with each other and the cranium. However, examination of both MCZ 8893 and SAM-PK-K1314 does indicate that the mandibles of these specimens are, in fact, slightly shorter than the cranium even when deformation is taken into account. As the presence of an elongate mandibular rhamphotheca seems unlikely, this may simply indicate that these taxa possessed an overbite, a feature that is accentuated by their procumbent premaxillary teeth. The single wear facet on the premaxillary tooth of the Kayenta specimen could easily have been produced by tooth-food wear: it is not necessary to invoke the presence of a beak.

Evidence for the presence of a small mandibular rhamphotheca in other taxa is equivocal. Many nutritive foramina are located around the rostral margin of the dentary in some specimens of Massospondylus and Plateosaurus, but these are not present in all individuals. Also, some of these foramina, where present, are too caudally situated to be associated with a terminal beak. Similarly, there is little evidence to support the presence of a premaxillary beak in the majority of basal sauropodomorph or prosauropod taxa (contra Sereno 1989, 1997): the tooth row extends to the rostral limit of the skull; no foramina are present on the margin of the premaxilla; and the premaxillary margin is generally smooth, lacking obvious rugosities (Gow et al. 1990; Yates 2003a). The only exception to this may be Riojasaurus, which possesses a rugose, raised surface at the rostral margin of the premaxilla that may have supported a beak (Bonaparte and Pumares 1995).

Postcranial evidence

The necks of basal sauropodomorphs and prosauropods are relatively long, with NL/TL ratios that range between 0·7 and 0·9 (Galton and Upchurch 2004; also see Table 2). This contrasts with the relatively short necks seen in basal saurischians and ornithischians, which have NL/TL ratios of approximately 0·4 [e.g. *Eoraptor* (Sereno et al. 1993); *Lesothosaurus* (Thulborn 1972)]. Prosauropods and basal sauropodomorphs possess ten cervical vertebrae (e.g. von Huene 1926; Cooper 1981; Yates 2003a), with the possible exception of *Riojasaurus*, in which there may have been only nine cervicals (Bonaparte and Pumares 1995). Cervicals were amphicoelous in all taxa.

Cooper (1981) suggested that the neck of *Massospondylus* was laterally inflexible and only capable of movement in the vertical plane. This was on the basis of his observations on the orientations of the cervical pre- and postzygapophyses, which were stated to '… articulate at an angle of 45 degrees to the horizontal …' (Cooper 1981, p. 720). Such an arrangement would indeed place severe limits on neck flexibility, effectively preventing mediolateral rotation around the intervertebral articulations as it would have been impossible for the pre- and postzygapophyses to slide past each other to a significant degree. However, this morphology is seen only in caudal cervicals (Cooper 1981; PMB, pers. obs.). In lateral view, the cranial cervicals of *Massospondylus* possess pre- and postzygapophyses that extend almost subparallel to the long axis of the cervical vertebrae, while in dorsal view, the zygapophyses project outwards at angles of approximately 10 degrees to the long axis of the neural arch (Cooper 1981; PMB, pers. obs.). In addition, the elongate prezygapophyses had a relatively large surface area. Consequently, as the pre- and postzygapophyses actually met at a much more shallow angle than envisaged by Cooper (1981), translational movements along the articular surfaces of the zygapophyses might have allowed a reasonable degree of both lateral and vertical movement while the cervicals remained in articulation. This combination of features suggests that although the base of the neck may have been relatively inflexible in *Massospondylus*, a greater range of movements may have been possible in the anterior region. The cervical ribs of *Massospondylus* are moderately long and extend caudally to overlap the succeeding cervical vertebra (Cooper 1981; PMB, pers. obs.). This may have placed some constraints on the lateral flexure of the neck, but as the overlap is not extensive is it unlikely to have prohibited movement. The necks of

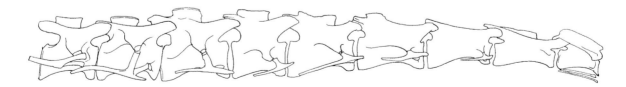

TEXT-FIG. 5. Cervical series of *Plateosaurus longiceps* in lateral view (after von Huene 1926). Scale bar represents 200 mm.

Plateosaurus (von Huene 1926; Moser 2003; Text-fig. 5), *Thecodontosaurus* (Kermack 1984; Yates 2003*a*) and *Riojasaurus* (Bonaparte and Pumares 1995) possessed a similar combination of vertebral characters; therefore, it is not unreasonable to conclude that basal sauropodomorphs and prosauropods could use their necks to feed over moderately wide horizontal and vertical ranges.

The function of the manus has been discussed by Galton (1971) and Cooper (1981). Both authors concluded that the manus was capable of grasping, but that digit I was not opposable. Galton (1971) suggested that digits II–IV, and particularly digits II and III, were capable of load-bearing during quadrupedal locomotion. Cooper (1981) disagreed, believing that a functionally didactyl manus would not have been able to bear such loads. Digits II–IV of smaller taxa (such as *Thecodontosaurus* and *Anchisaurus*) are more slender than in those in larger forms; this may simply be correlated with body size, or with the stance (bipedal or quadrupedal) that the animal habitually adopted (Galton 1971). Digit I was highly specialized and capable of hyperextension; it diverged (medially) from the other digits, and could not be used for grasping or load-bearing (Galton 1971; Cooper 1981). A large, trenchant ungual is present on digit I in all basal sauropodomorphs and prosauropods and may have had a variety of functions, which could have included a role in food gathering, such as grasping vegetation or dismembering carrion, or defence (Barrett 2000; Galton and Upchurch 2004).

Establishing the stance of basal sauropodomorphs and prosauropods has proved controversial: some authors have regarded most genera as facultative bipeds (e.g. Coombs 1978; Galton and Upchurch 2004; Bonnan and Senter 2007), while others have suggested quadrupedal habits (e.g. Christian and Preuschoft 1996; Langer 2003). This lack of consensus stems from the unusual mosaic of locomotory features seen in these animals. For example, most genera possess relatively short forelimbs, a feature usually associated with bipedal gaits (Galton 1970; Coombs 1978), and have FL/HL ratios of approximately 0·5–0·6, which are comparable with those of many bipedal and secondarily quadrupedal ornithischians (see Galton 1970; Coombs 1978, though note that the HL/TL ratios used herein are lower than those obtained by Galton, due to omission of metapodial length from the calculation of hindlimb length). However, a contradictory signal is obtained from consideration of HL/TL ratios. Bipedal animals tend to have relatively short trunks (with the result that the centre of mass, CM, is positioned closer to the acetabulum), but the HL/TL ratios of basal sauropodomorphs and prosauropods are intermediate with respect to those of obligate quadrupeds (such as sauropods) and obligate bipeds (such as small ornithopods and theropods: Galton 1970, 1976). Furthermore, although it has

been proposed that the cranial positioning of the gut and the possession of a gastric mill in basal sauropodomorphs would tend to pull the animal on to all fours (Norman and Weishampel 1991), calculations to ascertain the position of the CM in *Plateosaurus* (Henderson in press) indicate that a habitual bipedal posture was possible, as the CM was situated just cranial to the acetabulum. The relatively short forelimbs and trunk region of *Thecodontosaurus* have been used to support habitual bipedal habits for this animal (Kermack 1984); however, no associated fore- and hindlimbs or complete dorsal vertebral series are known for this taxon, so HL/TL and FL/HL ratios cannot be established with confidence (Langer 2003).

Biomechanical studies also provide conflicting results. The morphology of the shoulder, elbow and wrist joints in *Plateosaurus* and *Massospondylus* indicates that the range of forelimb motion was limited, preventing effective pronation of the manus, which in turn might have precluded quadrupedal locomotion (Bonnan and Senter 2007). Conversely, calculation of the bending moments along the length of the axial skeleton in *Plateosaurus* indicates a pattern of sustainable bending moments similar to those obtained from extant and extinct quadrupedal reptiles and mammals (Christian and Preuschoft 1996). Nevertheless, the latter analysis did not reject the possibility that facultative bipedalism may have been possible for short periods.

The dual-purpose nature of the manus, as a load-bearing and manipulative structure, supports suggestions of both facultative bipedality and quadrupedality in the majority of these forms (Galton 1971; Cooper 1981). Finally, some trackway evidence supports the notion that some basal sauropodomorphs and prosauropods might have walked quadrupedally, at least occasionally (Baird 1980; Thulborn 1990). Unfortunately, the recognition of bipedal trackways is difficult, owing to the similarity of basal sauropodomorph pedes to those of other tetrapods; consequently, it may not be possible to demonstrate that they moved bipedally using this line of evidence alone (Thulborn 1990).

As features associated with both bipedality and quadrupedality are present in the majority of basal sauropodomorphs and prosauropods, it seems reasonable to interpret these animals as either facultative bipeds or quadrupeds, with behaviours that encompassed both locomotory modes. To date, most work has concentrated on *Plateosaurus* and *Massospondylus*, but further biomechanical work on additional genera is obviously necessary. For example, *Riojasaurus* apparently deviates from the 'typical' condition seen in *Plateosaurus*: the former has relatively long forelimbs, with an FL/HL ratio approaching that of sauropods and an elongate trunk (Table 2: Galton and Upchurch 2004). These features, in combination with the large body size of *Riojasaurus* (up to 10 m in length: Bonaparte 1972), are suggestive of obligate quadrupedality (Galton and Upchurch 2004). Moreover, it is likely that the habitual stance of these early sauropodomorphs changed during ontogeny. Embryos of *Massospondylus* have relatively longer forelimbs than adult individuals, which may be indicative of quadrupedal locomotion (Reisz *et al.* 2005); the forelimbs of hatchling *Mussaurus* are also relatively long (Bonaparte and Vince 1979: Table 2). In addition, *Saturnalia* and *Anchisaurus* also possess relatively long forelimbs (see 'Discussion').

Browse heights varied with body size and with the ability to feed while in a bipedal stance (Bakker 1978). A fully grown *Plateosaurus*, in bipedal stance, could probably reach heights of around 3·0–3·5 m. A large quadrupedal form, such as *Riojasaurus*, may have been able to browse at similar levels.

Raath (1974) reported the occurrence of gastroliths in three *Massospondylus* skeletons from the Forest Sandstone (Early Jurassic) of Zimbabwe, each of which contained 19–50 polished and rounded stones in the abdominal cavity. Similar gastric mills have been reported in cf. *Ammosaurus* (Shubin *et al.* 1994) and *Sellosaurus* (Galton 1973*b*). Gastric mills would have been useful in the mechanical breakdown of food, particularly as basal sauropodomorphs and prosauropods lacked sophisticated chewing mechanisms.

FEEDING IN BASAL SAUROPOD DINOSAURS

General comments

Until recently, most of the available information on basal sauropods was gained from a handful of Early Jurassic taxa, principally *Vulcanodon* (southern Africa), *Barapasaurus* (India) and *Kotasaurus* (India) (e.g. McIntosh 1990). These taxa are known only on the basis of incomplete postcranial remains and occasional isolated teeth; consequently, key features of the sauropod feeding apparatus, such as the nature of the skull and neck, could not be investigated (Upchurch and Barrett 2000). Ghost lineages derived from phylogenetic analyses of saurischian and sauropodomorph interrelationships (e.g. Gauthier 1986; Upchurch 1995; Wilson and Sereno 1998; Sereno 1999) predicted that additional taxa should have been present during the Late Triassic and Early Jurassic, but remains of these animals proved elusive. However, the past 5 years have witnessed a dramatic improvement in this situation, with the discovery of new sauropods that appear to lie outside of Eusauropoda, including *Gongxianosaurus* (Early Jurassic, China: He *et al.* 1998), the 'Lufeng sauropod' (?Sinemurian, China: Barrett 1999) and *Isanosaurus* (Norian, Thailand: Buffetaut *et al.* 2000).

Moreover, reinterpretation of several 'prosauropods' has led to their recognition as non-eusauropod sauropods (Yates and Kitching 2003; Galton and Upchurch 2004; Yates 2004*a*; Upchurch *et al.* 2007), namely: *Antetonitrus* (Norian, South Africa), *Blikanasaurus* (Norian, South Africa), *Camelotia* (Rhaetian, UK), *Chinshakiangosaurus* (Lower Jurassic, China), *Jingshanosaurus* (?Hettangian, China), *Lessemsaurus* (Norian, Argentina) and *Melanorosaurus* (Norian, South Africa). All of these taxa provide significant new information on sauropodomorph phylogeny and on the sequence of character acquisition at the base of Sauropoda (e.g. Yates and Kitching 2003; Upchurch *et al.* 2007).

These recent discoveries indicate that sauropods lived alongside prosauropods and basal sauropodomorphs for approximately 45 million years and that sauropods gained a similar geographical distribution during this time (though Late Triassic–Early Jurassic sauropods have yet to be recovered from Antarctica, Australia and North America: Weishampel *et al.* 2004). Most authors have concluded that all sauropods were herbivorous quadrupeds (e.g. Coombs 1975; Dodson 1990; Upchurch *et al.* 2004; Wilson 2005); however, although this conclusion remains secure for eusauropods, newly recognized basal sauropod material may force a partial reappraisal of this view (see 'Discussion').

Craniodental anatomy

Although represented by limited amounts of material, the skulls of basal sauropods display greater structural variety than those of contemporary prosauropods and basal sauropodomorphs. Complete, or nearly complete, skulls are available for *Jingshanosaurus* and *Melanorosaurus*. The latter is currently under study by A. M. Yates (see Yates 2007); consequently, we have limited our comments on the skull of this taxon to existing published accounts (Welman 1999; Yates 2005). Hypodigms of other basal sauropod taxa (*Chinshakiangosaurus*, *Gongxianosaurus*, '*Kunmingosaurus*' and the 'Lufeng sauropod') include isolated tooth-bearing bones.

Cranial structure and function. The skull of *Jingshanosaurus* is identical to that of prosauropods in almost every respect (Text-fig. 6A). For example, the snout is elongate in lateral view, rostrally narrow (V-shaped) in dorsal view and the rostral end of the dentary forms a shallow, unexpanded symphysis. Unfortunately, it is not possible to determine the degree of cranial kinesis or the range of movement at the jaw joint on the basis of current data; however, the strong similarities between *Jingshanosaurus* and prosauropods suggest that cranial kinesis was absent and that jaw action was orthal, though this remains to be confirmed. The craniomandibular joint is ventrally offset with respect to the tooth row and the coronoid eminence is well developed, as in some prosauropods. The snout of *Melanorosaurus* is also V-shaped in dorsal view.

Other basal sauropods display a mosaic of features seen in more basal sauropodomorphs and eusauropods. In lateral view, the dentary of *Chinshakiangosaurus* tapers rostrally and terminates in a shallow, prosauropod-like symphysis. However, in dorsal view, the rostral portion of the dentary is bowed laterally; as a result, the snout of *Chinshakiangosaurus* would have been broad and U-shaped, as in eusauropods. In addition, *Chinshakiangosaurus* also possesses a prominent, vertical sheet of bone, termed the 'lateral plate' (*sensu* Upchurch 1995), which arises from the dorsolateral margins of the dentaries and ventrolateral margins of the premaxillae and maxillae, respectively, obscuring the bases of erupted tooth crowns in lateral view. This feature is unknown in prosauropods and *Jingshanosaurus*, but is present in all eusauropods (Upchurch 1995; Wilson 2002; Upchurch *et al.* 2004). It has been suggested that the lateral plate provided additional mechanical support to the bases of the tooth crowns, counteracting the outwardly directed forces on the teeth that would have been generated during feeding (Upchurch and Barrett 2000). Lateral plates are also present on the premaxilla of *Gongxianosaurus*, the maxilla of the 'Lufeng sauropod' and the dentary of '*Kunmingosaurus*' (Text-fig. 6B).

In addition to the lateral plate, the premaxilla of *Gongxianosaurus* possesses another eusauropod-like character: the elongate and subvertically orientated nasal process of the premaxilla. This contrasts with the situation in prosauropods, in which the nasal process is shorter and more horizontally directed (orientated at an angle of approximately 45 degrees to the horizontal in most cases). The structure of the nasal process in *Gongxianosaurus* suggests that the snout of basal sauropods was deeper than in prosauropods or basal sauropodomorphs. Also, the premaxilla of *Melanorosaurus* displays an incipient 'muzzle' (indicated by a slight inflexion of the nasal process of the premaxilla: Yates 2005), as in most eusauropods. Finally, the rostral end of the dentary in '*Kunmingosaurus*' is dorsoventrally expanded, producing a robust elongate symphysis that is significantly deeper than the remainder of the dentary ramus. This feature is absent in all prosauropods, but is present in eusauropods (e.g. Wilson and Sereno 1998; Upchurch *et al.* 2004). Unfortunately, no other information on cranial anatomy is available at present.

Dentition. As in basal sauropodomorphs and prosauropods, the tooth rows of basal sauropods extend for the entire length of the premaxilla/maxilla and dentary. This contrasts with the situation in sauropods more derived than *Shunosaurus*, in which the maxillary and dentary

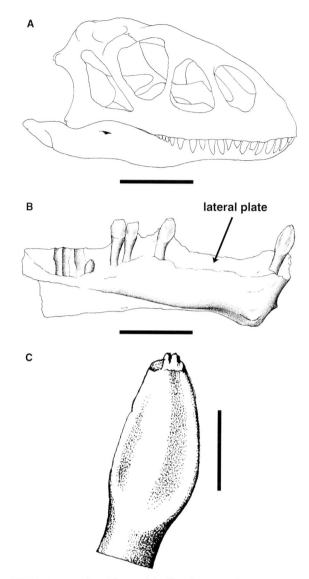

TEXT-FIG. 6. Cranial material of basal sauropods. A, reconstructed skull of *Jingshanosaurus xinwaensis* in right lateral view (based on Zhang and Yang 1994). B, right dentary of '*Kunmingosaurus wudingensis*' in medial view (based on Young 1966). C, tooth of '*Kunmingosaurus wudingensis*' in lingual view, showing the small number of apical denticles and the presence of the lingual concavity (based on Young 1966). Scale bars represent 100 mm in A, 50 mm in B and 10 mm in C.

tooth rows are abbreviated caudally and exhibit a corresponding reduction in tooth numbers (e.g. Wilson 2002; Upchurch *et al.* 2004). The complete dental formula of *Jingshanosaurus* included four premaxillary, 17–18 maxillary and 20–21 dentary teeth. The premaxillae of *Gongxianosaurus* possess four teeth; the maxilla of the 'Lufeng sauropod' contained a maximum of 16 teeth; and the dentaries of *Gongxianosaurus*, *Chinshakiangosaurus* and '*Kunmingosaurus*' contained 14–16, 19 and *c.* 20 teeth, respectively. These tooth counts fall towards the lower end of the range known in basal sauropodomorphs and prosauropods (see above): no sauropod possesses more than 21 maxillary or dentary teeth (Upchurch *et al.* 2004). Isolated teeth are also known in *Barapasaurus* and *Kotasaurus*.

In general, the teeth of basal sauropods are mesiodistally expanded with respect to the tooth crown, are arranged *en echelon* and decrease in height caudally, as in basal sauropodomorphs and prosauropods. The premaxillary and rostral maxillary teeth of *Jingshanosaurus* are strongly recurved, whereas those of the caudal maxilla and dentary lack recurvature. The teeth of *Melanorosaurus* are described as '... more theropod-like than those of *Thecodontosaurus* and *Plateosaurus*' (Welman 1999, p. 228), though it is not clear whether this statement refers to the presence of recurvature or the nature of the marginal serrations. Coarse marginal denticles are present on the dentary teeth of *Jingshanosaurus*, but are less prominent on the upper teeth. Marginal denticles are present on the unworn teeth of other basal sauropods where appropriate material is known (*Chinshakiangosaurus*, '*Kunmingosaurus*', *Barapasaurus*, *Kotasaurus* and *Gongxianosaurus*). In all cases (with the possible exception of *Jingshanosaurus*), the denticles are restricted to the apical half of the tooth crown and do not extend more basally (Text-fig. 6C); denticles are absent from some teeth of *Gongxianosaurus*.

The tooth enamel of almost all basal sauropods (including *Melanorosaurus*) exhibits a reticulate texture, as in eusauropods (cf. Wilson and Sereno 1998). This contrasts with the smooth enamel present in basal sauropodomorphs and prosauropods. Wrinkled enamel is present on the teeth of *Anchisaurus*, which is regarded as a basal sauropod by Yates (2004*a*). However, we regard this taxon as a prosauropod (see Upchurch *et al.* 2007), implying that the presence of wrinkled enamel may be convergent in *Anchisaurus* and sauropods.

The apices of basal sauropod teeth are inclined slightly lingually and the lingual surface of the tooth crown is mesiodistally concave. The latter feature (the 'lingual concavity'), in combination with a strongly convex labial crown surface, produces a D-shaped transverse cross-section through the crown and the 'spatulate' tooth morphology characteristic of sauropods (although this becomes modified in more derived taxa: Wilson and Sereno 1998; Upchurch 1998; Wilson 2002; Upchurch *et al.* 2004; Text-fig. 6C). Some basal sauropods (the 'Lufeng sauropod', '*Kunmingosaurus*', *Kotasaurus* and *Chinshakiangosaurus*) possess 'lingual ridges', which are apicobasally oriented and extend along the midline lingual concavity; these taxa also possess 'labial grooves' (apicobasally oriented furrows extending along the labial surface of the tooth crown). Labial grooves and lingual ridges are absent in prosauropods and basal sauropodomorphs, but are primitively present in all eusauropods

(though they are secondarily lost in some diplodocoids and advanced titanosaurs: Upchurch 1998; Wilson 2002; Upchurch *et al.* 2004). Unfortunately, the published descriptions of *Jingshanosaurus* and *Melanorosaurus* do not provide information on many of these tooth characters, so it is not clear if these features appeared in all sauropods or only in those forms more derived than *Jingshanosaurus* and *Melanorosaurus*.

Teeth referred to *Gongxianosaurus* possess small, subtriangular apical wear facets. The occurrence of similar wear facets on adjacent teeth suggests that they were produced by occlusion rather than tooth-food wear or breakage. This represents the earliest known evidence for tooth-tooth wear in Sauropoda, a feature previously thought to occur only in eusauropods (e.g. Upchurch and Barrett 2000). To date, tooth wear has not been reported in any other basal sauropod, but this may reflect the paucity of appropriate specimens rather than a genuine absence.

Soft anatomy. *Chinshakiangosaurus* possesses a marked ridge on the lateral surface of the dentary, similar to that of prosauropods, which has been interpreted as the site of attachment of a fleshy cheek (see above: Upchurch *et al.* in press). This feature is absent in all eusauropods (Upchurch and Barrett 2000), but its distribution among basal sauropods cannot be assessed owing to the lack of appropriate material. The elongate, highly ossified hyoid apparatus of *Melanorosaurus* indicates that at least one basal sauropod had a muscular tongue: the presence of a well-developed hyoid apparatus in many prosauropod and eusauropod taxa suggests that this feature characterized all sauropodomorphs. There is currently no evidence for a keratinous rhamphotheca in any basal sauropod or basal eusauropod.

Postcranial evidence

Unfortunately, no known basal sauropod specimen possesses a complete cervical series, limiting discussion of neck function. One example of relative neck length can be gained from *Jingshanosaurus*: the holotype specimen was discovered in articulation and, although the middle and posterior cervicals are badly weathered, their positions in the sequence can be determined from the remnants of the vertebrae and their disposition in the excavation. If it is assumed that the preserved positions of the cervical and cervicodorsal vertebrae approximate their positions *in vivo*, this yields an NL/TL ratio of 1·08, which is considerably higher than that of basal sauropodomorphs and prosauropods and closer to the range of relative neck lengths observed in eusauropods (NL/TL = 1·24–3·80: Upchurch and Barrett 2000). Many basal sauropods (*Jingshanosaurus*, *Melanorosaurus*, *Gong-*

xianosaurus, Antetonitrus, Camelotia and *Chinshakiangosaurus*) have amphicoelous centra, as in prosauropods and basal sauropodomorphs, but *Isanosaurus, Barapasaurus, Kotasaurus* and *Vulcanodon* possess opisthocoelous centra, as in eusauropods (Wilson and Sereno 1998; Upchurch 1998; Wilson 2002; Upchurch *et al.* 2004). The development of opisthocoely, with the presence of well-developed ball-and-socket joints between vertebrae, may have affected the degree of neck flexibility, though this suggestion requires testing. Where known (*Gongxianosaurus*), cervical ribs are short and do not extend beyond the caudal limit of the vertebra bearing them; consequently, they are unlikely to have hindered neck flexibility.

Complete manüs are known only in *Jingshanosaurus*, though some manual elements are present in *Vulcanodon* and *Antetonitrus*. The hands of *Jingshanosaurus* are very similar to those of prosauropods and basal sauropodomorphs. Metacarpal I has asymmetrical distal condyles, which would have permitted hyperextension and flexion of digit I, and digit I bears a trenchant ungual claw. However, in other respects the hand is sauropod-like (for example, many of the phalanges are subequal in length and width, as in sauropods, rather than being longer than wide, as in prosauropods). *Antetonitrus* possesses a prosauropod-like digit I; unfortunately, the rest of the manus is unknown in this taxon. In both taxa, it appears that the hand had other functions in addition to load-bearing, which may have been similar to those employed by prosauropods (see above). The metacarpals of *Vulcanodon* are much more elongate than those of *Jingshanosaurus*, *Antetonitrus* or prosauropods, a feature seen in eusauropods that contributes to the overall elongation of the forelimb (e.g. Coombs 1978). Digit V is enlarged in *Vulcanodon*, a feature that might represent its increased use as a load-bearing structure.

Jingshanosaurus and *Melanorosaurus* possess relatively short forelimbs and elongate trunks (Table 2), the same combination of features seen in the majority of prosauropods. This suggests that these taxa were facultatively bipedal rather than obligate quadrupeds (Van Heerden and Galton 1997). FL/HL ratios are much higher in all other basal sauropods (> 0·70: Table 2) and are indicative of obligate quadrupedality (Yates and Kitching 2003; Carrano 2005; Wilson 2005). The FL/HL ratios of these basal forms lie well within the range occupied by eusauropods (Upchurch and Barrett 2000; Wilson 2005). Eusauropods possess slightly lower HL/TL ratios than prosauropods, which also indicates a quadrupedal stance in these animals (Galton 1970, 1976). The shift to quadrupedality in sauropods more derived than *Melanorosaurus* correlates with increased body size and is also consistent with concurrent changes in manus structure (see above; Table 2; Carrano 2005).

Browse heights of basal sauropods would have overlapped those of larger prosauropods, with smaller taxa, such as *Jingshanosaurus* and *Melanorosaurus*, browsing at heights of up to 3 m. Large taxa, such as *Gongxianosaurus* and *Kotasaurus*, may have been capable of reaching vegetation 4–5 m above ground level. Late Triassic sauropods attained adult body sizes similar to those of contemporaneous prosauropods: prosauropods therefore can no longer be regarded as the first high-browsing herbivores as both clades radiated into this niche almost simultaneously (contra Bakker 1978).

DISCUSSION

Until recently a considerable morphological gap existed between basal sauropods and prosauropods/basal sauropodomorphs, obscuring the early evolution of Sauropoda (Wilson 2005). For example, many sauropod features, such as quadrupedal locomotion, were already well established in basal forms such as *Vulcanodon*, but the sequence of anatomical transformations that occurred could not be determined owing to the absence of intermediate taxa combining features seen in both sauropods and more primitive sauropodomorphs (Carrano 2005). However, discoveries of new basal sauropod taxa (e.g. *Isanosaurus* and *Antetonitrus*), the recognition that some 'prosauropods' are probably basal sauropods (e.g. *Camelotia* and *Lessemsaurus*) and re-examination of sauropodomorph phylogeny have reinvigorated the debates surrounding the origin of sauropods. Novel anatomical and phylogenetic data are shedding new light on the sequences of character acquisition for various sauropod functional complexes, such as locomotion (Yates and Kitching 2003) and feeding (Upchurch *et al.* 2007, in press). In particular, the recognition that *Jingshanosaurus* and *Melanorosaurus* are basal sauropods (see also Yates 2003*a*, 2004*a*; Yates and Kitching 2003) has several important implications for understanding the evolution of the sauropod *bauplan* (see below).

Consideration of the phylogenetic distribution of the varied feeding-related characters present in basal sauropodomorphs, prosauropods and sauropods reveals several broad trends in the evolution of sauropodomorph herbivory (Text-fig. 7: also Upchurch *et al.* in press), which can be categorized into five evolutionary 'grades':

Basal Sauropodomorpha. Primitively, all sauropodomorphs possess lanceolate, symmetrical, coarsely denticulate teeth that lack recurvature, high maxillary and dentary tooth counts, an orthal jaw action, akinetic skulls, a transversely narrow snout and fleshy cheeks. Postcranially, they share an elongated neck, a hyperextensible digit I on the manus bearing a trenchant claw, and a manus

TABLE 2. Postcranial features of sauropodomorph dinosaurs, including forelimb length : hindlimb length (FL/HL), hindlimb length : trunk length (HL/TL) and neck length : trunk length (NL/TL) ratios. Estimated total body length (BL) is also provided. Data taken from sources in Table 1. An asterisk (*) indicates BL estimates from Galton and Upchurch (2004). Two asterisks (**) indicate that forelimb length was calculated using ulna length.

Taxon	FL/HL	HL/TL	NL/TL	BL (m)
Basal sauropodomorphs				
Efraasia minor	0·60	0·78[1]	–	6·5*
Mussaurus patagonicus	0·71	–	–	3*[2]
Saturnalia tupiniquim	0·69**	0·80	–	1·5
Prosauropoda				
Anchisaurus polyzelus	0·69	0·60	–	2·5*
'Gyposaurus sinensis'	0·51**	0·79[3]	–	2·5
Lufengosaurus huenei	0·55	0·77	0·84	5
Massospondylus carinatus	0·51–0·71**[4]	–	–	5*
Plateosauravus cullingworthi	0·72			10*
Plateosaurus spp.	0·54	0·83	0·73	9*
Riojasaurus incertus	0·70–0·74	–	–	10
Yunnanosaurus huangi	0·49**	–	–	4·5
Basal (non-eusauropod) sauropods				
Antetonitrus ingenipes	0·84	–	–	8–10
Camelotia borealis	–	–	–	10*
Chinshakiangosaurus chunghoensis	–	–	–	12–13
Gongxianosaurus shibeiensis	0·70–0·75	–	–	14
Isanosaurus attavipachi	–	–	–	6·5
Jingshanosaurus xinwaensis	0·46	0·79	1·08[5]	8
Kotasaurus yamanpalliensis	0·79[6]	–	–	–
Melanorosaurus readi	0·63–0·64**[7]	–	–	7·5*
Vulcanodon karibaensis	0·77[8]	–	–	10

[1] Trunk length estimated on the basis of almost complete dorsal series.

[2] Adult length. It should be noted that the limb proportions of juveniles do not necessarily reflect those of adults (e.g. Reisz *et al.* 2005).

[3] Trunk length estimated on the basis of an almost complete dorsal series.

[4] Forelimbs of subadults and adults are relatively shorter than those of embryonic individuals (Reisz *et al.* 2005).

[5] See text for further explanation.

[6] Association between fore- and hindlimb elements requires confirmation. Previous estimates of NL/TL ratios in *Kotasaurus* (Upchurch and Barrett 2000) may have been based on a chimaera: associations between cervical vertebrae also require confirmation.

[7] Although an almost complete individual has been discovered it remains to be described (Galton *et al.* 2005). This ratio is based on measurements taken from two incomplete individuals whose remains were found mixed together; consequently, associations between the various limb elements require confirmation.

[8] Humeral and femoral lengths estimated (see Cooper 1984).

capable of grasping a variety of objects. Basal sauropodomorph taxa lack the following features: rhamphothecae; reticulate dental enamel; lateral plates on the tooth-bearing elements; a dorsoventrally expanded mandibular symphysis; and a precise occlusion. Most of these animals were small-bodied (< 3 m in adult body size). Unfortunately, lack of associated material prevents deduction of FL/HL ratios for some of these taxa. *Efraasia* has FL/HL and HL/TL ratios consistent with facultative bipedality whereas these metrics suggest that *Saturnalia* may have been quadrupedal. FL/HL proportions of *Mussaurus* fall within the range of quadrupedal taxa (Table 2), but as data are only available for hatchlings at present this result is difficult to interpret: limb proportions are known to vary ontogenetically, with embryonic and juvenile sauropodomorphs having much longer forelimbs than adults (Reisz *et al.* 2005). Nevertheless, outgroup comparisons and the prevalence of facultative bipedality in basal sauropodomorphs, prosauropods and basal sauropods indicate that bipedality is likely to have been the primitive condition for the clade as a whole. The combination of craniodental and postcranial character states present in these animals is consistent with a facultatively herbivorous or omnivorous diet (Barrett 2000).

Prosauropoda. Prosauropods retain all of the feeding-related features present in basal sauropodomorphs, but display a trend towards increased body size (ranging from

2·5 to 10 m in adult body length). The latter may have been an adaptation reflecting greater reliance on vegetation in the diet, though facultative omnivory may still have been important in some taxa (Barrett 2000; Yates 2004a). A larger body can accommodate a longer gut, facilitating slower passage times and permitting more time for enzymatic degradation: in addition, large size also results in a decrease in mass-specific metabolic rate, enabling subsistence on poor-quality, high-fibre vegetation (Farlow 1987). Increasing body size may have been particularly important in the evolution of obligate sauropodomorph herbivory as the anteriorly directed pubis would have precluded posterior elongation of the gut. FL/HL ratios indicate that the majority of prosauropods were facultative bipeds, though very large taxa (*Riojasaurus*) were probably quadrupedal.

Basalmost sauropods   (Jingshanosaurus, Melanorosaurus). FL/HL ratios indicate that the most basal sauropods were facultative bipeds; consequently, sauropods can no longer be regarded as an exclusively quadrupedal radiation. The majority of the character states that link *Jingshanosaurus* and *Melanorosaurus* with other sauropods relate to details of the axial and appendicular skeleton (Yates and Kitching 2003; Yates 2004a; Upchurch et al. 2007): most of the feeding-related characters present in these taxa are identical to those of prosauropods and basal sauropodomorphs. Known size ranges of the basal sauropods (6·5–10 m) overlap with those of prosauropods, suggesting similar browse heights. The retention of grasping hands, large manual unguals and recurved teeth suggests that at least some basal sauropods may have been facultatively omnivorous, though increased body sizes suggest that vegetation formed the majority of the diet (Barrett 2000). Some evidence (*Jingshanosaurus*) suggests that early sauropods had necks that were slightly more elongate than those of prosauropods/basal sauropodomorphs, permitting enlarged feeding envelopes, but neck elongation was not as marked as in eusauropods. Wrinkled tooth enamel characterizes all sauropods more derived than *Jingshanosaurus*, though the functional significance of this feature (if any) is unknown.

Non-eusauropod sauropods more derived than 'melanorosaurs'. Increases in relative forelimb length indicate that all of these taxa were quadrupedal. *Antetonitrus*, the most basal of these taxa, may have used the manus for grasping, but in more derived forms (*Vulcanodon*) the hand was primarily a weight-bearing structure. Various craniodental specializations of sauropods first appear in *Chinshakiangosaurus* and are retained in more derived taxa, including: spatulate tooth crowns with lingual ridges and labial grooves; reduced number of marginal denticles; a lateral plate (on the dentary, at least); and broader,

U-shaped jaws. Body size ranges from 8 to 13 m in these taxa, suggesting increased browse heights. Unfortunately, no information is available on neck length in these taxa. All of these features indicate obligate herbivory: the concurrent development of a broader snout, spatulate teeth and larger body size may be correlated with a shift to bulk-feeding on poor-quality vegetation (Upchurch et al. in press).

Eusauropoda. This grade is characterized by the appearance of additional craniodental and postcranial features associated with the processing and gathering of vegetation (Calvo 1994; Barrett and Upchurch 1995; Christiansen 1999, 2000; Upchurch and Barrett 2000). These features include: lateral plates on the premaxillae and maxillae (as well as the dentaries); development of occlusion; a reduction in the lengths of the maxillary and dentary tooth rows; dorsoventral expansion of the dentary symphysis; a decrease in snout length and an increase in snout height; marked increases in neck elongation (accomplished by the incorporation of additional cervicals: Upchurch 1994; Wilson and Sereno 1998); and numerous clade-specific features (Calvo 1994; Barrett and Upchurch 1995; Christiansen 1999, 2000; Upchurch and Barrett 2000; Sereno and Wilson 2005). Body size also continued to increase rapidly in some lineages (Yates 2004a; Carrano 2005). Cheeks were lost in eusauropods, perhaps reflecting larger oral gapes and/or involvement of the marginal dentition in food gathering, both of which may have been correlated with increased body size and necessary concurrent increases in assimilation rates (Upchurch et al. in press).

This survey demonstrates that many feeding-related characters previously thought to characterize Eusauropoda (e.g. Upchurch 1995, 1998; Wilson and Sereno 1998; Wilson 2002) were acquired in an incremental manner during the early stages of sauropod evolution (Text-fig. 7). *Jingshanosaurus*, *Melanorosaurus* and *Chinshakiangosaurus* help to bridge the morphological gap that previously existed between sauropods and more basal sauropodomorphs. Distributions of feeding-related characters on the phylogeny imply integration of increasing body size, quadrupedality and a shift towards obligate herbivory (as demonstrated by the development of skulls and dentitions with more 'herbivorous' adaptations, such as a U-shaped snout and lateral plate) in more derived members of the clade (Farlow 1987; Barrett 1998, 2000; Yates 2004a). This is suggestive of a macroevolutionary mode known as correlated progression (Thomson 1966, 1988), in which the development of one set of characters facilitates the origin and elaboration of related characters during the evolutionary history of a clade. Correlated progression posits that positive feedback loops reinforce the selective advantages of novel anatomical features or functions (Thomson

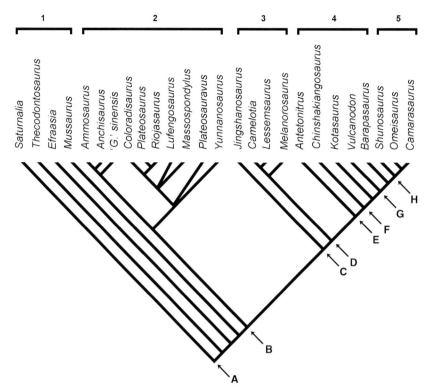

TEXT-FIG. 7. Feeding adaptations of sauropodomorphs mapped on to the phylogeny of Upchurch *et al.* (2007). Numbers along the top (1–5) refer to the various stages recognized in the evolution of sauropodomorph herbivory. Letters indicate nodes at which various innovations were acquired. A, Sauropodomorpha (lanceolate, coarsely denticulate teeth; cheeks; onset of neck elongation; grasping manus with hyperextensible, trenchant ungual). B, Anchisauria (onset of increasing body size). C, Sauropoda (further elongation of neck). D, unnamed clade (appearance of wrinkled tooth enamel). E, unnamed clade [quadrupedality (increased FL/HL ratio); further body size increase]. F, unnamed clade (dentary lateral plate; spatulate teeth with lingual concavity, lingual ridge and labial grooves; reduction in marginal denticles; U-shaped mandible). G, unnamed clade (manus adapted for weight-bearing). H, Eusauropoda (premaxillary and maxillary lateral plates; occlusion; expanded dentary symphysis; reduction in tooth count; loss of cheeks; increase in snout height, decrease in snout length; opisthocoelous cervical centra; further neck elongation; further body size increases; evolution of clade-specific feeding adaptations). See text for discussion.

1966, 1988; Kemp 1982; Lee 1996). For example, increased body size would lead to longer passage times through the gut, which in turn would allow greater reliance on coarse vegetation, which in turn would be favoured by selection for further adaptations to high-fibre herbivory (Barrett 1998; Yates 2004*a*). Correlated progression does provide a partial explanation for the development of sauropodomorph and ornithischian herbivory (Barrett 1997, 1998; Yates 2004*a*), such as the linkage between trends for large body size and more sophisticated craniodental feeding adaptations, but rigorous tests of this concept need to be developed in order to rule out (or at least identify) the possible influence of other macroevolutionary processes on the evolution of herbivory in these clades (e.g. character displacement, coevolution, competition).

Although the evolution of sauropod feeding was dominated by the elaboration of several major trends (cranial specializations to herbivory, increasing body size, quadrupedality and neck elongation), optimization of feeding characters on to the phylogeny of Upchurch *et al.* (2007) reveals high levels of homoplasy in the evolution of several important features: the same phenomena are encountered under both ACCTRAN and DELTRAN optimizations. For example, the evolution of exceptionally large body size (> 10 m) occurred on at least four independent occasions in Sauropodomorpha: in *Riojasaurus*, *Plateosauravus*, *Camelotia* and in the clade *Chinshakiangosaurus* + higher sauropods. In at least three of these cases (*Riojasaurus*, *Plateosauravus* and *Chinshakiangosaurus* + higher sauropods), this is correlated with the convergent acquisition of elongated forelimbs and presumed quadrupedality (see below). Increased body size in *Efraasia* also arose independently of the more general size increases that characterized Anchisauria. Interestingly, a possible reversal in the trend towards increased body size is also witnessed: *Melanorosaurus* is significantly smaller than other sauropod taxa. In general, it appears that sauropodomorphs were subject to strong selection pressures favouring size increase: as mentioned above, this may reflect their increasing reliance on poor-quality fodder and the need for more capacious guts.

Saturnalia, Anchisaurus, Plateosauravus, Riojasaurus and sauropods more derived than melanorosaurs were quadrupedal (on the basis of their high FL/HL ratios: Table 2). As these taxa are either distantly related to each other and/or are nested within clades that are primitively bipedal, each of these instances represents an independent acquisition of quadrupedal posture. A final example of homoplasy is found in the distribution of ventrally offset craniomandibular joints. Several previous analyses of prosauropod interrelationships (Galton 1990; Galton and Upchurch 2004) proposed that this feature was a synapomorphy uniting a clade consisting of *Efraasia, Coloradisaurus, Plateosaurus* and *Lufengosaurus*. However, this study (and those of Yates 2003*a*, 2004*a*) suggests that ventrally offset jaw joints appeared independently on several occasions within Sauropodomorpha, in *Sellosaurus*, Plateosauria and at the base of Sauropoda. Reversals to the primitive condition (jaw joint level with the maxillary tooth row) occur in *Riojasaurus* and *Massospondylus* (both in Plateosauria).

Gongxianosaurus has not yet been incorporated into a cladistic analysis: as a result, its phylogenetic position is uncertain (Upchurch *et al.* 2004). This is unfortunate, as many of its craniodental features (premaxillary lateral plate, high snout and presence of occlusion) are similar to those known in eusauropods (see above) and would represent some of the earliest appearances of these features in the fossil record. However, *Gongxianosaurus* lacks many of the synapomorphies that unite eusauropods and their close relatives, such as the presence of pleurocoels in the dorsal vertebrae and opisthocoelous cervical centra, and also retains several plesiomorphic features (e.g. an elongate caudolateral premaxillary process: He *et al.* 1998; Luo and Wang 1999). The combination of characters present in *Gongxianosaurus* currently suggests that it was a non-eusauropod more derived than *Chinshakiangosaurus*: establishing its phylogenetic position with more certainty may result in some of the craniodental features that currently define Eusauropoda moving down the basal sauropod lineage to define more inclusive clades.

A second Chinese taxon, *Yimenosaurus* (Fengjiahe Formation, Lower Jurassic, Yunnan: Bai *et al.* 1990), also has the potential to affect our scenarios of sauropodomorph craniodental evolution. Although originally described as a large prosauropod (up to 9 m in length: Bai *et al.* 1990; Galton and Upchurch 2004), several features (such as the retraction and enlargement of the external nares) indicate that it may be a basal sauropod. Unfortunately, the existing description does not provide enough anatomical data to incorporate this taxon into a phylogenetic analysis. Nevertheless, *Yimenosaurus* exhibits an interesting set of feeding-related features, including: high maxillary and dentary tooth counts; the possible presence of a premaxillary lateral plate; denticulate, spatulate teeth; and a ventrally offset jaw joint (Bai *et al.* 1990). Fuller descriptions of *Yimenosaurus* and *Gongxianosaurus* are likely to provide a great deal of new information on the early evolution of the sauropod feeding apparatus.

Archaeodontosaurus, from the Middle Jurassic (Bathonian) of Madagascar, is based on a partial right dentary containing several unerupted teeth (Buffetaut 2005). It possesses a lateral plate, a dorsoventrally expanded symphysis, a U-shaped mandible and teeth with wrinkled enamel. All of these characters support referral to Eusauropoda (Wilson 2002; Upchurch *et al.* 2004; see above). Buffetaut (2005) noted that the teeth retained a number of sauropodomorph symplesiomorphies, including coarse marginal denticles and the absence of a lingual concavity, reminiscent of the morphology present in prosauropods: this combination of primitive and derived character states led him to suggest that early sauropod feeding systems exhibited mosaic evolution. However, although the combination of dental and mandibular characters present in *Archaeodontosaurus* is currently unique, if this taxon is interpreted as a eusauropod the only unusual aspect of its anatomy relates to the lack of a lingual concavity: denticles are present on at least some teeth in many other eusauropods (including *Shunosaurus, Omeisaurus, Mamenchisaurus* and *Brachiosaurus*). Furthermore, a lingual concavity is also absent in *Cardiodon*, from the Middle Jurassic (Bathonian) of England, showing that this feature appeared in at least one other sauropod taxon (Owen 1840–45; Upchurch and Martin 2003). As the character states present in *Archaeodontosaurus* are otherwise entirely congruent with those in all other eusauropods, the absence of the lingual concavity is simply regarded as homoplasy herein, rather than evidence of mosaic evolution.

Basal sauropodomorphs/prosauropods and basal sauropods exhibited considerable geographical and temporal overlap: members of these clades/grades are frequently found in the same lithostratigraphical units (Weishampel *et al.* 2004). In addition, the trophic adaptations and body sizes of these two groups also overlapped, suggesting that they had similar ecological roles, although sauropods possessed more features indicative of obligate herbivory (see above). The species richness of prosauropods/basal sauropodomorphs declines during the Early Jurassic from its Late Triassic peak; conversely, sauropod diversity increases during the Early Jurassic (Barrett and Upchurch 2005). These observations have led to the suggestion that the taxa belonging to these two grades competed with each other for resources. Ultimately, the sauropods, with their more specialized feeding mechanisms, may have outcompeted prosauropods/basal sauropodomorphs, leading to the extinction of the latter (Barrett and Upchurch 2005).

CONCLUSIONS

Recent discoveries of basal sauropod material and reinterpretation of sauropodomorph phylogeny have provided many new insights into the evolution of sauropodomorph feeding and have permitted construction of a novel hypothesis to account for the evolution of herbivory in this clade. However, many aspects of the feeding complexes in these animals remain poorly understood, including neck function in basal sauropodomorphs/prosauropods and basal sauropods, and jaw mechanics in basal sauropods. In addition, although functional hypotheses have been proposed to account for the evolution of features such as the lateral plate and robust mandibular symphysis, these have yet to be tested using rigorous biomechanical approaches such as Finite Element Analysis (cf. Rayfield *et al.* 2001; Rayfield 2005). More detailed descriptions of key taxa (e.g. *Gongxianosaurus* and *Yimenosaurus*), application of quantitative biomechanical modelling approaches and continuing work on phylogenetic hypotheses, as well as the serendipitous discovery of new taxa, will all contribute to a more holistic view of feeding in this important dinosaur clade.

Acknowledgements. We thank the many people who provided access to the specimens included in this study: A. Arcucci (ULR), C. Chandler (YPM), Dong Zhi-Ming (IVPP), W.-D. Heinrich (MB), S. Kaal (SAM), C. Mehling (AMNH), M. A. Raath (BPI), C. Schaff (MCZ), W. Simpson (FMNH), Wang Xiao-Lin (IVPP), Wang Yuan (IVPP), Xu Xing (IVPP), Zhao Xi-Jin (IVPP), Zhou Xiao-Dan (NGMJ). R. Laws is thanked for producing the line drawings used in Text-figures 2A–B, 5 and 6. D. Maizels produced the original line art in Text-figure 2C. Our sincere thanks go to the Royal Society of London, NERC, The Jurassic Foundation, the Cambridge Philosophical Society and the Palaeontological Research Fund of the Natural History Museum for funding various parts of this work. We are grateful to M. F. Bonnan, D. M. Henderson and A. M. Yates for permission to cite their unpublished work, and E. Rayfield and R. Butler for discussion. Useful comments on an earlier version of this paper were received from D. B. Weishampel, M. Vickaryous and S. Modesto. Translations of Bai *et al.* (1990) and He *et al.* (1998) were produced by W. Downs and made available via the Polyglot Paleontologist website (http://ravenel.si.edu/palaeo/paleoglot/index.cfm).

REFERENCES

ATTRIDGE, J., CROMPTON, A. W. and JENKINS, F. A. Jr 1985. The southern African Liassic prosauropod *Massospondylus* discovered in North America. *Journal of Vertebrate Paleontology*, **5**, 128–132.

BAI ZI-QI, YANG JIE and WANG GUO-HUI 1990. *Yimenosaurus*, a new genus of Prosauropoda from Yimen County, Yunnan Province. *Yuxiwenbo (Yuxi Culture and Scholarship)*, **1**, 14–23. [In Chinese].

BAIRD, D. 1980. A prosauropod dinosaur trackway from the Navajo Sandstone (Lower Jurassic) of Arizona. 219–230. *In* JACOBS, L. L. (ed.). *Aspects of vertebrate history*. Museum of Northern Arizona Press, Flagstaff, AZ, 407 pp.

BAKKER, R. T. 1978. Dinosaur feeding behaviour and the origin of flowering plants. *Nature*, **274**, 661–663.

BARRETT, P. M. 1997. Correlated progression and the evolution of herbivory in the non-avian Dinosauria. *Journal of Vertebrate Paleontology*, **17** (Supplement to No. 3), 31A.

—— 1998. Herbivory in the non-avian Dinosauria. Unpublished PhD dissertation, University of Cambridge, 308 pp.

—— 1999. A sauropod dinosaur from the Lower Lufeng Formation (Lower Jurassic) of Yunnan Province, People's Republic of China. *Journal of Vertebrate Paleontology*, **19**, 785–787.

—— 2000. Prosauropods and iguanas: speculation on the diets of extinct reptiles. 42–78. *In* SUES, H.-D. (ed.). *Evolution of herbivory in terrestrial vertebrates: perspectives from the fossil record*. Cambridge University Press, Cambridge, 256 pp.

—— 2004. Sauropodomorph dinosaur diversity in the upper Elliot Formation (*Massospondylus* range zone: Lower Jurassic) of South Africa. *South African Journal of Science*, **100**, 501–503.

—— and UPCHURCH, P. 1994. Feeding mechanisms of *Diplodocus*. Gaia, **10**, 195–204.

—— —— 1995. Sauropod feeding mechanisms: their bearing on palaeoecology. 107–110. *In* SUN AI-LIN and WANG YUAN-QING (eds). *Sixth symposium on Mesozoic terrestrial ecosystems and biota. Short papers*. China Ocean Press, Beijing, 250 pp.

—— —— 2005. Sauropod diversity through time: possible macroevolutionary and palaeoecological implications. 125–156. *In* CURRY-ROGERS, K. A. and WILSON, J. A. (eds). *The sauropods: evolution and paleobiology*. University of California Press, Berkeley, CA, 349 pp.

—— —— and WANG XIAO-LIN 2005. Cranial osteology of *Lufengosaurus huenei* Young (Dinosauria: Prosauropoda) from the Lower Jurassic of Yunnan, People's Republic of China. *Journal of Vertebrate Paleontology*, **25**, 806–822.

—— —— ZHOU XIAO-DAN and WANG XIAO-LIN in press. The skull of *Yunnanosaurus huangi* Young (Dinosauria: Prosauropoda) from the Lower Lufeng Formation (Lower Jurassic) of Yunnan, China. *Zoological Journal of the Linnean Society*.

—— and YATES, A. M. 2006. New information on the palate and lower jaw of *Massospondylus* (Dinosauria: Sauropodomorpha). *Palaeontologia Africana*, **4**, 123–130.

BENTON, M. J., JUUL, L., STORRS, G. W. and GALTON, P. M. 2000. Anatomy and systematics of the prosauropod dinosaur *Thecodontosaurus antiquus* from the Upper Triassic of southwest England. *Journal of Vertebrate Paleontology*, **20**, 77–108.

BONAPARTE, J. F. 1972. Los tetrapodos del sector superior de la Formacion Los Colorados, La Rioja, Argentina (Triásico Superior). *Opera Lilloana*, **22**, 1–183.

—— 1978. *Coloradia brevis* n. g. et n. sp. (Saurischia, Prosauropoda), dinosaurio Plateosauridae de la Formacion Los Colorados, Triásico Superior de La Rioja, Argentina. *Ameghiniana*, **15**, 327–332.

—— and PUMARES, J. A. 1995. Notas sobre el primer craneo de *Riojasaurus incertus* (Dinosauria, Prosauropoda, Melanorosauridae) del Triásico Superior de La Rioja, Argentina. *Ameghiniana*, **34**, 341–349.

—— and VINCE, M. 1979. El hallazgo del primer nido de dinosaurios Triásicos, (Saurischia, Prosauropoda), Triásico Superior de Patagonia, Argentina. *Ameghiniana*, **16**, 173–182.

BONNAN, M. F. and SENTER, P. 2007. Were the basal sauropodomorph dinosaurs *Plateosaurus* and *Massospondylus* habitual quadrupeds? 139–155. *In* BARRETT, P. M. and BATTEN, D. J. (eds). *Evolution and palaeobiology of early sauropodomorph dinosaurs*. Special Papers in Palaeontology, **77**, 289 pp.

BUFFETAUT, E. 2005. A new sauropod dinosaur with prosauropod-like teeth from the Middle Jurassic of Madagascar. *Bulletin de la Société Géologique de France*, **176**, 467–473.

—— SUTEETHORN, V., CUNY, G., TONG HAIYAN, LE LOEUFF, J., KHANSUBHA, S. and JONG-AUTCHARIYAKUL, S. 2000. The earliest known sauropod dinosaur. *Nature*, **407**, 72–74.

CALVO, J. O. 1994. Jaw mechanics in sauropod dinosaurs. *Gaia*, **10**, 183–194.

CARRANO, M. T. 2005. The evolution of sauropod locomotion: morphological diversity of a secondarily quadrupedal radiation. 229–251. *In* CURRY-ROGERS, K. A. and WILSON, J. A. (eds). *The sauropods: evolution and paleobiology*. University of California Press, Berkeley, CA, 349 pp.

CHRISTIAN, A. and PREUSCHOFT, H. 1996. Deducing the body posture of extinct large vertebrates from the shape of the vertebral column. *Palaeontology*, **39**, 801–812.

CHRISTIANSEN, P. 1999. On the head size of sauropodomorph dinosaurs: implications for ecology and physiology. *Historical Biology*, **13**, 269–297.

—— 2000. Feeding mechanisms of the sauropod dinosaurs *Brachiosaurus*, *Camarasaurus*, *Diplodocus* and *Dicraeosaurus*. *Historical Biology*, **14**, 137–152.

COOMBS, W. P. Jr 1975. Sauropod habits and habitats. *Palaeogeography, Palaeoclimatology, Palaeoecology*, **17**, 1–33.

—— 1978. Theoretical aspects of cursorial adaptations in dinosaurs. *Quarterly Review of Biology*, **53**, 393–418.

COOPER, M. R. 1981. The prosauropod dinosaur *Massospondylus carinatus* Owen from Zimbabwe: its biology, mode of life and phylogenetic significance. *Occasional Papers of the National Museums and Monuments of Rhodesia, Series B, Natural Sciences*, **6**, 689–840.

—— 1984. A reassessment of *Vulcanodon karibaensis* Raath (Dinosauria: Saurischia) and the origin of Sauropoda. *Palaeontologia Africana*, **25**, 203–231.

CROMPTON, A. W. and ATTRIDGE, J. 1986. Masticatory apparatus of the larger herbivores during Late Triassic and Early Jurassic time. 223–236. *In* PADIAN, K. (ed.). *The beginning of the age of the dinosaurs*. Cambridge University Press, Cambridge, 378 pp.

—— and HIIEMÄE, K. 1969. How mammalian molar teeth work. *Discovery*, **5**, 23–34.

DODSON, P. 1990. Sauropod paleoecology. 402–407. *In* WEISHAMPEL, D. B., DODSON, P. and OSMÓLSKA,

H. (eds). *The Dinosauria*. University of California Press, Berkeley, CA, 733 pp.

DONG ZHI-MING 1992. *The dinosaurian faunas of China*. Springer-Verlag, Berlin, 188 pp.

FARLOW, J. O. 1987. Speculations about the diet and digestive physiology of herbivorous dinosaurs. *Paleobiology*, **13**, 60–72.

FIORILLO, A. R. 1998. Dental microwear patterns of the sauropod dinosaurs *Camarasaurus and Diplodocus*: evidence for resource partitioning in the Late Jurassic of North America. *Historical Biology*, **13**, 1–16.

GALTON, P. M. 1970. The posture of hadrosaurian dinosaurs. *Journal of Paleontology*, **44**, 464–473.

—— 1971. The prosauropod dinosaur *Ammosaurus*, the crocodile *Protosuchus*, and their bearing on the age of the Navajo Sandstone of north-eastern Arizona. *Journal of Paleontology*, **45**, 781–795.

—— 1973*a*. The cheeks of ornithischian dinosaurs. *Lethaia*, **6**, 67–89.

—— 1973*b*. On the anatomy and relationships of *Efraasia diagnostica* (Huene) n. gen., a prosauropod dinosaur (Reptilia: Saurischia) from the Upper Triassic of Germany. *Paläontologische Zeitschrift*, **47**, 229–255.

—— 1974. The ornithischian dinosaur *Hypsilophodon* from the Wealden of the Isle of Wight. *Bulletin of the British Museum (Natural History), Geology*, **25**, 1–152.

—— 1976. Prosauropod dinosaurs (Reptilia: Saurischia) of North America. *Postilla*, **169**, 1–98.

—— 1984*a*. Cranial anatomy of the prosauropod dinosaur *Plateosaurus*, from the Knollenmergel (Middle Keuper, Upper Triassic) of Germany. I. Two complete skulls from Trossingen/Württ. with comments on the diet. *Geologica et Palaeontologica*, **18**, 139–171.

—— 1984*b*. An early prosauropod dinosaur from the Upper Triassic of Nordwürttemberg, West Germany. *Stuttgarter Beiträge zur Naturkunde, Serie B*, **106**, 1–25.

—— 1985*a*. Cranial anatomy of the prosauropod dinosaur *Plateosaurus*, from the Knollenmergel (Middle Keuper, Upper Triassic) of Germany. II. All the cranial material and details of soft-part anatomy. *Geologica et Palaeontologica*, **19**, 119–159.

—— 1985*b*. Diet of prosauropods from the Late Triassic and Early Jurassic. *Lethaia*, **18**, 105–123.

—— 1985*c*. Cranial anatomy of the prosauropod dinosaur *Sellosaurus gracilis* from the Middle Stubensandstein (Upper Triassic) of Nordwürttemberg, West Germany. *Stuttgarter Beiträge zur Naturkunde, Serie B*, **118**, 1–39.

—— 1986. Herbivorous adaptations of Late Triassic and Early Jurassic dinosaurs. 203–221. *In* PADIAN, K. (ed.). *The beginning of the age of the dinosaurs*. Cambridge University Press, Cambridge, 378 pp.

—— 1990. Basal Sauropodomorpha–Prosauropoda. 320–344. *In* WEISHAMPEL, D. B., DODSON, P. and OSMÓLSKA, H. (eds). *The Dinosauria*. University of California Press, Berkeley, CA, 733 pp.

—— 1998*a*. The prosauropod dinosaur *Plateosaurus (Dimodosaurus) poligniensis* (Pidancet & Chopard, 1862) (Upper Triassic, Poligny, France). *Neues Jahrbuch für Geologie und Paläontologie, Abhandlungen*, **207**, 255–288.

—— 1998*b*. Saurischian dinosaurs from the Upper Triassic of England: *Camelotia* (Prosauropoda, Melanorosauridae) and *Avalonianus* (Theropoda, ?Carnosauria). *Palaeontographica Abteilung A*, **250**, 155–172, pls 24–27.

—— 2001. The prosauropod dinosaur *Plateosaurus* Meyer, 1837 (Saurischia: Sauropodomorpha; Upper Triassic). II. Notes on referred species. *Revue de Paléobiologie*, **20**, 435–502.

—— and BAKKER, R. T. 1985. The cranial anatomy of the prosauropod dinosaur '*Efraasia diagnostica*', a juvenile individual of *Sellosaurus gracilis* from the Upper Triassic of Nordwürttemberg, West Germany. *Stuttgarter Beiträge zur Naturkunde, Series B*, **117**, 1–15.

—— and UPCHURCH, P. 2004. Prosauropoda. 232–258. *In* WEISHAMPEL, D. B., DODSON, P. and OSMÓLSKA, H. (eds). *The Dinosauria*. Second edition. University of California Press, Berkeley, CA, 861 pp.

—— VAN HEERDEN, J. and YATES, A. M. 2005. Postcranial anatomy of referred specimens of the sauropodomorph dinosaur *Melanorosaurus* from the Upper Triassic of South Africa. 1–37. *In* TIDWELL, V. and CARPENTER, K. (eds). *Thunder-lizards: the sauropodomorph dinosaurs*. Indiana University Press, Bloomington, IN, 495 pp.

GAUTHIER, J. 1986. Saurischian monophyly and the origin of birds. *Memoirs of the Californian Academy of Sciences*, **8**, 1–55.

GOW, C. E., KITCHING, J. W. and RAATH, M. A. 1990. Skulls of the prosauropod dinosaur *Massospondylus carinatus* Owen in the collections of the Bernard Price Institute for Palaeontological Research. *Palaeontologia Africana*, **27**, 45–58.

GREAVES, W. S. 1998. The relative positions of the jaw joint and the tooth row in mammals. *Canadian Journal of Zoology*, **76**, 1203–1208.

HE XIN-LU, WANG CHANG-SHENG, LIU SHANG-ZHONG, ZHOU FENG-YUN, LIU TU-QIANG, CAI KAI-JI and DAI BING 1998. A new sauropod dinosaur from the Early Jurassic in Gongxian County, south Sichuan. *Acta Geologica Sichuan*, **18**, 1–6, pl. 1. [In Chinese, English abstract].

HENDERSON, D. M. in press. Burly gaits: centers of mass, stability and the trackways of sauropod dinosaurs. *Journal of Vertebrate Paleontology*.

HUENE, F. VON 1926. Vollständige osteologie eines plateosauriden aus dem schwäbischen Keuper. *Geologische und Palaeontologische Abhandlungen*, **15**, 139–179.

JAIN, S. L., KUTTY, T. S., ROY-CHOWDHURY, T. and CHATTERJEE, S. 1975. The sauropod dinosaur from the Lower Jurassic Kota Formation of India. *Proceedings of the Royal Society of London, Series B*, **188**, 221–228.

—— —— —— —— 1979. Some characteristics of *Barapasaurus tagorei*, a sauropod dinosaur from the Lower Jurassic of Deccan, India. 204–216. *In* LASKAR, B. and RAO, C. S. R. (eds). *IV international Gondwana symposium, Calcutta. Volume I*. Hindustan Publishing Corporation, Calcutta, 384 pp.

KEMP, T. S. 1982. *Mammal-like reptiles and the origin of mammals*. Academic Press, London, 363 pp.

KERMACK, D. 1984. New prosauropod material from South Wales. *Zoological Journal of the Linnean Society*, **82**, 101–117.

KING, G. M. and CLUVER, M. A. 1993. The aquatic *Lystrosaurus*: an alternative lifestyle. *Historical Biology*, **4**, 323–341.

LANGER, M. C. 2003. The pelvic and hind limb anatomy of the stem-sauropodomorph *Saturnalia tupiniquim* (Late Triassic, Brazil). *PaleoBios*, **23** (2), 1–40.

—— ABDALA, F., RICHTER, M. and BENTON, M. J. 1999. A sauropodomorph dinosaur from the Upper Triassic (Carnian) of southern Brazil. *Comptes Rendus de l'Academie des Sciences de Paris, Sciences de la Terre et des Planètes*, **329**, 511–517.

LEAL, L. A., AZEVEDO, S. A. K., KELLNER, A. W. A. and DA ROSA, Á. A. S. 2004. A new early dinosaur (Sauropodomorpha) from the Caturrita Formation (Late Triassic), Paraná Basin, Brazil. *Zootaxa*, **690**, 1–24.

LEE, M. S. Y. 1996. Correlated progression and the origin of turtles. *Nature*, **379**, 812–815.

LUCAS, S. G., HERNE, M. C., HECKERT, A. B., HUNT, A. P. and SULLIVAN, R. M. 2004. Reappraisal of *Seismosaurus*, a Late Jurassic sauropod dinosaur from New Mexico. *Geological Society of America, Abstracts with Programs*, **36** (5), 422.

LUO YAO-NAN and WANG CHANG-SHENG 1999. New discovery on dinosaur fossils from Early Jurassic, Sichuan, China. *Chinese Science Bulletin*, **44**, 2182–2188.

MAZZETTA, G. A., CHRISTIANSEN, P. and FARIÑA, R. 2004. Giants and bizarres: body size of some southern South American Cretaceous dinosaurs. *Historical Biology*, **16**, 71–83.

McINTOSH, J. S. 1990. Sauropoda. 345–401. *In* WEISHAMPEL, D. B., DODSON, P. and OSMÓLSKA, H. (eds). *The Dinosauria*. University of California Press, Berkeley, CA, 733 pp.

MOSER, M. 2003. *Plateosaurus engelhardti* Meyer, 1837 (Dinosauria: Sauropodomorpha) aus dem Feuerletten (Mittelkeuper; Obertrias) von Bayern. *Zitteliana (Reihe B, Abhandlungen der Bayerischen Staatssammlung für Paläontologie und Geologie)*, **24**, 3–186.

NORMAN, D. B. and WEISHAMPEL, D. B. 1991. Feeding mechanisms in some small herbivorous dinosaurs: processes and patterns. 161–181. *In* RAYNER, J. M. V. and WOOTTON, R. J. (eds). *Biomechanics in evolution*. Cambridge University Press, Cambridge, 273 pp.

OWEN, R. 1840–45. *Odontography, or a treatise on the comparative anatomy of the teeth: their physiological relations, mode of development and microscopic structure in the vertebrate animals. Parts I–III (two volumes)*. Hippolyte Balliere, London, xix + lxxiv + 665 pp. (Volume 1) and 37 pp., 168 pls (Volume 2).

PAUL, G. S. 1984. The segnosaurian dinosaurs: relics of the prosauropod–ornithischian transition? *Journal of Vertebrate Paleontology*, **4**, 507–515.

RAATH, M. A. 1972. Fossil vertebrate studies in Rhodesia: a new dinosaur (Reptilia: Saurischia) from near the Trias–Jurassic boundary. *Arnoldia*, **5** (30), 1–37.

—— 1974. Fossil vertebrate studies in Rhodesia: further evidence of gastroliths in prosauropod dinosaurs. *Arnoldia*, **7** (5), 1–7.

—— 1977. The anatomy of the Triassic theropod *Syntarsus rhodesiensis* (Saurischia: Podekesauridae) and a consideration of its functional biology. Unpublished PhD dissertation, Rhodes University, Grahamstown, 233 pp.

RAYFIELD, E. J. 2005. Using Finite-Element Analysis to investigate suture morphology: a case study using large carnivorous dinosaurs. *Anatomical Record*, **283A**, 349–365.

—— NORMAN, D. B., HORNER, C. C., HORNER, J. R., MAY SMITH, P., THOMASON, J. J. and UPCHURCH, P. 2001. Cranial design and function in a large theropod dinosaur. *Nature*, **409**, 1033–1037.

REISZ, R. R., SCOTT, D., SUES, H.-D., EVANS, D. C. and RAATH, M. A. 2005. Embryos of an Early Jurassic prosauropod dinosaur and their evolutionary significance. *Science*, **309**, 761–764.

SALGADO, L. and CALVO, J. O. 1997. Evolution of titanosaurid sauropods. II: the cranial evidence. *Ameghiniana*, **34**, 33–48.

SERENO, P. C. 1989. Prosauropod monophyly and basal sauropodomorph phylogeny. *Journal of Vertebrate Paleontology*, **9** (Supplement to No. 3), 38A.

—— 1991. *Lesothosaurus*, 'fabrosaurids', and the early evolution of Ornithischia. *Journal of Vertebrate Paleontology*, **11**, 168–197.

—— 1997. The origin and evolution of dinosaurs. *Annual Reviews of Earth and Planetary Sciences*, **25**, 435–489.

—— 1999. The evolution of dinosaurs. *Science*, **284**, 2137–2147.

—— and WILSON, J. A. 2005. Structure and evolution of a sauropod tooth battery. 157–177. *In* CURRY-ROGERS, K. A. and WILSON, J. A. (eds). *The sauropods: evolution and paleobiology*. University of California Press, Berkeley, CA, 349 pp.

—— FORSTER, C. A., ROGERS, R. R. and MONETTA, A. M. 1993. Primitive dinosaur skeleton from Argentina and the early evolution of Dinosauria. *Nature*, **361**, 64–66.

SHUBIN, N. H., OLSEN, P. E. and SUES, H.-D. 1994. Early Jurassic small tetrapods from the McCoy Brook Formation of Nova Scotia, Canada. 242–250. *In* FRASER, N. C. and SUES, H.-D. (eds). *In the shadow of the dinosaurs*. Cambridge University Press, Cambridge, 435 pp.

SIMMONS, D. J. 1965. The non-therapsid reptiles of the Lufeng Basin, Yunnan, China. *Fieldiana, Geology*, **15**, 1–93.

STEVENS, K. A. and PARRISH, J. M. 1999. Neck posture and feeding habits of two Jurassic sauropod dinosaurs. *Science*, **284**, 798–800.

—— —— 2005. Digital reconstructions of sauropod dinosaurs and implications for feeding. 178–200. *In* CURRY-ROGERS, K. A. and WILSON, J. A. (eds). *The sauropods: evolution and paleobiology*. University of California Press, Berkeley, CA, 349 pp.

SUES, H.-D., REISZ, R. R., HINIC, S. and RAATH, M. A. 2004. On the skull of *Massospondylus carinatus* Owen, 1854 (Dinosauria: Sauropodomorpha) from the Stormberg Group (Lower Jurassic) of South Africa. *Annals of the Carnegie Museum*, **73**, 239–257.

THOMSON, K. S. 1966. The evolution of the tetrapod middle ear in the rhipidistian-amphibian transition. *American Zoologist*, **6**, 379–397.

—— 1988. *Morphogenesis and evolution*. Oxford University Press, Oxford, 154 pp.

THULBORN, R. A. 1972. The post-cranial skeleton of the Triassic ornithischian dinosaur *Fabrosaurus australis*. *Palaeontology*, **15**, 29–60.

—— 1990. *Dinosaur tracks*. Chapman and Hall, London, 410 pp.

UPCHURCH, P. 1994. Sauropod phylogeny and palaeoecology. *Gaia*, **10**, 249–260.

—— 1995. The evolutionary history of sauropod dinosaurs. *Philosophical Transactions of the Royal Society of London, Series B*, **349**, 365–390.

—— 1998. The phylogenetic relationships of sauropod dinosaurs. *Zoological Journal of the Linnean Society*, **124**, 43–103.

—— and BARRETT, P. M. 2000. The evolution of sauropod feeding mechanisms. 79–122. *In* SUES, H.-D. (ed.). *Evolution of herbivory in terrestrial vertebrates: perspectives from the fossil record*. Cambridge University Press, Cambridge, 256 pp.

—— —— and DODSON, P. 2004. Sauropoda. 259–322. *In* WEISHAMPEL, D. B., DODSON, P. and OSMÓLSKA, H. (eds). *The Dinosauria*. Second edition. University of California Press, Berkeley, CA, 861 pp.

—— —— and GALTON, P. M. 2007. A phylogenetic analysis of basal sauropodomorph relationships: implications for the origin of sauropod dinosaurs. 57–90. *In* BARRETT, P. M. and BATTEN, D. J. (eds). *Evolution and palaeobiology of early sauropodomorph dinosaurs*. Special Papers in Palaeontology, **77**, 289 pp.

—— —— ZHAO XI-JIN and XU XING in press. A re-evaluation of *Chinshakiangosaurus chunghoensis* Ye *vide* Dong 1992 (Dinosauria, Sauropodomorpha): implications for cranial evolution in basal sauropod dinosaurs. *Geological Magazine*.

—— and MARTIN, J. 2003. The anatomy and taxonomy of *Cetiosaurus* (Saurischia, Sauropoda) from the Middle Jurassic of England. *Journal of Vertebrate Paleontology*, **23**, 208–231.

VAN HEERDEN, J. 1979. The morphology and taxonomy of *Euskelosaurus* (Reptilia: Saurischia; Late Triassic) from South Africa. *Navorsinge van die Nasionale Museum, Bloemfontein*, **4**, 21–84.

—— and GALTON, P. M. 1997. The affinities of *Melanorosaurus* – a Late Triassic prosauropod dinosaur from South Africa. *Neues Jahrbuch für Geologie und Paläontologie, Monatschefte*, **1997**, 39–55.

WEISHAMPEL, D. B., BARRETT, P. M., CORIA, R. A., LE LOEUFF, J., XU XING, ZHAO XI-JIN, SAHNI, A., GOMANI, E. M. P. and NOTO, C. R. 2004. Dinosaur distribution. 517–606. *In* WEISHAMPEL, D. B., DODSON, P. and OSMÓLSKA, H. (eds). *The Dinosauria*. Second edition. University of California Press, Berkeley, CA, 861 pp.

WELMAN, J. 1999. The basicranium of a basal prosauropod from the *Euskelosaurus* range zone and thoughts on the origin of dinosaurs. *Journal of African Earth Sciences*, **29**, 227–232.

WILSON, J. A. 2002. Sauropod dinosaur phylogeny: critique and cladistic analysis. *Zoological Journal of the Linnean Society*, **136**, 217–276.

—— 2005. Overview of sauropod phylogeny and evolution. 15–49. *In* CURRY ROGERS, K. and WILSON, J. A. (eds). *The sauropods: evolution and paleobiology*. University of California Press, Berkeley, CA, 349 pp.

—— and SERENO, P. C. 1998. Early evolution and higher-level phylogeny of the sauropod dinosaurs. *Memoir of the Society of Vertebrate Paleontology*, **5**, 1–68.

YADAGIRI, P. 1988. A new sauropod *Kotasaurus yamanpalliensis* from Lower Jurassic Kota Formation of India. *Records of the Geological Survey of India*, **116**, 102–127.

—— 2001. The osteology of *Kotasaurus yamanpalliensis*, a sauropod dinosaur from the Early Jurassic Kota Formation of India. *Journal of Vertebrate Paleontology*, **21**, 242–252.

YATES, A. M. 2003a. A new species of the primitive dinosaur *Thecodontosaurus* (Saurischia: Sauropodomorpha) and its implications for the systematics of early dinosaurs. *Journal of Systematic Palaeontology*, **1**, 1–42.

—— 2003b. The species taxonomy of the sauropodomorph dinosaurs from the Löwenstein Formation (Norian, Late Triassic) of Germany. *Palaeontology*, **46**, 317–337.

—— 2004a. *Anchisaurus polyzelus* (Hitchcock): the smallest known sauropod dinosaur and the evolution of gigantism among sauropodomorph dinosaurs. *Postilla*, **230**, 1–57.

—— 2004b. A definite prosauropod dinosaur from the Lower Elliot Formation (Norian: Upper Triassic) of South Africa. *Palaeontologia Africana*, **39**, 63–68.

—— 2005. The skull of the Triassic sauropodomorph, *Melanorosaurus readi*, from South Africa and the definition of Sauropoda. *Journal of Vertebrate Paleontology*, **25** (Supplement to No. 3), 132A.

—— 2007. The first complete skull of the Triassic dinosaur *Melanorosaurus* Haughton (Sauropodomorpha: Anchisauria). 9–55. *In* BARRETT, P. M. and BATTEN, D. J. (eds). *Evolution and palaeobiology of early sauropodomorph dinosaurs*. Special Papers in Palaeontology, **77**, 289 pp.

—— and KITCHING, J. W. 2003. The earliest known sauropod dinosaur and the first steps towards sauropod locomotion. *Proceedings of the Royal Society of London, Series B*, **270**, 1753–1758.

—— HANCOX, P. J. and RUBIDGE, B. S. 2004. First record of a sauropod dinosaur from the upper Elliot Formation (Early Jurassic) of South Africa. *South African Journal of Science*, **100**, 504–506.

YOUNG CHUNG-CHIEN 1941. A complete osteology of *Lufengosaurus huenei* Young (gen. et sp. nov.) from Lufeng, Yunnan, China. *Palaeontologica Sinica, Series C*, **7**, 1–53, pls 4–5.

—— 1942. *Yunnanosaurus huangi* Young (gen. et sp. nov.), a new Prosauropoda from the Red Beds at Lufeng, Yunnan. *Bulletin of the Geological Society of China*, **22**, 63–104.

—— 1948. Further notes on *Gyposaurus sinensis* Young. *Bulletin of the Geological Society of China*, **28**, 91–103, pls 1–5.

—— 1951. The Lufeng saurischian fauna. *Palaeontologica Sinica, Series C*, **13**, 1–96, pls 1–12.

—— 1966. On a new locality of the *Lufengosaurus* of Yunnan. *Vertebrata PalAsiatica*, **10**, 64–67. [In Chinese, English summary].

ZHANG YI-HONG and YANG ZHAO-LONG 1994. *A new complete osteology of Prosauropoda in Lufeng Basin, Yunnan, China*. Yunnan Publishing House of Science and Technology, Kunming, 100 pp., 24 pls. [In Chinese, English summary].

[Special Papers in Palaeontology 77, 2007, pp. 113–137]

THE PECTORAL GIRDLE AND FORELIMB ANATOMY OF THE STEM-SAUROPODOMORPH *SATURNALIA TUPINIQUIM* (UPPER TRIASSIC, BRAZIL)

by MAX C. LANGER*, MARCO A. G. FRANÇA* *and* STEFAN GABRIEL†

*FFCLRP, Universidade de São Paulo (USP), Av. Bandeirantes, 3900, Ribeirão Preto 14040–901, SP, Brazil; e-mails: mclanger@ffclrp.usp.br; marquinhobio@yahoo.com.br
†School of Biological and Chemical Sciences, Queen Mary University of London, Mile End Road, London E1 4NS, UK; e-mail: s.n.gabriel@qmul.ac.uk

Typescript received 24 February 2006; accepted in revised form 27 October 2006

Abstract: Description of the pectoral girdle (scapulocoracoid) and forelimb (humerus, radius and ulna) elements of two specimens of *Saturnalia tupiniquim*, a stem-sauropodomorph from the Upper Triassic Santa Maria Formation, southern Brazil, reveals a distinctive set of plesiomorphic, derived and unique traits, which shed light on the function and phylogenetic significance of these skeletal elements within early dinosaurs. Autapomorphic features of *S. tupiniquim* include, among others, an unusually long olecranon process of the ulna. Its function is still unclear, but it might have helped to sustain a quadrupedal gait, as inferred from the structure of the entire forearm.

Although less clear than previously suggested, some traits of *S. tupiniquim*, such as a long deltopectoral crest and a broad distal humeral end, are indicative of its sauropodomorph affinity. The taxon also bears several features previously regarded as autapomorphic of *Herrerasaurus ischigualastensis*, alluding to their broader distribution among basal dinosaurs. Variations within *S. tupiniquim* are mainly robustness-related and do not necessarily imply taxonomic distinctions.

Key words: *Saturnalia tupiniquim*, Dinosauria, Brazil, Triassic, pectoral girdle, forelimb, anatomy.

THE shoulder girdle and forelimb osteology of early dinosaurs is poorly known. Apart from the relatively abundant material referred to *Herrerasaurus ischigualastensis* (Reig 1963; Novas 1986; Brinkman and Sues 1987; Sereno 1993), and the still undescribed skeleton of *Eoraptor lunensis* (Sereno *et al.* 1993), most of the reported remains are incomplete. A nearly complete scapulocoracoid is part of the holotype of *Guaibasaurus candelariensis* (Bonaparte *et al.* 1999), but only scapula fragments and a dubious proximal humerus were assigned to *Staurikosaurus pricei* (Galton 2000; Bittencourt 2004). Within other putative Triassic dinosaurs, incomplete scapula and forelimb elements are among the material referred to *Saltopus elginensis* (von Huene 1910), *Spondylosoma absconditum* (Galton 2000) and *Agnosphitys cromhallensis* (Fraser *et al.* 2002).

In the austral summer of 1998, fieldwork conducted by the Museu de Ciências e Tecnologia, Pontifícia Universidade Católica do Rio Grande do Sul, collected three partial skeletons of a basal dinosaur in the red mudstone that typically crops out on the outskirts of Santa Maria (Text-fig. 1), in south Brazil (Langer *et al.* 1999; Langer 2005*a*). The material is only partially prepared, but a comprehensive description of the pelvis and hindlimb of *Saturnalia*

tupiniquim is available (Langer 2003). Among the other elements resulting from preparation of two of the skeletons are partial shoulder girdles and forelimbs, which are the subject of the present contribution.

Until now, because of the abundance of its material, *Herrerasaurus* has been the main basis on which the anatomy of the shoulder girdle and forelimb of basal dinosaurs was assessed. The constraint of using the condition in a single taxon, with its own set of derived and unique features, as almost the sole window on the plesiomorphic anatomy of a clade as diverse as the Dinosauria might lead to significant biases. The data available for *S. tupiniquim* is believed to alleviate this bias, adding morphological diversity to produce a better picture of the general anatomy of the shoulder girdle and forelimb of basal dinosaurs in general, and basal saurischians in particular.

Institutional abbreviations. BMNH, the Natural History Museum, London, UK; MACN, Museo Argentino de Ciencias Naturales 'Bernardino Rivadavia', Buenos Aires, Argentina; MB, Museum für Naturkunde, Berlin, Germany; MCN, Fundação Zoobotânica do Rio Grande do Sul, Porto Alegre, Brazil; MCP, Museu de Ciências e Tecnologia PUCRS, Porto Alegre, Brazil; PVL,

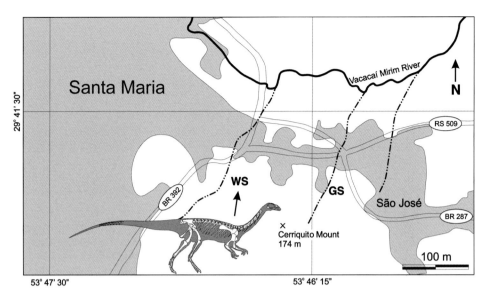

TEXT-FIG. 1. Map showing the fossil-bearing sites on the eastern outskirts of Santa Maria, Rio Grande do Sul, Brazil. Shaded parts indicate urban areas. GS: 'Grossesanga', type locality of *Staurikosaurus pricei* Colbert, 1970; WS: 'Waldsanga', type locality of *Saturnalia tupiniquim*.

Fundacíon Miguel Lillo, Tucumán, Argentina; PVSJ, Museo de Ciencias Naturales, San Juan, Argentina; QG, National Museum of Natural History, Harare, Zimbabwe; SMNS, Staatlisches Museum für Naturkunde, Stuttgart, Germany.

MATERIAL AND METHODS

The shoulder girdle and forelimb elements of the holotype of *Saturnalia tupiniquim* (MCP 3844-PV) include a nearly complete right scapulocoracoid, humerus and radius, and a right ulna lacking its distal third. Additional material belongs to one of the paratypes (MCP 3845-PV), and includes two nearly complete scapulocoracoids, a partial right humerus and the proximal portion of the right ulna. No carpals, metacarpals or phalanges have been recovered, and there is also no trace of any of the dermal elements of the shoulder girdle. If not explicitly mentioned, the described features and elements are shared by both specimens.

Directional and positional terms used herein are those defined in Clark (1993) and the dinosaur compendium of Weishampel *et al.* (2004). Considering the rather uncertain, although most probably oblique (Colbert 1989), orientation of the shoulder girdle in basal dinosaurs (Text-fig. 2), it is treated as vertical for descriptive purposes (Nicholls and Russell 1985). Accordingly, the coracoid lies ventral to its articulation with the scapula, whereas the scapula blade expands dorsally. This orientation is chosen, rather than one that more closely reflects avian anatomy (Ostrom 1974), because it is plesiomorphic for archosaurs (Romer 1956) and also more traditional

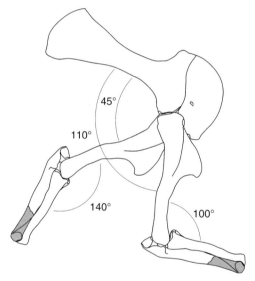

TEXT-FIG. 2. *Saturnalia tupiniquim*, Santa Maria Formation, Rio Grande do Sul, Brazil. Lateral view of the right pectoral girdle and partial forelimb reconstructed based mainly on MCP 3844-PV. Bones assembled in two different poses, corresponding to maximum angles of limb retraction and forearm extension, and limb protraction and forearm flexion. Shaded (see-through) area represents the missing distal part of the ulna, and scapulocoracoid long axis is at an angle of about 25 degrees to the horizontal.

(Romer 1966; Coombs 1978; Gauthier 1986). Regarding the forelimb, the arm and forearm are described with their long axes orientated vertically (Sereno 1993), and with the long axis of the elbow joint being orthogonal to the sagittal plane. This does not reflect their natural position (see Text-fig. 2), but should render the description easier to follow. Hence, the deltopectoral crest expands cranially from the humerus and the forearm moves caudally during extension.

The remains of *S. tupiniquim* are well preserved (Langer 2003) and lack evidence of major taphonomic distortions. This allows the recognition of various osteological traces left by the attachments of major muscle groups, and some insights on pectoral girdle and limb myology are presented here (Text-fig. 3). The tentative identification of the musculature corresponding to these traces represents inferences based on a 'phylogenetic bracket' approach (Felsenstein 1985; Bryant and Russell 1992; Witmer 1995; Hutchinson 2001). Crocodiles and birds are evidently the main elements of comparison, because they are the only extant archosaurs and the closest living relatives of *S. tupiniquim*.

SYSTEMATIC PALAEONTOLOGY

DINOSAURIA Owen, 1842
SAURISCHIA Seeley, 1887
EUSAURISCHIA Padian, Hutchinson and Holtz,1999
stem SAUROPODOMORPHA von Huene, 1932

Genus SATURNALIA Langer, Abdala, Richter and Benton,1999

Saturnalia tupiniquim Langer, Abdala, Richter and Benton, 1999
Text-figures 2–9, Tables 1–3

Referred specimens. The type series is composed of the holotype (MCP 3844-PV) and two paratypes (MCP 3845-PV and 3846-PV) (Langer 2003).

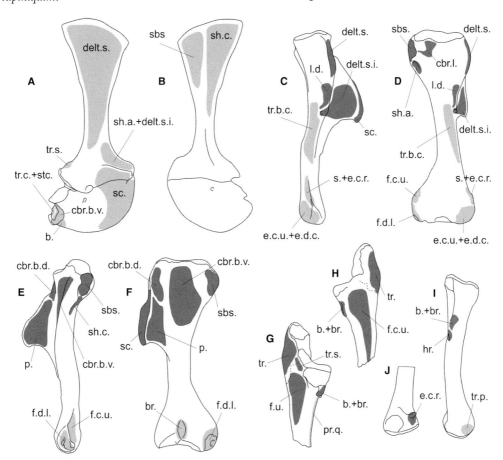

TEXT-FIG. 3. *Saturnalia tupiniquim*, Santa Maria Formation, Rio Grande do Sul, Brazil. Muscle attachment areas on A–B, right scapulocopracoid, C–F, humerus, G–H, partial ulna, I, radius, and J, distal part of radius, reconstructed based mainly on MCP 3844-PV, in lateral (A, C, G, J), medial (B, E, H–I), caudal (D), and cranial (F) views. Light shading indicates areas of muscle origin and dark shading their insertion areas. b., biceps; br., brachialis; cbr.b.d., coracobrachialis brevis pars dorsalis; cbr.b.v., coracobrachialis brevis pars ventralis; cbr.l., coracobrachialis longus; delt.s., deltoideous scapularis; delt.s.i., deltoideous scapularis inferior; e.c.r., extensor carpi radialis; e.c.u., extensor carpi ulnaris; e.d.c., extensor digitorum communis; f.c.u., flexor carpi ulnaris; f.d.l., flexor digitorum longus; f.u., flexor ulnaris; h., humeroradialis; l.d. latissimus dorsi; p. pectoralis; pr.q., pronator quadratus; s., supinator; sc., supracoracoideus; sbs., subscapularis; sh.a., scapulohumeralis anterior; sh.c., scapulohumeralis caudalis; stc., sternocoracoideus; tr., triceps tendon; tr.b.c., triceps brevis caudalis; tr.c., coracoidal head of triceps; tr.p., transvs. palmaris; tr.s., scapular head of triceps.

TABLE 1. Scapulocoracoid measurements of *Saturnalia tupiniquim* in millimetres. Brackets enclose approximate measurements, and inverted commas partial measurements taken from incomplete structures. Abbreviations: CGH, coracoidal glenoid dorsoventral height; CH, coracoid height on scapular axis; DSBB, distal scapula blade craniocaudal breadth; DSBT, distal scapula blade lateromedial thickness; MCGT, maximum coracoidal glenoid lateromedial thickness; MCL, maximum coracoid craniocaudal length; MCSB, maximum caput scapulae craniocaudal breadth; MSBB, minimum scapula blade craniocaudal breadth; MSBT, maximum scapula blade lateromedial thickness; MSGT, maximum scapula glenoid lateromedial thickness; MSL, maximum scapula length; SBL, scapula blade length; SGH, scapulaglenoid dorsoventral height; SPL, scapula prominence craniocaudal length.

	MCP 3844-PV (right)	MCP 3845-PV (right)	MCP 3845-PV (left
MSL	111	98	99
SBL	92	78	79
DSBB	41·5	39	–
MSBB	(14)	12·5	12·5
MSBT	8	6	7
DSBT	5	2	2·5
MCSB	45·5	43·5	45
SPL	18	23	22
SGH	7	10	9·5
MSGT	'12'	11·5	12
MCL	–	55	–
CH	33	23	'27'
CGH	10·5	8	7
MCGT	12·5	11	11·3

Type locality. All of the type series comes from the same locality: a private piece of land, no. 1845, on road RS-509; outskirts of the city of Santa Maria, on the north-western slope of Cerriquito Mount (Text-fig. 1). This is presumably the locality known as 'Waldsanga' (von Huene and Stahlecker 1931; Langer 2005*a*).

Horizon and age. Alemoa Local Fauna, *Hyperodapedon* Biozone (Barberena 1977; Barberena *et al.* 1985; Langer 2005*a*); Santa Maria 2 sequence (Zerfass *et al.* 2003); Alemoa Member, Santa Maria Formation, Rosário do Sul Group (Andreis *et al.* 1980); Late Triassic of the Paraná Basin. Based on comparisons with the Ischigualasto Formation (Rogers *et al.* 1993), the Alemoa Local Fauna can be given an early–middle Carnian age (Langer 2005*b*).

Revised diagnosis (based on pectoral girdle and forelimb elements only). A dinosaur that differs from other basal members of the group in a series of features, namely: oval pit on the caudal margin of the scapula blade, immediately dorsal to the glenoid border; central pit on the subglenoid fossa of the coracoid; oval excavation on the caudodistal corner of the lateral surface of the deltopectoral crest; marked fossa olecrani on the caudal surface of

the distal humeral end; greatly enlarged but partially hollow olecranon process of the ulna, with a separate ossification forming its proximocranial portion; pointed tuber on the craniolateral corner of the distal radius.

Comment. Most of these traits have also been identified in a few other dinosaurs (see descriptive section below) and so cannot be strictly defined as autapomorphic prior to assessing their phylogenetic distribution. These features might subsequently be shown to reflect either convergence or, most probably, excellent preservation of structures that are rarely preserved in the fossil record.

COMPARATIVE DESCRIPTION

Shoulder girdle

As is typical for dinosaurs, the shoulder girdle of *Saturnalia tupiniquim* (Text-figs 2–5, Table 1), includes a scapula and coracoid that are attached to each other by an immobile joint. They form a pair of long, lateromedially-flattened scapulocoracoids, which, as they follow the contour of the rib cage, are medially concave. The holotype right scapulocoracoid (Text-fig. 4) is complete except for the middle portion of the scapula blade and the cranioventral portion of the coracoid. The right scapulocoracoid of MCP 3845-PV (Text-fig. 5) is missing only a small central portion of the coracoid, while the left lacks the craniodorsal edge of the scapula blade and the cranial half of the coracoid. Although not preserved, clavicles and sternal plates were probably present, considering their occurrence in most dinosaur lineages (Bryant and Russell 1993; Padian 1997; Tykoski *et al.* 2002; Galton and Upchurch 2004*a*; Yates and Vasconcelos 2005).

The degree of fusion between the scapula and coracoid is similar in both holotype and paratype. The suture is clear in its caudal part, near the glenoid, where the coracoid seems to overlap the scapula laterally. Although partially fused cranially, the articulation is traceable for its entire length. Its caudal third extends cranioventrally as a nearly straight line from the glenoid until the level of the coracoid foramen, but deflects dorsally to project cranially as a slightly dorsally arched line. This defines a scapula margin that is more ventrally projected in the caudal portion, as commonly seen in basal dinosaurs (Bonaparte 1972; Welles 1984; Colbert 1989; Butler 2005; *Eoraptor*, PVSJ 514; *Guaibasaurus*, MCN 3844-PV; *Liliensternus*, HB R.1275; *Efraasia*, SMNS 17928). Adjacent to the scapulocoracoid suture, just dorsal to the coracoid foramen, is a bulging area that forms a marked tubercle in the holotype (Text-figs 4–5, ct). A similar structure also occurs in other dinosaurs (Ostrom 1974; Brinkman and Sues 1987; Butler 2005; *Efraasia*, SMNS 17928; see also Walker 1961), and is enlarged in some of them (Galton 1981, fig. 6A; *Liliensternus*, HB R.1275). This resembles, in shape and position, the ratite 'coracoid tuber' (Cracraft 1974), as figured for the ostrich by McGowan (1982, fig. 4E; 'acromial tuberosity'), which represents the origin of part of the deltoid musculature (Nicholls and Russell 1985). The whole scapulocoracoid junction of *S. tupiniquim* is bound by synchondral striations, more evident at the medial surface and laterally between the glenoid and the coracoid foramen. The scapular and coracoidal portions of the glenoid are nearly of the same size, but the latter projects further caudally. The scapulocoracoid is excavated at the cranial end of the articulation between the two bones. This is clearer in MCP 3845-PV, whereas a subtler concavity is present in the holotype. The significance of this excavation has been explored in the context of theropod phylogeny (Currie and Carpenter 2000; Holtz 2000; Holtz *et al.* 2004). Among basal dinosaurs, an excavation similar to that of *S. tupiniquim* is widespread (Colbert 1981; *Efraasia*, SMNS 17928; *Eoraptor*, PVSJ 514), and does not seem to bear an important phylogenetic signal (but see Tykoski and Rowe 2004).

Scapula. The scapula of *S. tupiniquim* is elongated, lateromedially flattened and arched laterally. It is formed of a slender dorsal blade and a basal portion (= caput scapulae; Baumel and Witmer 1993), along the ventral margin of which the coracoid articulates. The basal portion is composed of a lateromedially broad caudal column that extends onto the glenoid, and a plate-like cranial extension, the scapular prominence (='acromial process'; Nicholls and Russell 1985). This is convex medially, while

its concave lateral surface forms the 'preglenoid fossa' (Welles 1984; Madsen and Welles 2000) or 'subacromial depression' (Currie and Zhao 1993), which may have located part of the origin of the supracoracoid musculature (Coombs 1978; Nicholls and Russell 1985; Norman 1986; Dilkes 2000; Meers 2003). Dorsal to that, the 'preglenoid ridge' (Madsen and Welles 2000) extends caudally, but does not deflect ventrally as in forms with a deeper 'subacromial depression' (Madsen and Welles 2000; *Liliensternus*, HB R.1275; see also Brinkman and Sues 1987). Instead, the depression has a smooth caudal margin as in most basal dinosauromorphs. The 'preglenoid ridge' forms the entire dorsal margin of the acromion, which is thickened and striated, and was probably the origin of the m. scapulohumeralis anterior (Coombs 1978; Dilkes 2000). In addition, the acromial region represents the origin site for the avian mm. deltoideus major and deltoideus minor (George and Berger 1966; Vanden Berge 1975; McGowan 1982; Nicholls and Russell 1985), and the m. deltoideus clavicularis in crocodiles (Meers 2003; = m. deltoideus scapularis inferior: Nicholls and Russell 1985; Norman 1986) and some lizards (Romer 1922), and is probably also the origin for part of the deltoid musculature in *S. tupiniquim* (Text-fig. 3). The 'preglenoid ridge' is placed dorsal to the upper

TEXT-FIG. 4. *Saturnalia tupiniquim*, Santa Maria Formation, Rio Grande do Sul, Brazil; MCP 3844-PV. Photographs and outline drawings of right scapulocoracoid in A, lateral, B, caudal, C, medial, and D, cranial views. act, acrocoracoid tubercle; bo, origin of m. biceps; cf, coracoid foramen; ct, coracoid tuber; g, glenoid; hg, horizontal groove; msr, medial scapular ridge; pgf, preglenoid fossa; pgr, preglenoid ridge; sgb, subglenoid buttress; sgp, supraglenoid pit; sgr, subglenoid ridge; ss, striation of scapulocoracoid synchondrosis. Shaded areas indicate missing parts. Scale bars represent 20 mm.

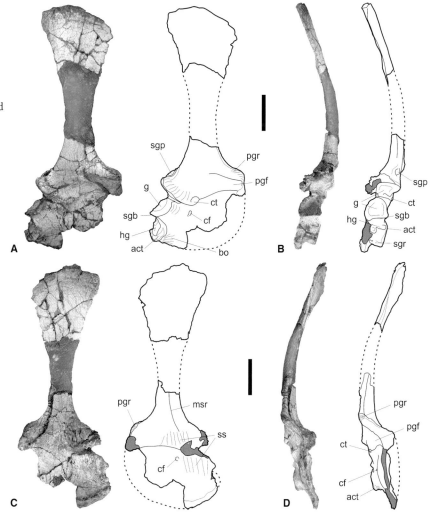

margin of the glenoid, a condition otherwise considered typical of herrerasaurs (Novas 1992), but also seen in other basal saurischians (Galton 1984; Rowe 1989; Raath 1990), although not in ornithischians (Owen 1863; Santa Luca 1980; Colbert 1981; Butler 2005).

Ventrally, the thickened caudal portion of the caput scapulae forms a subtriangular articulation with the coracoid and a broad glenoid. Dorsal to that, its caudal margin does not taper to a point, as does the majority of the scapula blade, but forms a flat caudomedially facing surface. The ridge that marks the lateral border of that surface is a ventral extension of the caudal margin of the blade, whereas the medial border is formed by a second ridge (Text-figs 4–5, msr) that extends along the medial surface of the blade, as seen also in *Herrerasaurus* (Sereno 1993). The distal part of this ridge may have separated the origins of the m. subscapularis cranially and the m. scapulohumeralis caudalis (= m. scapulohumeralis posterior, Dilkes 2000) caudally. Immediately dorsal to the glenoid border, an oval pit (Text-figs 4–5, sgp) lies at the end of the ridge extending from the caudal margin of the blade. A similar structure was reported for ornithomimosaurs (Nicholls and Russell 1985), *Heterodontosaurus* (Santa Luca 1980), and can be also seen in sauropods (e.g. *Barosaurus*: MB R.270.2 K34). This seems to represent the origin of a scapular branch of the m. triceps (Nicholls and Russell 1985; Brochu 2003), as this is usually immediately dorsal to the scapular part of the glenoid (George and Berger 1966; McGowan 1982) and often leaves a distinct scar (Meers 2003). Hence, in *S. tupiniquim* (Text-fig. 3), in contrast to the situation inferred for other dinosaurs (Borsuk-Bialynicka 1977; Coombs 1978; Norman 1986; Dilkes 2000), that muscle did not arise from the dorsal margin of the glenoid [i.e. the 'supraglenoid buttress' (Madsen and Welles 2000) or 'glenoid tubercle' (Norman 1986)]. Instead, the heavily striated surface, lateroventral to the oval pit, at the slightly projected laterodorsal border of the glenoid, is believed to represent an attachment area for ligaments of the shoulder joint. Indeed, this is the attachment site of the avian lig. scapulohumerale (Jenkins 1993) and the lepidosaurian caudodorsal ligament (Haines 1952). The scapular portion of the glenoid is ovoid in *S. tupiniquim* and meets the coracoid along its flat cranioventral margin, the medial part of which is more caudally projected. From that junction, its medial margin projects laterodorsally, and slightly caudally, while the lateral margin also projects caudodorsally, but diverges medially at its dorsal portion. As a consequence, the glenoid does not face strictly caudoventrally, but it is also directed somewhat laterally (MCP 3845-PV). This is typical of basal dinosaurs, although variants on the glenoid direction are seen in derived groups (Novas and Puerta 1997; Upchurch *et al.* 2004).

The scapula blade of the holotype lacks its middle portion, but has been safely reconstructed in length and shape based on the impression that the missing portion left in the matrix. It arches laterally, while the blade of MCP 3845-PV is more sinuous, with a straighter dorsal part, as seen in *Herrerasaurus* (Sereno 1993). Minor taphonomic distortions could, however, easily produce such a variation. In both specimens, the ventral portion of the blade is constricted to form the scapular neck, which is ovoid in cross-section. Dorsal to this, the bone becomes gradually thinner lateromedially, so that the distal end is plate-like. The blade also expands craniocaudally, so that its minimal breadth is less than half that of the dorsal margin. As in most basal dinosaurs (but see Welles 1984; Raath 1990), this expansion is neither abrupt nor restricted to the dorsal summit. A similar condition is seen in basal ornithischians (Owen 1863; Thulborn 1972; Santa Luca 1980), theropods (von Huene 1934; Rowe 1989) and sauropodomorphs (von Huene 1926; Benton *et al.* 2000), but not in *Herrerasaurus* (Sereno 1993) and more derived theropods (Currie and Zhao 1993; Currie and Carpenter 2000; Madsen and Welles 2000), in which the scapula blade is strap-shaped and does not expand much distally. Large areas of the scapula blade bear subtle longitudinal striations, which might correspond to origin areas of the m. subscapularis medially and the m. deltoideous scapularis laterally (McGowan 1982; Jenkins and Goslow 1983; Nicholls and Russell 1985; Dilkes 2000; Meers 2003). The dorsal margin of the blade is convex with sharp edges, and its more porous surroundings might indicate that it supported a cartilaginous extension (see Butler 2005).

Coracoid. The coracoid is a craniocaudally elongated element that is concave medially and convex laterally. Its cranial two-thirds are plate-like, with a subcircular cranioventral margin. Subtle craniocaudally directed striations are seen on its lateral surface, which seem to correspond to the origin of part of the m. supracoracoideus (Ostrom 1974; Coombs 1978; Nicholls and Russell 1985; Dilkes 2000). The coracoid thickens towards its caudal margin, and caudodorsally towards the glenoid. The scapular articulation is also cranially thin, and widens caudally, forming a subtriangular surface caudal to the 'coracoid tuber'. The coracoid foramen pierces the lateral surface of the bone well below the scapular articulation, and extends mediodorsally in an oblique fashion. In the holotype, the internal aperture is also below the scapular articulation (see also Santa Luca 1984; Sereno 1993), while in MCP 3845-PV it perforates the scapula-coracoid junction, forming a smooth excavation on the medioventral corner of the scapula (Text-fig. 5B), as reported for various basal dinosaurs (Norman 1986; Madsen and Welles 2000; Butler 2005).

The coracoidal portion of the glenoid is subrectangular and bears prominent lip-like lateral and caudal borders. The latter forms a delicate caudally projecting platform variously referred to as the 'horizontal' (Welles 1984), 'infraglenoid' (Kobayashi and Lü 2003) or 'subglenoid' (Madsen and Welles 2000) buttress (Text-figs 4–5, sgb). The flat to slightly concave humeral articulation faces almost entirely caudodorsally, and is not as laterally inclined as that of the scapula. Ventral to this, the coracoid bears a complex morphology. From near the lip-like caudolateral corner of the glenoid, but separated from it by a cleft, a short ridge extends cranioventrally along the lateral surface of the bone to meet a laterally extensive and craniodorsally to caudoventrally 'elongated tuber' (Text-figs 4–5, act). From the caudal end of that tuber, a blunt ridge projects medially forming a 'loop' (Text-figs 4–5, sgr) that reaches the medial margin of the bone. This supports a broad concave surface (= 'horizontal groove': Welles 1984, fig. 26b) with a deep pit at its centre, which is also seen in other archosaurs (Walker 1961; *Liliensternus*, HB R.1275).

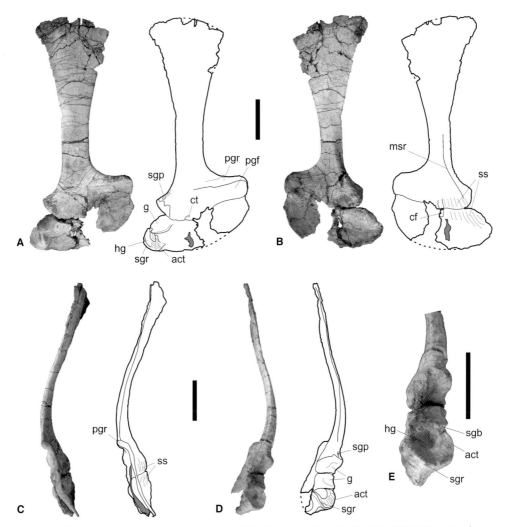

TEXT-FIG. 5. *Saturnalia tupiniquim*, Santa Maria Formation, Rio Grande do Sul, Brazil; MCP 3845-PV. A–D, photographs and outline drawings of right scapulocoracoid in A, lateral, B, medial, C, cranial, and D, caudal views. E, detail of glenoid area of right scapulocoracoid in caudal view. Abbreviations as in Text-figure 4. Shaded areas indicate missing parts. Scale bars represent 20 mm.

The above-mentioned 'elongated tuber' seems to be equivalent to a fainter ridge extending ventrally from the caudolateral corner of the glenoid of some archosaurs (Walker 1961, 1964; Long and Murry 1995; *Marasuchus*, PVL 3871), but a closer condition is shared by *Silesaurus* (Dzik 2003), *Guaibasaurus* (Bonaparte *et al.* 1999), *Eoraptor* (PVSJ 512) and basal sauropodomorphs (*Efraasia*, SMNS 17928), although the 'tuber' of the latter forms is often less expanded dorsally (Young 1941*a*, *b*, 1947; Moser 2003; Yates 2003; *Plateosaurus*, SMNS F65). This was referred to as the 'biceps tubercle' (Cooper 1981), whereas its ventral end was termed the 'caudolateral process of the coracoid' (Bonaparte 1972). The subglenoid part of the coracoid of basal theropods (Welles 1984; Madsen and Welles 2000; *Liliensternus*, HB R.1275; *Coelophysis rhodesiensis*, QG 1) also compares to that of *S. tupiniquim*, despite the suggestion of Holtz (2000) that the 'biceps tubercle' is more developed in *Dilophosaurus* and Coelophysidae than in 'prosauropods'. More derived theropods (Osmólska *et al.* 1972; Madsen 1976; Nicholls and Russell 1985; Makovicky and Sues 1998; Norell and Makovicky 1999; Brochu

2003) have a tuber placed further from the glenoid, and their dorsal 'concave surface' is more craniocaudally elongated. This follows an extension of the caudal process of the coracoid, as also seen in derived ornithischians (Gauthier 1986; Coria and Salgado 1996). Names applied to those structures vary: the tuber (= 'diagonal buttress': Welles 1984) has been termed 'coracoid' (Osmólska *et al.* 1972; Walker 1977; Norell and Makovicky 1999; Yates 2004) or 'biceps' (Ostrom 1974; Rowe 1989; Pérez-Moreno *et al.* 1994; Madsen and Welles 2000; Brochu 2003; Kobayashi and Lü 2003) tubercle, whereas the 'subglenoid fossa' (Norell and Makovicky 1999; Makovicky *et al.* 2005) seems to represent a caudally elongated version of the 'horizontal groove' (Welles 1984). In contrast, the coracoid of most ornithischians has a more plate-like subglenoid portion that apparently lacks those elements (Ostrom and McIntosh 1966; Colbert 1981; Forster 1990; Butler 2005; but see Janensch 1955; Santa Luca 1980).

The reconstruction of dinosaur coracoid musculature has been an issue of some debate (Ostrom 1974; Walker 1977), leading to the nomenclatural inconsistency seen above. In previous works, the origins of the m. biceps and m. coracobrachialis have been reconstructed according to two different patterns. Some authors (Ostrom 1974; Nicholls and Russell 1985; Dilkes 2000) favoured origins restricted to the subglenoid portion of the bone, while in other reconstructions (Borsuk-Bialynicka 1977; Coombs 1978; Norman 1986; Bakker *et al.* 1992; Carpenter and Smith 2001) these spread along most of the ventral half of the coracoid. Comparisons to the myology of ratites and crocodiles seem to favour the first hypothesis, given that those two muscles originate on the caudal portion of their coracoid (McGowan 1982; Meers 2003), and that the origin of the m. biceps is consistently ventral to that of the m. coracobrachialis. Indeed, the 'elongated tuber' of *S. tupiniquim* is suggested to accommodate the origin of the latter (Text-fig. 3), most probably its cranial (= brevis) branch, which may extend onto the 'concave surface'. This corresponds to the 'depression on the dorsal edge of the posterior coracoid process' where Nicholls and Russell (1985, p. 669) also placed the origin of the m. coracobrachialis brevis in *Struthiomimus*. In such forms, a caudal (= longus) branch of the m. coracobrachialis might originate from their elongated caudal coracoid process. This is lacking in *S. tupiniquim*, but the oval pit and medial part of its 'concave surface' can be related to the origin of a coracoidal head of the m. triceps (Norman 1986; Dilkes 2000; Brochu 2003). In birds, the impressio m. sternocoracoidei lies in this region of the bone (George and Berger 1966; McGowan 1982; Baumel and Witmer 1993; Vanden Berge and Zweers 1993), so the 'concave surface' may also represent the insertion of the eponymous muscle (Vanden Berge and Zweers 1993; = m. costocoracoideus, Meers 2003). The m. biceps, on the other hand, might have originated from a rugose bump ventral to the 'elongated tuber' (Text-fig. 4, bo). Accordingly, this could be tentatively considered equivalent to the 'coracoid' or 'biceps' tubercle of theropods, which is inferred to accommodate the origin of the m. biceps (Nicholls and Russell 1985; Brochu 2003; contra Walker 1977; Norell and Makovicky 1999), but was also considered to represent an 'artefact' of bone growth, related to the convergence of three muscle masses (Carpenter 2002). In any case, the entire 'elongated tuber' of *S. tupiniquim* resembles, in shape and position, the 'acrocoracoid tuberosity' of ratites (Parker 1891; McGowan 1982), which is related to the origin of the mm. coracobrachialis and biceps. Indeed, the origin of these muscles is often so intimately associated (McGowan 1982; Nicholls and Russell 1985) that the search for their exact origin in dinosaurs might prove very difficult.

Forelimb

Humerus. The humerus of 3845-PV (Text-fig. 7) lacks most of the deltopectoral crest and the lateral half of the proximal articulation, whereas only the centre of the deltopectoral crest and part of the medial tuberosity is missing in the holotype (Text-fig. 6). Manipulation of the humerus on the caudolaterally facing glenoid of *S. tupiniquim* reveals a resting pose (with scapulocoracoid positioned parasagittally) in which the bone is abducted about 20 degrees. It reaches maximal protraction and retraction of about 70 and 45 degrees relative to the long axis of the scapulocoracoid, respectively (Text-fig. 2), allowing an arm rotational

movement of 65 degrees. The humerus of the holotype is bowed cranially along its proximal two-thirds and caudally in its distal half, while that of MCP 3845-PV is somewhat straighter with the proximal half bent caudally at an angle of 20 degrees, and the distal end curved cranially. Both arrangements give the bone a sigmoid outline, as is typical of basal dinosaurs (Rauhut 2003), resulting in a permanent minor retraction. The relatively short humeral shaft connects lateromedially expanded distal and proximal ends, the margins of which are also craniocaudally expanded. The bone is therefore markedly waisted in cranial-caudal view, with a medial excavation extending through the entire length of the shaft, and a lateral one distal to the deltopectoral crest.

The proximal surface of the humerus is almost entirely occupied, except for its caudolateral and medial corners (Text-fig. 6E), by the humeral head (= caput articulare humeri; Baumel and Witmer 1993). This includes a broad, lateromedially elongated medial body, and a narrower lateral portion (= 'ectotuberosity': Welles 1984) that projects craniolaterally at an angle of about 35 degrees. As a result, the head has a 'bean-shaped' proximal outline that is caudolaterally rounded and excavated craniomedially. It articulated with the glenoid via a slightly caudally facing flat proximal surface, which is crossed by a shallow transverse groove and probably had a cartilaginous cover. Taken as a whole, the long axis of the humeral head forms an angle of approximately 30 degrees to that of the distal part of the bone, but the angle is merely 10 degrees if only the larger medial part of the head is considered. These account for the so called 'humeral torsion' (Raath 1969; Cooper 1981; Benton *et al.* 2000; Tykoski and Rowe 2004), which imposes a permanent 'supination' to the distal part of the bone. This is clearly seen in basal theropods (Welles 1984; *Coelophysis rhodesiensis*, QG 1; *Liliensternus*, HB R.1275) and 'prosauropods' (Moser 2003; Galton and Upchurch 2004a; *Riojasaurus*, PVL 3808), but is apparently more marked in the former group (Holtz 2000).

An indistinct trough separates the humeral head of *S. tupiniquim* from the medial/internal tuberosity (= tuberculum ventrale, Baumel and Witmer 1993; for alternative names, see Welles 1984; Nicholls and Russell 1985; Moser 2003). This corresponds to the insisura capitis humeri (Baumel and Witmer 1993), and is not as broad as in other basal dinosaurs (Raath 1969; Cooper 1981; Sereno 1993). The swollen and proximodistally elongated medial tuberosity (Text-fig. 7, mt) forms the medial margin of the proximal humerus (MCP 3845-PV), but does not rise proximally as in *Herrerasaurus* (Sereno 1993). It has a rugose texture that also enters the cranial surface of the bone, representing the insertion of the m. subscapularis (Ostrom 1969; Vanden Berge 1975; Coombs 1978; Meers 2003). The medial tuberosity gives rise to a sharp crista bicipitalis (Baumel and Witmer 1993; see Carpenter *et al.* 2005) extending distally along the medial corner of the humerus, the caudal surface of which might represent the insertion of the m. scapulohumeralis caudalis (Vanden Berge and Zweers 1993; Dilkes 2000). An oval pit is seen caudal to the crest (Text-fig. 7, ftp), which is comparable to the avian fossa pneumotricipitalis (Baumel and Witmer 1993). This was probably the origin of the medial head of the m. humerotriceps (= m. triceps brevis caudalis; Meers 2003) and the insertion of the m. scapulohumeralis anterior (Dilkes 2000; = m. scapulohumer-

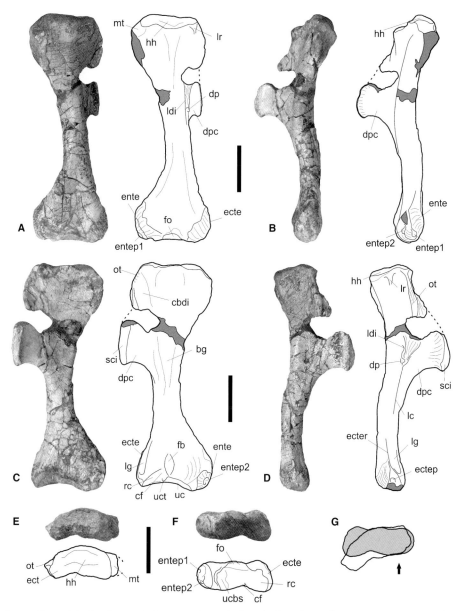

TEXT-FIG. 6. *Saturnalia tupiniquim*, Santa Maria Formation, Rio Grande do Sul, Brazil; MCP 3844-PV. A–F, photographs and outline drawings of right humerus in A, caudal, B, medial, C, cranial, D, lateral, E, proximal, and F, distal views. G, relative position of proximal and distal ends of the right humerus (arrow points caudally). bg, biceps gutter; cbdi, insertion of m. coracobrachialis brevis dorsalis; cf, cranial furrow; dp, deltoid pit; dpc, deltopectoral crest; ect, ectotuberosity; ecte, ectepicondyle; ectep, ectepicondyle pit; ecter, ectepicondyle ridge; ente, entepicondyle; entep1 and 2, entepicondyle pit 1 and 2; fb, fossa m. brachialis; fo, *fossa olecrani*; hh, humeral head; lc, lateral carina; ldi, insertion of m. latissimus dorsi; lg, ligament groove; lr, ligament ridge; mt, medial tuberosity; ot, outer tuberosity; rc, radial condyle; sci, insertion of m. supracoracoideus; uc, ulnar condyle; ucbs, ulnar condyle biconvex surface; uct; tubercle on ulnar condyle. Shaded areas indicate missing parts. Scale bars represent 20 mm.

alis cranialis, Vanden Berge 1975; see also Ostrom 1969). Lateral to that, the caudal surface of the proximal humerus forms a slightly concave smooth surface that somewhat continues to the 'capital groove'. This is laterally bound by a blunt ridge that defines a protruding lip-like border on the humeral head (MCP 3845-PV) and extends distally, in the direction of the ectepicondyle, as seen in *Scutellosaurus* (Colbert 1981, fig. 19a). The proximal part of the ridge is covered by a finely striated surface

(MCP 3844-PV) that forms a loop, extending medially until the base of the medial tuberosity (Text-fig. 3D), and possibly represents the insertion of m. coracobrachialis longus (Dilkes 2000; = caudalis, Vanden Berge 1975). More laterally, an ovoid depression (MCP 3845-PV) and a short, but rugose, ridge (MCP 3844-PV) mark the caudolateral margin of the humeral head and might be related to the caudodorsal ligaments of the shoulder joint (Haines 1952). On the cranial surface of the proximal humerus, a shallow excavation projects distally from the concavity of the humeral head. Its smooth and longitudinally striated surface extends medially, approaching the margin of the bone, and probably accommodated the insertion of the m. coracobrachialis brevis (Dilkes 2000; = cranialis, Vanden Berge 1975) pars ventralis (Meers 2003).

The lateral border of the proximal humerus is formed by a sharp ridge that expands from the craniolateral corner of the humeral head, at an angle of 45 degrees to the long axis of the distal end of the bone, and extends distally. Such a ridge is widespread among dinosaurs (Ostrom and McIntosh 1966; Cooper 1981; *Coelophysis rhodesiensis*, QG 1; *Liliensternus*, HB R.1275; *Plateosaurus*, SMNS F65; *Riojasaurus*, PVL 3808). Its rugose proximal portion ('outer tuberosity', Godefroit *et al.* 1998; 'greater tubercle', Madsen and Welles 2000; Carrano *et al.* 2002; tuberculum majus *sic*, Moser 2003) seems equivalent to the avian

tuberculum dorsale (Baumel and Witmer 1993), which receives the insertion of the m. deltoideus minor (Vanden Berge 1975; Vanden Berge and Zweers 1993). In 'reptiles', the m. deltoideus scapularis has a comparable insertion point lateral to the humeral head (Nicholls and Russell 1985; Dilkes 2000; Meers 2003), although it takes its origin from the scapula blade, while the m. deltoideus minor originates in the acromial area. Moreover, both the m. deltoideus scapularis and m. deltoideus minor lie deep to the m. deltoideus clavicularis and m. deltoideus major in crocodiles (Meers 2003) and birds (Vanden Berge 1975), respectively. In the case that they represent homologues, the shift of the origin of the m. deltoideus minor to a more proximal portion of the scapula might have been necessary if the muscle was to carry on acting as a forelimb abductor on the horizontally orientated avian scapulocoracoid. In *S. tupiniquim*, this ridge becomes less prominent distally and might correspond to the origin of the

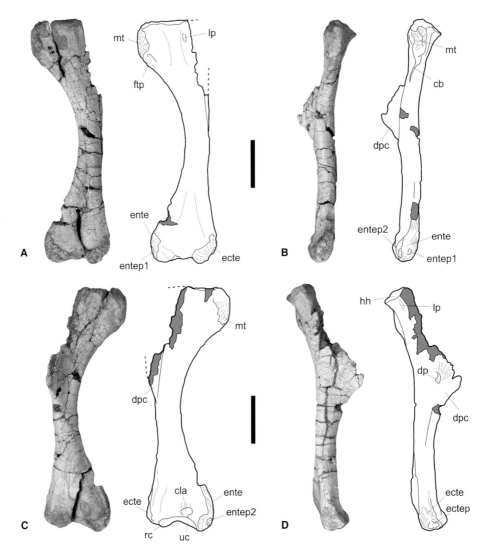

TEXT-FIG. 7. *Saturnalia tupiniquim*, Santa Maria Formation, Rio Grande do Sul, Brazil; MCP 3845-PV. Photographs and outline drawings of right humerus in A, caudal, B, medial, C, cranial, and D, lateral views. Abbreviations as in Text-figure 6 and: cb, crista bicipitalis; cla, attachment of collateral ligament; ftp, fossa tricipitalis; lp, ligament pit. Shaded areas indicate missing parts. Scale bars represent 20 mm.

m. triceps brevis cranialis, as described for crocodiles (Meers 2003). Its medial margin is heavily ornamented with pits and tubers, representing a likely insertion area for the m. latissimus dorsi (Borsuk-Bialynicka 1977; Dilkes 2000; Brochu 2003; Meers 2003). This portion of the ridge was described for tetanurans as related to the m. humeroradialis (Madsen 1976; Galton and Jensen 1979; Azuma and Currie 2000; Currie and Carpenter 2000; but see Carpenter and Smith 2001; Brochu 2003; Carpenter *et al.* 2005), but in *S. tupiniquim* that muscle probably had a more distal origin, near the margin of the deltopectoral crest. The distal end of the ridge under description is marked by an oval pit (Text-figs 6–7, dp) from which marked striae radiate proximocranially (MCP 3845-PV), partially representing the insertion of the deltoid musculature (see below). This is somewhat continuous (MCP 3844-PV) with an intermuscular line (= 'lateral carina'; Cooper 1981, fig. 26) that extends along the lateral margin of the shaft. In certain reconstructions (Norman 1986), a similar line outlines the boundary between the m. brachialis laterally, and a humeral branch of the m. triceps medially.

The deltopectoral crest is the most prominent element of the proximal humerus, but it is not continuous with the humeral head. It rises from the proximal part of the previously described ridge at an angle of 90 degrees to the long axis of the distal end of the humerus, and arches medially, before flaring laterally. This is a muted version of the medial inflection of the crest defined by Yates (2003) for some 'prosauropods' and is also seen in other members of the group (Cooper 1981), but not in theropods (Welles 1984; Madsen and Welles 2000; *Coelophysis rhodesiensis*, QG 1; *Liliensternus*, HB R.1275) or *Herrerasaurus*

TABLE 2. Right humerus measurements of *Saturnalia tupiniquim* in millimetres. Brackets enclose approximate measurements, and inverted commas partial measurements taken from incomplete structures. Abbreviations: DCL, deltopectoral crest length; DW, distal width across condyles; ET, entepicondyle maximum craniocaudal thickness; LDC1, length from proximal margin to apex of deltopectoral crest; LDC2, length from distal margin to distal base of deltopectoral crest; ML, maximum length; MPW, maximum proximal lateromedial width; MPT, maximum proximal craniocaudal thickness; MWPA, maximum lateromedial width of proximal articulation; RCT, radial condyle maximum craniocaudal thickness; SB, craniocaudal shaft breadth; UCT, ulnar condyle maximum craniocaudal thickness.

	MCP 3844-PV	MCP 3845-PV
ML	97	98
MPW	'32·5'	–
MWPA	28	–
MPT	14	11
DCL	33·5	–
LDC1	43	(47)
LDC2	50	51
SB	10·5	8·5
DW	33	28
ET	11·5	8
RCT	13	10
UCT	12	11

(MACN 18.060; but see Brinkman and Sues 1987). The crest in *S. tupiniquim* attains its maximum expansion and robustness near its distal margin, where it forms an angle of 60 degrees to the long axis of the distal end of the bone. In lateral view, it has a truncated distal end, with a hook-like cranial corner, but merges smoothly onto the shaft. Its flat to slightly bulging craniolateral margin is the inferred location of the insertion of the m. supracoracoideus (Vanden Berge 1975; Coombs 1978; Meers 2003), while its striated caudolateral surface (see also Charig and Milner 1997) represents the insertion of a muscle of the deltoid group that, judging by its position, seems to correspond to the avian m. deltoideus major (Vanden Berge 1975) and the m. deltoideus clavicularis (Meers 2003). The smooth distal portion of the craniomedial surface of the crest was occupied by the insertion of the m. pectoralis (Cooper 1981; Dilkes 2000; Meers 2003), while the m. coracobrachialis brevis dorsalis (Meers 2003) inserted proximally, on a shallow grove that extends onto the cranial margin of the ridge for the m. deltoideus minor (see above). Both insertion areas are medially separated from that of the m. coracobrachialis brevis ventralis by a faint ridge. Mediodistal to that, a well-developed 'biceps gutter' (Godefroit *et al.* 1998) crosses the cranial humeral surface longitudinally.

The humeral shaft has a subcircular cross-section, with a caudally flattened distal portion. This is continuous with the triceps fossae that extend distally as feeble excavations along the flat caudal surface of the distal end of the bone. The expanded distal humerus has well-developed and rugose epicondyles, although the ectepicondyle is not much expanded laterally, especially in MCP 3845-PV. It is barely separated from the radial condyle by a laterodistally facing cleft, and its more prominent element is a sharp longitudinal ridge that expands along the lateral corner of the bone. This is somewhat continuous with the 'lateral carina' (Cooper 1981), and might represent the origin area of the mm. supinator and extensor carpi radialis (McGowan 1982; Vanden Berge and Zweers 1993; Meers 2003). It forms the steep lateral margin of an elongated and distally deeper concavity that extends longitudinally on the cranial face of the ectepicondyle (Text-fig. 6, lg), and may represent the attachment area of dorsal collateral ligaments of the elbow joint (Baumel and Raikow 1993). Caudal to the lateral ridge, the caudolateral surface of the ectepicondyle has marked striations that surround a more distally placed pit (MCP 3845-PV), originally described as autapomorphic for *Herrerasaurus* (Sereno 1993). This whole area and pit are also probably related to the origin of extensor muscles such as the mm. extensor carpi ulnaris and extensor digitorum comunis (Dilkes 2000), and perhaps other elements (Meers 2003), including the m. ectepicondylo-ulnaris (Vanden Berge 1975; McGowan 1982). The entepicondyle corresponds mainly to a medially expanded rugose swelling, the caudolateral margin of which is separated from the ulnar condyle by a cleft. Its heavily striated caudal surface is continuous with a striated ridge (Text-fig. 7B) that extends proximally along the medial corner of the bone and probably corresponds to the origin of the m. flexor carpi ulnaris (Vanden Berge 1975; McGowan 1982; Meers 2003). The raised medial rim of that surface forms the caudal border of a longitudinally orientated ovoid pit that occupies the centre of a protruding area on the medial surface of the entepicondyle. This is paralleled by a similar depression placed cranial and slightly proximal to it, on a craniomedial extension of that protruding area. Comparable elements were described as unique for *Herrerasaurus* (Sereno 1993; see also Brinkman and Sues 1987, fig. 3D), and might correspond to the origin of flexor muscles such as the m. flexor digitorum longus (Dilkes 2000; Meers 2003). A steep border separates the rugose medial margin of the entepicondyle from its smooth and slightly concave cranial surface, which might have received the origin of pronator muscles (Vanden Berge 1975; McGowan 1982; Meers 2003). Laterodistal to this, a small rugose area (MCP 3845-PV) probably corresponds to the attachment of the ventral collateral ligament of the elbow joint (Baumel and Raikow 1993). Between the inner limits of the epicondyles, a large eye-shaped depression occupies the centre of the cranial surface of the distal humerus (MCP 3844-PV). This compares to the fossa m. brachialis (Baumel and Witmer 1993) of birds, which is the site of the humeral origin of the eponymous muscle in this group (George and Berger 1966; McGowan 1986). An avian-like origin for the m. brachialis was inferred for some dinosaurs (Cooper 1981; Moser 2003), whereas a condition more similar to that of crocodiles, with the muscle originating from the distal margin of the deltopectoral crest (Meers 2003), was reconstructed for others (Borsuk-Bialynicka 1977; Coombs 1978; Norman 1986). The frequent occurrence of a similar fossa in basal dinosaurs (Yates 2004, p. 14) seems to favour the first hypothesis.

The distal humeral articulation is lateromedially expanded, occupying about 70 per cent of the distal margin of the bone. On the whole, it faces slightly cranially and is gently concave, with ulnar and radial condyles equally projected distally. The radial condyle occupies the lateral two-fifths of the articulation area, and is nearly continuous with the ulnar facet, except for faint cranial and caudal furrows. It has steep caudal, lateral and craniolateral rims, but lacks a caudal ridge as described for *Herrerasaurus* (Sereno 1993). Its craniomedial margin merges smoothly into the cranial surface of the bone, so that the condyle is craniocaudally convex (MCP 3844-PV). Its distally upturned medial border gives the radial condyle a barely concave transverse outline, so that it can be described as saddle-shaped, as in *Herrerasaurus* (Sereno 1993). The cranial extension of the condyle is medially bound by an enlarged expansion of the cranial furrow (Text-fig. 6C, cf) that separates it from the ulnar articulation facet and also surrounds that facet proximally (MCP 3844-PV). It leads into the 'brachial fossa' and may represent a feeble version of the incisura intercondylaris (Baumel and Witmer 1993). The lateromedially elongated ulnar condyle occupies the medial and central parts of the distal humeral articulation. It is crossed by a craniocaudal groove, medial to which the condyle has a craniocaudally elongated biconvex surface. This is surrounded by well-developed lip-like borders, and articulated with a groove on the 'medial process' of the ulna. The transversely flat lateral part of the ulnar condyle abuts the base of the olecranon region. This is not restricted to the distal margin of the bone, but extends onto its cranial surface, where the lip-like border of the articulation ends laterally in a small tuberosity. The ulnar articulation also enters the caudal surface of the humerus, forming a rounded facet with rugose margin, which corresponds to the avian fossa olecrani (Baumel and Witmer 1993). The whole ulnar articulation is therefore markedly convex craniocaudally,

forming a saddle-shaped facet, as also seen, and originally considered unique to *Herrerasaurus* (Sereno 1993). Manipulation of the radius and ulna on the humeral condyles reveal that the elbow joint performed a basically fore-and-aft hinge movement, but some degree of pronation occurred during flexion. The forearm could attain maximal flexion and extension of about 100 and 140 degrees to the humeral long axis, respectively (Text-fig. 2). Indeed, the cranial projections of the distal condyles form a shallow 'cuboid fossa', suggesting that a reasonable degree of forearm flexion was possible (Bonnan 2003; Bonnan and Yates 2007).

Ulna. The recovered portion (proximal end and partial shaft) of the most complete (MCP 3844-PV) ulna of *S. tupiniquim* (Text-fig. 8A–E) accounts for about 70 per cent of the total length of the bone, as estimated based on the length of the complete radius. Its proximal end is composed of a broad body, the caudal half of which expands proximally to form the base of the olecranon, and the main portion of that process, which projects further proximally. In its entirety, the olecranon corresponds to 23 per cent of the estimated ulnar length. Such a large process is unusual for basal dinosaurs, but typical of some derived members of the group (Galton and Upchurch 2004b; Vickaryous *et al.* 2004; Senter 2005). In fact, the olecranon of *S. tupiniquim* is formed of what seems to be three separately ossified, but firmly attached portions. The subpyramidal stout portion that forms the base of the process is continuous with the rest of the ulna, and distinguished from the other parts by its smoother outer surface. Its cranial margin is slightly proximally orientated, and articulated with the lateral part of the ulnar condyle of the humerus and to the fossa olecrani, whereas the caudolateral surface has a scarred cranial portion, just proximal to the 'lateral process' (see also Santa Luca 1980) that might represent a separate insertion for the scapular branch of the m. triceps (Baumel and Raikow 1993; Baumel and Witmer 1993). This basal portion seems to correspond to the entire olecranon of most basal dinosaurs (Young 1941a, b, 1947; Bonaparte 1972; Galton 1973, 1974, 1976, 1981, 1984; Van Heerden 1979; Cooper 1981; Welles 1984; Forster 1990; Benton *et al.* 2000; Yates and Kitching 2003; Butler 2005; *Liliensternus*, HB R.1275) that is usually much shorter than distally broad.

In the above-mentioned forms, the olecranon has an often broad and rugose caudoproximally orientated flat surface kinked from the caudal margin of the ulna, which sets the proximal tip of the process apart from its caudal margin. In *S. tupiniquim*, this is covered by a proximally projected sheet of bone (Text-fig. 8, aoo1) that seems to have ossified independently from the rest of the ulna. As a result, the rounded caudal margin of the olecranon is nearly continuous with that of the ulnar shaft and its tip is more caudally placed. The flat medial and bowed caudolateral surfaces of that ossification are heavily marked by longitudinal striations that represent the insertion of the m. triceps tendon (Coombs 1978; Norman 1986; Dilkes 2000). That element tapers proximally from its broad base, whilst thin palisades project cranially from its medial and lateral margins, enveloping the proximal portion of the humeral articulation, to form a shallow trench (Text-fig. 8, ob) proximal to it. The elongated olecranon of some other basal dinosaurs (Raath 1969, 1990;

TABLE 3. Right epipodium measurements of *Saturnalia tupiniquim* in millimetres. Inverted commas enclose partial measurements taken from incomplete structures. Abbreviations: MiRPB, minimum radius proximal breadth; MRPB, maximum radius proximal breadth; OPB, olecranon process craniocaudal breadth; OPL, olecranon process length; PUL, preserved ulna length; RAWU, width of ulnar articulation for radius; RDB, radius distal craniocaudal breadth; RDW, radius distal lateromedial width; RL, radius length; RSW, radius mid-shaft lateromedial width; UPB, ulna proximal craniocaudal breadth; UPW, ulna proximal lateromedial width; USB, ulnar shaft craniocaudal breadth; USW, ulnar shaft lateromedial width.

	MCP 3844-PV	MCP 3845-PV
PUL	55	33
UPB	22	18
RAWU	18	15
UPW	14	12
OPL	15	'12'
OPB	11	–
USW	5	–
USB	8	–
RL	61	–
MRPB	17	–
MiRPB	9	–
RSW	6·6	–
RDW	12·5	–
RDB	13	–

Santa Luca 1980; Sereno 1993) might encompass an equivalent ossification. In *S. tupiniquim*, this element does not contribute to the humeral articulation, but its cranial surface holds another separate ossification. This is not preserved in MCP 3845-PV, neither was it reported in any other basal dinosaur of which we are aware. It has the shape of a medially compressed half-hemisphere, forming the cranial half of the olecranon, proximal to the humeral articulation. It also does not take part in the humeral articulation, but roofs the basin formed by the former ossification, and defines a proximally hollow olecranon process. This peculiar construction is reminiscent of that of the ulnar epiphysis of *Agama agama* figured by Haines (1969, fig. 29), the cranial surface of which has a non-ossified gap, occupied by un-eroded cartilage.

The homology of the proximal elements of the olecranon of *S. tupiniquim* is hard to deduce. They could be tentatively interpreted as ossifications of the triceps tendon (see Haines 1969, fig. 39), such as sesamoids like the ulnar patella of some reptiles (Haines 1969) and birds (Baumel and Witmer 1993). Most probably, however, these represent ossifications of a separate epiphyseal centre that, often in conjunction with tendon ossifications, co-ossify to the ulnar shaft to form a long olecranon, as seen in lizards (Haines 1969). Crocodiles lack discrete epiphyseal ossifications (Haines 1969), and their olecranon remains mainly cartilaginous (Brochu 2003), as also inferred for some fossil archosaurs (Romer 1956; Cooper 1981). In any case, given that an expanded olecranon is known in both preserved ulnae of *S. tupiniquim*,

and also in other finely preserved basal dinosaurs (see above), this morphology is not considered pathological, but typical of the taxon. Interestingly, a similar process is seen emanating from the proximal margin of the left ulna of one specimen of *Plateosaurus* from Halberstadt, Germany (HNM C mounted skeleton; Galton 2001, fig. 27). In this case, even if considered abnormal (the right ulna is typical of 'prosauropods'); this might correspond to a rarely ossified or preserved anatomical feature of basal dinosaurs. In other tetrapods, a similarly large olecranon is associated with a strong, but not necessarily fast, forearm extension (Coombs 1978; Fariña and Blanco 1996; Vizcaíno *et al.* 1999). This could be related to digging abilities, even if not connected to fossorial habits (Senter 2005). In this scenario, however, the olecranon would experience a significant stress, which does not seem to match its rather fragile construction in *S. tupiniquim*. Accordingly, the function of the large but hollow and thin-walled olecranon of that dinosaur is unclear.

Craniodistal to the olecranon, the humeral articulation expands cranially to form the 'medial' and 'lateral' processes (Godefroit *et al.* 1998), both of which bear lip-like outer borders that might represent attachment areas for ligaments of the elbow joint (Baumel and Raikow 1993; Meers 2003). The 'medial process' probably represents the attachment site for the posterior radioulnar ligament whereas the anterior radioulnar ligament would have attached to the 'lateral process' (Landsmeer 1983). The 'medial process' is more cranially projected, and separated from the more caudal portion of the humeral articulation by a medially deeper subtle groove. This corresponds to the avian 'ventral cotyle' (Baumel and Witmer 1993), which articulates with the medial part of the ulnar condyle of the humerus. Between the 'medial' and 'lateral' processes, the straight to slightly concave margin of the humeral articulation forms the proximal border of the radial articulation. This extends distally along the craniolateral surface of the ulna, especially on its medial part, forming a subtriangular flat area for the reception of the proximal head of the radius. The ridge-like proximal part of its caudal margin is sharper in MCP 3845-PV (Text-fig. 8F), forming a steep border that cranially bounds a concave area, which may represent the insertion of the m. flexor ulnaris (Text-fig. 3H). The distal portion of the articulation is lined medially by a rugose buttress (Text-fig. 8, bt), which might accommodate an ulnar insertion of the m. biceps, possibly coupled to that of the m. brachialis inferior (Norman 1986; Dilkes 2000). At this point, the ulna is subtriangular in cross-section, with a flat radial articulation, a rounded caudolateral surface that formed part of the insertion of the m. flexor ulnaris (Meers 2003; = m. ectepicondylus-ulnaris, McGowan 1982), and a flat to slightly concave medial surface. The latter is more excavated in MCP 3845-PV (Text-fig. 8, mia) and might represent the insertion area of either the m. brachialis (Ostrom 1969; Baumel and Raikow 1993) or, most likely, a branch of the m. flexor carpi ulnaris (Borsuk-Bialynicka 1977; Dilkes 2000).

The proximal portion of the ulnar shaft has a hemispherical cross-section (flat medially and round laterally), marked by cranial and caudal margins and a 'lateral crest' (Cooper 1981, fig. 31). Its cranial margin is formed of a subtle flat area expanding distally from the insertion of the m. biceps. It tapers along the shaft, and might represent the origin area of the m. pronator

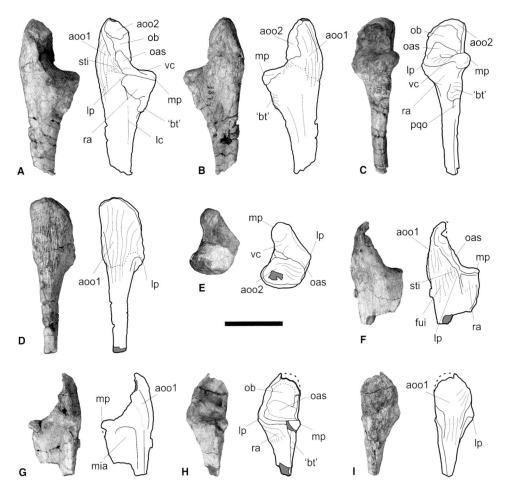

TEXT-FIG. 8. *Saturnalia tupiniquim*, Santa Maria Formation, Rio Grande do Sul, Brazil. Photographs and outline drawings of partial right ulnae. A–E, MCP 3944-PV in A, lateral, B, medial, C, cranial, D, caudal, and E, proximal views. F–I, MCP 3945-PV in F, lateral, G, medial, H, cranial, and I, caudal views. aoo1 and 2, additional olecranon ossifications 1 and 2; 'bt', 'biceps tubercle'; fui, insertion of m. flexor ulnaris; lc, lateral crest; lp, lateral process; mia, muscle insertion area; mp, medial process; oas, olecranon articular surface; ob, olecranon 'basin'; pqo, origin of m. pronator quadradus; ra, radius articulation; sti, insertion of m. triceps scapularis; vc, ventral cotyle. Shaded areas indicate missing parts. Scale bar represents 20 mm.

quadratus (Meers 2003). The 'lateral crest' extends distally from the 'lateral process', diminishing distally, so that the ulnar shaft is elliptical at its most distally preserved portion. Its medullar channel, which is also elliptical, occupies one-quarter of the craniocaudal and one-fifth of the lateromedial width of the bone. From what is preserved of the ulna, it is not possible to determine the medial displacement (Sereno 1993) or distal twisting (Benton *et al.* 2000) of its distal portion, but the whole bone is not caudally arched as is seen in some sauropodomorphs (Van Heerden 1979).

Radius. The radius of *S. tupiniquim* (Text-fig. 9) is composed of an elongated body and expanded proximal and distal ends. The latter is craniolaterally placed relative to the ulna and articulates with it via a flat caudomedial surface. The opposite margin is rounded, and the proximal end as a whole has a caudolaterally to craniomedially elongated ovoid outline. The proximal articulation surface has a shallow oblique depression extending lateromedially throughout its centre, which receives the radial condyle

of the humerus. The caudolateral and craniomedial corners of the distal margin are slightly upturned, but neither is particularly projected proximally as in some other basal dinosaurs (Santa Luca 1980; Sereno 1993; *Plateosaurus*, SMNS F65). The articular facet for the ulna is broad at the proximal margin of the radius and tapers distally, forming a subtriangular surface that extends for almost one-fifth of the length of the bone. Its rugose summit (Text-fig. 9A–B, mt) lies caudal to the distal end of a ridge that marks the craniomedial margin of the articulation. This is somewhat distal to the suggested ulnar insertion of the m. biceps, and possibly the m. brachialis inferior (see above), and might represent the radial insertion of these same muscles. It extends distally as an elongated rugose tuber until the craniomedial corner of the shaft, forming the so-called 'biceps tubercle' (Sereno 1993), which might represent the insertion of the m. humeroradialis (Brochu 2003; Meers 2003), and continues as a short faint ridge until the middle of the bone. A comparable, but not necessarily equivalent arrangement of medial elements in the proximal radius has been figured for other dinosaurs (von

Huene 1926; Galton 1974), but in comparison to *Herrerasaurus* (Sereno 1993), the 'biceps tubercle' is not so well marked, and that part of the bone is not medially kinked.

The radius twists along its body, as if the distal end suffered a counter-clockwise rotation of 90 degrees (from a proximal standpoint on the right side). This is inferred from both comparison with other dinosaurs and tracing the intermuscular lines along the shaft. This indicates, for example, that the cranial surface of the proximal part of the bone is continuous with the medial surface at its distal part. The caudomedial surface of the proximal radial shaft is still flat distal to the ulnar articulation. A faint ridge (Text-fig. 9, ril1) emerges from that surface, entering the distal half of the bone as a marked intermuscular line. This forms the caudomedial corner of the distal portion of the shaft, which is subquadratic in cross-section. Another intermuscular line (Text-fig. 9, ril2), also seen in *Herrerasaurus* (Sereno 1993, figs 7B, 8A), arises from the craniolateral surface of the proximal radius, becomes more distinct at the middle of the bone, reaching the craniomedial corner of its distal end. A less obvious line (Text-fig. 9, ril3) marks the craniolateral corner of the distal shaft and is somewhat continuous to a faint ridge extending distally from the caudolateral margin of the proximal end of the bone (see also Sereno 1993, fig. 8B).The more rounded caudolateral corner of the distal shaft is aligned to the ventral margin of the flat caudolateral surface of the proximal radius. Distal to the 'biceps tubercle' the ulnar shaft is slightly bowed laterally, especially on its proximal part, but not to the extent seen in *Herrerasaurus* (Sereno 1993).

The radius has an expanded distal end, the perimeter of which is heavily ornate with tubers and grooves. Its cross-section is subtrapezoidal, formed by a broader cranial, a narrower caudal, and oblique lateral and medial surfaces. The cranial surface is flat, but slightly concave laterally, where an inverted extension of the radiale articulation bounds its distal margin. This excavation may correspond to the avian sulcus tendinosus (Baumel and Witmer

1993), which is occupied by tendons of the extensor muscles of the wrist joint. Medial to that, a bulging area occupies the craniomedial corner of the distal radius, stretching caudally along its medial surface (Text-fig. 9, la). A similar rugose element was figured for *Hypsilophodon* (Galton 1974, fig. 40, x), and its relative position seems to correspond to the attachment for the avian lig. radio-radiocarpale craniale (Baumel and Raikow 1993). Laterally, the distal end of the radius has a smooth cranial surface that distally and caudally surrounds a pointed tubercle, marking the craniolateral corner of the bone. A similar tubercle was reported for *Heterodontosaurus* (Santa Luca 1980) and related to the m. extensor carpi radialis. Alternatively, this might correspond to an insertion of the pronator musculature (Vanden Berge 1975; McGowan 1982; Meers 2003), which assists in flexing the forearm. Caudal to that, the ovoid articular facet for the ulna (Sereno 1993) and/or ulnare (Santa Luca 1980) occupies the lateral surface of the distal radius, and also expands into its caudolateral corner. That facet is proximally bound by marked longitudinal striations, possibly related to the lig. radioulnare. This may have also extended into a groove (= depressio ligamentosa: Baumel and Witmer 1993; Brochu 2003), on the rounded caudal surface of the distal radius, just medial to the aforementioned articulation. Medial to that, another rugose bulging area (Text-fig. 9, dfo) might correspond to the origin of a digit flexor (Carpenter and Smith 2001), perhaps the m. transvs. palmaris (Meers 2003). All but the craniomedial corner of the distal surface of the radius is occupied by the articulation with the radiale. This has an ovoid shape, with the long axis nearly perpendicular to that of the proximal end of the bone. The articulation is almost flat, but dimly concave medially and convex laterally. The entire distal end of the radius is laterally kinked, so that the distal surface forms an angle of 70 degrees to the long axis of the shaft.

Given that the hand and distal ulna of *S. tupiniquim* are unknown, the orientation of its manus and the relative position

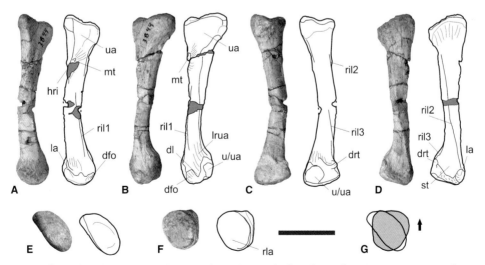

TEXT-FIG. 9. *Saturnalia tupiniquim*, Santa Maria Formation, Rio Grande do Sul, Brazil; MCP 3844-PV. A–F, photographs and outline drawings of right radius in A, medial, B, caudal, C, lateral, D, cranial, E, proximal, and F, distal views. G, relative position of proximal and distal ends of the right radius (arrow points cranially). dfo, origin of digit flexor muscle; dl, depressio ligamentosa; drt, distal radius tubercle; hri, insertion of m. humeroradialis; la, ligament attachment; lrua, attachment of lig. radioulnaris; mt, medial tuber; ril1, 2 and 3, radial intermuscular lines 1, 2 and 3; rla, radiale articulation; st, sulcus tendinosus; ua, ulna articulation; u/ua, ulna/ulnare articulation. Shaded areas indicate missing parts. Scale bar represents 20 mm.

of its radius and ulna cannot be positively established. However, it is possible to infer that, based on the position of the ulnar articulation facets on the radius and manipulation of the preserved parts of its epipodium, the twisting of the radius would allow it to cross over the ulna cranially, so that its distal end would be craniomedially placed relative to that bone. In this tentative scenario, the palmar surface of the manus would be directed caudomedially, and not medially to craniomedially as suggested for most saurischians (Sereno 1993; Carpenter 2002; Bonnan 2003; Senter and Robins 2005), although not for most sauropods (Bonnan 2003), which possess a pronated manus. Indeed, the twisting of the radius is not as clear in other basal saurischians (*Herrerasaurus*, PVSJ 373, 407; *Plateosaurus*, SMNS F65) as it is in *S. tupiniquim*, and their wrist joints seem only slightly pronated relative to the proximal radius and ulna. Yet, a radius-ulna crossing is seen in *Stormbergia* (Butler 2005), and a distal radius-ulna articulation similar to that inferred for *S. tupiniquim* was described for other taxa (Thulborn 1972, fig. 7J; Welles 1984, p. 129; Norman 1986, fig. 76C), indicating that the inferred arrangement is not unlikely for a basal dinosaur.

If the wrist was pronated in *S. tupiniquim*, as proposed here, that pronation must have been permanent, given that an active rotation of its radius relative to the ulna is prevented by their flat proximal articulation (see Carpenter 2002; Senter and Robins 2005). That pronation would enforce, at least partially, the caudal orientation of the palmar surface of the manus, which could be fully achieved by means of a minor abduction of the forelimb, as given by the regular articulation of the shoulder joint (see above). This implies that *S. tupiniquim* would be able to face its hand towards the ground, so that the forelimb could tentatively sustain a quadrupedal locomotion, as also inferred on the basis of hindlimb anatomy (Langer 2003). In this context, the enlarged olecranon of *S. tupiniquim* might have been needed to hold the body in a semierect (humerus abducted) position via forearm extension. Yet, this function also does not match its somewhat fragile construction, as previously discussed. Knowledge of manus anatomy is clearly necessary before fully establishing the role, if any, of the forelimb in the locomotion of *S. tupiniquim*. In any case, it would be important to determine whether its forearm construction represents the plesiomorphic saurischian condition, shared even by fully bipedal basal members of the group, or if it is linked to the reacquisition of a quadrupedal gait in an animal that is on the threshold between being an obligate biped and a facultative quadruped (Langer 2003). Indeed, as discussed by Bonnan (2003; see also Bonnan and Yates 2007), a shift in the position of the entire radius, not only of its distal portion, apparently characterizes the transition of facultative to obligatory quadrupedalism among sauropodomorphs.

INFERENCES ON EARLY DINOSAUR PHYLOGENY

Except for the manus, which is particularly important in the characterization of Saurischia (Gauthier 1986; Langer 2004), the anatomy of the pectoral girdle and forelimb has been scarcely considered in phylogenetic studies of

basal dinosaurs. The latter elements represent the source of approximately 12 per cent of the characters used by Holtz (2000) and Yates (2004) in their phylogenies, 5 per cent of characters in Carrano *et al.* (2002), Rauhut (2003), Langer (2004), Tykoski and Rowe (2004) and Galton and Upchurch (2004a), and less than 2 per cent in the 'basal Dinosauria' section of Sereno's (1999) analysis. Yet, some of these characters are central to the definition of certain key hypotheses of relationships, such as herrerasaur-theropod affinity (Sereno *et al.* 1993), as well as dinosaur (Novas 1996) and 'prosauropod' monophyly (Sereno 1999). Various characters of phylogenetic significance have already been discussed in the descriptive section of this paper. Here, based on the pectoral girdle and forelimb elements described for *Saturnalia tupiniquim*, the status of various related morphological characters proposed in the literature is evaluated. A numerical phylogenetic study has not been carried out, but this reassessment of previously used characters can be incorporated into further studies.

Variations in the length and shape of the scapula blade have been coded differently in cladistic studies of basal dinosaurs. These attempted to define how elongate the scapula, or its blade, is (Holtz 2000, character 211; Carrano *et al.* 2002, character 97; Rauhut 2003, character 132; Yates 2004, character 113; Butler 2005, character 46), how distally expanded and/or constricted the middle of the blade is (Gauthier 1986; Novas 1992; Holtz 2000, character 212; Carrano *et al.* 2002, character 96; Rauhut 2003, character 133; Yates 2004, character 114; Tykoski and Rowe 2004, character 106), or a combination of these conditions (Sereno 1999). Even if somewhat shorter in *Eoraptor* (PVSJ 512), the scapula and/or scapula blade of most basal dinosaurs are equally long in comparison to the breadth of the caput scapulae, and that relation does not seem to bear an important phylogenetic signal. Regarding the second parameter, *S. tupiniquim* shares with most basal dinosaurs a clearly expanded distal blade. On the contrary, *Herrerasaurus* (Sereno 1993) and, to a lesser extent, *Eoraptor* (PVSJ 512) have less expanded blades. As stated by several authors (Holtz 2000; Carrano *et al.* 2002; Rauhut 2003; Langer 2004; contra Sereno *et al.* 1993; Sereno 1999), however, this is not considered to support a theropod-herrerasaur affinity, given that the scapula blade of most basal theropods is also expanded. Instead, it most probably represents an apomorphy of *Herrerasaurus*, and perhaps *Eoraptor*, which is convergently acquired by more derived theropods (Rauhut 2003; Tykoski and Rowe 2004). The condition in *Staurikosaurus* is ambiguous; its blade was either considered strap-like (Novas 1992) or distally expanded (Sereno 1993; Galton 2000; Langer 2004), but never with a strong basis. Considering that the identification of the putative proximal scapular (Bittencourt 2004) and humeral (Sereno 1993;

Galton 2000) fragments are disputed, there is no element of the pectoral girdle or forelimb left with which to compare its incomplete distal scapula. Yet, the craniocaudal length of that element is subequal to that of proximal trunk centra, while in other basal dinosaurs (Owen 1863; Galton 1973; Santa Luca 1980; Welles 1984; Colbert 1989), *Herrerasaurus* included (PVSJ 373; measurements in Novas 1993; Sereno 1993), the distal scapula blade is at least 1·5 times longer. Indeed, this suggests that the distal end of the scapula blade of *Staurikosaurus* is not expanded.

Somewhat related to the distal expansion of the scapula blade are characters dealing with the curvature of its caudal margin (Tykoski and Rowe 2004, character 107; Yates 2004, character 114). The holotype scapula blade of *S. tupiniquim* has an evenly curved caudal margin, whereas in MCP 3845-PV, especially on the left side, the proximal part of the blade is straighter. Indeed, this character has a somewhat erratic distribution: putatively distantly related forms (Owen 1863; Young 1942; Bonaparte 1972; Colbert 1981; Welles 1984; Butler 2005) have a curved caudal margin, whereas a straighter margin is more common among basal dinosaurs. This may be the case for the entire margin of the blade (Sereno 1993). Alternatively, it may be caudally curved only near its distal tip, which is typical of some coelophysoids (Tykoski and Rowe 2004), but is also seen in other basal forms (von Huene 1926; Young 1947; Thulborn 1972; Santa Luca 1980; Cooper 1981; Madsen and Welles 2000; *Efraasia*, SMNS 17928; *Eoraptor*, PVSJ 512; *Guaibasaurus*, MCN 3844-PV). Ultimately, we believe that such a highly variable character contributes only a poor phylogenetic signal to understanding the relationships of basal dinosaurs.

Forelimb length relations have been discussed in the context of theropod (Holtz 2000; Holtz *et al.* 2004), sauropodomorph (Yates 2003; Galton and Upchurch 2004a) and ornithischian (Butler 2005) evolution. The humerus plus radius length in *S. tupiniquim* is just over half that of the femur plus tibia, a condition typical of basal dinosaurs, including *Guaibasaurus* (Bonaparte *et al.* 2006), and dinosaur outgroups (Sereno and Arcucci 1994; Benton 1999; but see Dzik 2003). A longer arm plus forearm is commonly seen among sauropodomorphs (Cooper 1981), whereas the reverse is often the case for theropods (Raath 1969; Welles 1984) and ornithischians (Peng 1992; Butler 2005). On the contrary, the length of the humerus relative to the scapula does not vary greatly among basal dinosaurs. These are usually subequal, or the scapula is slightly longer (Owen 1863; Raath 1969; Galton 1973; Welles 1984; Colbert 1989; Butler 2005; *Eoraptor*, PVSJ 514), while a significantly longer humerus is seen in the outgroups to Dinosauria (Bonaparte 1975; Benton 1999; see also Dzik 2003). Moreover, the humerus of *S. tupiniquim* is approximately 60 per cent of the femoral length, a low ratio compared to that of various 'prosauropods' (Yates 2004). This is, however, comparable to the condition of *Guaibasaurus* (Bonaparte *et al.* 2006) and dinosaur outgroups, while *Herrerasaurus* and *Eoraptor* share with basal theropods a humerus of about half the length of the femur (Novas 1993; Langer 2004; Langer and Benton 2006). The extension of the deltopectoral crest was used to define dinosaur and 'prosauropod' monophyly (Sereno 1999) and that of *S. tupiniquim* extends for about 45 per cent of the humeral length. Although typically dinosaurian, this length is over the usual proportional length seen in basal theropods (*Liliensternus*, HB R.1275; *Coelophysis rhodesiensis*, QG 1) and ornithischians (Santa Luca 1980; Colbert 1981; Butler 2005), but below that of most basal sauropodomorphs (Galton and Upchurch 2004a; Yates 2004). For this character, *S. tupiniquim* apparently possesses an intermediate condition in the basal dinosaur-sauropodomorph transition, as might also be the case for *Thecodontosaurus*. Fissure-fill deposits from the British Rhaetian have yielded many isolated humeri attributed to that taxon, including gracile and robust morphotypes that vary in the length of the deltopectoral crest (Galton 2005) between general dinosaur and sauropodomorph conditions. Other basal dinosaurs such as *Herrerasaurus* (MACN 18.060) and *Eoraptor* (PVSJ 514) have shorter crests (Langer and Benton 2006), but that of *Guaibasaurus* (Bonaparte *et al.* 2006) falls right into the typical 'prosauropod' range of lengths. Sereno (1999) also considered a deltopectoral crest forming an angle of 90 degrees to the long axis of the distal end of the bone as diagnostic of 'prosauropods'. Although we agree with Yates (2003) on the susceptibility of this character to taphonomic distortion, most 'prosauropods' show the derived condition (Galton and Upchurch 2004a), as is also the case for most basal theropods (Welles 1984; *Coelophysis rhodesiensis*, QG 1; *Liliensternus*, HB R.1275). On the contrary, *S. tupiniquim* shares with *Herrerasaurus* (MACN 18.060) and basal ornithischians (Santa Luca 1980; *Lesothosaurus*, BMNH RUB17), a crest forming a lower angle to the distal end of the bone.

Sereno (1993) defined four autapomorphies for *Herrerasaurus ischigualastensis* based on the humeral anatomy, namely: prominent medial tuberosity, separated by a groove from the head; circular pit on the ectepicondyle; prominent entepicondyle with cranial and caudal depressions; and a saddle-shaped ulnar condyle. It should be clear from the description of *S. tupiniquim* that the taxon possesses the three latter features, while the first is shared by some basal theropods (*Coelophysis rhodesiensis*, QG 1). Indeed, autapomorphies of fossil taxa that belong to poorly understood groups should be defined with caution. These might represent widespread conditions, unknown in putatively related forms given their poor preservation, and may not represent specific anatomical features of the taxon in question.

The forearm anatomy has been poorly explored in the literature dealing with basal dinosaur phylogeny. The reduction of the olecranon process of the ulna was considered in the context of sauropod (Upchurch 1998; Wilson and Sereno 1998) and theropod (Rauhut 2003) evolution, and its unusually large size in *S. tupiniquim* might bear a phylogenetic signal. Yet, it is unclear whether the proximal portion of its olecranon is homologous to that preserved in most basal dinosaurs, and so direct comparison may be misleading. It is likely that it corresponds to those of *Herrerasaurus*, *Heterodontosaurus* and one *Coelophysis rhodesiensis* morphotype, but if so, its significance is jeopardized by both the phylogenetic distance between these taxa and the variation within *C. rhodesiensis*. This suggests that the condition was independently acquired or, most probably, only preserved as such in those particular forms. In the latter case, the presence of a large olecranon among basal dinosaurs would be phylogeneticaly meaningless. Additionally, Holtz (2000) considered 'prosauropods' distinct from most basal theropods (*Herrerasaurus* included) for their more expanded and concave radial articular facet on the ulna. Yet, that articulation in *S. tupiniquim* is not markedly different from those of *Herrerasaurus* and other basal saurischians (*Plateosaurus*, SMNS F65; *C. rhodesiensis*, QG 1), whereas an excavated proximal ulna better characterizes more derived sauropodomorphs (Wilson and Sereno 1998; Bonnan 2003; Yates and Kitching 2003; Galton and Upchurch 2004*a*; Yates 2004; Bonnan and Yates 2007). The brachial-antebrachial length relation has also been discussed in the context of sauropodomorph (Yates 2004; Galton and Upchurch 2004*a*) and theropod (Sereno *et al.* 1998; Holtz 2000; Rauhut 2003; Tykoski and Rowe 2004) evolution. In basal dinosaurs, forearm reduction is apomorphic (Langer 2001, 2004; Yates 2003; Langer and Benton 2006), and *S. tupiniquim* shares with most forms, *Guaibasaurus* (Bonaparte *et al.* 2006) included, a derived condition in which the radius accounts for less than 70 per cent of the humeral length. On the contrary, a radius to humerus length ratio of more than 0·8 can be inferred for *Eoraptor* (PVSJ 514) and *Herrerasaurus* (PVSJ 373, 407; MACN 18.060), and also characterizes the outgroups to Dinosauria (Bonaparte 1975; Benton 1999; Dzik 2003).

VARIATION IN *SATURNALIA TUPINIQUIM*

MCP 3845-PV and the holotype of *S. tupiniquim* differ in various details, the more significant of which are related to the greater robustness of the latter. The humeri (Table 2), femora, tibiae, fibulae and metatarsals (Langer 2003) of both specimens are nearly identical in length, suggesting that the animals were equivalent in size. The long bones of MCP 3845-PV have, however, thinner walls and smaller articulation areas, and its scapulocoracoid is relatively smaller. Indeed, the humerus of the holotype has a shaft that is 12 per cent broader, a distal end that is 20 per cent wider, and a deltopectoral crest that is approximately 50 per cent broader at the base of its distal margin. Its proximal ulna is also broader, both transversely (10 per cent) and craniocaudally (20 per cent), although the olecranon process is equivalent in length to that of MCP 3845-PV. Similarly, although the femoral shafts of both specimens are almost equally thick, that of the holotype has a 65 per cent wider bone-wall (measured in absolute terms at the same portion of the shaft), and a 20 per cent broader distal end. Likewise, the tibial shaft of the holotype is only slightly wider than that of the paratype, but its bone-wall is about 60 per cent wider, and its maximal proximal breadth is 30 per cent greater. The proximal fibula of the holotype is also 10 per cent wider craniocaudally and its metatarsals II–IV are about 20 per cent wider proximally. In that context, variations on the distal width of the humerus are important because it has been used for phylogenetic inferences (Langer 2001, 2004; Yates 2003, 2004). In *S. tupiniquim*, the distal width represents 35 per cent of the length of the holotype humerus, but only 29 per cent of that of MCP 3845-PV. Such variation does not take place in any particular part of the distal humerus, but all parts (condyles and epicondyles) are involved. Those ratios are, respectively, above and below the threshold proposed to discriminate basal sauropodomorphs from other basal dinosaurs. Indeed, whereas almost no basal dinosaur has a humeral distal end corresponding to more than one-third of the length of the bone (Langer and Benton 2006), this is typical of basal sauropodomorphs, although it is reversed in various derived members of the group (Yates 2004). Indeed, the condition of *S. tupiniquim* might be intermediate between the plesiomorphic state and the distally broader humerus of sauropodomorphs, but the detected variation prevents a more precise definition. Similar ranges of variation, but below the one-third boundary, were reported for coelophysoids (Raath 1990; Tykoski and Rowe 2004), and may also apply to dinosaurs with ratios above that limit (see Bonaparte 1972).

In addition to the disparity in robustness, other morphological differences, especially regarding the pectoral girdle and forelimb, were recognized in the specimens of *S. tupiniquim* discussed. The scapular prominence, at the cranioventral part of the bone, is more expanded in MCP 3845-PV than in the holotype, representing, respectively, 44 and 33 per cent of the craniocaudal length of the caput scapulae. Also, the acromion forms a steeper angle to the main axis of the blade in MCP 3845-PV (nearly 90 vs. 65 degrees in the holotype), whereas the 'preglenoid ridge' of the holotype is broader, and its cranial tip overhangs slightly laterally, as seen in *Stormbergia* (Butler 2005, fig. 10a). Yates (2004, character 112) codes *S. tupiniquim* as sharing a 'long acromion' with some sauropodomorphs and *Herrerasaurus*, but the variable length of its scapular prominence jeopardizes such an assumption. Furthermore, he considered that *S. tupiniquim* also shares with *Herrerasaurus* and some sauropodomorphs an acromion that forms an angle of 65 degrees or more to the scapula blade. Although this is true for both discussed specimens of *S. tupiniquim*, MCP 3845-PV approaches the atypical condition of *Herrerasaurus* more than that of any basal member of the sauropodomorph lineage, including the holotype of *S. tupiniquim*. Indeed, Novas (1992; see also Galton 2000) defined an acromion forming a right angle to the scapula blade as diagnostic for Herrerasauridae. Alternatively, more recent accounts have considered this condition to be either absent (Sereno 1993) or indeterminate (Bittencourt 2004) for *Staurikosaurus*, and hence 'autapomorphic' for *Herrerasaurus*. Yet, the record of a comparable morphology in the paratype of *S. tupiniquim* indicates that this condition is not unique to *Herrerasaurus*, but more widespread among basal dinosaurs. Furthermore, the variation in *S. tupiniquim* challenges the phylogenetic significance of this character.

The 'coracoid tuber' (Text-figs 4–5, ct) at the scapula-coracoid junction is more marked in the holotype, whereas the subglenoid part of the coracoid also varies between the two specimens of *S. tupiniquim*. In MCP 3845-PV the 'subglenoid buttress' (Text-figs 4–5, sgb) does not reach the medial margin of the bone but gives rise to a crest-like caudoventral projection of the caudomedial corner of the glenoid, which expands as a ridge to bound medially the 'subglenoid fossa' (Text-fig. 5E). As a consequence, this fossa faces slightly laterally, and not strictly caudodorsally as in the holotype, and the lateral outline of the coracoid is not excavated ventral to the glenoid. On the contrary, the 'subglenoid buttress' of the holotype expands medially, and separates the caudomedial corner of the glenoid from the subglenoid part of the bone (Text-fig. 4A). The former lacks a well-developed ridge-like caudal expansion, and the caudal margin of the coracoid is excavated ventral to the glenoid. Yates (2004, character 116) suggested that a 'flat caudoventrally facing surface between the glenoid and the coracoid tubercle' was missing in theropods and some sauropodomorphs. This is the condition observed in MCP 3845-PV and apparently also in some ornithischians (Colbert 1981), theropods (Welles 1984; *Liliensternus*, HB R.1275), and *Thecodontosaurus* (Yates 2003), in which the caudal expansion of the glenoid disrupts the medial extension of that surface. In contrast, that surface roofs the entire cranial portion of the subglenoid fossa in MCP 3844-PV, some basal theropods (*Coelophysis rhodesiensis*, QG 1), ornithischians (Butler 2005), sauropodomorphs (Young

1941*a*; *Plateosaurus*, SMNS F65; *Riojasaurus*, PVL 3808), and possibly *Herrerasaurus* (Brinkman and Sues 1987). Accordingly, the variation in *S. tupiniquim* and the erratic distribution of this character among basal dinosaurs jeopardizes its phylogenetic informativeness.

Despite major differences in robustness, the humeri of MCP 3844-PV and 3845-PV are rather alike in other aspects. Some structures are better defined in the latter, such as most excavations on the cranial and caudal surfaces of both proximal and distal ends of the bone. Raath (1990) mentioned that the humeri of the robust morphotypes of *Coelophysis rhodesiensis* have distal condyles with more pronounced rims, but this is not clear in *S. tupiniquim*. Likewise, Raath (1990) noted variations in the development of the olecranon processes in ulnae attributed to *C. rhodesiensis*, but no relationship to either of the morphotypes was drawn. As discussed above, the process is equally elongated in both specimens of *S. tupiniquim*, while the absence of an extra ossification roofing of the proximal part of the olecranon in MCP 3845-PV is probably owing to lack of preservation. In the hind limb (Langer 2003), the external tibial condyle of the holotype of *S. tupiniquim* is placed cranial to the internal condyle, as in *Marasuchus*, *Pseudolagosuchus*, most 'prosauropods' and basal ornithischians. In MCP 3845-PV, on the other hand, this element lies in the caudolateral corner of the proximal tibia, as in *Herrerasaurus*, *Staurikosaurus*, *Pisanosaurus* and basal theropods, so that its morphology seems apomorphic among basal dinosaurs (Langer and Benton 2006). On the contrary, the distal tibia of MCP 3845-PV seems plesiomorphic for its almost subtriangular cross-section, which approaches the condition in *Marasuchus* (Bonaparte 1975) and basal ornithischians (Thulborn 1972). These lack the strong caudolateral corner seen in the holotype and most saurischians. Likewise, the non-vertical caudal border of the astragalar ascending process of the paratype represents a primitive feature for dinosaurs, because it is present in basal dinosauromorphs and basal ornithischians (Novas 1989, 1996).

The robustness differences between the two specimens of *S. tupiniquim* can be related to sex, age or phylogeny, or some combination of these parameters. Horner *et al.* (2000) reported that older specimens of *Maiasaura* present relatively thinner bone-walls, because of the expansion of their medullary cavity. In this case, the endosteal margin spreads diffusely into the deep cortex, and the distinction between the medullar cavity and the bone-wall is not clear-cut. Such ageing signals are absent in *S. tupiniquim*, the long bones of which posses a discrete limit between these tissue layers. In this case, the thicker bone-walls of the holotype seem to be the result of more extensive appositional growth that, along with its broader articulations, would reflect an extended development, rather than the reverse. Sexual dimorphism, with a robust more developed gender, could explain those differences. Yet, it is noteworthy that femora of both specimens of *S. tupiniquim* have a well-developed trochanteric shelf, the presence of which was regarded as sexually dimorphic in *Coelophysis rhodesiensis* (Raath 1990). On the other hand, if individual age is to be considered as the source of the robustness differences of *S. tupiniquim*, the similar size of the two specimens implies that the juvenile, or subadult, would already have virtually reached adult size. This suggests a quick growth in length during early ontogenetic stages (for a review of this subject, see Erickson 2005), with subsequent appositional growth increasing the thickness of the bone-walls and some skeletal parts.

If the variation identified in *S. tupiniquim* is at least partially driven by phylogeny, one has to admit the chance of taxonomic distinction between the two specimens in question, and search for morphological differences not associated *a priori* with developmental constraints. Indeed, some of their differences are putatively significant in the context of basal dinosaur evolution, standing for plesiomorphic or derived states (see above). Nevertheless, the two specimens share unique features, so that if future work defines *Saturnalia* as a congregation of different taxa, these will probably form a clade. As argued by Raath (1990) for *Coelophysis rhodesiensis*, the study of intraspecific variation is important to define whether states of a potential phylogenetically informative character fall within the range of variation of a single taxon. In this scenario, the paratype would simply not show certain features that define a fully developed individual of *S. tupiniquim*.

Acknowledgements. MCL thanks Drs Paul Barrett and Tim Fedak for the invitation to present this paper at the 'Evolution of giants' symposium. We are indebted to DNPM-RS and Drs Jeter Bertoleti, Martha Richter and Cláudia Malabarba (PUCRS, Porto Alegre) for allowing the study of the specimens of *Saturnalia* under their responsibility. We also thank the following for permission to examine specimens and help during the course of our work: Judith Babot and Jaime Powell (PVL); José Bonaparte (MACN); Sandra Chapman and Angela Milner (BMNH); Jorje Ferigolo and Ana Maria Ribeiro (MCN); Alex Kellner (Museu Nacional, Rio de Janeiro), Ricardo Martinez (Universidad Nacional de San Juan), Bruce Rubidge (Bernard Price Institute for Palaeontological Research, Johannesburg); Rainer Schoch (SMNS); and Dave Unwin (MB). Reviews by Drs Matthew Bonnan, Phil Senter and Paul Barrett greatly improved the manuscript. The work of MAGF in this project was granted by an MSc scholarship from the Brazilian agency FAPESP.

REFERENCES

ANDREIS, R. R., BOSSI, G. E. and MONTARDO, D. K. 1980. O Grupo Rosário do Sul, (Triássico) no Rio Grande do Sul. *Anais do XXXI Congresso Brasileiro de Geologia, Camboriú*, **2**, 659–673.

AZUMA, Y. and CURRIE, P. J. 2000. A new carnosaur (Dinosauria: Theropoda) from the Lower Cretaceous of Japan. *Canadian Journal of Earth Sciences*, **37**, 1735–1753.

BAKKER, R. T., KRALIS, D., SEIGWARTH, J. and FILLA, J. 1992. *Edmarka rex*, a new gigantic theropod dinosaur from the middle Morrison Formation, Late Triassic of the Como Bluff outcrop region, with comments on the evolution of the chest region and shoulder in theropods and birds, and a discussion of the five cycles of origin and extinction among giant dinosaurian predators. *Hunteria*, **2**, 1–24.

BARBERENA, M. C. 1977. Bioestratigrafia preliminar da Formação Santa Maria. *Pesquisas*, **7**, 111–129.

—— ARAÚJO, D. C. and LAVINA, E. L. 1985. Late Permian and Triassic tetrapods of southern Brazil. *National Geographic Research*, **1**, 5–20.

BAUMEL, J. J. and RAIKOW, R. J. 1993. Arthrologia. 133–187. *In* BAUMEL, J. J. (ed.). *Handbook of avian anatomy: nomina anatomica avium*. Publications of the Nuttall Ornithological Club, Cambridge, 779 pp.

—— and WITMER, L. M. 1993. Osteologia. 45–132. *In* BAUMEL, J. J. (ed.). *Handbook of avian anatomy: nomina anatomica avium*. Publications of the Nuttall Ornithological Club, Cambridge, 779 pp.

BENTON, M. J. 1999. *Scleromochlus taylori* and the origin of dinosaurs and pterosaurs. *Philosophical Transactions of the Royal Society of London, Series B*, **354**, 1423–1446.

——JULL, L., STORRS, G. W. and GALTON, P. M. 2000. Anatomy and systematics of the prosauropod dinosaur *Thecodontosaurus antiquus* from the Upper Triassic of southwestern England. *Journal of Vertebrate Paleontology*, **20**, 77–108.

BITTENCOURT, J. S. 2004. Revisão descritiva e posicionamento filogenético de *Staurikosaurus pricei* Colbert 1970 (Dinosauria, Theropoda). Unpublished MSc Thesis, Universidade Federal do Rio de Janeiro, Rio de Janeiro, 158 pp.

BONAPARTE, J. F. 1972. Los tetrápodos del sector superior de la Formacion Los Colorados, La Rioja, Argentina (Triássico superior). *Opera Lilloana*, **22**, 1–183.

—— 1975. Nuevos materiales de *Lagosuchus talampayensis* Romer (Thecodontia – Pseudosuchia) y su significado en el origin de los Saurischia. Chañarense inferior, Triasico Medio de Argentina. *Acta Geologica Lilloana*, **13**, 5–90.

—— BREA, G., SCHULTZ, C. L. and MARTINELLI, A. G. 2006. A new specimen of *Guaibasaurus candelariensis* (basal Saurischia) from the Late Triassic Caturrita Formation of southern Brazil. *Historical Biology*, **18** (3), 1–10.

——FERIGOLO, J. and RIBEIRO, A. M. 1999. A new early Late Triassic saurischian dinosaur from Rio Grande do Sul State, Brazil. 89–109. *In* TOMIDA, Y., RICH, T. H. and VICKERS-RICH, P. (eds). *Proceedings of the Second Gondwanan Dinosaur Symposium*. National Science Museum Monographs, **15**, 296 pp.

BONNAN, M. F. 2003. The evolution of manus shape in sauropod dinosaurs: implications for functional morphology, forelimb orientation, and phylogeny. *Journal of Vertebrate Paleontology*, **23**, 595–613.

—— and YATES, A. M. 2007. A new description of the fore-limb of the basal sauropodomorph *Melanorosaurus*: implications for the evolution of pronation, manus shape and quadrupedalism in sauropod dinosaurs. 157–168. *In* BARRETT, P. M. and BATTEN, D. J. (eds). *Evolution and palaeobiology of early sauropodomorph dinosaurs*. Special Papers in Palaeontology, **77**, 289 pp.

BORSUK-BIALYNICKA, M. 1977. A new camarasaurid sauropod *Opistocelicaudia skarzynskii* gen. n., sp. n. from the Upper Cretaceous of Mongolia. *Palaeontologia Polonica*, **37**, 5–63.

BRINKMAN, D. B. and SUES, H.-D. 1987. A staurikosaurid dinosaur from the Upper Triassic Ischigualasto Formation of Argentina and the relationships of the Staurikosauridae. *Palaeontology*, **30**, 493–503.

BROCHU, C. A. 2003. Osteology of *Tyrannosaurus rex*: insights from a nearly complete skeleton and high-resolution computed tomographic analysis of the skull. *Memoir of the Society of Vertebrate Palaeontology*, **7**, 1–138.

BRYANT, H. N. and RUSSELL, A. P. 1992. The role of phylogenetic analysis in the inference of unpreserved attributes of extinct taxa. *Philosophical Transactions of the Royal Society of London, Series B*, **337**, 405–418.

—— ——1993. The occurrence of clavicles within Dinosauria: implications for the homology of the avian furcula and the utility of negative evidence. *Journal of Vertebrate Paleontology*, **13**, 171–184.

BUTLER, R. J. 2005. The 'fabrosaurid' ornithischian dinosaurs of the upper Elliot Formation (Lower Jurassic) of South Africa and Lesotho. *Zoological Journal of the Linnean Society*, **145**, 175–218.

CARPENTER, K. 2002. Forelimb biomechanics of nonavian theropod dinosaurs in predation. 59–76. *In* GUDO, M., GUTMANN, M. and SCHOLZ, J. (eds). *Concepts of functional engineering and constructional morphology*. Senckenbergiana Lethaea, **82**, 372 pp.

—— and SMITH, M. 2001. Forelimb osteology and biomechanics of *Tyrannosaurus rex*. 90–116. *In* TANKE, D. H. and CARPENTER, K. (eds). *Mesozoic vertebrate life*. Indiana University Press, Bloomington, IN, 352 pp.

—— MILES, C., OSTROM, J. H. and CLOWARD, K. 2005. Redescription of the small maniraptoran theropods *Ornitholestes* and *Coelurus* from the Upper Jurassic Morrison Formation of Wyoming. 49–71. *In* CARPENTER, K. (ed.). *The carnivorous dinosaurs*. Indiana University Press, Bloomington, IN, 371 pp.

CARRANO, M. T., SAMPSON, S. D. and FORSTER, C. A. 2002. The osteology of *Masiakasaurus knopfleri*, a small abelisauroid (Dinosauria: Theropoda) from the Late Cretaceous of Madagascar. *Journal of Vertebrate Paleontology*, **22**, 510–534.

CHARIG, A. J. and MILNER, A. C. 1997. *Baryonyx walkeri*, a fish-eating dinosaur from the Wealden of Surrey. *Bulletin of the Natural History Museum (Geology Series)*, **53**, 11–70.

CLARK., A. C. Jr 1993. Termini situm et directionem partium corporis indicantes. 1–6. *In* BAUMEL, J. J. (ed.). *Handbook of avian anatomy: nomina anatomica avium*. Publications of the Nuttall Ornithological Club, Cambridge, 779 pp.

COLBERT, E. H. 1970. A saurischian dinosaur from the Triassic of Brazil. *American Museum Novitates*, **2405**, 1–39.

—— 1981. A primitive ornithischian dinosaur from the Kayenta Formation of Arizona. *Bulletin of the Museum of Northern Arizona*, **53**, 1–61.

—— 1989. The Triassic dinosaur *Coelophysis*. *Bulletin of the Museum of Northern Arizona*, **57**, 1–160.

COOMBS, W. P. Jr 1978. Forelimb muscles of the Ankylosauria (Reptilia; Ornithischia). *Journal of Paleontology*, **52**, 642–657.

COOPER, M. R. 1981. The prosauropod dinosaur *Massospondylus carinatus* Owen from Zimbabwe: its biology, mode of life and phylogenetic significance. *Occasional Papers of the National Museums and Monuments of Rhodesia, Series B (Natural Sciences)*, **6**, 689–840.

CORIA, R. A. and SALGADO, L. 1996. A basal iguanodontian (Ornithischia: Ornithopoda) from the Late Cretaceous of South America. *Journal of Vertebrate Paleontology*, **16**, 445–457.

CRACRAFT, J. 1974. Phylogeny and evolution of the ratite birds. *Ibis*, **116**, 494–521.

CURRIE, P. J. and CARPENTER, K. 2000. A new specimen of *Acrocanthosaurus atokensis* (Theropoda, Dinosauria) from the Lower Cretaceous Antlers Formation (Lower Cretaceous, Aptian) of Oklahoma, USA. *Geodiversitas*, **22**, 207–246.

—— and ZHAO, XI-JIN 1993. A new carnosaur (Dinosauria, Theropoda) from the Jurassic of Xinjiang, People's Republic of China. *Canadian Journal of Earth Sciences*, **30**, 2037–2081.

DILKES, D. W. 2000. Appendicular myology of the hadrosaurian dinosaur *Maiasaura peeblesorum* from the Late Cretaceous (Campanian) of Montana. *Transactions of the Royal Society of Edinburgh: Earth Sciences*, **90**, 87–125.

DZIK, J. 2003. A beaked herbivorous archosaur with dinosaur affinities from the early Late Triassic of Poland. *Journal of Vertebrate Paleontology*, **23**, 556–574.

ERICKSON, G. M. 2005. Accessing dinosaur growth patterns: a microscopic revolution. *Trends in Ecology and Evolution*, **20**, 677–684.

FARIÑA, R. A. and BLANCO, R. E. 1996. *Megatherium*, the stabber. *Proceedings of the Royal Society of London, Series B*, **263**, 1725–1729.

FELSENSTEIN, J. 1985. Phylogenies and the comparative method. *The American Naturalist*, **125**, 1–15.

FORSTER, C. A. 1990. The postcranial skeleton of the ornithopod dinosaur *Tenontosaurus tilletti*. *Journal of Vertebrate Paleontology*, **10**, 273–294.

FRASER, N. C., PADIAN, K., WALKDEN, G. M. and DAVIS, A. L. 2002. Basal dinosauriform remains from Britain and the diagnosis of Dinosauria. *Palaeontology*, **45**, 79–95.

GALTON, P. M. 1973. On the anatomy and relationships of *Efraasia diagnostica* (Huene) n. gen., a prosauropod dinosaur (Reptilia: Saurischia) from the Upper Triassic of Germany. *Paläontologische Zeitschrift*, **47**, 229–255.

—— 1974. The ornithischian dinosaur *Hypsilophodon* from the Wealden of the Isle of Wight. *Bulletin of the British Museum of Natural History (Geology)*, **25**, 1–152.

—— 1976. Prosauropod dinosaurs (Reptilia: Saurischia) of North America. *Postilla*, **169**, 1–98.

—— 1981. *Dryosaurus*, a hypsilophodontid dinosaur from the Upper Jurassic of North America and Africa. Postcranial skeleton. *Paläontologische Zeitschrift*, **55**, 271–312.

—— 1984. An early prosauropod dinosaur from the Upper Triassic of Nordwürttenberg, West Germany. *Stuttgarter Beiträge zur Naturkunde, Serie B (Geologie und Paläontologie)*, **106**, 1–25.

—— 2000. Are *Spondylosoma* and *Staurikosaurus* (Santa Maria Formation, Middle–Upper Triassic, Brazil) the oldest known dinosaurs? *Paläontologische Zeitschrift*, **74**, 393–423.

—— 2001. The prosauropod dinosaur *Plateosaurus* Meyer, 1837 (Saurischia: Sauropodomorpha; Upper Triassic). II. Notes on the referred species. *Revue de Paléobiologie*, **20**, 435–502.

—— 2005. Basal sauropodomorph dinosaur taxa *Thecodontosaurus* Riley & Stutchbury, 1836, *T. antiquus* Morris, 1843 and *T. caducus* Yates, 2003: their status re humeral morphs from the 1834 Fissure Fill (Upper Triassic) in Clifton, Bristol, UK. *Journal of Vertebrate Paleontology*, **23** (Supplement to No. 3), 61A.

—— and JENSEN, J. A. 1979. A new large theropod dinosaur from the Upper Jurassic of Colorado. *Brigham Young University, Geological Studies*, **26**, 1–12.

—— and UPCHURCH, P. 2004*a*. Prosauropoda. 232–258. *In* WEISHAMPEL, D. B., DODSON, P. and OSMÓLSKA, H. (eds). *The Dinosauria*. Second edition. University of California Press, Berkeley, CA, 861 pp.

—— and UPCHURCH, P. 2004*b*. Stegosauria. 343–362. *In* WEISHAMPEL, D. B., DODSON, P. and OSMÓLSKA, H. (eds). *The Dinosauria*. Second edition. University of California Press, Berkeley, CA, 861 pp

GAUTHIER, J. A. 1986. Saurischian monophyly and the origin of birds. 1–55. *In* PADIAN, K. (ed.). *The origins of birds and the evolution of flight*. Memoirs of the Californian Academy of Sciences, **8**, 98 pp.

GEORGE, J. C. and BERGER, A. J. 1966. *Avian myology*. Academic Press, New York, NY, and London, 500 pp.

GODEFROIT, P., DONG ZHI-MING, BULTYNCK, P., LI HONG and FENG LU 1998. Sino-Belgian Cooperative Program. Cretaceous dinosaurs and mammals from Inner Mongolia: (1) New *Bactrosaurus* (Dinosauria: Hadrosauroidea) material from Iren Dabasu (Inner Mongolia, P.R. China). *Bulletin de l'Institut Royal des Science Naturelles de Belgique, Sciences de la Terre*, **68** (Supplement), 3–70.

HAINES, R. W. 1952. The shoulder joint of lizards and the primitive reptilian shoulder mechanism. *Journal of Anatomy*, **86**, 412–422.

—— 1969. Epiphyses and sesamoids. 81–115. *In* GANS, C. BELLAIRS, A. d'A. and PARSONS, T. S. (eds). *Biology of the Reptilia, morphology A, volume 1*. Academic Press, New York, NY, 263 pp.

HOLTZ, T. R. Jr 2000. A new phylogeny of the carnivorous dinosaurs. 5–61. *In* PÉREZ-MORENO, B. P., HOLTZ, T. R. Jr, SANZ, J. L. and MORATALLA, J. J. (eds). *Aspects of theropod paleobiology*. Gaia, **15**, 403 pp.

—— MOLNAR, R. E. and CURRIE, P. J. 2004. Basal tetanurae. 71–110. *In* WEISHAMPEL, D. B., DODSON, P. and OSMÓLSKA, H. (eds). *The Dinosauria*. Second edition. University of California Press, Berkeley, CA, 861 pp.

HORNER, J. R., DE RICQLÈS, A. and PADIAN, K. 2000. Long bone histology of the hadrosaurid dinosaur *Maiasaura peeblesorum*: growth dynamics and physiology based on an ontogenetic series of skeletal elements. *Journal of Vertebrate Paleontology*, **20**, 115–129.

HUENE, F. VON 1910. Ein primitiver Dinosaurier aus der mittleren Trias von Elgin. *Geologische und Palaeontologische Abhandlungen, Neue Folge*, **8**, 317–322.

—— 1926. Vollständige Osteologie eines Plateosauriden aus dem schwäbischen Keuper. *Geologische und Palaeontologische Abhandlungen, Neue Folge*, **15**, 139–179.

—— 1932. Die fossile Reptil-Ordnung Saurischia, ihre Entwicklung und Geschichte. *Monographien zur Geologie und Palaeontologie, Series 1*, **4**, 1–361.

—— 1934. Ein neuer Coelurosaurier in der thüringschen Trias. *Palaeontologische Zeitschrift*, **16**, 145–168.

—— and STAHLECKER, R. 1931. Geologische Beobachtungen in Rio Grande do Sul. *Neues Jahrbuch für Mineralogie, Geologie und Paläontologie, B*, **65**, 1–82.

HUTCHINSON, J. R. 2001. The evolution of hindlimb anatomy and function in theropod dinosaurs. Unpublished PhD thesis, University of California, Berkeley, CA, 415 pp.

JANENSCH, W. 1955. Der Ornithopode *Dysalotosaurus* der Tendaguruschichten. *Palaeontographica (Supplement 7)*, **3**, 105–176.

JENKINS, F. A. Jr 1993. The evolution of the avian shoulder joint. *American Journal of Science*, **293A**, 253–267.

—— and GOSLOW, G. E. Jr 1983. The functional anatomy of the shoulder of the savannah monitor lizard (*Varanus exanthematicus*). *Journal of Morphology*, **175**, 195–216.

KOBAYASHI, Y. and LÜ JUN-CHANG 2003. A new ornithomimid dinosaur with gregarious habits from the Late Cretaceous of China. *Acta Palaeontologica Polonica*, **48**, 235–259.

LANDSMEER, J. M. F. 1983. The mechanism of forearm rotation in *Varanus exanthematicus*. *Journal of Morphology*, **175**, 119–130.

LANGER, M. C. 2001. *Saturnalia Tupiniquim* and the early evolution of dinosaurs. Unpublished PhD thesis, University of Bristol, Bristol, 415 pp.

—— 2003. The sacral and pelvic anatomy of the stem-sauropodomorph *Saturnalia tupiniquim* (Late Triassic, Brazil). *Paleobios*, **23**, 1–40.

—— 2004. Basal Saurischia. 25–46. *In* WEISHAMPEL, D. B., DODSON, P. and OSMÓLSKA, H. (eds). *The Dinosauria*. Second edition. University of California Press, Berkeley, CA, 861 pp.

—— 2005*a*. Studies on continental Late Triassic tetrapod biochronology. I. The type locality of *Saturnalia tupiniquim* and the faunal succession in south Brazil. *Journal of South American Earth Sciences*, **19**, 205–218.

—— 2005*b*. Studies on continental Late Triassic tetrapod biochronology. II. The Ischigualastian and a Carnian global correlation. *Journal of South American Earth Sciences*, **19**, 219–239.

—— and BENTON, M. J. 2006. Early dinosaurs: a phylogenetic study. *Journal of Systematic Palaeontology*, **4**, 309–358.

—— ABDALA, F., RICHTER, M. and BENTON, M. J. 1999. A sauropodomorph dinosaur from the Upper Triassic (Carnian) of southern Brazil. *Comptes Rendus de l'Académie des Sciences, Paris, Sciences de la Terre et des Planètes*, **329**, 511–517.

LONG, J. A. and MURRY, P. A. 1995. Late Triassic (Carnian and Norian) tetrapods from the southwestern United States. *Bulletin of the New Mexico Museum of Natural History and Science*, **4**, 1–254.

MADSEN, J. H. 1976. *Allosaurus fragilis*: a revised osteology. *Bulletin of the Utah Geology and Mineralogy Survey*, **109**, 3–163.

—— and WELLES, S. P. 2000. *Ceratosaurus* (Dinosauria, Theropoda); a revised osteology. *Utah Geological Survey, Miscellaneous Publications*, **00-2**, 1–80.

MAKOVICKY, P. J. and SUES, H.-D. 1998. Anatomy and phylogenetic relationships of the theropod dinosaur *Microvenator celer* from the Lower Cretaceous of Montana. *American Museum Novitates*, **3240**, 1–27.

—— APESTEGUÍA, S. and AGNOLÍN, F. L. 2005. The earliest dromaeosaurid theropod from South America. *Nature*, **437**, 1007–1011.

McGOWAN, C. 1982. The wing musculature of the Brown Kiwi *Apteryx australis mantelli* and its bearing on ratite affinities. *Journal of Zoology*, **197**, 173–219.

—— 1986. The wing musculature of the Weka (*Gallirallus australis*), a flightless rail endemic to New Zealand. *Journal of Zoology*, **210**, 305–346.

MEERS, M. B. 2003. Crocodylian forelimb musculature and its relevance to Archosauria. *Anatomical Record, Part A*, **274A**, 891–916.

MOSER, M. 2003. *Plateosaurus engelhardti* Meyer, 1837 (Dinosauria: Sauropodomorpha) aus dem Feuerletten (Mittelkeuper; Obertrias) von Bayern. *Zitteliana*, **24**, 3–186.

NICHOLLS, E. L. and RUSSELL, A. P. 1985. Structure and function of the pectoral girdle and forelimb of *Struthiomimus altus* (Theropoda: Ornithomimidae). *Palaeontology*, **28**, 643–677.

NORELL, M. A. and MAKOVICKY, P. J. 1999. Important features of the dromaeosaurid skeleton II: information from newly collected specimens of *Velociraptor mongoliensis*. *American Museum Novitates*, **3282**, 1–45.

NORMAN, D. B. 1986. On the anatomy of *Iguanodon atherfieldiensis* (Ornithischia: Ornithopoda). *Bulletin de l'Institut Royal des Sciences Naturelles de Belgique: Sciences de la Terre*, **56**, 281–372.

NOVAS, F. E. 1986. Un probable terópodo (Saurischia) de la Formación Ischigualasto (Triásico superior), San Juan, Argentina. *Actas del IV Congreso Argentino de Paleontologia y Estratigrafia. Mendoza*, **2**, 1–6.

—— 1989. The tibia and tarsus in Herrerasauridae (Dinosauria, *incertae sedis*) and the origin and evolution of the dinosaurian tarsus. *Journal of Paleontology*, **63**, 677–690.

—— 1992. Phylogenetic relationships of the basal dinosaurs, the Herrerasauridae. *Palaeontology*, **35**, 51–62.

—— 1993. New information on the systematics and postcranial skeleton of *Herrerasaurus ischigualastensis* (Theropoda: Herrerasauridae) from the Ischigualasto Formation (Upper Triassic) of Argentina. *Journal of Vertebrate Paleontology*, **13**, 400–423.

—— 1996. Dinosaur monophyly. *Journal of Vertebrate Paleontology*, **16**, 723–741.

—— and PUERTA, P. F. 1997. New evidence concerning avian origins from the Late Cretaceous of Patagonia. *Nature*, **387**, 390–392.

OSMÓLSKA, H., MARYANSKA, T. and BARSBOLD, R. 1972. A new dinosaur, *Gallimimus bullatus* n. gen., n. sp.

(Ornithomimidae) from the Upper Cretaceous of Mongolia. *Palaeontologia Polonica*, **27**, 103–143.

OSTROM, J. H. 1969. Osteology of *Deinonychus antirrhopus*, an unusual theropod from the Lower Cretaceous of Montana. *Bulletin of the Peabody Museum of Natural History*, **30**, 1–165.

—— 1974. The pectoral girdle and forelimb function of *Deinonychus* (Reptilia: Saurischia): a correction. *Postilla*, **165**, 1–11.

—— and McINTOSH, J. S. 1966. *Marsh's dinosaurs*. Yale University Press, New Haven, CT, 388 pp.

OWEN, R. 1842. Report on British fossil reptiles. Part II. *Reports of the British Association for the Advancement of Science*, **11**, 60–204.

—— 1863. A monograph of the fossil Reptilia of the Liassic formations. Second part. *Scelidosaurus harrisonii* continued. *Palaeontographical Society Monographs*, **14**, 1–26.

PADIAN, K. 1997. Pectoral girdle. 530–536. *In* CURRIE, P. J. and PADIAN, K. (eds). *Encyclopedia of dinosaurs*. Academic Press, San Diego, CA, 869 pp.

—— HUTCHINSON, J. R. and HOLTZ, T. R. Jr 1999. Phylogenetic definitions and nomenclature of the major taxonomic categories of the carnivorous Dinosauria (Theropoda). *Journal of Vertebrate Paleontology*, **19**, 69–80.

PARKER, T. J. 1891. Observations on the anatomy and development of *Apteryx*. *Philosophical Transactions of the Royal Society of London, Series B*, **182**, 25–134.

PENG GUANG-ZHAO 1992. Jurassic ornithopod *Agilisaurus louderbacki* (Ornithopoda: Fabrosauridae) from Zigong, Sichuan, China. *Vertebrata Palasiatica*, **30**, 39–53. [In Chinese, English abstract].

PÉREZ-MORENO, B. P., SANZ, J. L., BUSCALLONI, A. D., MORATALLA, J. J., ORTEGA, F. and RASSKIN-GUTMAN, D. 1994. A unique multi-toothed ornithomimosaur dinosaur from the Lower Cretaceous of Spain. *Nature*, **370**, 363–367.

RAATH, M. A. 1969. A new coelurosaurian dinosaur from the Forest Sandstone of Rhodesia. *Arnoldia*, **4** (28), 1–25.

—— 1990. Morphological variation in small theropods and its meaning in systematics: evidence from *Syntarsus rhodesiensis*. 91–105. *In* CARPENTER, K. and CURRIE, P. J. (eds). *Dinosaur systematics. Approaches and perspectives*. Cambridge University Press, Cambridge, 334pp.

RAUHUT, O. W. M. 2003. The interrelationships and evolution of basal theropod dinosaurs. *Special Papers in Palaeontology*, **69**, 1–213.

REIG, O. A. 1963. La presencia de dinosaurios saurisquios en los 'Estrados de Ischigualasto' (Mesotri·sico superior) de las Provincias de San Juan y La Rioja (Republica Argentina). *Ameghiniana*, **3**, 3–20.

ROGERS, R. R., SWISHER, C. C. III, SERENO, P. C., MONETTA, A. M., FORSTER, C. A. and MARTÍNEZ, R. N. 1993. The Ischigualasto tetrapod assemblage (Late Triassic, Argentina) and ^{40}Ar/^{39}Ar dating of dinosaurs origins. *Science*, **260**, 794–797.

ROMER, A. S. 1922. The locomotor apparatus of certain primitive and mammal-like reptiles. *Bulletin of the American Museum of Natural History*, **46**, 517–606.

—— 1956. *Osteology of the reptiles*. University of Chicago Press, Chicago, IL, 772 pp.

—— 1966. *Vertebrate paleontology*. Third edition. University of Chicago Press, Chicago, IL, 468 pp.

ROWE, T. 1989. A new species of the theropod dinosaur *Syntarsus* from the Early Jurassic Kayenta Formation of Arizona. *Journal of Vertebrate Paleontology*, **9**, 125–136.

SANTA LUCA, A. P. 1980. The postcranial skeleton of *Heterodontosaurus tucki* (Reptilia, Ornithischia) from the Stormberg of South Africa. *Annals of the South African Museum*, **79**, 15–211.

—— 1984. Postcranial remains of Fabrosauridae (Reptilia: Ornithischia) from the Stormberg of South Africa. *Palaeontologia Africana*, **25**, 151–180.

SEELEY, H. G. 1887. On the classification of the fossil animals commonly named Dinosauria. *Proceedings of the Royal Society of London*, **43**, 165–171.

SENTER, P. 2005. Function in the stunted forelimbs of *Mononykus olecranus* (Theropoda), a dinosaurian anteater. *Paleobiology*, **31**, 373–381.

—— and ROBINS, J. H. 2005. Range of motion in the forelimb of the theropod dinosaur *Acrocanthosaurus atokensis*, and implications for predatory behaviour. *Journal of Zoology*, **266**, 307–318.

SERENO, P. C. 1993. The pectoral girdle and forelimb of the basal theropod *Herrerasaurus ischigualastensis*. *Journal of Vertebrate Paleontology*, **13**, 425–450.

—— 1999. The evolution of dinosaurs. *Science*, **284**, 2137–2147.

—— and ARCUCCI, A. B. 1994. Dinosaurian precursors from the Middle Triassic of Argentina: *Marasuchus lilloensis*, gen. nov. *Journal of Vertebrate Paleontology*, **14**, 53–73.

—— BECK, L., DUTHEIL, D. B., GADO, B., LARSSON, H. C. E., LYON, G. H., MARCOT, J. D., RAUHUT, O. W. M., SADLEIR, R. W., SIDOR, C. A., VARRICCHIO, D. D., WILSON, G. P. and WILSON, J. A. 1998. A long-snouted predatory dinosaur from Africa and the evolution of spinosaurids. *Science*, **282**, 1298–1302.

—— FORSTER, C. A., ROGERS, R. R. and MONETTA. A. M. 1993. Primitive dinosaur skeleton from Argentina and the early evolution of the Dinosauria. *Nature*, **361**, 64–66.

THULBORN, R. A. 1972. The post-cranial skeleton of the Triassic ornithischian dinosaur *Fabrosaurus australis*. *Palaeontology*, **15**, 29–60.

TYKOSKI, R. S. and ROWE, T. 2004. Ceratosauria. 47–70. *In* WEISHAMPEL, D. B., DODSON, P. and OSMÓLSKA, H. (eds). *The Dinosauria*. Second edition. University of California Press, Berkeley, CA, 861 pp.

—— FORSTER, C. A., ROWE, T., SAMPSON, S. D. and MUNYIKWA, D. 2002. A furcula in the coelophysid theropod *Syntarsus*. *Journal of Vertebrate Paleontology*, **22**, 728–733.

UPCHURCH, P. 1998. The phylogenetic relationships of sauropod dinosaurs. *Zoological Journal of the Linnean Society*, **124**, 43–103.

—— BARRETT, P. M. and DODSON, P. 2004. Sauropoda. 259–322. *In* WEISHAMPEL, D. B., DODSON, P. and OSMÓLSKA, H. (eds). *The Dinosauria*. Second edition. University of California Press, Berkeley, CA, 861 pp.

VAN HEERDEN, J. 1979. The morphology and taxonomy of *Euskelosaurus* (Reptilia: Saurischia; Late Triassic) from South Africa. *Navorsinge van die Nasionale Museum*, **4**, 21–84.

VANDEN BERGE, J. C. 1975. Aves myology. 1802–1848. *In* GETTY, R. (ed.). *Sisson and Grossman's The anatomy of the domestic animals. Volume 2*. Fifth edition. W.B. Saunders, Philadelphia, PA, 1881 pp.

—— and ZWEERS, G. A. 1993. Myologia. 189–247. *In* BAUMEL, J. J. (ed.). *Handbook of avian anatomy: nomina anatomica avium*. Publications of the Nuttall Ornithological Club, Cambridge, 779 pp.

VICKARYOUS, M. K., MARYANSKA, T. and WEISHAMPEL, D. B. 2004. Ankylosauria. 363–392. *In* WEISHAMPEL, D. B., DODSON, P. and OSMÓLSKA, H. (eds). *The Dinosauria*. Second edition. University of California Press, Berkeley, CA, 861 pp.

VIZCAÍNO, S. F., FARIÑA, R. A. and MAZZETTA, G. 1999. Ulnar dimensions and fossoriality in armadillos and other South American mammals. *Acta Theriologica*, **44**, 309–320.

WALKER, A. D. 1961. Triassic reptiles from the Elgin area: *Stagonolepis, Dasygnathus* and their allies. *Philosophical Transactions of the Royal Society of London, Series B*, **244**, 103–204.

—— 1964. Triassic reptiles from the Elgin area: *Ornithosuchus* and the origin of carnosaurs. *Philosophical Transactions of the Royal Society of London, Series B*, **248**, 53–134.

—— 1977. Evolution of the pelvis in birds and dinosaurs. 319–358. *In* ANDREWS, S. M., MILES, R. S. and WALKER, A. D. (eds). *Problems in vertebrate evolution*. Linnean Society Symposia, **4**, 411 pp.

WEISHAMPEL, D. B., DODSON, P. and OSMÓLSKA, H. 2004. Introduction. 1–3. *In* WEISHAMPEL, D. B., DODSON, P. and OSMÓLSKA, H. (eds). *The Dinosauria*. Second edition. University of California Press, Berkeley, CA, 861 pp.

WELLES, S. P. 1984. *Dilophosaurus wetherilli* (Dinosauria, Theropoda). Osteology and comparisons. *Palaeontographica A*, **185**, 85–180.

WILSON, J. A. and SERENO, P. C. 1998. Early evolution and higher-level phylogeny of sauropod dinosaurs. *Memoir of the Society of Vertebrate Paleontology*, **5**, 1–68.

WITMER, L. M. 1995. The extant phylogenetic bracket and the importance of reconstructing soft tissues in fossils. 19–33. *In* THOMASON, J. J. (ed.). *Functional morphology in vertebrate paleontology*. Cambridge University Press, Cambridge, 293 pp.

YATES, A. M. 2003. A new species of the primitive dinosaur *Thecodontosaurus* (Saurischia: Sauropodomorpha) and its implications for the systematics of early dinosaurs. *Journal of Systematic Palaeontology*, **1**, 1–42.

—— 2004. *Anchisaurus polyzelus* (Hitchcock): the smallest known sauropod dinosaur and the evolution of gigantism among sauropodomorph dinosaurs. *Postilla*, **230**, 1–58.

—— and KITCHING, J. W. 2003. The earliest known sauropod dinosaur and the first steps towards sauropod locomotion. *Proceedings of the Royal Society of London, Series B*, **270**, 1753–1758.

—— and VASCONCELOS, C. C. 2005. Furcula-like clavicles in the prosauropod dinosaur *Massospondylus*. *Journal of Vertebrate Paleontology*, **25**, 466–468.

YOUNG CHUNG-CHIEN 1941*a*. A complete osteology of *Lufengosaurus huenei* Young (gen. et sp. nov.). *Palaeontologia Sinica, Series C*, **7**, 1–53.

—— 1941*b*. *Gyposaurus sinensis* Young (sp. nov.), a new Prosauropoda from the Upper Triassic beds at Lufeng, Yunnan. *Bulletin of the Geological Society of China*, **21**, 205–252.

—— 1942. *Yunnanosaurus huangi* Young (gen. et sp. nov.), a new Prosauropoda from the Red Beds at Lufeng, Yunnan. *Bulletin of the Geological Society of China*, **22**, 63–104.

—— 1947. On *Lufengosaurus magnus* (sp. nov.) and additional finds of *Lufengosaurus huenei* Young. *Palaeontologia Sinica, Series C*, **12**, 1–53.

ZERFASS, H., LAVINA, E. L., SCHULTZ, C. L., GARCIA, A. G. V., FACCINI, U. F. and CHEMALE, F. Jr 2003. Sequence stratigraphy of continental Triassic strata of southernmost Brazil: a contribution to southwestern Gondwana palaeogeography and palaeoclimate. *Sedimentary Geology*, **161**, 85–105.

[Special Papers in Palaeontology 77, 2007, pp. 139–155]

WERE THE BASAL SAUROPODOMORPH DINOSAURS *PLATEOSAURUS* AND *MASSOSPONDYLUS* HABITUAL QUADRUPEDS?

by MATTHEW F. BONNAN* *and* PHIL SENTER†

*Department of Biological Sciences, Western Illinois University, Macomb, IL 61455, USA; e-mail: MF-Bonnan@wiu.edu (author to whom correspondence should be addressed)

†Department of Math and Science, Lamar State College at Orange, 410 Front Street, Orange, TX 76630, USA; e-mail: philsenter@lsco.edu

Typescript received 3 February 2006; accepted in revised form 5 July 2006

Abstract: The basal sauropodomorph dinosaurs *Plateosaurus* and *Massospondylus* are often portrayed as habitual quadrupeds that were facultatively bipedal. Surprisingly, the functional morphology of their forelimbs has rarely been considered when reconstructing their locomotor habits. If *Plateosaurus* and *Massospondylus* were efficient, habitual quadrupeds we predict that the manus would have been pronated such that it produced a caudally directed force in parallel with the pes. We articulated and manipulated the forelimbs of *Plateosaurus*, *Massospondylus* and several extant outgroup taxa (*Varanus*, *Alligator*, *Anser* and *Struthio*) using a standardized protocol. Moreover, we compared our results with previously published estimates of forelimb movement in saurischian outgroup taxa from Theropoda and Sauropoda and with the basal sauropodomorph/sauropod *Melanorosaurus*. Our results indicate that the range of motion in the forelimbs of *Plateosaurus* and *Massospondylus* did not allow efficient, habitual quadrupedal locomotion. The range of humeral flexion and abduction is limited and the articular surfaces of the radius and ulna orient the palmar surfaces of the manus medially in semi-supination. Active or passive pronation of the manus was not possible and the manus could not function in a dynamically similar way to the pes for efficient quadrupedal locomotion. Our results also rule out specialized forms of quadrupedal locomotion, such as the knuckle-walking gait of some mammals. We suggest that most known 'prosauropod' trackways were probably not made by animals such as *Plateosaurus* or *Massospondylus*, but the ichnotaxon *Otozoum* may have been created by animals similar to these taxa. Furthermore, we show that trunk and limb ratios do not yield consistent results and should not be used solely to determine posture. Although these two taxa probably assumed a quadrupedal posture as hatchlings, we show that the morphological orientations of the forelimb elements remained consistent across ontogeny, precluding efficient, quadrupedal locomotion at any age. As with theropods, forelimb use in basal sauropodomorphs is difficult to reconstruct and interpret. We speculate that the forelimb could have aided in acquiring vegetation or defence in *Plateosaurus* and *Massospondylus* only if these animals reared or assumed a tripodal posture.

Key words: *Plateosaurus*, *Massospondylus*, Sauropodomorpha, Dinosauria, locomotion, functional morphology.

P LATEOSAURUS and *Massospondylus* are two well-known sauropodomorphs that have been reconstructed as animals capable of walking both bipedally and quadrupedally (see Galton 1990 and Galton and Upchurch 2004 for an overview). However, deciphering the posture of basal sauropodomorphs poses numerous difficulties because of the phylogenetic and functional ambiguities surrounding these dinosaurs. Phylogenetically, these taxa may, or may not, constitute the monophyletic sister-group to the Sauropoda (Wilson and Sereno 1998; Wilson 2002; Yates and Kitching 2003; Galton and Upchurch 2004; Upchurch *et al.* 2004; Yates 2007). Alternatively, some taxa, such as *Melanorosaurus*, may either be part of a paraphyletic 'Prosauropoda' close to the common ancestor of sauropods (Yates and Kitching 2003; Bonnan

and Yates 2007) or may be basal members of Sauropoda (Upchurch *et al.* 2004), depending on the definition of Sauropoda that is adopted. Functionally, because of marked differences in limb and vertebral morphology among the basal sauropodomorphs, the habitual posture of so-called 'prosauropod' dinosaurs has been difficult to constrain. Many of these dinosaurs are reconstructed as facultative bipeds, whereas others may have been habitually quadrupedal (Bonaparte 1969; Baird 1980; Paul 1987; Galton 1990; Thulborn 1990; Christian and Preuschoft 1996; Van Heerden 1997; Galton and Upchurch 2004; Bonnan and Yates 2007).

A diversity of data, including trackways (Baird 1980; Thulborn 1990), trunk length to hindlimb length ratios (Galton 1971, 1976, 1990; Van Heerden 1997) and vertebral dimensions (Christian and Preuschoft 1996), have previously been utilized to reconstruct basal sauropodomorph posture and locomotion. Surprisingly, with the exception of manus function (Galton 1971; Cooper 1981), discussions of basal sauropodomorph locomotion have rarely considered the functional morphology and anatomy of their forelimbs. Here, we examine the forelimb articulations and morphology of *Plateosaurus* and *Massospondylus* using the Extant Phylogenetic Bracket (EPB) approach (Witmer 1995) to clarify the role that the forelimb of these two dinosaurs may have played in locomotion.

Despite the current flux in our understanding of non-sauropod (basal) sauropodomorph relationships, most current phylogenetic analyses consistently divide these taxa into two major groups that generally correspond to previously inferred posture. One group, labelled by some as Plateosauria (e.g. Yates and Kitching 2003; Galton and Upchurch 2004), is a monophyletic clade that includes *Plateosaurus*, *Massospondylus* and other forms considered in previous analyses to be habitual bipeds (Galton 1971, 1976, 1990; Christian and Preuschoft 1996; Van Heerden 1997). By contrast, the other grouping, which Galton and Upchurch (2004) regarded as a monophyletic clade (Melanorosauridae), includes taxa such as *Melanorosaurus* and *Riojasaurus* that have been regarded as at least habitual quadrupeds (Bonaparte 1969; Van Heerden 1997; Bonnan and Yates 2007; although see Van Heerden and Galton 1997 for contrasting views on *Melanorosaurus*).

In this context, we selected *Plateosaurus* and *Massospondylus* for comparative examination because: (1) they currently occupy relatively stable phylogenetic positions within Sauropodomorpha; (2) they have been used extensively as examples of sauropodomorphs that were capable of both quadrupedal and bipedal locomotion; and (3) they were not closely related to other taxa, such as *Melanorosaurus*, which either share a close phylogenetic relationship with sauropods or are basal sauropods. *Plateosaurus* and *Massospondylus* are well-known taxa represented by abundant material, including almost complete ontogenetic series (Galton and Upchurch 2004; Reisz *et al.* 2005; Sander and Klein 2005), and their anatomy is often treated as a basal sauropodomorph 'template' (Van Heerden 1997; Galton and Upchurch 2004).

We focused in particular on anatomical features in the forelimbs related to pronation of the forearm and manus because of its bearing on locomotor ability and quadrupedal posture. Certainly, animals that walk bipedally may occasionally assume quadrupedal postures or move for short distances on all-fours (e.g. kangaroos: Hildebrand and Goslow 2001). However, if *Plateosaurus* and *Massospondylus* habitually walked quadrupedally, the forelimb would have had to function as a propulsive organ. In most extant quadrupeds, the manus is pronated such that the palmar surfaces of the digits and/or metacarpals

contact the ground. Pronation allows the manus to produce a caudally directed propulsive force that parallels the actions of the pes in a parasagittal plane (Bonnan 2003). In most quadrupedal mammals, permanent pronation is accomplished by crossing the radius over the ulna (permanently or actively), which subsequently orients the palmar surface of the manus caudally (Hildebrand and Goslow 2001; Bonnan 2003). Whereas tetrapods with non-parasagittal gaits pronate their manus without active or permanent radius cross-over, mammals and dinosaurs share parasagittal limbs that moved in dynamically similar ways (Carrano 1998). Therefore, we predict that if *Plateosaurus* and *Massospondylus* employed habitual quadrupedal locomotion, their forelimbs should display anatomical evidence compatible with pronation, which would have allowed the manus to produce a caudally directed force paralleling the action of the pes.

To avoid awkward phrases such as 'basal (non-sauropod) sauropodomorphs', we shall hereafter refer to *Plateosaurus* and *Massospondylus* as 'prosauropods'. We emphasize that our results and inferences apply strictly to these two 'prosauropods' only, and not to sauropodomorphs in general.

Institutional abbreviations. AMNH, American Museum of Natural History, New York; BP, Bernard Price Institute of Palaeontology and Palaeoanthropology, Johannesburg; FMNH, Field Museum of Natural History, Chicago; NM, National Museum, Bloemfontein, Free State, South Africa; NMZ, National Museum of Zimbabwe, Bulawayo; OMNH, Oklahoma Museum of Natural History, Norman; SAM, Iziko South African Museum, Cape Town; USNM, United States Museum of Natural History, Washington, DC; WIU, Western Illinois University, Macomb; ZDM, Zigong Dinosaur Museum, Zigong.

Abbreviations on text-figures. c, coracoid; c2, c3, c4, manus unguals; c1, pollex claw; clp, craniolateral process of the ulna; cmp, craniomedial process of the ulna; h, humerus; r, radius; s, scapula; u, ulna.

MATERIAL AND METHODS

Material and manipulation

We examined the complete and undistorted scapulocoracoids and proximal forelimbs of the 'prosauropods' *Plateosaurus engelhardti* (AMNH 2409) and *Massospondylus carinatus* (SAM-PK-K5135). Observations and measurements derived from these specimens were compared with other *Plateosaurus* and *Massospondylus* forelimbs to rule out inferring abnormal movements owing to individual variation. To polarize the primitive and derived characteristics of the forelimb movements reported here, we examined the range of motion in skeletonized forelimbs in several extant outgroup taxa: two monitor lizards (*Varanus* sp. FMNH 31338 and USNM 290873), an American alligator (*Alligator mississippiensis*: WIU NSFG2), a goose (*Anser domesticus*: WIU NSFG3) and an ostrich (*Struthio camelus*: WIU NSFG1). We also compared our results with previously published data on forelimb movements or morphology in the saurischian outgroups of these two 'prosauropods': *Herrerasaurus* (Sereno 1993; Bonnan 2001, 2003), *Coelophysis* (Carpenter 2002), *Dilophosaurus* (Welles 1984), *Allosaurus* (Carpenter 2002), *Acrocanthosaurus* (Senter and Robins 2005), *Deinonychus* (Gishlick 2001), *Melanorosaurus* (Bonnan and Yates 2007), and *Apatosaurus* and other neosauropods (Bonnan 2001, 2003, 2004).

Examination of *Plateosaurus* and *Massospondylus* involved articulation and manipulation of the scapulocoracoid, humerus and antebrachium using previously published methods (Carpenter 2002; Senter and Robins 2005). Elements were either supported from beneath (by metal rods padded with foam rubber) and clamped to chemistry ring stands or carefully posed on foam and cloth. At the glenoid, the humerus was posed in three positions: maximum extension, maximum abduction and maximum flexion/depression (Text-figs 1–2). At the elbow, the humerus and antebrachium were posed with the elbow in full extension and flexion (Text-figs 1–2). Where necessary, elements were fastened together as needed with plastic chicken wire, plastic-coated wire twists, modelling clay or masking tape. The scapulocoracoids of the extant taxa are oriented differently from those of *Plateosaurus*, *Massospondylus* and other saurischian dinosaurs owing to their different locomotor modes. To simplify comparisons and maintain uniformity in the range of motion among the extant taxa, the scapulocoracoids and forelimbs have been rotated in the figures presented here to match the orientation of the *Plateosaurus* and *Massospondylus* glenoid.

Inferring limb movements

As in previous studies, the edges of articular surfaces were presumed to represent the limits of possible motion (Gishlick 2001; Carpenter 2002; Senter 2005; Senter and Robins 2005). We acknowledge that the absence of cartilage and other soft tissues poses a problem when attempting to place realistic constraints on forelimb movements in dinosaurs. In particular, the absence of secondarily ossified epiphyses in dinosaur long bones indicates that an unknown portion of the articular surface is lost during fossilization. In fact, the so-called 'epiphyses' of dinosaur long bones are actually the terminal ends of the metaphysis capped with the calcified cartilage that in life underlay the living and unpreserved articular cartilage (Haines 1939, 1952, 1969; Reid 1997; Horner *et al.* 2001;

Chinsamy-Turan 2005). This has led to concerns about the accuracy and reliability of inferring dinosaur long bone functional morphology. For example, Holliday *et al.* (2001) showed that 10–20 per cent of limb bone size, as well as major articular surface features, are severely truncated or lost during skeletonization in extant archosaurs. By extension, they suggested that an appreciable amount of morphological data may be absent from dinosaur skeletons (Holliday *et al.* 2001).

Although the loss of articular cartilage poses problems, it does not follow that no constraints can be placed on inferences of limb function. In dinosaurs, as in crocodilians and turtles, the calcified cartilage at the end of the metaphysis was the region of long bone growth and is often called a 'growth plate'. It is instructive to think about the 'growth plate' as a region of calcified cartilage sandwiched between the articular cartilage above and the metaphysis below. In this sense, the preserved, calcified cartilage 'growth plate' on the ends of dinosaur long bones followed the general shape of the genuine articular surface (Reid 1997). This differs from the situation in mammals and lizards, in which a secondary centre of ossification forms in the articular cartilage and the 'growth plate' does not reflect the shape of the overlying cartilage (Reid 1997; Chinsamy-Turan 2005). The *Varanus* specimens used here were adults with ossified epiphyses, so post-mortem loss of articular cartilage did not result in the loss of the epiphysis itself.

Therefore, although the thickness of dinosaur long bone articular cartilage is difficult to determine, we infer, on the basis of outgroup comparison, that the articular cartilages capping the bones in the shoulder and elbow joints of dinosaurs are reflected by the contours of the calcified cartilages at the ends of the metatphyses. Moreover, although soft tissue constraints on joint morphology cannot be precisely reconstructed, simplification of the complex mechanics of the forelimb using bone-to-bone articulations (1): can provide useful maxima and minima for forelimb movements; (2) can reveal relative differences in basic forelimb movements among the examined taxa; and (3) is an approach that can be replicated. We emphasize that we are reporting general rather than specific forelimb movements in *Plateosaurus* and *Massospondylus*. These methods are employed here to investigate whether pronation or regular quadrupedal locomotion were plausible given the anatomy of their forelimbs.

RESULTS

Extant taxa

We will not reiterate the basic forelimb movements of the extant outgroup taxa, which are described sufficiently in the literature [e.g. varanids (Haines 1952; Jenkins and Goslow 1983; Landsmeer 1983, 1984); crocodilians (Zug 1974; Reilly and Elias 1998); volant birds (Muybridge 1957; Jenkins *et al.* 1988; Poore *et al.* 1997)]. Instead, we focus on shared movements among the extant taxa from which we infer the plesiomorphic functional morphology for dinosaur forelimbs, including those of the two 'pro-sauropods'.

Manual articulation and manipulation of the humeral head with the glenoid shows that, despite differences in the maximum range of motion among the extant taxa, flexion at the glenohumeral joint is less well developed than extension in the extant outgroup taxa (Text-fig. 1). Flexion at the glenohumeral joint was most constrained in the birds *Anser* and *Struthio*. As might be predicted, the range of motion at the glenoid allows a substantial amount of rotational movement in *Varanus* and *Alligator*, relating to their 'sprawling' gait (e.g. Jenkins and Goslow 1983; Landsmeer 1983, 1984; Meers 2003). Rotational movements related to flight were also possible in *Anser*, as predicted from previous studies of volant bird taxa (e.g. Jenkins *et al.* 1988; Proctor and Lynch 1993; Poore *et al.* 1997). With the exception of *Struthio*, all extant outgroup taxa display a large range of abduction related either to a sprawling gait (e.g. *Varanus, Alligator*) or flight

(e.g. *Anser*). For *Struthio*, abduction movements were more limited, agreeing with wing manipulations in a freshly dead ostrich specimen (MFB, pers. obs.).

The elbow can flex to a strongly acute angle in all extant outgroup taxa except *Struthio*, where flexion is more limited (Text-fig. 2). Active pronation of the radius about the ulna, as in some mammals, was not possible in any of the above-mentioned extant outgroup taxa. In the archosaur taxa, the oblong head of the radius, the relatively flat ulna-radius articulations, and the parallel orientations of the radius and ulna throughout their lengths prevent the radius from crossing the ulna. In wet specimens of *Alligator, Anser* and *Struthio*, it was possible to slide the radius distally in relation to the ulna, producing the parallel, piston-like movements previously reported for birds (Liem *et al.* 2001). In birds, these movements act on the carpus, allowing ulnar deviation of the manus, a movement used in flight (Proctor and Lynch 1993; Liem *et al.* 2001). In *Alligator*, the piston-like movement of the radius on the ulna appears to bring about ulnar deviation. One of us (MFB) observed that as the radius slides distally, it presses against the lateral carpals, which may also cause ulnar deviation of the manus. However, further investigations and cineradiography are required to substantiate this proposed mechanism.

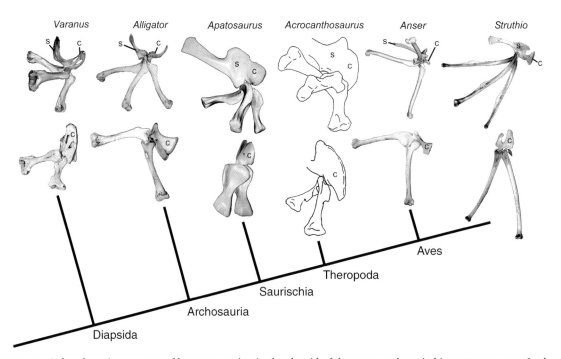

TEXT-FIG. 1. Inferred maximum range of humerus motion in the glenoid of the extant and saurischian outgroup taxa. In the top row, all humeri, scapulae and coracoids are oriented in lateral view, showing maximum range of humeral flexion and extension. In the bottom row, all humeri, scapulae and coracoids are oriented in cranial view, showing maximum range of humeral abduction and adduction. For consistency and ease of comparison, forelimbs of *Varanus* (FMNH 3133), *Alligator* (WIU NSFG2), *Anser* (WIU NSFG3) and *Struthio* (WIU NSFG1) are oriented in the same plane as that of the dinosaurs. The *Apatosaurus* forelimb is a scale-model by Phil Platt discussed in Bonnan (2003). The illustrations of *Acrocanthosaurus* are modified from Senter and Robins (2005). Not to scale.

In *Varanus*, a unique relationship between the humerus, radius and ulna allows manus pronation. In this taxon, independent, lateral rotation of both the radius and the ulna against the humeral condyles cause the distal ends of these bones to cross and separate (Landsmeer 1983). This motion is transferred through the carpal series, which in turn 'flips' the manus into a pronated orientation (see Landsmeer 1983, 1984 for more details). As Landsmeer (1983, 1984) described, this type of pronation is possible in *Varanus* because: (1) the radius and ulna have independent articulations with their respective condyles on the humerus; (2) a radial facet on the ulna allows the proximal end of the radius to slide and rotate laterally with the ulna; (3) the complex morphology of the carpus transfers the rotational movements of the radius and ulna to the manus; and (4) a system of ligaments between the radius, ulna and carpals augment and refine these movements. We were able to replicate the movements of the radius and ulna described by Landsmeer (1983, 1984) in a skeletonized *Varanus* specimen (USNM 290873).

Saurischian outgroups to Plateosaurus and Massospondylus

To polarize further the plesiomorphic forelimb movements of *Plateosaurus* and *Massospondylus*, we utilized previously published studies describing forelimb movements in their saurischian outgroups: theropods, sauropods and the sauropodomorph *Melanorosaurus*. In basal theropods, ceratosaurs, tetanurans and maniraptorans, maximum flexion at the glenohumeral joint is reduced, with the humerus assuming a subvertical position, whereas at maximum extension the humerus assumes a subhorizontal orientation relative to the glenoid (Welles 1984; Sereno 1993; Bonnan 2001, 2003; Gishlick 2001; Carpenter 2002; Senter and Robins 2005: Text-fig. 1). These results are similar to those reported for the above-mentioned extant taxa, bolstering the inference that the plesiomorphic range of extension at the glenohumeral joint in saurischian dinosaurs was far greater than flexion. In contrast to the extant taxa, non-maniraptoran theropods display limited abduction, and rotational movements within the glenoid appear reduced or absent (Sereno 1993; Bonnan 2001, 2003; Carpenter 2002; Senter and Robins 2005). An exception appears in maniraptorans and other derived coelurosaurs where the humerus can be abducted to a subhorizontal position (Jenkins *et al.* 1988; Poore *et al.* 1997; Gishlick 2001; Carpenter 2002; Senter and Robins 2005).

In most theropods, elbow flexion is less acute than in the extant taxa, whereas elbow extension can approach, but not achieve, 180 degrees (Welles 1984; Sereno 1993;

Gishlick 2001; Carpenter 2002; Senter and Robins 2005: Text-fig. 2). In many theropods, such as *Allosaurus*, the presence of a cuboid fossa on the cranial aspect of the humerus would have been capable of accepting the head of the radius during flexion (Bonnan 2001, 2003). The presence of this feature on many theropod humeri suggests that although elbow flexion was less acute than in the extant outgroup taxa, a significant amount of flexion was still possible. In derived theropods such as *Deinonychus*, the antebrachium can be flexed to a strongly acute angle (Jenkins *et al.* 1988; Poore *et al.* 1997; Gishlick 2001; Carpenter 2002; Senter and Robins 2005). As with the extant taxa, the radius was incapable of crossing the ulna to achieve pronation, and the radius cannot be rotated actively about the ulna to produce manus pronation in these theropods.

Pronation and supination were probably absent or reduced in the earliest saurischians (e.g. *Herrerasaurus*: Sereno 1993; Bonnan 2001, 2003), but it is not clear whether the radius could move in a piston-like fashion against the ulna, as in extant archosaurs. In *Herrerasaurus*, such a movement is probably precluded because the distal ends of the antebrachial elements are pitted and scarred, suggesting that they were held together by ligaments or other soft tissues (Sereno 1993). Furthermore, the carpal series of most theropods is fairly short and does not display the arrangement observed in *Alligator*, which allows ulnar deviation and possibly pronation (see above). If proximodistal movements of the radius did occur, it appears most likely that ulnar deviation of the manus, rather than its pronation, would result. In fact, in maniraptorans such as *Deinonychus*, ulnar deviation was apparently the primary movement of the manus (Ostrom 1969; Sereno 1993; Gishlick 2001). The condyles of varanid humeri are distinctly separated, allowing pronation through independent movements of the antebrachial elements, and the ulna possesses a radial facet that allows the radius to rotate against it: these features are absent in theropods. Overall, it appears that the neutral position of the manus was semi-supinated in most theropods: the palms faced medially, or medioventrally, in a prayer-like orientation (Carpenter 2002; Senter and Robins 2005).

Forelimb osteology and trackway data show that most sauropods held their forelimbs in a columnar posture with the manus pronated to within 5–55 degrees of the direction of travel (Bonnan 2001, 2003). Movement of the humerus at the glenoid was more restricted than in theropods: nevertheless, similar, generalized patterns of forelimb movements emerge. Flexion at the glenohumeral joint is reduced compared with extension, although some sauropods (e.g. *Apatosaurus*) were apparently capable of more humeral flexion than theropods (Bonnan 2001: Text-fig. 1). Abduction of the humerus and rotational movements within the glenoid were apparently very limited (Bonnan 2001, 2003, 2004). It is clear that active pronation was impossible in sauropods because the radial head was angular and cradled within a deep radial fossa on the cranial aspect of the ulna (Upchurch 1998; Wilson and Sereno 1998; Bonnan 2003). Distally, the ends of the radius and ulna were scarred and rugose, which suggests that a ligamentous network greatly immobilized these elements (Hatcher 1902; Bonnan 2001, 2003). The relatively flat and undivided humeral condyles, the lack of a cuboid fossa, and the gently concave or convex surfaces of the radius and ulna, respectively, suggest elbow flexion was relatively reduced in sauropods (Upchurch 1998; Wilson and Sereno 1998; Bonnan 2003: Text-fig. 2). As in proboscideans, complete extension of the forearm against the humerus was possible and manus pronation was permanent (Wilson and Sereno 1998).

Disagreement regarding how sauropods pronated their manus without crossing their radius over their ulna as in graviportal mammals has continued for over a century (see Bonnan 2003 for a synopsis). Most recently, Wilson and Sereno (1998) and Wilson and Carrano (1999) suggested that the radius and ulna assumed typical

saurischian orientations in sauropods, with the radius positioned lateral to the ulna proximally and cranial to the ulna distally. They inferred that in this arrangement the manus was permanently supinated relative to the direction of travel (Wilson and Sereno 1998; Wilson and Carrano 1999). However, using a scale model of *Apatosaurus*, Bonnan (2003) demonstrated that the radius has shifted cranially and slightly medially in relation to the ulna proximally and distally in sauropods. Bonnan (2003) showed that the craniolateral process of the ulna, a well-known synapomorphy of Sauropoda, was a morphological 'signal' for this change in radius orientation in sauropods. Bonnan (2003) was only able to replicate the range of manus pronation in sauropod trackways with the antebrachial bones in this derived orientation, and could not do so following the proposals of Wilson and Sereno (1998) and Wilson and Carrano (1999). Bonnan (2003) concluded that manus pronation in sauropods was correlated with the development of a craniolateral process on the proximal part of the ulna.

The forelimb osteology of the basal sauropodomorph *Melanorosaurus* may help resolve this disagreement and

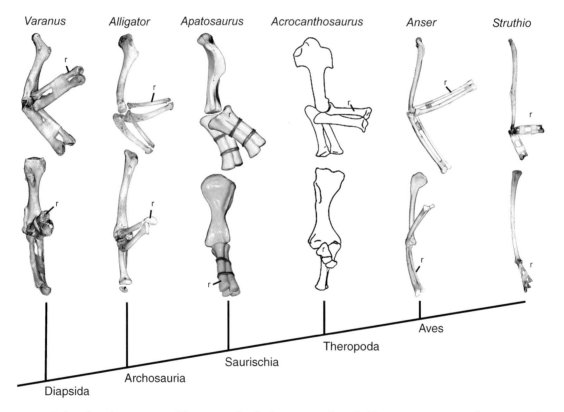

TEXT-FIG. 2. Inferred maximum range of forearm motion in the extant and saurischian outgroup taxa. In the top row, the humeri, radii and ulnae are oriented in lateral view, showing the maximum range of antebrachial flexion and extension. In the bottom row, all humeri, radii and ulnae are oriented in cranial view. For consistency and ease of comparison, forelimbs of *Varanus* (FMNH 3133), *Alligator* (WIU NSFG2), *Anser* (WIU NSFG3) and *Struthio* (WIU NSFG1) are oriented in the same plane as that of the dinosaurs. The *Apatosaurus* forelimb is a scale-model by Phil Platt discussed in Bonnan (2003). The illustrations of *Acrocanthosaurus* are modified from Senter and Robins (2005). Not to scale.

shows anatomical features that presage a permanent 'switch' to full quadrupedality in sauropods. The forelimb of *Melanorosaurus* shares some characteristics with those of *Plateosaurus* and *Massospondylus*, but other aspects of its forelimb anatomy diverge from these two taxa. Overall, the morphology of the humerus shares many similarities to those of *Plateosaurus* and *Massospondylus*. Proximally, the subconvex humeral head resembles those of *Plateosaurus* and *Massospondylus*, and we infer that its articulation and movement within the scapulocoracoid glenoid could not have differed much from these taxa. Muscle insertion landmarks, such as the deltopectoral crest, differ from the other 'prosauropods' in being less pronounced (Galton and Upchurch 2004; Galton *et al.* 2005), but the overall dimensions of the humerus and these landmarks still fall within the range reported for other basal sauropodomorphs (Bonnan and Yates 2007). Distally, a shallow cuboid fossa is present, and distal condyles are present but not well developed.

The radius and ulna not only assume a sauropod-like morphology in *Melanorosaurus*, but these elements also articulate with one another as they do in the derived condition reported for sauropods by Bonnan (2003). The radius is cradled within a radial fossa on the cranial face of the ulna proximally, and the ulna bears a craniolateral process that is identical to those reported for all described sauropod ulnae (Bonnan and Yates 2007). The radius is a straight element that has shifted to a cranial and slightly medial orientation proximally and distally in relation to the ulna. In the *Melanorosaurus* specimen NM QR3314, the forelimb is preserved in articulation *in situ*, clearly showing this change in radius and ulna morphology and orientation. These observations suggested to Bonnan and Yates (2007) that *Melanorosaurus* had developed some degree of manus pronation. Unlike *Plateosaurus* and *Massospondylus*, the distal ends of the radius and ulna bear roughened areas that may indicate their binding by ligamentous tissues (Bonnan and Yates 2007), which is another sauropod-like feature.

In the articulated specimen NM QR3341 carpal bones are not present. Bonnan and Yates (2007) suggested that the completeness and articulation of the specimen indicated that the carpal elements had failed to ossify, even though the individual was quite large (over 4 m long). An ontogenetic delay in carpus ossification may be typical of basal sauropodmorphs, as indicated by a juvenile *Massospondylus*, well over hatchling size (BP/1/4376), with an articulated forelimb and manus that lacks an ossified carpus (see below). Although the manus of *Melanorosaurus* is relatively short, it still falls within the range of most sauropodomorphs (Bonnan and Yates 2007). The presence of a divergent pollex claw and the retention of claws on digits II and III

bear many resemblances to those of other 'prosauropods' (Bonnan and Yates 2007). The *in situ* preservation of the manus in specimen NM QR3314 shows that proximally the metacarpus assumed the gently arched shape characteristic of other non-sauropod saurischians (Wilson and Sereno 1998; Bonnan 2003; Bonnan and Yates 2007). This morphology contrasts with the columnar and semicircular arrangement of metacarpals in most known sauropods (Wilson and Sereno 1998; Wilson 2002; Bonnan 2003; Upchurch *et al.* 2004).

Plateosaurus *and* Massospondylus

Based on the preceding descriptions of forelimb anatomy in the outgroup taxa, we infer the following movements to be plesiomorphic for *Plateosaurus* and *Massospondylus*. At the glenoid, extension at the glenohumeral joint was more pronounced than flexion of the humerus. Although a large degree of abduction and rotational movements were originally possible at the glenohumeral joint, abduction and rotation were later reduced in saurischian dinosaurs. The condition of increased humeral abduction and rotation observed in certain maniraptorans is probably derived. Elbow flexion, although not as acute as in birds, crocodilians and varanids, was still present in saurischian dinosaurs. In theropods, the presence of a cuboid fossa may have augmented this movement, and elbow extension was well developed but the forearm was not capable of complete extension against the humerus. Because of their columnar limb posture, the forearms of sauropods were less flexible but capable of complete extension against the humerus. Active pronation was plesiomorphically reduced or absent in saurischians, although piston-like movements of the radius observed in extant archosaurs could have contributed to some ulnar deviation of the manus. The absence of several anatomical specializations in the forelimb of the saurischian outgroup taxa precludes the specialized movements of the radius and ulna described for *Varanus*. Permanent pronation in sauropods appears to have occurred via the evolution of a derived antebrachial morphology.

The ranges of movements inferred for *Plateosaurus* and *Massospondylus* did not vary significantly from one another. Thus, the movements and articulations described here apply to both taxa. At the glenoid, as in other saurischian dinosaurs and the extant outgroup taxa, the humerus is incapable of flexing much past vertical (Text-fig. 3). Extension at the glenohumeral joint is well developed and the humerus can assume a subhorizontal orientation at its maximum range. Humeral abduction is limited when compared with theropods or the extant outgroup taxa, but the range of abduction is greater than that reported for sauropods.

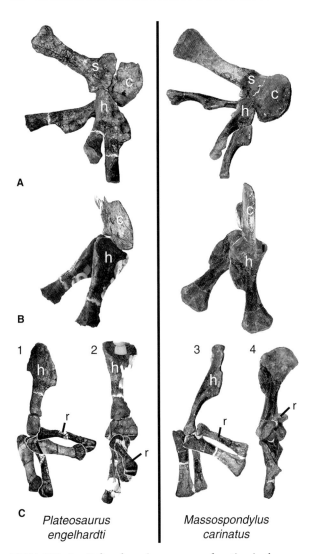

A

B

1   2   3   4

C

*Plateosaurus
engelhardti*

*Massospondylus
carinatus*

TEXT-FIG. 3. Inferred maximum range of motion in the basal sauropodomorphs *Plateosaurus engelhardti* (AMNH 2409) and *Massospondylus cariantus* (SAM-PK-K5135). A, maximum humeral extension and flexion at the glenoid in lateral view. B, maximum humeral abduction and adduction at the glenoid in cranial view. C, maximum antebrachial extension and flexion in 1, 3, lateral view and 2, 4, cranial view. Not to scale.

Flexion and extension at the elbow in both *Plateosaurus* and *Massospondylus* falls within the range reported previously for most non-maniraptoran theropods (Text-fig. 3). The presence of a cuboid fossa that could accommodate the radial head on the humeri of both *Plateosaurus* and *Massospondylus* supports a range of forearm flexion at least as acute as those reported for theropods. As with theropods, the forearm has a fair range of extension, but cannot open to 180 degrees. This condition contrasts with that of sauropods in which flexion is greatly reduced and forearm extension is complete.

As in both the extant outgroup taxa and the saurischian sister-groups, active pronation of the radius about the ulna is precluded in 'prosauropods'. The oblong radial head and flat articulations between the radius and ulna prevent long-axis rotation of the former about the latter (see also Bonnan 2003). The radius and ulna parallel one another throughout their lengths, and the radius does not cross the ulna as in some quadrupedal mammals. Distally, the radius remains craniolateral, whereas the end of the ulna retains its caudal position. The ulna does not have a craniolateral process as in sauropods. As in all tetrapods, digit I aligns with the radius and digit V aligns with the ulna (see Bonnan 2003). Given these constraints, the manus is in a semi-supinated or 'prayer-like' orientation with the palmar surfaces facing medially or slightly ventromedially as in theropods (see also Bonnan 2003). As in other saurischians, the anatomical specializations observed in varanid lizards for independent movement of the radius and ulna are lacking. The humeral condyles of these two 'prosauropods' are less well developed than those of theropods but more developed than those of sauropods, and there is no distinct separation between the condyles. Moreover, as in theropods, the ulna lacks a radial facet.

It is possible that a piston-like action of the radius against the ulna could have occurred in the two 'prosauropod' taxa, but evidence for such a mechanism is lacking. Moreover, two observations preclude a significant carpal contribution to manus pronation in *Plateosaurus* and *Massospondylus*. First, without exception, all examined *Plateosaurus* and *Massospondylus* forelimbs articulated *in situ* show no evidence of manus pronation: digit I is always aligned with the radius and digit V is always aligned with the ulna (Text-figs 4–5). The manus is never significantly rotated or otherwise unaligned with the antebrachium in naturally articulated specimens, including juveniles (Text-figs 4–5). Secondly, in adult specimens the carpal bones form a narrow band of distal, ossified elements that may have capped the proximal ends of the metacarpals and/or articulated proximally with cartilaginous carpal elements. In *Herrerasaurus*, Sereno (1993) suggested that the two-tiered arrangement of its carpals may have contributed to additional wrist flexion and slight wrist rotation. However, the short, quadrangular morphology and flattened articular surfaces of 'prosauropod' carpals appear to limit their contribution to rotational movements such as manus pronation and supination. A similar condition is also noted in neosauropod carpals (Bonnan 2003). Therefore, it appears unlikely that movements of the carpal bones would have rotated the manus to such an extent that the orientation observed in naturally articulated forelimbs would be significantly altered. In juvenile specimens, no carpals are reported and these elements are presumably present as cartilage condensations.

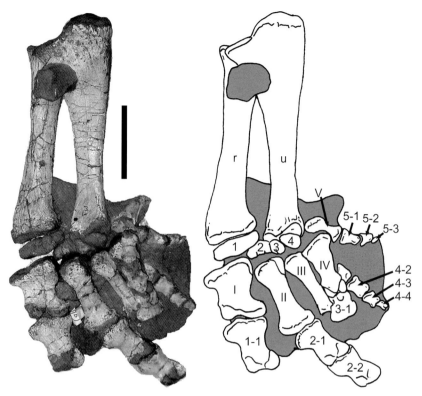

TEXT-FIG. 4.   Articulated *Plateosaurus* forearm and manus (SMNS F61) *in situ*, palmar view with interpretive drawing. Note that the radius and ulna do not cross and that the manus is aligned in the same orientation as the antebrachium. The carpal bones are arranged in a narrow band and their quadrangular morphologies prevent any significant rotational or twisting motions. Phalanx III-1 has rotated out of position with metacarpal III. 1, 2, 3, 4, carpal bones; 1-1, phalanx I-1; 2-1, 2-2, phalanges of digit II; 3-1, phalanx III-1; 4-1, 4-2, 4-3, phalanges of digit IV; 5-1, 5-2, 5-3, phalanges of digit V; for other abbreviations, see text. Roman numerals indicate digit numbers. Grey shading indicates matrix. Scale bar represents 10 cm.

DISCUSSION

The forelimbs of *Plateosaurus* and *Massospondylus* retain many of the plesiomorphic features of the outgroup taxa. At the glenoid, the humerus could not flex much beyond vertical, whereas the humerus could be extended to a nearly horizontal orientation. Movements of abduction and rotation at the glenoid were more limited than the plesiomorphic condition, but not as constrained as in sauropods. Elbow flexion and extension were similar to results reported for most theropods, wherein flexion and extension were well developed but not as extensive as in certain extant outgroup taxa. A significant degree of forearm flexion at the elbow is also indicated by the presence of a cuboid fossa on the humerus. Active pronation via rotation of the radius about the ulna is precluded by the morphology and articulations of the antebrachial bones. Moreover, the morphology of the carpus and the *in situ* articulated forelimbs of *Plateosaurus* and *Massospondylus* specimens further suggest that pronation was limited or absent in these dinosaurs. Based on outgroup comparison and manual manipulation of the forelimbs of these taxa, the most parsimonious orientation of the manus was in a

semi-supinated position with the palmar surface directed medially or slightly ventromedially. This orientation matches the condition observed in theropods.

Our results, combined with previously published observations of saurischian outgroup taxa and our own examination of extant outgroup taxa, suggest that the forelimbs of *Plateosaurus* and *Massospondylus* lacked adaptations for efficient, habitual quadrupedal locomotion. It is clear that the forelimb of *Plateosaurus* and *Massospondylus* could not operate in a dynamically similar way to the hindlimb. The articular shapes of the glenoid and humeral head would have prevented the humerus from achieving much flexion past vertical and this would not allow the forelimb to swing cranially and caudally in a fashion similar to the hindlimb, as in mammals (Hildebrand and Goslow 2001; Liem *et al.* 2001). Moreover, the orientation of the manus relative to the direction of travel would be nearly orthogonal and this would have prevented the manus from producing a caudally directed force if locomotion were quadrupedal.

It is also doubtful that movements of the scapulocoracoid could contribute to humeral flexion or cranial translation of the forelimb as the scapula does in many quadrupedal mammals, in which it is a free element that often lacks a connection to the sternum and is suspended entirely by soft tissues (Hildebrand and Goslow 2001; Liem *et al.* 2001). By contrast, the scapulocoracoid of dinosaurs, including *Plateosaurus* and *Massospondylus*, articulated directly with the sternum as in extant archosaurs and lepidosaurs. In lepidosaurs and crocodilians, movement of the coracoid within the sternal joint increases rotational movements associated with a sprawling gait, and allows some cranial translation (Jenkins and Goslow 1983; Meers 2003; MFB pers. obs.), but not to the extent observed in quadrupedal mammals. In birds, the articulation of the coracoid is nearly perpendicular to the long axis of the sternum. This articulation allows a lateral sliding motion during wing beats (Jenkins *et al.* 1988), but rotational movements or cranial translation are essentially non-existent. All of these observations suggest that movements of the scapulocoracoid were restricted in *Plateosaurus* and *Massospondylus* and did not contribute significantly to cranial translation of the forelimb. Moreover, the presence of clavicles arising from the acromion region and bridging the scapulocoracoids in *Massospondylus* and possibly other 'prosauropods' (Yates and Vasconcelos 2005) suggests that no mammal-like movements of that element were possible.

Many sprawling tetrapods do not cross the radius over the ulna, but pronate the manus by virtue of the horizontal orientation of their humerus. In this position, the lateral condyle of the humerus can be positioned cranially or craniomedially in relation to the medial condyle, which subsequently positions the radius craniomedial to the ulna, resulting in manus pronation. Could *Plateosaurus* and *Massospondylus* have been habitual quadrupeds in which the forelimbs operated in a sprawling or semisprawling orientation as suggested for some ceratopsians (Johnson and Ostrom 1995; Dodson 1996)? It is significant to note that in *Plateosaurus* and *Massospondylus*, although the plesiomorphic range of humeral flexion and extension were retained, the amount of inferred abduction at the glenoid was reduced. Thus, unlike varanids, crocodilians and even some ceratopsians, these 'prosauropods' did not have a large range of abduction movements. This would have limited the range of mobility of the forelimb if it was utilized in a sprawled or semisprawled orientation.

Knuckle-walking adaptations in Plateosaurus *and* Massospondylus?

Giant anteaters (*Myrmecophaga tridactyla*) and some great apes (*Gorilla* and *Pan*) utilize a specialized type of forelimb locomotion called 'knuckle-walking' (Orr 2005). In these mammals, the middle phalanges are flexed and bear weight, whereas the metacarpophalangeal joints are 'locked' in extension to prevent wrist instability. Although the manus remains pronated in most knuckle-walkers, some chimpanzees are capable of knuckle-walking with a semi-supinated, medially directed manus (Orr 2005). Thus, in rare cases, a non-pronated forelimb can be used for propulsion if knuckle-walking is possible.

Knuckle-walking requires stability at the carpus and manus to prevent hyperextension. Orr (2005) has shown that, despite their distant phylogenetic relationships, giant anteaters and great apes share a suite of convergent osteological features in their antebrachium and manus, including: (1) dorsal extension of the radius over the proximal part of the carpus; (2) prevention of hyperextension of the proximal phalanges by a ridge at the dorsal edge of the phalangeal articular surface of each weight-bearing metacarpal; and (3) the heads of weight-bearing metacarpals are wider dorsally than on their palmar aspects. In chimpanzees, caudally directed forces can presumably be generated during semi-supinated knuckle-walking because of the large amount of ulnar deviation allowed at the wrist. In hominoids, the shortened end of the styloid process of the ulna, coupled with a meniscus-filled space between this element and the proximal carpal row, 'enhances' such ulnar deviation (Lewis 1969; Orr 2005).

None of the specializations observed in knuckle-walking quadrupedal mammals is present in *Plateosaurus*, *Massospondylus* or any other saurischian dinosaur. The radius of *Plateosaurus* and *Massospondylus* does not extend dorsally over the proximal portion of the carpus, no ridges are present on the distal ends of the metacarpals, and there is no significant dimensional difference between the dorsal and palmar aspects of the metacarpals. Moreover, ulnar deviation of the kind observed in chimpanzees and other hominoids was not possible given the short, block-like carpals of *Plateosaurus* and *Massospondylus*. These 'prosauropods' also lack the space for a large meniscus between the ulna and the proximal carpals. Thus, it is very unlikely that *Plateosaurus*, *Massospondylus* or other saurischian dinosaurs utilized knuckle-walking.

'Prosauropod' manus prints and quadrupedal locomotion

Several Triassic manus prints have been attributed to 'prosauropods', but these designations have been problematic (see Rainforth 2003 and Wilson 2005 for overviews and several examples). Several manus prints attributed to 'prosauropods', especially those of the ichnotaxon *Tetrasauropus*, exhibit foreshortened digits, a compact outline, and well-developed pronation that suggests a closer resemblance to the manus of basal sauropods (see Wilson 2005, fig. 5). In other cases, 'prosauropod' manus tracks exhibit large pollex imprints with shortened lateral digits (Baird 1980; Thulborn 1990; Lockley 1991; Lockley and Hunt 1995). Although it is not clear whether these ichnotaxa represent 'prosauropods' or basal sauropods (see Wilson 2005) or even non-dinosaur archosaurs (Rainforth 2003), they do not resemble the expected morphology of a manus print that would be generated by *Plateosaurus* or *Massospondylus*.

As shown by Galton (1990), the articular surfaces of the pollex claw, phalanx I-1 and the distal end of metacarpal I would have acted to abduct and direct the pollex medially in *Plateosaurus*. This would have resulted in this digit being held off the ground if used quadrupedally, presumably leaving no impression in trackways. The manus morphology of *Massospondylus* is essentially similar to that of *Plateosaurus* (see Cooper 1981; Galton 1990; Text-fig. 5), and we would expect a similar manus posture in this taxon. Moreover, we would predict that the second and third digits of 'prosauropods' like *Plateosaurus* and *Massospondylus* would have left more elongate impressions instead of the foreshortened impressions of ichnotaxa attributed to 'prosauropods'. It is also unlikely that these manus prints are the result of knuckle-walking because this posture is osteologically untenable in these two taxa (see above).

However, other sauropodmorphs with a manus morphology that differs from *Plateosaurus* and *Massospondylus* may be 'responsible' for these trackways. For example, the manus of *Melanorosaurus* is more foreshortened than those of *Plateosaurus* and *Massospondylus*, would have been at least partially pronated when this animal walked quadrupedally, and bears blunted claws on its first three digits (Bonnan and Yates 2007). The trackways of *Navahopus* and *Tetrasauropus* certainly bear resemblances to the manus of such a sauropodomorph (see Lockley 1991 for trackway illustrations of *Navahopus*; Wilson 2005 for trackway illustrations of *Tetrasauropus*; but see Rainforth 2003 for contrasting views on *Tetrasauropus*). Moreover, basal sauropods such as *Antetonitrus* may have retained a manus similar in morphology to *Melanorosaurus* and could also fit the 'profile' of these trackways (see Yates and Kitching 2003 for an inferred reconstruction of the *Antetonitrus* manus).

We find it significant that most of the reported manus prints attributed to Late Triassic and Early Jurassic sauropodomorphs are relatively pronated. Specifically, trackmakers such as *Navahopus* and *Tetrasauropus* show well-pronated manus prints that fall within the range of lateral rotation of sauropods (5–55 degrees: Bonnan 2003; see above). Given the osteological and functional constraints described here for the forelimbs of *Plateosaurus* and *Massospondylus*, it is doubtful that either of these taxa could make such manus prints. We would expect much more laterally out-turned manus prints with no pollex marks and elongate digit impressions if either of these two 'prosauropods' walked quadrupedally.

It is intriguing therefore that a thorough analysis of the mostly bipedal ichnotaxon *Otozoum* by Rainforth (2003) shows: (1) it was made by a 'prosauropod' based on analyses of saurischian pedal skeletons; (2) it occurs in strata where sauropodomorphs such as *Massospondylus* are common; and (3) one set of footprints shows laterally rotated, semi-supinated manus prints without a pollex claw impression. These manus prints appear only once in the trackway and parallel one another (Rainforth 2003, fig. 4). Significantly, Rainforth (2003) concluded that these manus prints were made when the bipedal *Otozoum* animal stopped walking and placed both of its hands down before continuing on. The supinated orientation of the manus, the absence of a pollex impression and the presence of four digit impressions in this *Otozoum* trackway resemble the morphology predicted for *Massospondylus* or *Plateosaurus* based on our results. The single set of manus prints in the trackway would also support our contention that habitual quadrupedal locomotion was not the usual form of locomotion in these 'prosauropods', with the animal placing its manus down only when stopped or resting. Certainly, we cannot draw definitive inferences from a single trackway. However, the *Otozoum* ichnotaxon does agree with our inferences regarding locomotor posture and manus orientation in *Plateosaurus* and *Massospondylus*, and it is the closest in morphology and orientation to what we have described and inferred for these taxa.

Trunk and limb ratios

Although the limb and trunk proportions of theropods and sauropods demonstrate obligate bipedality in the former and obligate quadrupedality in the latter, the proportions of many basal sauropodomorphs fall somewhere in between. The ratio of trunk length to hindlimb length of obligate bipedal dinosaurs falls within a range of 1 : 1·22–1 : 1·9, whereas those of obligate quadrupeds have a reported range of 1 : 0·69–1 : 0·9 (Galton 1970; Galton and Upchurch 2004). Trunk to hindlimb ratios of most 'prosauropods', including *Plateosaurus* and *Massospondylus*, give 'intermediate' ratios of 1 : 0·95–1 : 1·15 (Galton and Upchurch 2004). It is notable that *Riojasaurus*, a taxon morphologically similar to *Melanorosaurus*, has a trunk length to hindlimb length ratio of 1 : 0·71 and falls just within the range reported for quadrupedal dinosaurs (Galton 1970; although see Van Heerden and Galton 1997 for a contrasting perspective). Moreover, Christian and Preuschoft (1996) predicted that the sagittal bending moment about the vertebral column would be greatest over the limb girdles that were supporting the majority of the body mass during locomotion. In various bipeds they found a single peak in the vertebral sagittal bending

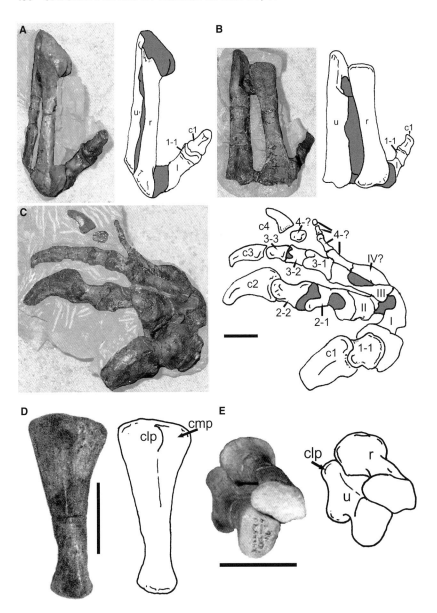

TEXT-FIG. 5. *In situ* articulation of a small juvenile *Massospondylus* (BP/1/4376) forearm showing the parallel and uncrossed relationships of the radius and ulna. A, cranial view with interpretive drawing. B, lateral view with interpretive drawing. C, medioventral view of manus with interpretive drawing. D–E, small juvenile radius and ulna of *Apatosaurus* sp. (OMNH 01289): D, in craniolateral view showing the craniolateral process on the ulna; E, in distal view when articulated with the ulna, the radius assumes a craniomedial orientation. Note that in the *Massospondylus* juvenile, digit I remains aligned with the radius, and that this relationship leads to a semi-supinated manus even in a juvenile specimen. In the *Massospondylus* figures, the matrix in which the bones are contained has been artificially 'lightened' in order to highlight the orientations of the bones. Question marks denote uncertainty about element identity. Roman numerals indicate digit numbers. Grey shading indicates matrix. Scale bar for *Massospondylus* forearm and manus represents 10 mm. Scale bars for sauropod juvenile represent 5 cm.

moment above the pelvic girdle, whereas two peaks were reported in quadrupeds above each girdle (Christian and Preuschoft 1996). *Plateosaurus* does exhibit a small peak at the shoulder in sagittal bending moment about the vertebral column, which suggested to Christian and Preuschoft (1996) that this taxon was at least facultatively quadrupedal.

Forelimb length to hindlimb length ratios also show most basal sauropodomorphs falling somewhere between obligate bipedalism and quadrupedalism (Galton and Upchurch 2004). However, our comparison of the ratio of humerus length to femur length in several specimens of *Plateosaurus* and *Massospondylus* with those of other selected sauropodomorphs shows a surprising amount of overlap (see Table 1). Over a range of sauropodomorph taxa (*Plateosaurus*, *Massospondylus*, *Vulcanodon*, *Shunosaurus*,

Apatosaurus) the humerus is approximately 60 per cent of the length of the femur (see Table 1). By contrast, taxa such as *Melanorosaurus*, *Antetonitrus* and *Omeisaurus* have humeri approximately 71–90 per cent the length of the femur (see Table 1). A somewhat different picture emerges when the ratio forelimb length (humerus + ulna + metacarpal II) to hindlimb length (femur + tibia + metatarsal III) is compared (see Table 2). In this case, the forelimb of *Plateosaurus* is nearly 75 per cent the length of the hindlimb, whereas that of *Massospondylus* is just greater than 50 per cent. *Antetonitrus* has the longest relative forelimb length at 84 per cent of the length of the hindlimb. By contrast, the relative lengths of the forelimb in *Melanorosaurus*, *Shunosaurus* and *Apatosaurus* are approximately 62–65 per cent of the hindlimb length, falling very close to their humerus : femur ratio. We suggest the foreshortened

TABLE 1. Ratio of maximum humerus length vs. maximum femur length in specimens of *Plateosaurus*, *Massospondylus* and other selected sauropodomorphs.

Taxon	Specimen	Humerus length	Femur length	Ratio
Plateosaurus	SMNS 53537	402	636	0·63
Plateosaurus	SMNS F48	428	716	0·60
Plateosaurus	SMNS Graburg 1932	350	596	0·59
Plateosaurus	SMNS 13200	401	620	0·65
Massospondylus	SAM-PK-K5135	220	350	0·63
Massospondylus	BP/1/4998	180	314	0·57
Melanorosaurus	NM QR3144	430	610*	0·71
Antetonitrus	BP/1/4952	707	795	0·89
Vulcanodon	NMZ QG24†	700	1100	0·64
Shunosaurus	ZDM T5401†	670	1200	0·56
Omeisaurus	ZDM T5701†	1080	1310	0·82
Apatosaurus	CM 3018	1103	1730	0·63

Note that the humerus : femur ratios of both *Plateosaurus* and *Massospondylus* overlap those of *Vulcanodon*, *Shunosaurus* and *Apatosaurus*, but are lower than those of *Melanorosaurus*, *Antetonitrus* or *Omeisaurus*.
* Estimate.
† Measurements and ratios from Yates and Kitching (2003). All measurements are in millimetres.

TABLE 2. Ratios of forelimb length (humerus + ulna + metacarpal II) vs. hindlimb length (femur + tibia + metatarsal III) in *Plateosaurus*, *Massospondylus* and selected sauropodomorph taxa.

Taxon	Specimen	Forelimb length	Hindlimb length	Ratio
Plateosaurus	SMNS 13200	1031	1386	0·744
Massospondylus	SAM-PK-K 5135	424*	794	0·534
Melanorosaurus	NM QR3314	752	1115†	0·674
Antetonitrus	BP/1/4952	1253	1498	0·840
Shunosaurus	ZDM T5401‡	1268	2055	0·617
Apatosaurus	CM 3018	1931	3035	0·640

Note that the forelimb of the selected *Plateosaurus* specimen is approximately 74 per cent of the length of the hindlimb, whereas that of the selected *Massospondylus* specimen is only about half the length of the hindlimb.
* Estimated from metacarpal III because metacarpal II was missing.
† Estimated from a weathered femur.
‡ Measurements from Yates and Kitching (2003). All measurements are in millimetres.

manus and ulna of these taxa accounts for the negligible difference in ratios.

Overall, these ratios show that among several sauropodomorphs, including *Plateosaurus* and *Massospondylus*, the relative size of the humerus is fairly consistent irrespective of inferred posture. Moreover, there is no consistent picture regarding relative humerus size among the examined taxa: a basal sauropodomorph/sauropod (*Melanorosaurus*), a basal sauropod (*Antetonitrus*) and a eusauropod (*Omeisaurus*) all have a relatively long humerus, whereas two 'prosauropods' (*Plateosaurus*, *Massospondylus*), a basal sauropod (*Vulcanodon*), a eusauropod (*Shunosaurus*) and a neosauropod (*Apatosaurus*) have shorter humeri of the same relative size. When the total forelimb length : total hindlimb length ratio is compared among the selected taxa, there is again no clear association of this result with posture. *Plateosaurus* has a relatively longer forelimb than that of *Massospondylus*, which is much shorter, *Antetonitrus* has the relatively longest forelimb of the examined taxa, and *Melanorosaurus*, *Shunosaurus* and *Apatosaurus* have forelimbs of relatively similar length.

Trunk : limb ratios for various sauropodomorphs appear more consistent across a range of saurischian taxa than do the limb ratios reported here. For most basal sauropodomorphs, the trunk is relatively shorter than in obligate quadrupedal taxa but longer than that reported for bipedal taxa (Galton and Upchurch 2004). Thus, *Plateosaurus* and *Massospondylus* have a relative trunk length that falls somewhere between known bipeds and quadrupeds. If the trunk length is assumed to act like a cantilever extending from the hindlimb, this ratio by itself is suggestive of at least some quadrupedal locomotion in *Plateosaurus* and *Massospondylus*. However, we caution about over-reliance on this ratio alone for illuminating habitual posture in sauropodomorphs. We suggest that the ratio of trunk length : limb length be considered in the context of other metrics, functional morphology and phylogeny for a more complete picture.

The overlap and variation in the two sets of limb ratios reported here suggests that such data are not always reliable indicators of posture. Moreover, the relative differences in the forelimb and hindlimb may also be associated with deep phylogenetic or functional trends in limb development among sauropodomorphs, saurischians, dinosaurs generally or even within the crown-group Archosauria. Such ratios are too coarse to distinguish between locomotor, developmental or phylogenetic influences on relative limb length. We suggest that, as with trunk and limb ratios, without the inclusion of functional and phylogenetic data, such metrics by themselves are not reliable for illuminating habitual posture and locomotion in sauropodomorph taxa.

Embryonic limb posture and development in sauropodomorphs

Reisz *et al.* (2005) recently reported *Massospondylus* embryos with limb proportions similar to those of quadrupedal dinosaurs. Their data show that throughout ontogeny the forelimbs of *Massospondylus* become relatively shorter compared with the femur. Reisz *et al.* (2005) inferred from changes in the relative proportions of the limbs, axial skeleton and skull that this taxon was initially quadrupedal as a hatchling with adult bipedalism resulting from appendicular heterochrony (Reisz *et al.* 2005). They further suggested that a quadrupedal limb posture might be plesiomorphic for all sauropodomorphs (Reisz *et al.* 2005). Although it is not known how widespread this ontogenetic pattern may be, it should be noted that the limb proportions of the *Mussaurus* embryos reported by Bonaparte and Vince (1979) are similar to those of the embryonic *Massospondylus*. Furthermore, given the relationship of *Plateosaurus* to *Massospondylus*, parsimony suggests that a similar developmental trajectory was probably present in the former taxon as well.

The data reported by Reisz *et al.* (2005) are significant in the context of interpreting the development of basal sauropodomorph locomotor posture. It is clear that hatchling *Massospondylus* (and perhaps *Plateosaurus*) would have assumed a quadrupedal posture. However, as Reisz *et al.* (2005) suggested, the relatively large heads and small pectoral and pelvic girdle elements would have made efficient locomotion as a quadruped difficult. Furthermore, because a relatively large amount of epiphyseal cartilage was probably present in the joints of hatchling *Massospondylus* (Reisz *et al.* 2005), we suggest that this would have further limited efficient quadrupedal locomotion. These data may indicate that hatchlings of *Massospondylus* (and perhaps *Plateosaurus*) may have been altricial (Reisz *et al.* 2005).

In the *Massospondylus* embryos as well as juvenile specimens there is no sign that the forelimb was capable of pronation. A juvenile *Massospondylus* forearm and manus preserved articulated *in situ* shows clearly that the radius was craniolateral to the former element (Text-fig. 5). Unlike sauropods and *Melanorosaurus*, the ulna lacks a craniolateral process proximally and the radius does not assume a more craniomedial orientation. Even the smallest juvenile sauropod forelimbs clearly show these morphological features (Text-fig. 5), and this suggests that 'prosauropods' such as *Massospondylus* had not developed the sauropod 'solution' to pronation and quadrupedalism. This is significant because such morphology reveals functional data that cannot be conveyed by standard limb ratios. As we have cautioned, such metrics must be used in combination with functional and phylogenetic data to ascertain their relevance.

Taking the unchanging relationships of the antebrachial elements into account, the picture that emerges for *Massospondylus* locomotion is rendered more complex: during ontogeny, despite changing limb proportions, no morphological change occurred in the relationships or inferred functional morphology of the forelimb. Owing to

the fragile nature or embedded condition of available juvenile specimens, we cannot articulate and manipulate juvenile *Massospondylus* forelimbs as we did for the adults. However, based on our analysis of adult sauropodomorphs, other saurischians and extant outgroup taxa, and given the functional constraints we have elaborated on, we can be reasonably confident that the manus of hatchling and juvenile *Massospondylus* (and perhaps *Plateosaurus*) were not pronated.

We are not suggesting that a semi-supinated, laterally rotated manus precludes quadrupedal locomotion in hatchling and juvenile *Massospondylus* (and perhaps *Plateosaurus*). Instead, we infer that juveniles would have utilized a different locomotor pattern when they did move quadrupedally. Such a pattern may have involved a more horizontal orientation of the humerus, as in some ceratopsians (Johnson and Ostrom 1995). If the forelimb was instead held closer to a sagittal plane perhaps the manus was simply a semi-supinated 'prop' that allowed the forelimb to support the cranial end of the hatchling as it moved. Whatever the mode of locomotion in juvenile *Plateosaurus* and *Massospondylus* 'prosauropods', it is clear from the consistent morphological relationships of their antebrachium and manus at all sizes that efficient, habitual quadrupedal locomotion was unlikely.

Forelimb use

The robust nature of basal sauropodomorph forelimbs suggests they were used actively. Yet, the limited range of humeral flexion observed in *Plateosaurus* and *Massospondylus* makes it difficult to determine how the forelimb was used. In some respects our results parallel the frustrations surrounding the reconstruction of forelimb function in theropod dinosaurs.

Based on the results reported here (see also Galton and Upchurch 2004), the inability of the humerus to flex much beyond vertical at the glenoid implies that it is unlikely the manus could have transferred food directly to the mouth. This suggests a number of possibilities, including a more flexible cervical series that could bring the mouth to the manus, or that food was only acquired orally. However, we doubt that the inability of the manus to reach the mouth precludes the role of the forelimb in food acquirement in *Plateosaurus* or *Massospondylus*. Instead, our results suggest that the forelimbs and manus of these taxa could have been used on objects located directly cranial and/or ventral to the animal when the animal was standing bipedally with the trunk oriented horizontally. *Plateosaurus* and *Massospondylus* probably fed on lower level plants (< 1 m) as well as taller vegetation (Galton and Upchurch 2004), and may have used their cranioventrally oriented forelimbs to rake or uproot ground-level plants. It has also been suggested that some basal sauropodomorphs could rear tripodally (see Galton and Upchurch 2004 for an overview). In such a position with the pectoral region elevated to face more cranially, the reach and use of the forelimbs may have allowed *Plateosaurus* and *Massospondylus* to pull down branches to reach higher vegetation (see Galton and Upchurch 2004). Furthermore, many of these food-acquiring activities would certainly have benefited from a large, grasping pollex claw.

The limited flexion of the forelimb in *Plateosaurus* and *Massospondylus* also suggests that in a typical bipedal stance the manus and pollex claws could not be used effectively in defence or intraspecific combat. That these sauropodomorphs could not use their forelimbs for defensive swatting would be surprising as such behaviour has been proposed to explain the function of the enlarged pollex claw (Galton 1971, 1976; Bakker 1986). One could speculate that the forelimbs would be useful in grappling movements between conspecifics. These types of forelimb movements are observed in fighting male kangaroos (Dawson 1995) or in the precopulatory contests of red kangaroos (Croft 1981). Although such movements seem unlikely in a typical saurischian bipedal pose given our results, if *Plateosaurus* and *Massospondylus* were to stand tripodally or rear, the forelimbs and large pollex could then be lifted and engaged in defence or grappling. Such behaviours are difficult to test but suggest to us that further functional investigations, especially into tripodal rearing, will yield interesting results for *Plateosaurus*, *Massospondylus* and other sauropodomorph dinosaurs.

We are surprised that our data do not support efficient, regular habitual quadrupedal locomotion, and are intrigued by the relative limit of movement that appears possible in the forelimbs of *Plateosaurus* and *Massospondylus*. The osteology of both 'prosauropods' shows that the manus could not be pronated and that the forelimb could not be utilized efficiently as a locomotor organ. Unlike sauropods, there is no evidence of pronation through morphological changes in the antebrachium. Moreover, specialized movements such as knuckle-walking were not possible given the forelimb morphology of these taxa. Overall, our data suggest manus pronation was not well developed in *Plateosaurus* and *Massospondylus*, and thus efficient, habitual quadrupedal locomotion was unlikely in these taxa. We are also intrigued by the unchanging morphological associations of the forelimb elements throughout ontogeny in *Plateosaurus* and *Massospondylus* despite significant changes in body size and relative element forelimb length. The implications of a non-pronated manus in the presumably quadrupedal juveniles of *Massospondylus* and *Plateosaurus* suggest that such a posture was not conducive to active locomotion on all-fours even at the earliest stages of development. Ultimately, the forelimbs of *Plateosaurus* and *Massospondylus* require us to re-examine our hypotheses and theories surrounding the evolution and diversity of sauropodomorph locomotion.

Acknowledgements. For help with various aspects of this project, we acknowledge the following for access to material: S. Bell, C. Mehling, M. Norell and R. Edwards (AMNH); M. Raath, A. Yates and B. Rubridge (BP); J. Ladonski, A. Resetar, H. Voris and M. Kearney (FMNH); R. Nuttall and J. Botha (NM); R. Cifeli and M. Wedel (OMNH); R. Smith, A. Chinsamy, E. Butler and S. Kaal (SAM); R. Schoch and staff (SMNS); and R. Crombie and A. Wynn (USNM). R. Elsey (Rockefeller Wildlife Refuge, Louisiana) and R. Wilhite (Louisiana State University, Baton Rouge) provided MFB with *Alligator* specimens. Partial funding was provided through an NSF grant 0093929, received by J. M. Parrish and K. Stevens on which MFB is an outside collaborator, a University Research Council grant 3-30135 from Western Illinois University (MFB), and National Geographic grant 7713-04 awarded to Adam Yates on which MFB is a participant-recipient. The Jurassic Foundation provided travel funds for PS. We thank P. Barrett, J. Wilson and D. Henderson for their comments, which improved the quality of the final manuscript. We are especially grateful to P. Barrett and A. Yates for their assistance with 'fine-tuning' our knowledge of sauropodomorph phylogeny. Finally, we thank P. Barrett and T. Fedak for organizing this special volume.

REFERENCES

BAIRD, D. 1980. A prosauropod dinosaur trackway from the Navajo Sandstone (Lower Jurassic) of Arizona. 219–230. *In* JACOBS, L. L. (ed.). *Aspects of vertebrate history.* Museum of Northern Arizona Press, Flagstaff, AZ, 407 pp.

BAKKER, R. T. 1986. *The dinosaur heresies.* Kensington, New York, NY, 481 pp.

BONAPARTE, J. F. 1969. Comments on early saurischians. *Zoological Journal of the Linnean Society*, **48**, 471–480.

—— and VINCE, M. 1979. El hallazgo del primer nido de dinosaurios triásicos (Saurischia, Prosauropoda), Triásico Superior de Patagonia, Argentina. *Ameghiniana*, **16**, 173–182.

BONNAN, M. F. 2001. The evolution and functional morphology of sauropod dinosaur locomotion. Unpublished PhD dissertation, Northern Illinois University, 722 pp.

—— 2003. The evolution of manus shape in sauropod dinosaurs: implications for functional morphology, forelimb orientation and sauropod phylogeny. *Journal of Vertebrate Paleontology*, **23**, 595–613.

—— 2004. Morphometric analysis of humerus and femur shape in Morrison sauropods: implications for functional morphology and paleobiology. *Paleobiology*, **30**, 444–470.

—— and YATES, A. M. 2007. A new description of the forelimb of the basal sauropodomorph *Melanorosaurus*: implications for the evolution of pronation, manus shape and quadrupedalism in sauropod dinosaurs. 157–168. *In* BARRETT, P. M. and BATTEN, D. J. (eds). *Evolution and palaeobiology of early sauropodomorph dinosaurs.* Special Papers in Palaeontology, **77**, 289 pp.

CARPENTER, K. 2002. Forelimb biomechanics of nonavian theropod dinosaurs in predation. *Senckenbergiana Lethaea*, **82**, 59–76.

CARRANO, M. T. 1998. Locomotion in non-avian dinosaurs: integrating data from hindlimb kinematics, *in vivo* strains, and bone morphology. *Paleobiology*, **24**, 450–469.

CHINSAMY-TURAN, A. 2005. *The Microstructure of dinosaur bone: deciphering biology with fine-scale techniques*. The John Hopkins University Press, Baltimore, MD, 195 pp.

CHRISTIAN, A. and PREUSCHOFT, H. 1996. Deducing the body posture of extinct large vertebrates from the shape of the vertebral column. *Palaeontology*, **39**, 801–812.

COOPER, M. R. 1981. The prosauropod dinosaur *Massospondylus carinatus* Owen from Zimbabwe: its biology, mode of life, and phylogenetic significance. *Occasional Papers of the National Museums and Monuments of Rhodesia, Series B, Natural Sciences*, **6**, 689–840.

CROFT, D. B. 1981. Behaviour of red kangaroos, *Macropus rufus* (Desmarest, 1822) in northwestern New South Wales, Australia. *Australian Mammalogy*, **4**, 5–58.

DAWSON, T. J. 1995. *Kangaroos: biology of the largest marsupials*. Comstock Publishing Associates, Ithaca, NY, 162 pp.

DODSON, P. 1996. *The horned dinosaurs*. Princeton University Press, Princeton. NJ, 392 pp.

GALTON, P. M. 1970. The posture of hadrosaurian dinosaurs. *Journal of Paleontology*, **44**, 464–473.

—— 1971. The prosauropod dinosaur *Ammosaurus*, the crocodile *Protosuchus*, and their bearing on the age of the Navajo Standstone of northeastern Arizona. *Journal of Paleontology*, **45**, 781–795.

—— 1976. Prosauropod dinosaurs of North America. *Postilla*, **169**, 1–98.

—— 1990. Basal Sauropodomorpha–Prosauropoda. 320–344. *In* WEISHAMPEL, D. B., DODSON, P. and OSMÓLSKA, H. (eds). *The Dinosauria*. University of California Press, Berkeley, CA, 733 pp.

—— and UPCHURCH, P. 2004. Prosauropoda. 232–258. *In* WEISHAMPEL, D. B., DODSON, P. and OSMÓLSKA, H. (eds). *The Dinosauria*. Second edition. University of California Press, Berkeley, CA, 861 pp.

—— VAN HEERDEN, J. and YATES, A. M. 2005. Postcranial anatomy of referred specimens of the sauropodomorph dinosaur *Melanorosaurus* from the Upper Triassic of South Africa. 1–37. *In* TIDWELL, V. and CARPENTER, K. (eds). *Thunder-lizards: the sauropodomorph dinosaurs*. University of Indiana Press, Bloomington, IN, 495 pp.

GISHLICK, A. D. 2001. The function of the manus and forelimb of *Deinonychus antirrhopus* and its importance for the origin of avian flight. 301–318. *In* GAUTHIER, J. and GALL, L. F. (eds). *New perspectives on the origin and early evolution of birds*. Yale Peabody Museum, New Haven, CT, 613 pp.

HAINES, R. W. 1939. A revision of the movements of the forearm in tetrapods. *Journal of Anatomy*, **73**, 211–233.

—— 1952. The shoulder joint of lizards and the primitive reptilian shoulder mechanism. *Journal of Anatomy*, **86**, 412–422.

—— 1969. Epiphyses and sesamoids. 81–115. *In* GANS, C., BELLAIRS, A. d'A. and PARSONS, T. S. (eds). *Biology of the Reptilia, morphology A*. Volume 1. Academic Press, New York, NY, 373 pp.

HATCHER, J. B. 1902. Structure of the forelimb and manus of *Brontosaurus*. *Annals of the Carnegie Museum*, **1**, 356–376.

HILDEBRAND, M. and GOSLOW, G. E. 2001. *Analysis of vertebrate structure*. Fifth edition. John Wiley and Sons Inc, New York, NY, 373 pp.

HOLLIDAY, C. M., RIDGELY, R. C., SEDLMAYR, J. C. and WITMER, L. M. 2001. The articular cartilage of extant archosaur limb bones: implications for dinosaur functional morphology and allometry. *Journal of Vertebrate Paleontology*, **21** (Supplement to No. 3), 62A.

HORNER, J. R., PADIAN, K. and RICQLES, A. D. 2001. Comparative osteohistology of some embryonic and perinatal archosaurs: developmental and behavioral implications for dinosaurs. *Paleobiology*, **27**, 39–58.

JENKINS, F. A. Jr and GOSLOW, G. E. Jr 1983. The functional anatomy of the shoulder of the Savannah Monitor Lizard (*Varanus exanthematicus*). *Journal of Morphology*, **175**, 195–216.

—— DIAL, K. P. and GOSLOW, G. E. Jr 1988. A cineradiographic analysis of bird flight: the wishbone in starlings is a spring. *Science*, **241**, 1495–1498.

JOHNSON, R. E. and OSTROM, J. H. 1995. The forelimb of *Torosaurus* and an analysis of the posture and gait of ceratopsian dinosaurs. 205–218. *In* THOMASON, J. J. (ed.). *Functional morphology in vertebrate paleontology*. Cambridge University Press, New York, NY, 277 pp.

LANDSMEER, J. M. F. 1983. The mechanism of forearm rotation in *Varanus exanthematicus*. *Journal of Morphology*, **175**, 119–130.

—— 1984. Morphology of the anterior limb in relation to sprawling gait in *Varanus*. *Symposium of the Zoological Society of London*, **52**, 27–45.

LEWIS, O. J. 1969. The hominoid wrist joint. *American Journal of Physical Anthropology*, **30**, 251–268.

LIEM, K. F., BEMIS, W. E., WALKER, W. F. and GRANDE, L. 2001. *Functional anatomy of the vertebrates*. Third edition. Harcourt College Publishers, New York, NY, 703 pp.

LOCKLEY, M. G. 1991. *Tracking dinosaurs: a new look at an ancient world*. Cambridge University Press, Cambridge, 238 pp.

—— and HUNT, A. G. 1995. *Dinosaur tracks and other fossil footprints of the western United States*. Columbia University Press, New York, NY, 388 pp.

MEERS, M. 2003. Crocodylian forelimb musculature and its relevance to Archosauria. *Anatomical Record, Part A*, **274**, 891–916.

MUYBRIDGE, E. 1957. *Animals in motion*. Dover Publications, New York, NY, 74 pp.

ORR, C. M. 2005. The knuckle-walking anteater: a convergence test of adaptation for purported knuckle-walking features of the African Hominidae. *American Journal of Physical Anthropology*, **128**, 639–658.

OSTROM, J. H. 1969. Osteology of *Deinonychus antirrhopus*, an unusual theropod from the Lower Cretaceous of Montana. *Bulletin of the Yale Peabody Museum of Natural History*, **30**, 1–165.

PAUL, G. S. 1987. The science and art of restoring the life appearances of dinosaurs and their relatives: a rigorous how-to guide. 4–49. *In* CZERKAS, S. J. and OLSON, E. C. (eds). *Dinosaurs past and present.* Volume 2. Los Angeles County Museum of Natural History, Los Angeles, CA, 149 pp.

POORE, S. O., SÁNCHEZ-HALMAN, A. and GOSLOW, G. E. Jr 1997. Wing upstroke and the evolution of flapping flight. *Nature,* **387,** 799–802.

PROCTOR, N. S. and LYNCH, P. J. 1993. *Manual of ornithology: avian structure and function.* Yale University Press, New Haven, CT, 340 pp.

RAINFORTH, E. C. 2003. Revision and re-evaluation of the Early Jurassic dinosaurian ichnogenus *Otozoum. Palaeontology,* **46,** 803–838.

REID, R. E. H. 1997. How dinosaurs grew. 403–413. *In* FARLOW, J. O. and BRETT-SURMAN, M. K. (eds). *The complete dinosaur.* Indiana University Press, Bloomington, IN, 752 pp.

REILLY, S. M. and ELIAS, J. A. 1998. Locomotion in *Alligator mississippiensis:* kinematic effects of speed and posture and their relevance to the sprawling-to-erect paradigm. *Journal of Experimental Biology,* **201,** 2559–2574.

REISZ, R. R., SCOTT, D., SUES, H.-D., EVANS, D. C. and RAATH, M. A. 2005. Embryos of an early prosauropod dinosaur and their evolutionary significance. *Science,* **309,** 761–764.

SANDER, P. M. and KLEIN, N. 2005. Developmental plasticity in the life history of a prosauropod dinosaur. *Science,* **310,** 1800–1802.

SENTER, P. 2005. Function in the stunted forelimbs of *Mononykus olecranus* (Theropoda), a dinosaurian anteater. *Paleobiology,* **31,** 373–381.

—— and ROBINS, J. H. 2005. Range of motion in the forelimb of the theropod dinosaur *Acrocanthosaurus atokensis,* and implications for predatory behaviour. *Journal of Zoology,* **266,** 307–318.

SERENO, P. C. 1993. The pectoral girdle and forelimb of the basal theropod *Herrerasaurus ischigualastensis. Journal of Vertebrate Paleontology,* **13,** 425–450.

THULBORN, T. 1990. *Dinosaur tracks.* Chapman and Hall, London, 410 pp.

UPCHURCH, P. 1998. The phylogenetic relationships of sauropod dinosaurs. *Zoological Journal of the Linnean Society,* **124,** 43–103.

—— BARRETT, P. M. and DODSON, P. 2004. Sauropoda. 259–322. *In* WEISHAMPEL, D. B., DODSON, P. and OSMÓLSKA, H. (eds). *The Dinosauria.* Second edition. University of California Press, Berkeley, CA, 861 pp.

VAN HEERDEN, J. 1997. Prosauropods. 242–263. *In* FARLOW, J. O. and BRETT-SURMAN, M. K. (eds). *The complete dinosaur.* Indiana University Press, Bloomington, IN, 752 pp.

—— and GALTON, P. M. 1997. The affinities of *Melanorosaurus,* a Late Triassic prosauropod dinosaur from South Africa. *Neues Jahrbuch für Geologie und Paläontologie, Abhandlungen,* **1997,** 39–55.

WELLES, S. P. 1984. *Dilophosaurus wetherilli* (Dinosauria, Theropoda). Osteology and comparisons. *Palaeontographica Abteilung A,* **185,** 85–180.

WILSON, J. A. 2002. Sauropod dinosaur phylogeny: critique and cladistic analysis. *Zoological Journal of the Linnean Society,* **136,** 217–276.

—— 2005. Integrating ichnofossil and body fossil records to estimate locomotor posture and spatiotemporal distribution of early sauropod dinosaurs: a stratocladistic approach. *Paleobiology,* **31,** 400–423.

—— and CARRANO, M. T. 1999. Titanosaurs and the origin of 'wide-gauge' trackways: a biomechanical and systematic perspective on sauropod locomotion. *Paleobiology,* **25,** 252–267.

—— and SERENO, P. C. 1998. Early evolution and higher-level phylogeny of sauropod dinosaurs. *Memoir of the Society of Vertebrate Paleontology,* **5,** 1–68.

WITMER, L. M. 1995. The Extant Phylogenetic Bracket and the importance of reconstructing soft tissues in fossils. 19–33. *In* THOMASON, J. J. (ed.). *Functional morphology in vertebrate paleontology.* Cambridge University Press, New York, NY, 293 pp.

YATES, A. M. 2007. The first complete skull of the Triassic dinosaur *Melanorosaurus* Haughton (Sauropodomorpha: Anchisauria). 9–55. *In* BARRETT, P. M. and BATTEN, D. J. (eds). *Evolution and palaeobiology of early sauropodomorph dinosaurs.* Special Papers in Palaeontology, **77,** 289 pp.

—— and KITCHING, J. W. 2003. The earliest known sauropod dinosaur and the first steps towards sauropod locomotion. *Proceedings of the Royal Society of London, Series B,* **270,** 1753–1758.

—— and VASCONCELOS, C. C. 2005. Furcula-like clavicles in the prosauropod dinosaur *Massospondylus. Journal of Vertebrate Paleontology,* **25,** 466–468.

ZUG, G. R. 1974. Crocodilian galloping: a unique gait for reptiles. *Copeia,* **1974,** 550–552.

[Special Papers in Palaeontology 77, 2007, pp. 157–168]

A NEW DESCRIPTION OF THE FORELIMB OF THE BASAL SAUROPODOMORPH *MELANOROSAURUS*: IMPLICATIONS FOR THE EVOLUTION OF PRONATION, MANUS SHAPE AND QUADRUPEDALISM IN SAUROPOD DINOSAURS

by MATTHEW F. BONNAN* *and* ADAM M. YATES†

*Department of Biological Sciences, Western Illinois University, Macomb, IL 61455, USA; e-mail: MF-Bonnan@wiu.edu (author to whom correspondence should be addressed)

†Bernard Price Institute for Palaeontological Research, University of the Witwatersrand, Private Bag 3, Johannesburg 2050, South Africa; e-mail: yatesa@geosciences.wits.ac.za

Typescript received 3 February 2006; accepted in revised form 18 September 2006

Abstract: The evolution of a quadrupedal limb posture is characteristic of the earliest sauropod dinosaurs and involved secondarily modifying a non-supporting forelimb into a pronated support column with a semicircular metacarpus. *Melanorosaurus readi* is a basal sauropodomorph phylogenetically close to the earliest sauropods, and the morphology of its forelimb sheds additional light on the origins of the unique manus shape of sauropods and the initial stages of manus pronation. We describe the osteology of a complete forelimb of *Melanorosaurus* (NM QR3314), as well as partial referred specimens (SAM-PK-K3449, SAM-PK-K3532), and show that the forelimb elements of this taxon comprise a mosaic of basal sauropodomorph and basal sauropod characteristics. The humerus retains the plesiomorphic morphology of basal sauropodomorphs. However, like sauropods, the forearm of *Melanorosaurus* clearly shows the development of a proximal craniolateral process on its ulna and a shift in the position of the radius to a more cranial orientation relative to the ulna. The manus of *Melanorosaurus* was not a semicircular colonnade as in sauropods: instead its metacarpals were arranged closer to the orientation more typical of theropods and basal sauropodomorphs. A recurved, medially divergent pollex claw and straighter but blunter claws on digits II and III were present. We suggest that the characteristic U-shaped manus of eusauropods and neosauropods may have resulted from mosaic evolution. The forelimb morphology of *Melanorosaurus* suggests that pronation of the manus occurred early in basal sauropods through a change in antebrachial morphology, but that changes to the morphology of the manus followed later in eusauropods, perhaps related to further manus pronation and improved stress absorption in the metacarpus. Thus, we conclude that changes to antebrachial morphology and manus morphology were not temporally linked in sauropods and constitute separate phylogenetic events.

Key words: *Melanorosaurus*, Sauropodomorpha, Sauropoda, pronation, manus, functional morphology, evolution.

SAUROPOD dinosaurs were the largest terrestrial herbivores, but their early history and their transition from bipedalism to quadrupedalism have remained obscure. Current phylogenetic hypotheses suggest that sauropods, like other dinosaurs, were descendants of a medium-sized bipedal ancestor (Wilson 2002; Upchurch *et al.* 2004). Although recent data from Reisz *et al.* (2005) suggest that some basal sauropodomorphs (e.g. *Massospondylus*) had a quadrupedal posture as hatchlings and switched to a bipedal posture during later ontogeny, it is not clear how widespread this pattern was (although similar proportions are reported for *Mussaurus*; Bonaparte and Vince 1979). What is clear is that the ancestor of all sauropodomorphs

was a biped (Wilson 2002; Upchurch *et al.* 2004), and that quadrupedalism in all sauropodomorph taxa is probably secondarily derived, regardless of the ontogenetic appearance of this feature.

The obligate quadrupedal posture of sauropod forelimbs required pronation of the manus such that it operated in a craniocaudal plane similar to that of the pes (Bonnan 2003). Pronation of the manus was apparently accomplished by a shift in the relationship of the antebrachial bones with each other and the humerus. In both the earliest recognized sauropods (e.g. *Antetonitrus*) and the most derived sauropod taxa (e.g. *Opisthocoelicaudia*), the radius articulates with the ulna proximally such that it is cranial and medial in relation to the latter element (Bonnan 2003; Yates and Kitching 2003). Evidence of this shift is apparent in the development of a novel craniolateral process on the ulna that would have articulated with the lateral portion of the distal articular surface of the humerus (Bonnan 2003). The new craniolateral process and primitive craniomedial process of the ulna 'cradle' are the radius in its derived, craniomedial orientation. Because the shaft of the radius and ulna parallel one another throughout their length and do not cross, this results distally in a pronated manus (Bonnan 2003). However, the point at which this morphological shift occurred and its relationship to manus shape remain unresolved.

Bonnan (2003) suggested that the unique, arched colonnade arrangement of the metacarpus in eusauropods evolved in tandem with the morphological changes in the radius and ulna that allowed manus pronation. He further suggested that, if this hypothesis were correct, the basal-most sauropods would have a pronated, U-shaped manus (Bonnan 2003). Within this context, the morphology of the antebrachium and manus in basal sauropodomorph dinosaurs closest to the earliest sauropods may shed additional light on the first appearance of obligate quadrupedalism and manus pronation. Moreover, the forelimb morphology of such dinosaurs may reveal whether manus shape and pronation were linked or separate events in sauropod evolution, which may have significant implications for sauropod phylogeny.

The relationships of the basal sauropodomorphs to one another and their relationship to sauropods are currently in a state of flux. The hypothesis that all sauropodomorphs outside the clade of 'typical sauropods' form a monophyletic group called the Prosauropoda (e.g. Galton and Upchurch 2004) has not been supported by several recent comprehensive analyses of sauropodomorph relationships (Pol and Powell 2007; Upchurch *et al.* 2007; Yates 2007, in press). Instead it seems that classic 'prosauropods' such as *Plateosaurus* and *Massospondylus* are relatively distant from sauropods, while some large-bodied forms such as *Camelotia*, *Blikanasaurus* and *Melanorosaurus* may lie close to the ancestry of sauropods, or may even be basal members of that clade. For the purposes of this paper, Sauropoda is defined as the most inclusive clade including *Saltasaurus* but not *Melanorosaurus* (Yates in press). Among these large taxa, *Melanorosaurus readi* is morphologically significant. It is known from relatively complete remains and its forelimb osteology is represented by specimens from more than one individual (Galton *et al.* 2005). Moreover, one relatively complete specimen of *Melanorosaurus* retains a nearly complete manus (NM QR3314). However one defines Sauropoda, the completeness of *Melanorosaurus* and its phylogenetic closeness to the earliest known sauropods (e.g. *Antetonitrus, Vulcano-*

don; Yates 2003, in press; Yates and Kitching 2003) make it an ideal taxon with which to examine the morphological evolution of the forelimb.

Institutional abbreviations. BPI, Bernard Price Institute for Palaeontological Research, University of the Witwatersrand, Johannesburg, South Africa; CM, Carnegie Museum of Natural History, Pittsburgh, Pennsylvania, USA; NM, National Museum, Bloemfontein, Free State, South Africa; SAM, South African Museum, Iziko Museums of Cape Town; SMNS, Staatliches Museum für Naturkunde, Stuttgart, Germany.

Abbreviations on text-figures. c1, pollex claw; c2, claw II; c3, claw III; clp, craniolateral process of the ulna; cmp, craniomedial process of ulna; cub, cuboid fossa of humerus; dpc, deltopectoral crest of humerus; h, humerus; hh, humeral head; i1, phalanx I-1; ii1?, inferred phalanx II-1; iii1, phalanx III-1; iii2, phalanx III-2; iv1, phalanx IV-1; v1, phalanx V-1; of, olecranon fossa; ole, olecranon region of ulna; r, radius; u, ulna.

MATERIAL AND METHODS

Our analysis involved the examination of a nearly complete specimen of *Melanorosaurus readi* (NM QR3314) and other forelimb bones assigned to this taxon (SAM-PK-K3449, SAM-PK-K3532). The well-preserved, but disarticulated, postcranial bones of NM 1551 were not used in the description because the forelimb elements (humerus and ulna) are damaged and, in the case of the ulna, probably pathologically deformed. For further discussion of the cranial and postcranial morphology of *Melanorosaurus*, especially NM 1551, we refer readers to Galton *et al.* (2005) and Yates (2007).

Three humeri assigned to *Melanorosaurus* were examined (NM QR3314, SAM-PK-K3449, SAM-PK-K3532) and are referred to below. Of the three humeri, two were complete with some damage to the deltopectoral region (NM QR3314, SAM-PK-K3449), whereas the other (SAM-PK-K3532) was incomplete but retained a well-preserved proximal end. Referral of the two South African Museum specimens to *Melanorosaurus* is uncertain, however. The more complete specimen (SAM-PK-K3532) comes from a series of bones collected at Rooi Nek (all catalogued together) that does not contain any bones that are clearly diagnostic of *M. readi* and may represent a mixed assemblage of sauropodomorph bones. However, we tentatively regard the humerus as referable to *M. readi* for two reasons. First, it does not match the morphology of the known humeri of other sauropodomorph taxa from the lower Elliot Formation. *Plateosauravus cullingworthi* has a strongly sinuous deltopectoral crest that occupies less than 50 per cent of the length of the humeral shaft, while *Antetonitrus ingenipes* has a strong paramarginal sulcus on the lateral distal surface of the deltopectoral crest. Second, the morphology of the SAM-PK-K3532 humerus matches that of NM QR3314 well. The proximal end of the humerus included among the holotype bones (SAM-PK-K3449) was, according to Haughton (1924), found loose and downslope from the main accumulation of bones. This specimen shows a strong paramarginal groove on the deltopectoral crest and suggests that this specimen is an isolated bone belonging to a small *Antetonitrus*.

A complete right radius and ulna were present for two specimens (NM QR3314, SAM-PK-K3449) and an additional left ulna was available from specimen NM QR3314. All elements described here were complete and relatively undamaged. In NM QR3314, the right radius, ulna and humerus were preserved *in situ* in a flexed orientation, providing evidence for the natural articulations of these elements. The additional left ulna from this specimen, and the disarticulated right radius and ulna of SAM-PK-K3449, provided opportunities to examine these elements independently. Specimen NM QR3314 is particularly valuable to our analysis as it retains a nearly complete manus (although no ossified carpals were present).

We also recorded a number of linear morphometric measurements of the humerus, radius, ulna and metacarpus of *Melanorosaurus* specimens for comparative purposes. Identical measurements were taken from specimens of *Plateosaurus* (SMNS G1932), *Massospondylus* (SAM-PK-K3394), *Antetonitrus* (BPI 4952) and *Apatosaurus* (CM 3018).

DESCRIPTION OF THE FORELIMB

Humerus

Overall, the shape of the humerus is typical of many saurischian dinosaurs, with expanded proximal and distal ends and a constricted midshaft (Galton and Upchurch 2004; Text-figs 1–2). Proximally, the head of the humerus is subconvex in cranial outline. The deltopectoral crest has a low, hatchet shape in lateral profile, which is similar to, but not as developed as, those of other basal sauropodomorphs (Galton and Upchurch 2004; Galton *et al.* 2005). The ratio and bivariate plot of maximum humerus

TEXT-FIG. 1. Articulated skeleton of *Melanorosaurus readi* (NM QR3314). A, D, details of the *in situ* relationships of the humerus, radius, ulna and manus. B, E, caudolateral view of the humerus, radius and ulna in articulation: note that the caudal portion of the lateral condyle of the humerus is in alignment with the craniolateral process of the ulna. C, F, lateral views of the humerus and antebrachium *in situ*. Scales are in millimetres; roman numerals indicate digit identities.

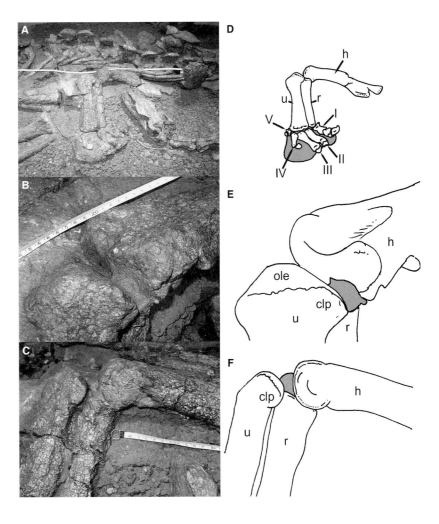

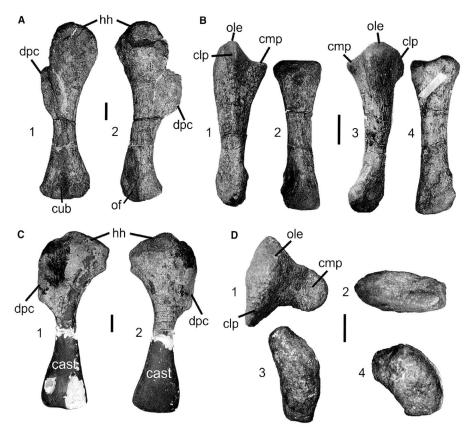

TEXT-FIG. 2. Disarticulated forelimb elements associated with *Melanorosaurus readi* (SAM-PK-K3449 and SAM-PK-K3532). A, right humerus (SAM-PK-K3532) in cranial (1) and caudal views (2). B, right ulna and radius in cranial (1–2) and caudal (3–4) views. C, right humerus possibly belonging to *Antetonitrus ingenipes* (SAM-PK-K3449) in cranial (1) and caudal (2) views: note that only the proximal end of the humerus is preserved in this specimen and distal end is casted. D, ulna and radius in proximal (1–2) and distal (3–4) views: note the strong craniolateral process on the ulna 1). All scale bars represent 5 cm.

length against deltopectoral crest length shows clearly that *Melanorosaurus* falls within the range of other basal sauropodomorphs (Text-fig. 3A; Table 1). At the distal end of the humerus, a shallow cuboid fossa is present, and distal condyles are present but not well developed. This distal morphology is similar to that of other basal sauropodomorphs (Bonnan 2003; Galton and Upchurch 2004), and differs from sauropods in which distinct condyles and a cuboid fossa are both absent (Bonnan 2003). The proximal and distal ends of the humerus bear the characteristic roughened and rugose texture of the calcified cartilage 'growth plate' between the metaphysis and unpreserved epiphysis of many dinosaurs and other archosaurs (Haines 1969; Chinsamy-Turan 2005). The texture of the calcified cartilage is reminiscent of other basal sauropodomorphs, but not as deep and crenulated as that of sauropods (Wilson and Sereno 1998). From the data presented in Table 1, it is evident from the overall dimensions of the humerus in *Melanorosaurus* that it falls closer to those of *Plateosaurus* and *Massospondylus* than to basal sauropods such as *Antetonitrus* (Text-fig. 3A–B).

TABLE 1. Comparative measurements of the humerus in *Melanorosaurus* and several outgroup taxa. All measurements in millimetres. H1, maximum length; H2, proximal breadth; H3, deltopectoral crest length; H4, craniocaudal midshaft width; H5, mediolateral midshaft width; H6, distal breadth; DPC ratio, ratio of deltopectoral crest length to H1.

Taxon	H1	H2	H3	H4	H5	H6	DPC ratio
Melanorosaurus NM QR3314	430	–	153	55	80	115	0·36
Melanorosaurus NM 1551	451	–	240	–	–	140	0·53
Melanorosaurus SAM-PK-K 3449	495	136	254	58	66	150	0·51
Plateosaurus SMNS G1932	350	146	180	60	50	122	0·51
Massospondylus SAM-PK-K 3394	207	70	115	27	33	78	0·56
Antetonitrus BPI 4952	707	308	420	97	110	226	0·59
Apatosaurus CM 3018	1103	566	482	168	215	415	0·44

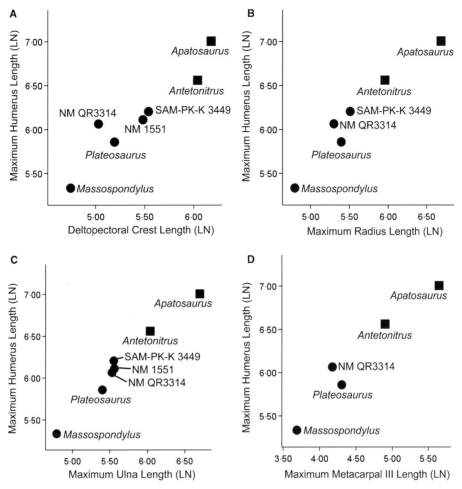

TEXT-FIG. 3. Bivariate plots of natural log-transformed measures of maximum humerus length against various forelimb bones and landmarks. In all of the plots, maximum humerus length (MHL) is indicated on the *y*-axis. A, MHL vs. deltopectoral crest length. B, MHL vs. maximum radius length. C, MHL vs. maximum ulna length. D, MHL vs. maximum length of metacarpal III. *Melanorosaurus* specimens (which are listed in Tables 1–3) are indicated by their specimen numbers. Note that *Melanorosaurus* plots closest to the trajectory of sauropods for the radius and ulna, whereas it groups more closely with *Plateosaurus* and *Massospondylus* for measurements of the humerus and metacarpal III.

Radius and ulna

The antebrachial elements have some of the most sauropod-like characteristics observed in *Melanorosaurus*. The radius is a straight element with a suboval head, a gently concave proximal articular surface, a flattened but slightly convex distal end, and a concave and rugose distolateral edge that articulated with the ulna (Text-fig. 4). The distolateral concavity continues onto the caudal face of the radius. These rugose and concave regions of the radius bear a striking resemblance to the distal end of this element in sauropods (Bonnan 2003). Hatcher (1902), Gilmore (1936), Wilson and Sereno (1998), Bonnan (2003) and others have attributed this morphology in sauropods to ligamentous attachments of the radius to the ulna, limiting its movement. We suggest, based on the morphology displayed on both of the radii examined here (NM QR3314, SAM-PK-K3449), that the radius of *Melanorosaurus* was similarly bound to the distal end of the ulna by ligamentous tissues.

Proximally, the ulna possesses both the standard craniomedial process typical of saurischians and an additional craniolateral process characteristic of sauropod ulnae (Bonnan 2003). These processes together cradle the radius proximally in a radial fossa, and this condition is identical to that observed in both basal sauropods such as *Antetonitrus* and *Vulcanodon* as well as more derived sauropod taxa (Bonnan 2003; Yates and Kitching 2003; see Text-fig. 5). This morphology is distinctive from many other basal sauropodomorphs, where the proximal end of the ulna is triangular in shape (Galton and Upchurch 2004). As reported for sauropod ulnae

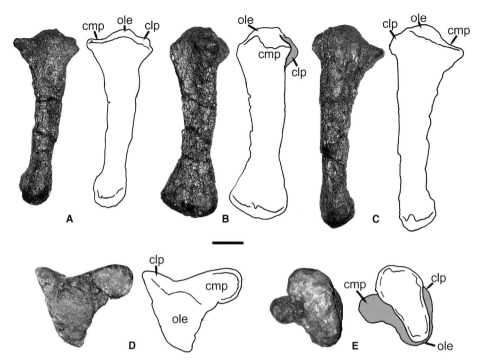

TEXT-FIG. 4. Left ulna of *Melanorosaurus readi* (NM QR3314) in A, cranial, B, lateral, C, caudal, D, proximal and E, distal views with associated line drawings. Note in view D the prominent craniolateral process of the ulna. Compare this figure to Text-figure 5, showing the proximal ends of the ulnae of saurischians, sauropodomorphs and sauropods. Scale bar represents 5 cm.

(Upchurch 1998; Upchurch *et al.* 2004), the craniomedial process is gently concave whereas the craniolateral process is distinctly convex (Text-figs 2, 4). However, unlike most sauropods, the olecranon process of the ulna is fairly prominent (Text-figs 2, 4). The distal end of the ulna bears a radial fossa reminiscent of that in sauropods (Bonnan 2003), into which the radius articulated. As with the radius, the distal end of the ulna bears rugose texturing that may correspond to ligamentous tissues binding this element to the radius.

Table 2 shows that the dimensions of these elements fall within the range of other basal sauropodomorphs in terms of their overall size. However, in the plots of

humerus length vs. that of the radius and ulna (Text-fig. 3B–C), the *Melanorosaurus* specimens lie on a trajectory closer to that of the sauropods (e.g. *Antetonitrus*, *Apatosaurus*) than to *Plateosaurus* or *Massospondylus*.

Manus

No trace of the carpus is preserved in NM QR3314. This cannot be attributed to post-mortem loss because the manus remained in complete articulation with the antebrachium. Neither can the loss be attributed to loss during collection because the manus remains in a slab of matrix

TABLE 2. Comparative measurements of the antebrachium in *Melanorosaurus* and several outgroup taxa. All measurements in millimetres. Radius measurements: R1, maximum length; R2, proximal breadth; R3, craniocaudal midshaft width; R4, mediolateral midshaft width; R5, distal breadth. Ulna measurements: U1, maximum length; U2, proximal breadth; U3, craniocaudal midshaft width; U4, mediolateral midshaft width; U5, distal breadth. R ratio, ratio of radius vs. humerus maximum length; U ratio, ulna vs. humerus maximum length.

Taxon	R1	R2	R3	R4	R5	U1	U2	U3	U4	U5	R ratio	U ratio
Melanorosaurus NM QR3314	200	–	28	–	–	252	60	50	–	85	0·47	0·59
Melanorosaurus NM 1551	–	–	–	–	–	260	–	–	–	–		0·58
Melanorosaurus SAM-PK-K 3449	247	72	28	41	83	258	76	40	46	82	0·50	0·52
Plateosaurus SMNS G1932	220	60	26	30	40	222	60	32	32	54	0·63	0·63
Massospondylus SAM-PK-K 3394	121	32	14	19	40	120	58	15	20	40	0·58	0·58
Antetonitrus BPI 4952	386	120	23	62	114	418	140	52	54	115	0·55	0·59
Apatosaurus CM 3018	800	237	108	107	241	815	353	141	150	161	0·73	0·74

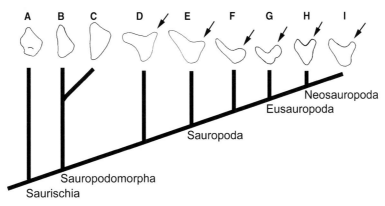

TEXT-FIG. 5. Proximal outlines of the ulnae of basal saurischian, sauropodomorph and sauropod taxa: cranial is up whereas caudal is down; arrows indicate the presence of a craniolateral process. Phylogenetic relationships of taxa based on Yates and Kitching (2003), Upchurch *et al.* (2004) and Galton *et al.* (2005). A, *Herrerasaurus*. B, *Plateosaurus*. C, *Massospondylus*. D, *Melanorosaurus*. E, *Antetonitrus*. F, *Vulcanodon*. G, *Omeisaurus*. H, *Apatosaurus*. I, *Camarasaurus*. Not to scale. Outlines of A–B and F–I after Bonnan (2003).

that can be fitted to the antebrachium. The only plausible explanation is that the carpal elements had failed to ossify, despite the fact that the individual was over 4 m long and was at least half-grown. This is not the only evidence that the carpus ossified very late in the ontogeny of early sauropodomorphs. A juvenile *Massospondylus* well over hatching size (BP/1/4376) also includes an articulated forelimb and manus that lacks any trace of an ossified carpus, whereas it is certainly present in adult specimens (Cooper 1981; BP/1/4934).

The manus is short relative to that of most known basal sauropodomorphs, but still does not fall into the range observed in sauropods (see Table 3 and Text-fig. 6). Metacarpal I is a proximodistally short, robust element that is wide mediolaterally. Although features of the distal articular surface of metacarpal I in specimen NM QR3314 are partially obscured by its articulation with phalanx I-1, it is clear that this element had the bevelled, medially directed ginglymus characteristic of other basal sauropodomorphs (Galton and Upchurch 2004). The proximolateral edge of metacarpal I partially overlaps the proximal articular surface of metacarpal II, and a proximolateral fossa on the first metacarpal may have enabled this articulation *in vivo*. The shape and relationship of

metacarpal I to metacarpal II bear striking similarities to those reported for metacarpal I in *Antetonitrus* (Yates and Kitching 2003) and those in some neosauropods, such as *Apatosaurus* (Bonnan 2001, 2003).

There is a single phalanx (I-1) that articulates obliquely with metacarpal I, and in the specimen described here it is orientated such that its medial face has rotated dorsally (Text-fig. 6). This element is fractured into two pieces at its midshaft. Phalanx I-1 is approximately similar in size to the first metacarpal and the distal ginglymus is well developed with prominent colateral ligament fossae (Text-fig. 6). The pollex claw is blunt, but recurved, and its proximal end articulates deeply into the articular groove on phalanx I-1 (Text-figs 6–7). In medial view, a subtle trace of an ungual nutrient groove is present. In its articulation with phalanx I-1, the medial surface of the pollex ungual has been rotated dorsally and its overall orientation is medially directed. The orientation and morphology of digit I are similar to that of other basal sauropodomorphs.

Metacarpal II is the longest element in the metacarpus in *Melanorosaurus*, a pattern that resembles that of basal sauropodomorphs such as *Plateosaurus* and *Massospondylus* (Table 2; Text-figs 3D, 6, 8). In contrast, metacarpal

TABLE 3. Comparative lengths of the metacarpals in *Melanorosaurus* and several outgroup taxa. All measurements in millimetres. Roman numerals refer to their respective digits. MCIII ratio, ratio of metacarpal III vs. humerus maximum length; MCII/III ratio, ratio of metacarpal II vs. metacarpal III maximum length.

Taxon	I	II	III	IV	V	MCIII ratio	MCII/III ratio
Melanorosaurus NM QR3314	55	70	65	50	35	0·15	1·08
Melanorosaurus SAM-PK-K 3449	–	–	–	–	–	–	–
Melanorosaurus NM 1551	–	–	–	–	–	–	–
Plateosaurus SMNS G1932	70	88	74	60	34	0·21	1·19
Massospondylus SAM-PK-K 3394	32	42	40	32	20	0·19	1·05
Antetonitrus BPI 4952	88	127	134	–	–	0·19	0·95
Apatosaurus louisae CM 3018	270	272	283	261	222	0·26	0·96

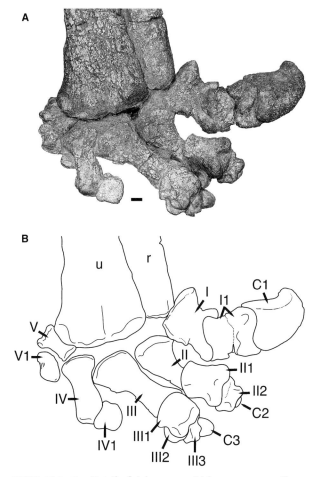

TEXT-FIG. 6. Detail of right manus, *Melanorosaurus readi* (NM QR3314), in A, dorsal view with B, associated line drawing. The manus contains a unique mosaic of basal sauropodomorph and sauropod features. Basal sauropodomorph features include a medially divergent, recurved pollex claw, claws on the second and third digits, and a gently arched metacarpus. Sauropod features include the foreshortened metacarpus and phalanges on digits II–IV. Scale bar represents 5 cm.

III is the longest metacarpus element in both basal sauropods (e.g. *Antetonitrus*) and many neosauropods (e.g. *Apatosaurus*) (Table 3; Text-fig. 3D; see also Bonnan 2001 for other metacarpal measurements of neosauropods). The distal condyles of metacarpal II are largely hidden underneath the dislocated first phalanx but it is evident from a ventral view of the block that the distal end was flared, giving metacarpal II a similar shape to that of *Antetonitrus* (Yates and Kitching 2003). The presence of deep, distal colateral ligament pits are observed on all subsequent metacarpals.

The first phalanx of digit II is dislocated and lies on the dorsal surface of metacarpal II. It is followed by a distinctively smaller non-terminal phalanx which articulates distally with a claw (Text-figs 6, 8). As with phalanx I-1, deep colateral ligament pits are present distally on the

non-terminal phalanges. The ungual that articulates with phalanx II-2 is blunt and slightly recurved, and is preserved in a ventriflexed orientation in which it is curled beneath the manus (Text-figs 6, 8). This orientation is suggestive of a fair degree of claw flexibility, as shown for many basal sauropodomorphs (e.g. Galton 1990), but without being able to manipulate these bones this conclusion must be viewed as tentative.

Metacarpal III is slightly shorter than metacarpal II (Table 3) and is an elongate element with gently expanded proximal and distal ends (Text-figs 6, 8). Digit III contains three non-terminal phalanges with well-developed articular surfaces and colateral ligament pits (Text-figs 6, 8) that strongly decrease in size distally. A claw articulates distally with phalanx III-3 that is blunt and straighter than that of the pollex, but similar in morphology to claw II (Text-figs 6, 8). As with claw II, this ungual is also flexed.

Metacarpal IV is significantly shorter than metacarpal III (Table 3) and has shifted proximally out of its natural articulation with the latter (Text-fig. 6). A single, small and subtriangular phalanx IV-1 is present that bears some similarities in overall morphology to the terminal phalanges of many sauropod dinosaurs (Wilson and Sereno 1998; Bonnan 2001). No distinct articular surface is present distally and colateral ligament pits are absent. Metacarpal V is the shortest element in the metacarpus (Table 3) and has rotated out of articulation with the manus such that its distal end is directed dorsolaterally (Text-fig. 6). Despite its small size, its distal end bears distinct condyles and a prominent intercondylar groove. A small, paddle-shaped element lies near the medial side of metacarpal V that we interpret as phalanx V-1. Phalanx

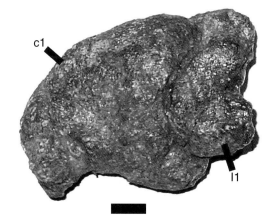

TEXT-FIG. 7. Medial view of pollex claw of *Melanorosaurus readi* (NM QR3314) in articulation with the distal half of phalanx I-1. Note the short and compact morphology of phalanx I-1 and the dull and non-trenchant shape of the pollex claw. The overall morphology is closer to that observed in sauropods than in basal sauropodomorphs such as *Plateosaurus* or *Massospondylus*. Scale bar represents 1 cm.

V-1 is wider proximally than it is distally, and has poorly developed articular surfaces (Text-fig. 6). No claws were present on digits IV and V, and this pattern parallels that of other known basal sauropodomorphs (Galton and Upchurch 2004).

Because the manus of *Melanorosaurus readi* NM QR3314 is preserved *in situ*, we could not individually articulate the bones of the metacarpus to determine their precise relationships to one another. However, based on the morphology described here and the basic orientation of the elements to one another, it is clear that proximally the metacarpus assumed the gently arched shape characteristic of theropods and basal sauropodomorphs (Wilson and Sereno 1998; Bonnan 2003). This morphology contrasts with that of most known sauropods in which the metacarpals are arranged in a semicircular colonnade (Wilson and Sereno 1998; Bonnan 2003; Upchurch *et al.* 2004).

The manus is flattened into a semi-supinated orientation with its palmar surface facing medially (Text-fig. 6), but it is not clear if this was its natural orientation. In most theropods and basal sauropodomorphs, the manus was orientated into a semi-supinated, prayer-like orientation because the radius did not cross over the ulna and remained lateral throughout its length (Bonnan 2003; Senter and Robins 2005; Bonnan and Senter 2007). The possession of a triradiate proximal ulna and the more craniomedial orientation of the radius in *Melanorosaurus* suggest to us that the manus was at least partially pronated, and that its palmar surface was orientated mostly caudally in life.

DISCUSSION

Forelimb morphology of Melanorosaurus: *one step before basal sauropods?*

The forelimb elements of *Melanorosaurus* comprise a mosaic of basal sauropodomorph and basal sauropod characteristics. The dimensions and basic morphology of the humerus resemble those of many basal sauropodo-

morphs and the presence of a cuboid fossa suggests flexion of the antebrachium was at least as well developed as in taxa such as *Plateosaurus* and *Massospondylus* (see also Bonnan 2003; Bonnan and Senter 2007). In contrast, the radius and ulna show many sauropod features, including the appearance of a craniolateral process on the proximal ulna. The manus displays a mixture of basal sauropodomorph and sauropod features. In one sense, aspects of the manus approach the derived condition seen in eusauropods, including: a foreshortened hand that is less than 38 per cent of the length of the humerus + radius and the reduced size of the unguals and penultimate phalanges of digits II and III. In eusauropods the latter bones are lost altogether (Wilson and Sereno 1998).

Other aspects of the hand retain distinctly plesiomorphic, 'prosauropod-like', states. The semicircular colonnade that typifies the manus of most known sauropods is absent. Instead, the metacarpals form a gentle arch with the more linear arrangement characteristic of other saurischians (see Bonnan 2003). The presence of three manus claws and especially the medially divergent, recurved pollex claw are plesiomorphic features common to most basal sauropodomorphs. This condition suggests to us that the manus of *Melanorosaurus* was not devoted solely to weight-bearing, as it is in sauropods. Although the use of the forelimb in quadrupedal locomotion in basal sauropodomorphs has been somewhat speculative and difficult to substantiate (see Bonnan and Senter 2007), there is little doubt that the forelimb functioned in some other capacity when these animals assumed a bipedal posture (e.g. Galton 1990; Galton and Upchurch 2004). Whatever the precise function of the forelimb, we suggest that *Melanorosaurus* was capable of using its forelimb in non-locomotor behaviours.

Moreover, the presence of a foreshortened manus with three claws could conceivably produce manus prints similar to those of the Triassic ichnotaxon *Tetrasauropus*, a trackmaker of uncertain sauropodomorph phylogenetic affinities. Wilson (2005) suggested convincingly that the manus prints ascribed to *Tetrasauropus* may have been

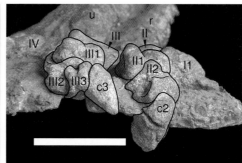

TEXT-FIG. 8. Detail of ventral view of manus, *Melanorosaurus readi* (NM QR3314). Note the flexed orientation of the claws on digits II and III and the pronounced articular ginglymi and colateral ligament pits on the phalanges. Scale bars represent 5 cm.

produced by a basal sauropod and not a typical basal sauropodomorph. We suggest, based on the manus morphology, that another plausible trackmaker of the *Tetrasauropus* tracks could be an animal such as *Melanorosaurus*. The appearance of other derived, sauropod-like character states in different parts of the skeleton, such as the skull, sacrum, pes and tail, are interpreted here as evidence for a closer relationship between *Melanorosaurus* and Sauropoda than most other basal sauropodomorphs, rather than simple convergence relating to large size and quadrupedalism (Yates 2007). In total, the osteological features observed in *Melanorosaurus* suggest that this taxon represents one of the closest outgroups to sauropods among the basal sauropodomorphs. This is significant because if this hypothesis is upheld then *Melanorosaurus* informs us of the sequence in which derived character states were acquired on the line leading to Sauropoda and eventually Eusauropoda.

Functional implications: mosaic evolution of manus pronation and morphology?

Bonnan (2003) suggested that the initial evolution of a U-shaped manus in most sauropods did not result from graviportal considerations, but as a result of correlated progression within the forelimb. He suggested that pronation was the driving selective pressure underlying manus shape: the radius of all known sauropods apparently shifted craniomedially in relation to the ulna. Bonnan (2003) argued that a shift in the position of the radius in relation to the ulna rotated the manus into pronation and subsequently altered the path of the digital arch during embryonic development, thereby forming a strongly arched metacarpal colonnade. This hypothesis was supported by morphological features (e.g. the development of a craniolateral process on the proximal articular surface of the ulna, a curvature created by the articulating distal ends of the radius and ulna which match the U-shaped metacarpus), which appear in all known sauropods with complete forelimb material. Bonnan (2003) proposed that when a complete forelimb of a basal sauropod was described, these morphological features and a U-shaped manus would be present.

The forelimb of *Melanorosaurus* partially supports the Bonnan (2003) hypothesis: the ulna has developed a craniolateral process which cradles the radius, as in all known sauropods including early forms such as *Antetonitrus* (Yates and Kitching 2003) and *Vulcanodon* (Raath 1972; Cooper 1984). The radius of *Melanorosaurus* has also apparently shifted in position from a plesiomorphic craniolateral orientation proximally to a more cranial and slightly medial region relative to the ulna (Text-figs 1–2). This is especially apparent in NM QR3314 in which the

caudolateral portion of the humerus would have articulated with the craniolateral process of the ulna, not the head of the radius, when the forearm was extended (Text-fig. 1).

However, the manus has not developed the U-shape characteristic of known eusauropods. Instead, as described above, the metacarpus forms a flatter shape and gentle arch more reminiscent of theropods and basal sauropodomorphs. This would appear to falsify the functional linkage between manus shape and forelimb morphology proposed by Bonnan (2003): the change in the digital arch was apparently not directly linked to a shift in radius orientation.

The forelimb osteology of *Melanorosaurus* suggests that a reversion to at least facultative quadrupedal locomotion and manus pronation occurred prior to the emergence of basal sauropod taxa. Functionally, the manus of many quadrupedal tetrapods is pronated because this orientation produces a caudally directed propulsive force that parallels the actions of the pes (Bonnan 2003; Bonnan and Senter 2007). In most quadrupedal mammals, the radius physically crosses the ulna to pronate the manus (Hildebrand and Goslow 2001). The forelimb of *Melanorosaurus* shows that, as Bonnan (2003) suggested, manus pronation in sauropods was apparently accomplished by an initial shift in the orientation of the radius, although this change alone may have only partially pronated the manus. The more characteristic U-shaped manus of later eusauropods and neosauropods may have developed as a function of improving pronation. In other words, with the manus already partially pronated in a close sister-group, such as the melanorosaurids, any change in the configuration and relationships of the metacarpals in sauropods would further pronate the manus, allowing it to function more effectively in a craniocaudal plane. As the evolution of a U-shaped manus also seems to be correlated with increased body size and mass (Wilson and Sereno 1998), selection for morphological changes correlated with alleviating weight-bearing stresses (e.g. more columnar metacarpal orientation and a U-shaped arrangement) may have further 'refined' manus shape in eusauropods.

Thus, a U-shaped manus in sauropods probably arose after pronation and was perhaps influenced by selective pressure to refine pronation and adopt a stress-alleviating morphology. In this respect, *Melanorosaurus* provides a crucial window into the sequence of these changes. The appearance of a morphologically sauropod-like ulna and radius in *Melanorosaurus* suggests that these features are correlated with reversion to quadrupedal posture and manus pronation. Later in the larger eusauropods and neosauropods, and perhaps basal sauropods, weight-support may have had a more direct influence on whatever selective pressures drove the development of the

U-shaped metacarpus. Ultimately, the shift to a U-shaped manus was not temporally linked with changes in the antebrachium, and was an evolutionary event that occurred after the appearance of both the sauropod clade and obligate quadrupedalism, but probably before the origination of the gigantic eusauropods.

Acknowledgements. We acknowledge the following for their assistance with specimens in their care: M. Raath and B. Rubidge (Bernard Price Institute of Palaeontology and Palaeoanthropology); R. Nuttall and J. Botha (National Museum, Bloemfontein); R. Smith, A. Chinsamy, E. Butler and S. Kaal (South African Museum, Iziko Museums of Cape Town); and R. Schoch and staff (Staatliches Museum für Naturkunde). Funding for this project was provided to both authors through a National Geographic grant award (7713-04). Additional funding was provided through an NSF grant 0093929, received by J. M. Parrish and K. Stevens on which MFB is an outside collaborator, and a University Research Council grant 3-30135 from Western Illinois University to MFB. We thank P. Barrett, J. Farlow and an anonymous reviewer for their comments, which improved the quality of the final manuscript. We are also grateful to P. Barrett for the quick turn-around and efficient editing of our manuscript, and sincerely thank him and Tim Fedak for organizing this special volume.

REFERENCES

BONAPARTE, J. F. and VINCE, M. 1979. El hallazgo del primer nido de dinosaurios triásicos (Saurischia, Prosauropoda), Triásico Superior de Patagonia, Argentina. *Ameghiniana*, **16**, 173–182.

BONNAN, M. F. 2001. The evolution and functional morphology of sauropod dinosaur locomotion. Unpublished PhD dissertation, Northern Illinois University, DeKalb, 000 pp.

—— 2003. The evolution of manus shape in sauropod dinosaurs: implications for functional morphology, forelimb orientation and sauropod phylogeny. *Journal of Vertebrate Paleontology*, **23**, 595–613.

—— and SENTER, P. 2007. Were the basal sauropodomorph dinosaurs *Plateosaurus* and *Massospondylus* habitual quadrupeds? 139–155. *In* BARRETT, P. M. and BATTEN, D. J. (eds). *Evolution and palaeobiology of early sauropodomorph dinosaurs.* Special Papers in Palaeontology, **77**, 289 pp.

CHINSAMY-TURAN, A. 2005. *The microstructure of dinosaur bone: deciphering biology with fine-scale techniques.* John Hopkins University Press, Baltimore, MD, 195 pp.

COOPER, M. R. 1981. The prosauropod dinosaur *Massospondylus carinatus* Owen from Zimbabwe: its biology, mode of life, and phylogenetic significance. *Occasional Papers of the National Museums and Monuments of Rhodesia, B. Natural Sciences*, **6**, 689–840.

—— 1984. A reassessment of *Vulcanodon karibaensis* Raath (Dinosauria: Saurischia) and the origin of the Sauropoda. *Palaeontologia Africana*, **25**, 203–231.

GALTON, P. M. 1990. Basal Sauropodomorpha – Prosauropoda. 320–344. *In* WEISHAMPEL, D. B., DODSON, P. and OSMÓLSKA, H. (eds). *The Dinosauria.* First edition. University of California Press, Berkeley, CA, 733 pp.

—— and UPCHURCH, P. 2004. Prosauropoda. 232–258. *In* WEISHAMPEL, D. B., DODSON, P. and OSMÓLSKA, H. (eds). *The Dinosauria.* Second edition. University of California Press, Berkeley, CA, 733 pp.

—— VAN HEERDEN, J. and YATES, A. M. 2005. Postcranial anatomy of referred specimens of the sauropodomorph dinosaur *Melanorosaurus* from the Upper Triassic of South Africa. 1–37. *In* TIDWELL, V. and CARPENTER, K. (eds). *Thunder-lizards: the sauropodomorph dinosaurs.* University of Indiana Press, Bloomington, IN, 495 pp.

GILMORE, C. W. 1936. Osteology of *Apatosaurus* with special reference to specimens in the Carnegie Museum. *Memoirs of the Carnegie Museum*, **10**, 175–300.

HAINES, R. W. 1969. Epiphyses and sesamoids. 81–115. *In* GANS, C., BELLAIRS, A. d'A. and PARSONS, T. S. (eds). *Biology of the Reptilia, morphology A.* Volume 1. Academic Press, New York, NY, 373 pp.

HATCHER, J. B. 1902. Structure of the forelimb and manus of *Brontosaurus*. *Annals of the Carnegie Museum*, **1**, 356–376.

HAUGHTON, S. H. 1924. The fauna and stratigraphy of the Stormberg Series. *Annals of the South African Museum*, **12**, 323–497.

HILDEBRAND, M. and GOSLOW, G. 2001. *Analysis of vertebrate structure.* Fifth edition. John Wiley and Sons, New York, NY, 373 pp.

POL, D. and POWELL, J. E. 2007. New information on *Lessemsaurus sauropoides* (Dinosauria, Sauropodomorpha) from the Upper Triassic of Argentina. 223–243. *In* BARRETT, P. M. and BATTEN, D. J. (eds). *Evolution and palaeobiology of early sauropodomorph dinosaurs.* Special Papers in Palaeontology, **77**, 289 pp.

RAATH, M. A. 1972. Fossil vertebrate studies in Rhodesia: a new dinosaur (Reptilia: Saurischia) from near the Trias-Jurassic boundary. *Arnoldia*, **5**, 1–37.

REISZ, R. R., SCOTT, D., SUES, H.-D., EVANS, D. C. and RAATH, M. A. 2005. Embryos of an early prosauropod dinosaur and their evolutionary significance. *Science*, **309**, 761–764.

SENTER, P. and ROBINS, J. H. 2005. Range of motion in the forelimb of the theropod dinosaur *Acrocanthosaurus atokensis*, and implications for predatory behaviour. *Journal of Zoology*, **266**, 307–318.

UPCHURCH, P. 1998. The phylogenetic relationships of sauropod dinosaurs. *Zoological Journal of the Linnean Society*, **124**, 43–103.

—— BARRETT, P. M. and DODSON, P. 2004. Sauropoda. 259–322. *In* WEISHAMPEL, D. B., DODSON, P. and OSMÓLSKA, H. (eds). *The Dinosauria.* Second edition. University of California Press, Berkeley, CA, 733 pp.

—— —— and GALTON, P. M. 2007. A phylogenetic analysis of basal sauropodomorph relationships: implications for the origin of sauropod dinosaurs. 57–90. *In* BARRETT, P. M.

and BATTEN, D. J., (eds). *Evolution and palaeobiology of early sauropodomorph dinosaurs.* Special Papers in Palaeontology, **77**, 289 pp.

WILSON, J. A. 2002. Sauropod dinosaur phylogeny: critique and cladistic analysis. *Zoological Journal of the Linnean Society*, **136**, 217–276.

—— 2005. Integrating ichnofossil and body fossil records to estimate locomotor posture and spatiotemporal distribution of early sauropod dinosaurs: a stratocladistic approach. *Paleobiology*, **31**, 400–423.

—— and SERENO, P. C. 1998. Early evolution and higher-level phylogeny of sauropod dinosaurs. *Memoirs of the Society of Vertebrate Paleontology*, **5**, 1–68.

YATES, A. M. 2003. A new species of the primitive dinosaur *Thecodontosaurus* (Saurischia: Sauropodomorpha) and its implications for the systematics of early dinosaurs. *Journal of Systematic Palaeontology*, **1**, 1–42.

—— 2007. The first complete skull of the Triassic dinosaur *Melanorosaurus* Haughton (Sauropodomorpha, Anchisauria). 9–55. *In* BARRETT, P. M. and BATTEN, D. J. (eds). *Evolution and palaeobiology of early sauropodomorph dinosaurs.* Special Papers in Palaeontology, **77**, 289 pp.

—— in press. Solving a dinosaurian puzzle: the identity of *Aliwalia rex* Galton. *Historical Biology.*

—— and KITCHING, J. W. 2003. The earliest known sauropod dinosaur and the first steps towards sauropod locomotion. *Proceedings of the Royal Society of London, Series B*, **270**, 1753–1758.

[Special Papers in Palaeontology 77, 2007, pp. 169–206]

BONE HISTOLOGY AND GROWTH OF THE PROSAUROPOD DINOSAUR *PLATEOSAURUS ENGELHARDTI* VON MEYER, 1837 FROM THE NORIAN BONEBEDS OF TROSSINGEN (GERMANY) AND FRICK (SWITZERLAND)

by NICOLE KLEIN* *and* P. MARTIN SANDER†

*Naturhistorisches Forschungsinstitut, Humboldt-Universität zu Berlin, Institut für Paläontologie, Invalidenstraße 43, D-10115 Berlin, Germany; e-mail: Nicole.Klein@museum.hu-berlin.de

†Institut für Paläontologie, Universität Bonn, Nußallee 8, D-53115 Bonn, Germany; e-mail: martin.sander@uni-bonn.de

Typescript received 6 February 2006; accepted in revised form 21 October 2006

Abstract: Bones of the prosauropod dinosaur *Plateosaurus engelhardti* from the Norian localities of Trossingen (Germany) and Frick (Switzerland) were sampled for palaeohistological study. Both localities yielded rich material of adult individuals with body sizes of 4·80–10 m. Altogether 50 bones from *c.* 27 individuals were sampled using different methods (coring, cross-sectioning, analysis of existing fracture surfaces). The cortex of most bones consists of the laminar fibro-lamellar complex typical for dinosaurs, indicating fast growth rates. The laminar fibro-lamellar complex in *Plateosaurus engelhardti* is always cyclically interrupted by lines of arrested growth (LAGs). In some bones, the laminar fibro-lamellar complex changed in the exterior cortex to lamellar-zonal bone, grading peripherally into avascular lamellar bone. Thus, growth was clearly determinate in *P. engelhardti*, but in contrast to all other dinosaur species for which this is known, the final size of fully grown individuals covers a very broad size range (6·5–10 m). Additionally, growth in *P. engelhardti* may have been influenced by environmental conditions, as suggested by irregular growth cycle thickness and differences in growth rates. Both reflect a more reptilian growth pattern than is known from any other dinosaur, agreeing with the basal phylogenetic position of *P. engelhardti*. Nevertheless, in *Plateosaurus* the foundations for very fast growth and the evolution of gigantism in sauropods are already in place.

Key words: Prosauropoda, bone histology, skeletochronology, growth pattern, life history.

THE only direct way to obtain information about past events in the life of an individual dinosaur is from the growth record preserved in its hard tissues (Castanet *et al.* 1993; Castanet 1994; Curry 1999; Sander 1999, 2000; Erickson 2005). Bone histology in particular may provide information about aspects of palaeobiology including individual age, age at sexual maturity, growth rates and growth patterns, sexual dimorphism, and cycles of activity and reproduction. From these data, information can be deduced regarding physiology and metabolic rate, ontogeny, evolution and phylogeny (Castanet *et al.* 1993; Chinsamy 1993*a*, *b*, 1994; Erickson and Tumanova 2000; Horner *et al.* 2000; Sander *et al.* 2004; Chinsamy-Turan 2005; Erickson 2005). Skeletochronology is a standard method for ageing extant fishes and ectothermic vertebrates (de Ricqlès *et al.* 1991; Castanet *et al.* 1993; Chinsamy 1994) and is almost the only way for assessing age in extinct vertebrates (Castanet 1994). The method is based on counting the cyclically deposited growth marks pre-

served in bone microstructure. Terminology used herein follows Francillon-Vieillot *et al.* (1990).

Plateosaurus engelhardti, from the Norian Knollenmergel beds (Löwenstein and Trossingen formations, Keuper) of central Europe, is of great interest for bone histological research. Firstly, its relatively early (Late Triassic) occurrence allows assessment of the basic question of whether the high growth rates of Jurassic and Cretaceous dinosaurs (Curry 1999; Horner *et al.* 1999, 2000; Sander 1999, 2000; Erickson and Tumanova 2000; Sander and Tückmantel 2003; Sander *et al.* 2004; Erickson 2005) evolved gradually or were present from the time of their origin. Secondly, as *Plateosaurus* is a prosauropod dinosaur, it belongs to the sister-group of the giant sauropods (Galton and Upchurch 2004). Thus, histological study of *Plateosaurus engelhardti* can contribute to our understanding of how sauropods were able to grow to gigantic body sizes from the Late Triassic onward. Finally, *Plateosaurus engelhardti* is well suited for histological study as several dozen complete and incomplete individuals of the same species are known from different *Plateosaurus* mass accumulations.

This paper aims to describe and document the bone histology of *Plateosaurus engelhardti*. The main focus is on the bone tissues and the kinds of vascularization present in different bones and at different sampling locations within single bones. Histological differences within samples of single individuals and between different elements of the skeleton, as well as histological differences between samples from two different localities (Trossingen in south-western Germany and Frick in northern Switzerland), will be documented and discussed. Growth rates, counts of growth marks and extrapolation methods are used for ageing the adult individuals. On the basis of these data, life history parameters such as longevity, possible age at sexual maturity, growth determination, age at growth cessation and sexual dimorphism are deduced for *Plateosaurus engelhardti*. Bone histology can also address the issue of a possible correlation between age and body size/body mass. Finally, the growth strategy and life history data of *Plateosaurus engelhardti* will be compared briefly with that of other prosauropods, sauropods and some recent tetrapods.

Institutional abbreviations. AMNH, American Museum of Natural History, New York, USA; IFG, Institut für Geowissenschaften, Universität Tübingen, Germany; IPB, Institut für Paläontologie, Universität Bonn, Germany; MSF, Municipal Sauriermuseum Frick, Frick, Cantone of Aargau, Switzerland; NAA, Cantonal Museum of Natural History 'Naturama', Aarau, Cantone of Aargau, Switzerland; PIMUZ, Paläontologisches Institut und Museum der Universität Zürich, Zürich, Switzerland; SMA, Sauriermuseum, Aathal near Zürich, Switzerland; SMNS, Staatliches Museum für Naturkunde, Stuttgart, Germany.

BONE HISTOLOGY AND GROWTH

Results from (and problems of) tetrapod skeletochronology

Interpretation of growth marks involves various problems (Reid 1997*a*). Consequently, the basic premise of most works dealing with skeletochronology, the annual deposition of the lines of arrested growth (LAGs) or of any other cyclical growth marks, has been discussed extensively. Following the work of several recent authors (Zug *et al.* 1986; Castanet and Smirina 1990; Francillon-Vieillot *et al.* 1990; Reid 1990, 1997*a*; de Ricqlès *et al.* 1991; Castanet *et al.* 1993; Chinsamy 1994; Horner *et al.* 1999; Erickson and Tumanova 2000; Horner and Padian 2004; Erickson 2005), growth marks and LAGs are interpreted as having been formed annually in the current study. This premise is supported by studies on recent reptiles of

known age (Castanet *et al.* 1993; Castanet 1994) and has also been supported using oxygen isotope geochemistry (Tütken 2004). According to Reid (1997*a*) growth lines in dinosaurs are rather common. Horner *et al.* (1999) pointed out that LAGs, and therefore cyclical growth, are a plesiomorphic feature for vertebrates and suggested that these structures had no particular physiological meaning as they appear in several orders of mammals (see also Klevezal 1996; Sander and Andrassy in press). Chinsamy and Hillenius (2004) argued for a more functional significance of growth marks, suggesting that their formation is strongly dependent on exogenous conditions.

One problem in using bone histology for age estimates is that each skeleton undergoes substantial histological changes during ontogeny, owing to longitudinal growth, changes in shape, reproductive activity and fatigue repair (Erickson and Tumanova 2000). Additionally, every bone of a skeleton has its own ontogenetic history (Horner *et al.* 1999; A. de Ricqlès pers. comm. 2001; Starck and Chinsamy 2002; Chinsamy-Turan 2005). Growth rate and bone tissue can vary within a single bone during ontogeny, as well as in different bones of a single individual (de Ricqlès *et al.* 1991; Chinsamy 1993*b*; Starck and Chinsamy 2002). Thus, growth stops at somewhat different stages in the varying bone types (Chinsamy 1994) as a result of morphological growth allometries. Some bones, for example, grow asymmetrically, and growth cycles may be restricted to the parts where growth was slowest (Reid 1997*b*). This means that the same bone sampled in different regions, as well as different kinds of bones from a single individual, may present a different bone microstructure and therefore a different growth history, as expressed by a varying number of LAGs (de Ricqlès 1983; Chinsamy 1993*a*; Horner *et al.* 2000) or by the deposition of different bone types. On one hand, this makes standardized sampling locations for each bone type necessary (de Ricqlès 1983; Reid 1990; Chinsamy 1993*a*, *b*; Sander 2000). Conversely, it is necessary to sample different kinds of bone from an individual to get the most complete growth record.

Owing to resorption from the medullary cavity outwards and remodelling of bone, which both start early in ontogeny and vary in extent from bone to bone and from group to group, the preserved growth record in most bones is incomplete. Most studies (e.g. Chinsamy 1990, 1993*a*, *b*; Sander 1999, 2000; Erickson and Tumanova 2000; Horner *et al.* 2000; Horner and Padian 2004; Chinsamy-Turan 2005) deal with long bones like femora and humeri, allowing histological comparison between the various tetrapods studied. However, long bones of subadult and adult individuals usually have a relatively large medullary cavity and, consequently, an incomplete growth record that makes it necessary to estimate the number of resorbed cycles by extrapolation (Chinsamy 1993*a*; Reid 1997*a*). A further problem is that many tetrapods grow at different rates during different ontogenetic stages (Varricchio 1993; Chinsamy 1994; Curry 1999; Erickson and Tumanova 2000; Horner *et al.* 2000; Sander 2000; Horner and Padian 2004). It is therefore necessary to sample different ontogenetic stages of one taxon to get the most complete growth record with the minimum amount of error. Unfortunately, this is seldom possible due to the rarity of appropriate fossil material.

Growth and physiology in dinosaurs

Comparison of histological data shows that, although very diverse, dinosaurs had a relatively uniform bone microstructure (fibro-lamellar bone), presumably inherited from a common ancestor. Most dinosaurs also had determinate growth, as indicated histologically by a clear cessation of growth (e.g. the external fundamental system, EFS; see Horner *et al.* 2001; Erickson 2005: also called outer circumferential layer, OCL; see Chinsamy-Turan 2005). However, not all dinosaurs grew at the same fast growth rates, leading to different life histories. For example, some dinosaurs grew continuously and rapidly, as shown by highly vascularized fibro-lamellar bone, and their growth rates are comparable with those of large, fast-growing mammals (Reid 1997*a*; Erickson *et al.* 2001) and precocial birds (Erickson *et al.* 2001). Other dinosaurs possess less vascularized fibro-lamellar bone, sometimes in combination with lamellar-zonal bone, indicating slower growth within a cyclical growth pattern, as seen in modern reptiles. Moreover, some dinosaurs exhibit both growth patterns in different parts of their skeletons (Reid 1997*a*; Erickson *et al.* 2001).

As a result of the various studies on dinosaur bone histology, it has been demonstrated that most dinosaurs grew at higher growth rates than extant reptiles, but not all grew as fast as extant mammals and birds. Thus, dinosaurs show a combination of bone tissue types and growth rates that is not present in any extant tetrapod group and is therefore not readily comparable. Consequently, dinosaur physiology seems to have been unique and different from that of modern tetrapods (Reid 1990, 1997*a*). Bone histology is not directly correlated with thermoregulation, nevertheless it does give insights into bone depositional rate, growth rate and growth strategy (Reid 1990; Chinsamy 1994; Chinsamy-Turan 2005; Erickson 2005).

Previous work on prosauropod bone histology

Several early works on bone histology mention prosauropods, especially *Plateosaurus*, but most of them are purely descriptive (Seitz 1907; Gross 1934). Enlow and Brown (1957) noted the similarity of the bone tissue of *Plateosaurus* with that of several mammalian groups, predominately artiodactyls, but did not study *Plateosaurus* firsthand. Currey (1962) used bone histology to suggest that prosauropods had a physiology similar to that of large herbivorous mammals, and de Ricqlès (1968) was the first to recognize the importance of prosauropod histology for sauropod evolution. Counting growth cycles to obtain information about life history was carried out by Reid (1990) and Chinsamy (1993*a*). All previous authors agree that the primary tissue in prosauropod long bones consists of laminar fibro-lamellar bone (although earlier authors used a different terminology).

Chinsamy (1993*a*) provided the first comprehensive work on prosauropod bone histology, based on transverse sections from a growth series of femora (12·0–44·0 cm long) of the South African taxon *Massospondylus carinatus*, which is closely related to *Plateosaurus* (Galton and Upchurch 2004). Although the growth rate in *Massospondylus* clearly decreased with increasing age, growth did not stop but continued indefinitely at a decreased rate (Chinsamy 1993*a*). Chinsamy (1993*a*) reported an indeterminate growth strategy from this and concluded that *Massospondylus* displayed an intermediate physiological level between ectotherms and endotherms (Chinsamy 1993*a*).

Foelix (1997, figs 4, 5; 1999, fig. 10) illustrated thin-sections of *Plateosaurus* ribs from the *Plateosaurus* bone-bed at Frick, Switzerland. He documented growth rings in the outer parts of the rib samples and suggested periodic growth for *Plateosaurus*.

Other histological work on prosauropods was conducted on the basal taxon *Thecodontosaurus antiquus* from the Upper Triassic of south-west England (Cherry 2002). Cherry (2002) studied long bones, phalanges and ribs. All sections showed laminar fibro-lamellar bone, but LAGs occurred only in the rib.

THE PROSAUROPOD DINOSAUR *PLATEOSAURUS*

The genus *Plateosaurus* belongs to the Prosauropoda, which together with the Sauropoda forms the Sauropodomorpha (von Huene 1932). Galton and Upchurch (2004) regarded the Prosauropoda as a monophyletic clade that forms the sister taxon of sauropods. *Plateosaurus* first appears in the upper middle Keuper (Norian) and disappears within the Rhaetian. In central Europe, *Plateosaurus* remains have mainly been found in the Stubensandstein and the Knollenmergel Beds of late Norian or early Rhaetian age in south-west Germany and stratigraphically equivalent marls and mudstones in Bavaria, France and

Switzerland (Galton 2001). Detailed osteological descriptions of this material have been produced by various authors (von Huene 1926; Galton 1985, 1986, 1990, 2001; Weishampel and Westphal 1986; Van Heerden 1997; Galton and Upchurch 2004).

Plateosaurus bonebeds of central Europe

Plateosaurus is known from many skeletons, several of which are complete, making this animal one of the best-known Late Triassic dinosaurs. The most important localities are Halberstadt in central Germany, Trossingen in south-west Germany and Frick in northern Switzerland. The fossils at Halberstadt were found in the Trossingen Formation, the fossils at Trossingen are from the Löwenstein Formation (not from the overlying Trossingen Formation according to Brenner and Villinger 1981) and the fossils at Frick are from a unit called the 'Obere Bunte Mergel'. These three localities were called '*Plateosaurus* bonebeds' by Sander (1992) as the same taphonomic processes were involved in their origin and as they occupy nearly identical stratigraphical positions. *Plateosaurus* bonebeds are currently thought to have been formed when the animals were mired in mud on a floodplain, followed by mummification or carcass disarticulation by theropod scavengers and weathering (Sander 1992). However, these do not necessarily represent mass mortality events as synchrony of miring cannot be established (Sander 1992). *Plateosaurus* is usually the most frequent terrestrial vertebrate in the beds in which it occurs.

Each of these accumulations yielded a large number of fragmentary and complete skeletons. For example, remains of 35 individuals were discovered at Halberstadt (Jaekel 1913–14; Sander 1992). During three main excavation periods the remains of *c.* 54 individuals were recovered at Trossingen (Sander 1992). The *Plateosaurus* remains from Frick belong to no fewer than 20 individuals, but new material continues to be found. Unfortunately, no juveniles or small individuals of *Plateosaurus engelhardti* are currently known from the bone-beds or from any other central European locality (Sander 1992; Moser 2003). Remains of juvenile *Plateosaurus* specimens from the Feuerletten beds of Ellingen/Bavaria that were mentioned by Wellnhofer (1993) are now referred to theropods (Moser 2003).

Taxonomy of Plateosaurus *from the central European bone-beds*

The genus *Plateosaurus* was erected by von Meyer in 1837 when he described fragmentary remains, most notably a femur, from the Feuerletten beds of Bavaria as *P. engelhardti* (von Meyer 1837). In the initial descriptions of the prosauropods from the *Plateosaurus* bone-beds of Trossingen and Halberstadt, none of the material was referred to this species, but numerous new species and even genera were erected, often on the basis of incomplete skeletons (Jaekel 1913–14; von Huene 1932). The *Plateosaurus* remains from the locality of Frick have only been the subject of a brief study by Galton (1985, 1986) who referred them to the type species, *P. engelhardti*. Beginning with these studies until the year 2000, all authors held the view that the bone-beds at Trossingen, Halberstadt and Frick contain only this species of *Plateosaurus* (Galton 1985, 1986, 1990; Weishampel and Westphal 1986; Weishampel and Chapman 1990; Sander 1992). Differences in osteology within one locality and between the different localities are small and were explained by intraspecific variations or possibly by sex differences. Although there is still agreement that only a single species of *Plateosaurus* is present in Trossingen and Frick (Galton 2001; A. Yates, pers. comm. 2002, 2003; Moser 2003; Galton and Upchurch 2004), the discussion concerning the correct name of this species is continuing. Galton (2001) argued that the material from the *Plateosaurus* bonebeds represents a different species from the Bavarian type material of *P. engelhardti*. The earliest available name for this species would be *P. longiceps* Jaekel (1913-14), based on material from Halberstadt (Galton 2001; Galton and Upchurch 2004). Moser (2003) and Yates (2003), on the other hand, have continued to refer all material from Trossingen, Frick and Halberstadt to *P. engelhardti*. Our own unpublished work indicates that the Frick material does indeed consist only of a single species, with a few small but consistent differences from the Trossingen material (see also Galton 2001). These may represent differences at the population level. We follow Moser (2003) and Yates (2003, 2004) in calling the *Plateosaurus* material from the localities of Trossingen and Frick *Plateosaurus engelhardti* because we feel that the differences between the type material of *P. engelhardti* from Bavaria and that of *P. longiceps* are insufficient for specific separation.

Size, body mass, and variation in Plateosaurus

Plateosaurus engelhardti is a medium-sized to large prosauropod dinosaur, with a maximum length of approximately 10 m (Weishampel 1984). Moser (2003) argued that this estimate is too high because it was based on bones enlarged by clay swelling. However, as the current study will show, the histological record of *Plateosaurus* bones from Trossingen and Frick is intact and provides no evidence for any significant diagenetic changes in bone size (both bone length and thickness). Based on the skeletal reconstruction in Weishampel and Westphal (1986) femur length is approximately 10 per cent of overall body length (Sander 1992). The reconstruction in Weishampel and Westphal (1986) was based on the mounted skeletons of two medium-sized individuals (SMNS 13200, femur length 66·0 cm, and IFG uncatalogued, femur length 58·5 cm). The IFG specimen is the one figured in Weishampel and Westphal (1986, pl. 1, specimen on the right).

Sander (1992) estimated the body mass of *Plateosaurus* by using a scale model based on skeleton SMNS 13200 and calculated that an individual of 8 m total length weighed 2179 kg (Table 1). Seebacher (2001) determined the body mass using a graphical reconstruction of the complete skeleton AMNH 6810 (Galton 1986) and high-order polynomial equations, yielding a body weight of 1072 kg for an individual 6·5 m in length.

Weishampel and Chapman (1990) performed multivariate morphometric analyses on femora of *Plateosaurus*, which led them to conclude that two morphs could be distinguished which may have been sexual dimorphs and that each sex may have possessed a different locomotory regime. Gow *et al.* (1990) proposed sexual differences in the skull of *Plateosaurus*, but Galton (2000) contradicted this as the skull differences were not consistent with femoral differences. Galton (1997, 2000, p. 236) described these differences as 'probably individual variations'. Variation in the postcranial skeleton of the prosauropods *Thecodontosaurus antiquus* and *Melanorosaurus readi* may also be the result of sexual dimorphism, as Galton (1997) suggested, or of allometry and individual variation (Benton *et al.* 2000).

Plateosaurus engelhardti bones examined for the current study show considerable variation. Among the Trossingen material, bones of a similar length show some slight morphological variations in size and shape, mainly of the proximal epiphyses and shaft diameter, leading to the recognition of robust and gracile morphs. These differences could not be identified consistently between complete or nearly complete skeletons, but were only apparent in single elements. Owing to the similar size range of the bones studied, allometric effects can be dismissed. This leaves the possibility of two different species occurring at Trossingen, which seems unlikely because there are no really significant differences in morphology. Moser (2003) believed that the high morphological variability of *Plateosaurus* bones from Trossingen and elsewhere is the result of diagenetic effects (palaeopedogenesis) and not of a biological nature, though the current histological study does not corroborate this. Consequently, we prefer the explanation that the observed differences in the samples are a result of either sexual dimorphism or individual variation.

MATERIAL

Specimens studied

Owing to their relatively simple appositional growth, long bones like the femur, tibia, fibula and humerus were used for studying growth and skeletochronology. Nevertheless, in the present study, flat bones like the pubis and scapula were also included. Other bone types such as vertebrae,

TABLE 1. DME on the basis of the mass estimates available in the literature. Masses in italic type were taken from Sander (1992) and Seebacher (2001).

Specimen	Femur length (cm)	Estimated body mass (kg) based on Sander (1992)	Estimated body mass (kg) based on Seebacher (2001)	Estimated body mass (kg) on the basis of the mean value for the largest femur (cm)
femur 1	50·0	523	482	515
femur 2	56·5	763	708	753
femur 3	59·0	872	804	832
femur 4	59·5	894	825	872
femur 5	62·5	1046	954	991
femur 6	63·5	1090	997	1030
femur 7	65·0	1177	*1072*	1109
femur 8	72·0	1591	1458	1545
femur 9	74·0	1721	1587	1664
femur 10	76·0	1874	1715	1783
femur 11	77·5	1983	1812	1902
SMNS 13200	80·0	*2179*	1994	2100
femur 12	81·0	2266	2080	2179
femur 13	99·0	4140	3784	3962

ribs, pedal phalanges and ischia were sampled in limited numbers for comparative purposes. Only *Plateosaurus* material from Trossingen and Frick, and material kept in the SMNS from other Knollenmergel localities, was included in the present study. Bones from the Trossingen and Frick localities were all measured and partially sampled: those from other localities were only measured. These data were used for comparison and to provide an overview of the general size range of bones and skeletal proportions. Most of the studied material from Trossingen is housed in the SMNS, but some bones were also sampled and measured in the IFG. Material from the excavations in Frick is mainly housed in the MSF and NAA; some measurements were taken from a specimen at the SMA.

Plateosaurus engelhardti *from Trossingen*

The fossils from the first excavation at Trossingen and *Plateosaurus* remains from other localities are fully accessioned in the collections of the SMNS, but those from the third Trossingen excavation (conducted by Seemann in 1932) is only partially accessioned. Almost all of this material was prepared soon after excavation, and broken bones were repaired and partially reconstructed with plaster. The material was not soaked in any stabilizing liquids or resins. Field numbers given by Seemann during the 1932 excavation range from 1 to 65 and represent the order in which the specimens were discovered (Seemann 1932, 1933). A single field number may either represent an assemblage of bones from several individuals, a complete or nearly complete individual, or a few isolated bones. During preparation, it became clear that some of the field numbers incorporating only a few bones could be assigned to individuals represented by other field numbers. Specimens 3, 6 and 54 represent turtle remains. Specimen 1 and some others that are currently missing were presumably lost during World War II (in 1944: Ziegler 1986; R. Wild, pers. comm. 2001). Before selection of material for histological examination, all of the material surviving in the SMNS and IFG was inventoried. On the basis of this inventory, the minimum number of individuals (MNI) was determined for each field number by counting long and flat bones of similar sizes and assigning them to right and left categories. The total MNI for Trossingen is 54 individuals. Field numbers 5, 10, 20, 36, 37, 45, 50, 53 and 65 were fully accessioned to the SMNS collection, and field number 33 was prepared *in situ* and is on exhibit (Ziegler 1988, fig. 17). Field numbers 8, 14, 15, 21, 27, 28, 29, 48 and 61 remain stored in large crates and are referred to by their field numbers alone (e.g. 'SMNS F8'). The material in the crates provided the main sample for the present histological research, while the fully accessioned material was only used for measure-

ments. A few bones from the IFG were sampled. This material was excavated by von Huene (in 1921 and 1923) and was mechanically prepared and stabilized with shellac (Weishampel and Westphal 1986). In total, 39 bones from Trossingen (Table 2) were sampled histologically, representing a minimum of 16 individuals. For some individuals, several bones were sampled allowing comparisons of the growth record in different bones of the same individual (SMNS F14, F29, F48). To check the consistency of the growth record in a single bone, different samples were taken from varied locations.

Plateosaurus engelhardti *from Frick*

The *Plateosaurus* material from Frick was also prepared, repaired and partially reconstructed with plaster. Material from excavations prior to 1995 is inventoried in the collection of the MSF. The remains of *Plateosaurus* exhibited in the SMA are on permanent loan from the MSF. The material from the 1995 excavation is not yet fully prepared and is stored in the NAA. Some of the material from Frick was stabilized with cyanoacrylate glue in the field and during preparation. An inventory of the long and flat bones was produced before sampling. The MNI excavated at Frick is approximately 20, and new specimens are being discovered as mining of the dinosaur-bearing mudstone progresses. Most of the Frick material is on exhibit and is not accessible for sampling. The bones from Frick fall into the lower size range of the medium-sized *Plateosaurus engelhardti* individuals from Trossingen (Text-fig. 1). Eleven bones from Frick were sampled for histological research (Table 2), representing an MNI of 11. None of the Frick individuals could be sampled from more than one bone. In some bones, samples were taken at different locations to check for the consistency of the growth record.

To identify sampled bones uniquely without a collection number, the locality, the original field numbers, bone length and body side of each specimen are listed in Table 2. The sampled bones were given consecutive numbers from the smallest to the largest for this study (Table 2).

Body size of Plateosaurus *from Frick and Trossingen*

On the basis of the femur/body length ratio of 1 : 10 (Sander 1992), the *Plateosaurus engelhardti* material from Trossingen consists of individuals with a total body length from approximately 4·80 m (scapula 1, calculated femur length 48·0 cm; see 'Methods') to 9·90 m (femur 13). Text-figure 1A is a size class histogram based on the femora from Trossingen and Frick. Two peaks are

TABLE 2. Sampled bones of *P. engelhardti* from the Trossingen and Frick localities arranged according to kind of bone and bone length. TS, thin-section; PS, polished section; CS, cross-section; FS, fracture surface; NM, not measurable.

Specimen	Total length (cm)	Kind of sample	Body side	Locality	Specimen number
femur 1	c. 50·0	TS, PS, CS	right	Trossingen	SMNS F 29
femur 2	56·5	TS, PS, CS	right	Frick	NAA F 88-B130
femur 3	59·0	TS, PS	left	Frick	NAA uncatalogued
femur 4	59·5	TS, PS	right	Trossingen	SMNS F 29 A
femur 5	62·5	FS	right	Trossingen	SMNS F 27
femur 6	63·5	TS, PS, FS	left	Trossingen	SMNS F 14 A
femur 7	72·0	TS, PS, FS	left	Trossingen	SMNS F 29 A
femur 8	74·0	FS	left	Trossingen	SMNS F 8
femur 9	74·0	TS, PS	left	Trossingen	IFG 192.1
femur 10	76·0	FS	right	Trossingen	SMNS F 48-1
femur 11	77·5	TS, PS	left	Trossingen	SMNS F 48-1
femur 12	81·0	TS, PS	left	Trossingen	SMNS F 27
femur 13	99·0	TS, PS	left	Trossingen	IFG uncatalogued
tibia 1	c. 51·0	TS, PS	right	Frick	MSF MSFM-1
tibia 2	51·0	CS	right?	Frick	NAA F 88/B70
tibia 3	51·0	FS	right	Trossingen	SMNS F 15
tibia 4	≫ 52·0	TS, PS	left	Frick	MSF MSFM-2
tibia 5	53·0	TS, PS	right	Frick	NAA A 9
tibia 6	55·0	TS, PS, FS	right	Trossingen	SMNS F 14 A
tibia 7	59·0	TS, PS, FS	left	Trossingen	SMNS F 29 A
tibia 8	66·0	FS	right	Trossingen	SMNS F 48-2
fibula 1	46·5	FS	left	Trossingen	SMNS F 29 B
fibula 2	52·5	FS	right	Trossingen	SMNS F 14
fibula 3	53·0	FS	right	Trossingen	SMNS F 15
fibula 4	57·5	FS	left	Trossingen	SMNS F 29 B
fibula 5	c. 59·0	FS	right?	Trossingen	SMNS F 8
fibula 6	59·0	FS	right	Trossingen	SMNS F 48-2
humerus 1	NM	TS	right?	Trossingen	IFG 11921
humerus 2	41·0	FS	left	Trossingen	SMNS F 14 A
humerus 3	43·5	TS, PS	left	Trossingen	SMNS F 29 A
humerus 4	c. 43·5	FS	right	Trossingen	SMNS F 61 B
humerus 5	44·5	TS, CS	right	Frick	NAA F 88/B640
humerus 6	53·0	TS	right	Trossingen	IFG uncatalogued
scapula 1	36·5	TS, PS	left	Frick	MSFM uncatalogued
scapula 2	≫ 39·0	CS	right?	Frick	NAA uncatalogued
scapula 3	≫ 42·0	CS	left	Trossingen	SMNS F 29
scapula 4	47·0	TS, PS, FS	right	Trossingen	SMNS F 14 B
scapula 5	48·0	TS, PS, FS	left	Trossingen	SMNS F 29 B
scapula 6	49·5	TS, PS	left	Trossingen	SMNS F 29 A
pubis 1	NM	TS, PS	?	Frick	NAA uncatalogued
pubis 2	48·0	CS	left	Trossingen	SMNS F 29
pubis 3	50·0	TS, PS, FS	left	Trossingen	SMNS F 14 B
pubis 4	52·5	TS, PS	right	Trossingen	SMNS F 29 B
pubis 5	53·5	TS, PS, FS	left	Trossingen	SMNS F 29 B
vertebra 1	dorsal vertebra (10·8)	TS, PS		Trossingen	IFG RW 12
vertebra 2	dorsal vertebra (10·7)	TS, PS		Trossingen	IFG RW 14
rib 1	rib	CS		Trossingen	SMNS F 29

TABLE 2. Continued.

Specimen	Total length (cm)	Kind of sample	Body side	Locality	Specimen number
rib 2	rib	TS, PS		Frick	MSF F88/738
phalanx	phalanx (NM)	TS, PS		Trossingen	IFG 8159
ischium	ischium (43·0)	TS, PS, FS		Trossingen	SMNS F 27

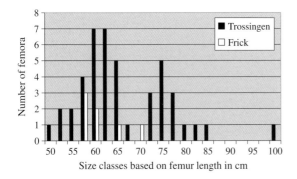

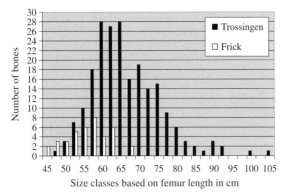

TEXT-FIG. 1. Size class histograms based on femur length for Trossingen and Frick. A, size class histogram of femora from Trossingen and Frick; 44 femora from Trossingen that correspond to an estimated minimum number of individuals (MNI) of 37, and seven femora from Frick that correspond to an estimated MNI of five, are included in this histogram. B, size distribution of all measured bones from Trossingen and Frick, scaled to femur size; all long and flat bones, 214 bones from Trossingen and 39 from Frick, are included. The MNI in Trossingen is around 54 and around 20 in Frick. Scaling to femur length based on bone ratios in Table 3.

observable, one between 57·5 and 67·5 cm (this includes all but one femur from Frick) and a second one between 72·5 and 80·0 cm femur length. Apart from the higher sample number, the peaks in the present histogram correspond well with the histograms published by Weishampel and Westphal (1986, fig. 11) and Sander (1992, fig. 16). A second histogram (Text-fig. 1B) includes all available long and flat bones. Most individuals from

Trossingen and all of those from Frick belonged to medium-sized individuals. Really large individuals of *Plateosaurus engelhardti* are very rare; only four femora from Trossingen attained a length > 80 cm, and none of this size is known from Frick. Medium-sized individuals thus make up the majority (about 75 per cent) of the *Plateosaurus engelhardti* individuals found in Trossingen. No bone from either locality indicates an individual smaller than 4·80 m, which is probably owing to a taphonomic filter (Sander 1992). This also means that the sample from Trossingen and Frick does not include juveniles.

Preservation and diagenesis

The fossils from the Trossingen locality are generally well preserved, with only minor crushing. However, the bones are brittle and fragile due to the carbonate-rich mudstone in which they are embedded and their unusual diagenesis. The mudstone has infiltrated the bones, filling even minor pores, but there is little precipitation of minerals in the pores. Microcracks and fractures extend through the bones, but they do not follow any pattern and extend radially or parallel to the bone surface. In many cases, the cracks follow the LAGs, sometimes resulting in an enlargement and splitting of the LAGs. Cementation and precipitation of minerals in vascular spaces is generally rare. Fossil bone apatite is sometimes substituted by calcite, mostly along the microcracks. However, these diagenetic effects do not affect the general interpretation of bone histology and the resulting swelling of the bone is less than 1 mm. This observation is contrary to that of Moser (2003) who postulated large size increases for the Trossingen bones through palaeopedogenesis. He envisaged the mechanism to be the expansion of the clay minerals in the pores of the bone. However, he did not provide evidence from thin-sections in support of this hypothesis.

The material from Frick is generally better preserved than the material from Trossingen as it is not as brittle and fragile, although it has suffered more crushing. Infiltration of greenish to greyish fine-grained sediment into the bone is common, as in Trossingen, but the bones from Frick are not as sensitive to fracturing. Microcracks are very rare in the Frick material. In some specimens the medullary cavity is crushed and/or parts of the cortex are displaced by crushing. The purported effects of palae-opedogenesis (Moser 2003) are not documented in the sample. Few diagenetic effects are observable at the histological level. The primary apatite is sometimes substituted by calcite and in some regions the bone microstructure has been destroyed by crushing. However, these diagenetic effects do not affect interpretation of the histology and growth record. The Frick samples are very light-coloured in thin-section, and features of the bone tissue and LAGs are often more difficult to identify in thin-sections than in the Trossingen samples. Conversely, the Frick material is easier to polish than the Trossingen material.

METHODS

Osteology, measurements and mass estimates

A detailed osteological study of the appendicular skeleton was conducted by the authors in concert with the histological research in order to confirm that only one species of *Plateosaurus* occurs in Trossingen and in Frick, and that this species is very similar in both localities, if not the same. Furthermore, we searched for indications of possible sexual dimorphism in bones from both localities. Osteological observations are less reliable for the Frick bones than for those from Trossingen because of strong crushing. Standard measurements like complete bone length, width and length of the epiphyses, and shaft circumferences were taken from all *Plateosaurus engelhardti* bones in the SMNS, IFG, MSF, NAA and SMA collections to obtain the largest database possible. Measurements of total bone length provide the most important information because they were used in scaling of the specimens to femur length and therefore to the total body length of the individual. Body length, in turn, is used for constructing growth curves and mass estimates. Furthermore, measurements from complete bones allowed estimation of the dimensions of incomplete bones. To determine proportions in the postcranial skeleton of *Plateosaurus*, the most complete individuals were used and a ratio, based on an average length for each of the bones in relation to the length of the femur (Table 3), was calculated. The ratios for material from Frick and Trossingen were determined separately. The data for Trossingen are based on several individuals (SMNS F5, F8, F14, F37, F48, 53537) and on the cast of SMNS 13200. The mean values from the measurements of the different individuals are used. The data from Frick are based on

TABLE 3. Ratios of selected postcranial bones to femur length in per cent.

	Trossingen	Frick (MSF 23)
Tibia/femur	80	81·5
Fibula/femur	80	92
Humerus/femur	62	68
Pubis/femur	81·5	not available
Scapula/femur	68	76

a single individual, which is exhibited at the MSF (MSF 23). During the collection of these data it became clear that the measurements for bones of the right and left body halves of a single skeleton, as well as the ratios of bones between different skeletons, can differ considerably, suggesting high intraspecific variability. The ratios obtained from the complete skeleton from Frick are conspicuously higher than the mean ratios of the Trossingen material (Table 3).

Mass estimates were obtained using developmental mass extrapolation (DME) as described by Erickson and Tumanova (2000). The cube of the femoral length of each specimen was taken, and two mass estimates for each individual were computed (Table 1): one as a percentage of the mass estimates of *Plateosaurus* provided by Seebacher (2001) and the other as a percentage of the mass estimate of Sander (1992). The two estimates for each individual differed remarkably little. To obtain the average of these mass estimates for each individual, the mean of the two values for the largest femur (femur 13) was set at 100 per cent, and the masses of the smaller individuals were calculated as percentages of the mass of the largest femur (Table 1). This approach was chosen for ease of calculation over the approach of calculating the mean of the two values for each bone.

Sampling

Whole bone cross-sections. Based on earlier studies on dinosaur long-bone histology and general principles of bone growth, the middle of the shaft is recognized as the best place to obtain the most complete growth record (de Ricqlès 1983; Chinsamy 1993a, b; Erickson and Tumanova 2000; Horner et al. 2000; Sander 2000). This is owing to the predominately appositional growth of the shaft and the location of the neutral zone in this region (Sander 2000). The best coring site is defined by a combination of the thickest available cortex with clear and numerous growth cycles and minimum remodelling (i.e. minimum erosion moving outward from the medullary cavity and minimum secondary bone tissue). These conditions are usually encountered in the mid-shaft region, but vary slightly from bone to bone as a result of different growth histories and loads experienced by the bones *in vivo*. To determine the best coring site around the circumference of the mid-shaft region, whole bone cross-sections were cut from a femur, pubis, scapula and rib from Trossingen (SMNS F 29), and from a femur, tibia, scapula and humerus from Frick (single bones from the NAA). These bones do not belong to single individuals, but represent several specimens. Before cutting, the mid-shaft region was stabilized in a girdle of synthetic resin. The sample was obtained by placing two parallel cuts across the shaft, resulting in a slice a few centimetres thick. After sawing, both sides of the slice were impregnated with a synthetic, low-viscosity resin (Araldite 2020 A/B, Huntsman, Cambridge), finely ground, and polished to a high gloss.

Coring. The sampling method developed by Sander (1999, 2000) was followed herein. In order to minimize damage to the sampled bone, a core of approximately 10–15 mm diameter is drilled with a diamond-studded coring bit into the cortex of the narrowest mid-shaft area. In humeri and femora this is approximately in the middle of the shaft (Text-fig. 2A). Tibial cores were taken somewhat distal to the mid-shaft region, and the best sampling location for the fibula was slightly proximal to this area as in both cases the narrowest place is not exactly in the middle of the diaphysis (Text-fig. 2A). To ensure the capture of the longest possible growth record the cores in long bones were always drilled through the cortex into the medullary region. The cores and cross-sections from the pubes were usually taken in the proximal to mid-shaft region, as near as possible to the lateral edge of the bone (Text-fig. 2B). This bone site contains the thickest available cortex and also the thickest primary bone tissue. The cores and cross-sections of the scapulae were taken from the middle of the narrowest mid-shaft region, which contains the least disturbed primary bone tissue (Text-fig. 2B), as shown by examination of fractures and cross-sections. Because the scapula and pubis are flat bones, cores can easily be drilled through the entire element.

The drill is usually lubricated with water but, depending on the bone material, oil was occasionally used to avoid swelling of the clay minerals. In some cases there was no lubrication (dry drilling). The coring bit is attached to a normal electrical drill with adjustable speed. For the brittle *Plateosaurus* material, low speed and low torque settings were necessary. A small plasticine dam was built around the drilling hole, into which the lubricant was poured. This protected the bones from excessive moisture and guarantees constant lubrication during drilling. The drill was mounted in a simple drill press. The foot of the press is weighted down with a dish of sand in

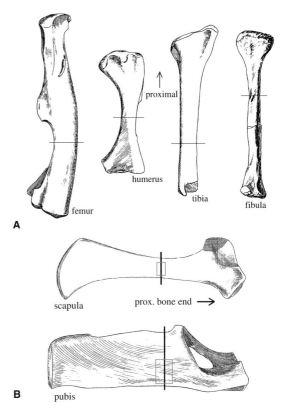

TEXT-FIG. 2. The sampling location in long and flat bones. A, sampling location in long bones is around the narrowest mid-shaft region, here marked by a line; right femur and right humerus in posterior view, right tibia and right fibula in medial view; bone drawings from von Huene (1926); not to scale. B, sampling location in pubis and scapula, marked by a rectangle; right pubis in dorsal view, left scapula in lateral view; bone drawing from von Huene (1926); not to scale.

which the bone is supported securely. The resulting cores were embedded in synthetic resin (Araldite 2020 A/B, Huntsman, Cambridge) and then cut longitudinally, perpendicular to the long axis of the whole bone. One half of the core was processed into a thin-section and the other half into a polished section. Thin-sectioning followed standard procedures for petrographic thin-sections.

Existing fracture surfaces. The *Plateosaurus* long bones from Trossingen and Frick show many fractures. Most of these run across the narrow mid-shaft region, which is the most interesting for histological research. The fracture surfaces are not covered by vein minerals, owing to the lack of diagenetic cements, and are similar to the cut whole bone cross-sections. The fracture surfaces provide a full overview of the growth record in this bone region. Gross histological structures, such as LAGs and degree of vascularization, can be observed in these surfaces with the naked eye and observations on these fractures complement those made on the cores and sections.

Methods of histological and skeletochronological study

The histological database consists of 50 specimens (33 long bones, 11 flat bones, six other bones) from both localities. All sections and surfaces were examined by standard light microscopic techniques. Bones with good fracture surfaces were studied under a binocular microscope and the growth marks were traced with a camera lucida. In addition, the fracture surfaces were photographed. Growth marks were later added to the digital images using standard graphics software. The advantage of the digital images is that they show whole cross-sections, providing an overview about the growth record without any further damage to the bone. However, owing to the limited resolution of the images and the binocular microscope, some histological details (e.g. bone tissue types) cannot be distinguished and LAGs cannot be identified with the same certainty as in thin-sections, particularly in the exterior cortex where growth cycles may become very close in fully grown individuals. Consequently, the number of LAGs counted on only such a fracture surface is commonly low and a greater error must be assumed than for the sections. The thin-sections were viewed under a compound microscope in transmitted light, both normal and polarized. At high magnifications, bone cell lacunae, bone types, and primary and secondary osteons were visible. The bone cross-sections and core polished sections were studied under a binocular microscope to obtain a good overview of the growth record and to determine the distances between LAGs, differences in the degree of vascularization, and the type and distribution of remodelling. Drawings were made of all sections with a camera lucida, and the visible LAGs were drawn in and counted. In contrast to the material from Trossingen, polish lines are very common in the Frick material. Polish lines are made visible to the naked eye or with a binocular microscope by tilting the section surface to create a reflection or by using bright-field illumination under a compound microscope (Sander 2000). In the samples from Frick, the polish lines generally correspond to LAGs. LAGs can be observed with the binocular microscope as fine lines, usually appearing in the middle of the polish line.

The current study shows that a combination of various photographic, microscopic and sectioning techniques is important in order to obtain the most complete understanding of the growth record. Because of the mode of preservation and colour of the material, the polished sections often gave better results in observing growth marks than thin-sections. The thin-sections were better suited for studying histological features such as bone tissue types or remodelling processes at higher magnifications.

LAG counts. Two forms of growth marks were used: LAGs and, in the Frick samples, polish lines that contain LAGs. Because of the lack of juveniles and the large medullary cavity of *Plateosaurus* long bones, extrapolation of resorbed LAGs is necessary. Consequently, the current study employs three different LAG counts. The visible LAG count corresponds to the actual LAG count observed in the samples. The extrapolated or resorbed LAG count is a calculation of the number of LAGs lost to resorption owing to expansion of the medullary cavity. The total LAG count is the sum of both, the visible and resorbed or extrapolated LAG count. LAG counts are assumed to correspond to annual growth cycles as discussed before.

Extrapolation method. In long bones, resorption and remodelling of the inner cortex mainly occurs around the margins of the medullary cavity, removing evidence of early growth cycles. Consequently, LAG counts in sub-adult and adult individuals are always incomplete and the lack of young *Plateosaurus* individuals makes extrapolation of the number of resorbed growth cycles necessary. However, extrapolation increases the margin of error in the skeletochronological analysis.

For *Massospondylus carinatus*, Chinsamy (1993a) estimated the original number of LAGs/growth cycles by using the thickness of the initial growth cycles of the smallest and presumably youngest specimens. de Ricqlès (1983) and Chinsamy (1993a) noted that it is necessary to consider the diameter of the inner remodelled region in order to obtain a rough estimate of the number of resorbed cycles. Reid (1997a) referred to Ferguson *et al.* (1982) who estimated the number of lost cycles in living *Alligator mississippiensis* by extrapolating from the thickness of the remaining cycles. Zug *et al.* (1986) developed a formula to reconstruct the lost marks of skeleton growth in living *Caretta caretta*. They estimated the total number of marks of skeletal growth by subtracting the average cross-sectional diameter of hatchling bones from half of the cross-sectional diameter of the bone in question and dividing the resulting value by the average width of the preserved growth marks (Zug *et al.* 1986, p. 13).

The extrapolation method used herein attempts to incorporate the aforementioned methods by adapting them to the specific requirements of the available *Plateosaurus* material. On the basis of the assumption that the cycle distances in juvenile individuals were greater than in older individuals (Chinsamy 1993a, p. 320), the greatest available cycle distance in each sampled bone was used as the basis for extrapolation. The distance between the centre of the medullary cavity and the first visible LAG (LAG 1) was measured and divided by the greatest distance between any two adjacent LAGs in the specimen (i.e. the widest growth cycle). In general, the greatest distance was that between visible LAG 1 and LAG 2. Nevertheless, distances between some of the outer LAGs were sometimes greater than observed between the innermost LAGs because of the irregular growth of

Plateosaurus. The result of the calculation is the number of maximally lost LAGs by remodelling and resorption. 'Maximally' because the resorbed cycles must have been wider than the widest preserved ones. This is because of the decrease in average cycle thickness with increasing age as generally observed in skeletochronology. The large medullary cavity and the lack of juvenile specimens for calibration will thus lead to an overestimate of the number of resorbed LAGs.

This extrapolation method works only for the samples of whole cross-sections or if the diameter of the medullary cavity is known. Extrapolation from thin-sections or polished sections cut from cores requires reconstruction of the missing part of the medullary cavity. This was done on the basis of the average percentage value of the cross-sectional diameter that the medullary cavity occupies in different elements. In the pubis, the ratio of cortex thickness to medullary cavity diameter varies strongly, depending on which location in the cross-section the measurements are taken. Consequently, in the pubis, measurements were taken from the proximolateral edge of the bone only, because of the greater mediodistal extent of the medullary cavity in this area. Thus, with the exception of pubis samples, cross-sections and fracture surfaces were measured on four surfaces of the bone (caudal, medial, cranial and lateral), whereas in pubic cross-sections only the dorsal and ventral surfaces were measured. Unfortunately, it was not always possible to measure all bone surfaces owing to the varying quality of the growth record within each element. The growth record may be generally unclear, disturbed by resorption and remodelling, or incomplete because of preservational factors. Additionally, the resolution of the fracture surfaces is not of the same quality on all surfaces of a bone. Thus, the visible and resorbed LAG number usually differs along the circumference of a single bone.

A further error is introduced by measuring from the centre of the medullary cavity because this includes the diameter of the bone at hatching, which has to be corrected. Owing to the lack of *Plateosaurus* hatchlings, hatchling femur diameter needs to be estimated. On the basis of *Mussaurus* (Carpenter 1999, p. 206, table 11.3) and *Massospondylus* (Reisz *et al.* 2005), the femoral diameter of a *Plateosaurus* hatchling must have been only a few millimetres. Consequently, the error introduced by not including the diameter of the bone at hatching is likely to be very small and will have no major influence on the extrapolation results. It should be remembered that extrapolation is not an exact method for reconstructing LAGs: results only indicate the probable number of resorbed LAGs for each bone sample and should be seen as approximations, and are likely to be overestimates of the true age of the animals.

HISTOLOGICAL DESCRIPTIONS

Long bones

Sampling location. As in most other dinosaurs, the femur is the longest bone in *Plateosaurus*. The cortex is approximately 2–3 cm thick. Diaphysis diameters, measured craniocaudally, are 5·8–8·9 cm. Most femoral cross-sections exhibit the thickest cortex on the caudal side, making this the best sampling location. Generally, a thin zone of remodelling surrounds the well-defined boundary of the medullary cavity, which is conspicuously large and often partially filled by cancellous bone. The average size of the medullary cavity is approximately 58 per cent of the shaft diameter (calculated craniocaudally). It is noteworthy that neither the cores from the caudal surface nor femoral cross-sections show any remodelling trace of the fourth trochanter, although the samples were often taken from only a few centimetres distal to it.

The cortex of the tibia is not as thick as that of the femur, reaching up to 2·5 cm. The tibial diameters, measured mediolaterally, are between 3·7 and 5·8 cm. Most cross-sections show the thickest cortex at the craniomedial surface. As in the femora, the medullary cavity is relatively large and forms (on average) approximately 49 per cent of the shaft diameter (calculated mediolaterally). Generally, the well-defined boundary of the medullary cavity is surrounded by a clear remodelling zone, which in some regions reaches up to the middle of the cortex. However, this remodelling is light and does not completely overprint the primary bone tissue. Extensive regions of cancellous bone were not observed in any tibiae. Cortex thickness in the fibula varies between 0·5 and 1·4 cm. In most samples, the cranial part of the bone contains the thickest cortex. The whole diameter, measured mediolaterally, ranges between 2·3 and 3·5 cm. On average, the medullary cavity occupies approximately 52 per cent of the shaft diameter (measured craniocaudally). Fibula 2, which is fractured into five pieces, yields information about the change in bone tissues along the diaphysis. The middle shaft region shows the least remodelling and only the caudomedial part of fibula possesses cancellous bone. The medullary cavity has a clear boundary in the middle shaft, but this becomes blurred at the proximal and distal extremities. Remodelling and the amount of cancellous bone increase proximally and distally, finally surrounding the medullary cavity completely. Only a thin layer of primary tissue remains in the cortex of the proximalmost and distalmost shaft regions.

In comparison with the femur and tibia, the humeral cortex is relatively thin, ranging between 0·5 and 1·2 cm in thickness. The whole bone diameters, measured craniocaudally, were between 4·0 and 5·0 cm. In all humerus samples, the medial part of bone contains the thickest cortex. As in the other long bones, the medullary cavity is large, accounting for 55 per cent of the shaft diameter (calculated craniocaudally). In most sampled humeri, a relatively large cancellous bone area surrounds the medullary cavity, but in two samples (humeri 3 and 4), no cancellous bone is present on the medial side.

Primary bone tissues and vascular density. The most widespread primary bone tissue in all *Plateosaurus* long bones is a laminar fibro-lamellar complex combined with LAGs (Text-fig. 3A, E). This makes up an estimated 75 per cent of the primary bone tissue. In spite of its cyclical growth, the laminar fibro-lamellar complex with LAGs shows a relatively high vascular density and therefore represents fast growing bone tissue. The vascular canals build a three-dimensional network. The primary vascularization pattern is laminar and consists mainly of circular and longitudinal vascular canals. In some samples, a few radial and reticular vascular canals occur in the fibro-lamellar complex, mainly in the inner cortex areas. The number of longitudinal vascular canals increases from the inner to the outer cortex. In general, the vascular density decreases gradually within a growth cycle and also from the inner to the outer cortex. The osteocyte lacunae in the fibro-lamellar tissue are large and rounded. Lamina density averages 8 laminae/mm in *Plateosaurus*.

The second regularly appearing bone tissue is lamellar-zonal bone, which is usually restricted to the exterior cortex (Text-fig. 3B, F) (with the exception of humerus 6). The vascular canals are mainly longitudinal in the lamellar-zonal bone. The osteocyte lacunae in this tissue are flattened. Lamellar bone also occurs regularly in the annuli of well-developed growth marks. In some specimens the lamellar-zonal bone grades outwards into avascular bone with closely spaced growth marks. This tissue represents an EFS (= OCL of Chinsamy-Turan 2005).

In two femoral samples (femora 9 and 12), the inner cortex consists of a very highly vascularized laminar bone tissue within a broad cycle with very thick laminae and large vascular canals (Text-fig. 3C, G). The individual vascular canals and laminae are much thicker than in the 'normal' fibro-lamellar complex of *Plateosaurus engelhardti*. The tissue appears in the first and second visible cycles. The inner part of the first cycle, close to the medullary cavity, and the later cycles consist of the normal laminar fibro-lamellar complex.

Several samples (femur 2, tibiae 2 and 7, fibula 2, vertebra 1 and ischium) show an additional bone type (Text-fig. 3D, H). The bone tissue is still the fibro-lamellar complex, but the vascularization is quite different. It consists of densely spaced, parallel radial vascular canals. Although this bone tissue type shows such high vascularization, which indicates a very rapid growth rate, the tissue is deposited cyclically and is delimited by normal LAGs. Owing to the predominance of radial vascular canals in this tissue, we refer to it as radial fibro-lamellar bone tissue (RFB), herein. It is similar to, but much more highly vascularized than, the 'radiating' fibro-lamellar bone tissue described by Francillon-Vieillot *et al.* (1990, fig. 14G). The cross-sections of the bones in question show that RFB is locally restricted, but it appears on different bone surfaces in the different samples and can be intercalated between cycles of laminar fibro-lamellar complex as well as appearing after cycles of lamellar-zonal bone. In femur 2 the RFB was deposited caudally in three cycles following a distinct cycle of lamellar-zonal bone in avascular tissue. As in the femur sample, the RFB in both tibia samples (tibiae 2 and 7) is restricted caudally and always in several cycles. It is unusual for such a highly vascularized tissue to occur consistently in the outer or outermost cortical areas. As documented in four of the five fracture surfaces along fibula 2, the RFB seems to be restricted to the mid-shaft region and disappears in the proximal and distal areas of the diaphysis: it is not evenly distributed along the whole shaft. In the latter sample, the RFB is restricted to the lateral surface and two cycles are present. In tibia 2, the RFB is not quite as highly vascularized as in tibia 7. However, in tibia 7, the RFB was deposited after a distinct layer of avascular lamellar-zonal bone, as in the femora. This sequence indicates that after deposition of bone tissue had nearly stopped, rapid growth started again and continued at a very high rate, at least in restricted areas. In tibia 7, the RFB was laid down directly after fibro-lamellar bone, which means that the normal growth rate had accelerated again at a relatively late time in growth history. These RFB cycles seem to correspond to regular growth cycles because they end in normal LAGs. On the opposite side of the same cross-section, the RFB cycles may be deposited as normal cycles of lamellar-zonal avascular bone tissue or as a laminar fibro-lamellar complex. Surprisingly, RFB seems to occur only in single bones of a skeleton and does not appear regularly in all bones of a single skeleton (Tables 4–5).

Secondary osteons and remodelling. All long bone samples share a generally low degree of remodelling. The major remodelling features are erosional cavities of varying diameters. These are connected to the medullary cavity and are usually concentrated on one bone surface. Many erosional cavities show some secondary bone deposition. Towards the medullary cavity the erosional cavities are transformed to a thin zone of mostly cancellous bone. The transition from cancellous bone at the margin of the medullary cavity via the erosional bone to the primary cortex is gradual and irregular. Secondary osteons occur in the cortex in all long bones. Where present, they are usually sparsely distributed throughout the cortex. In

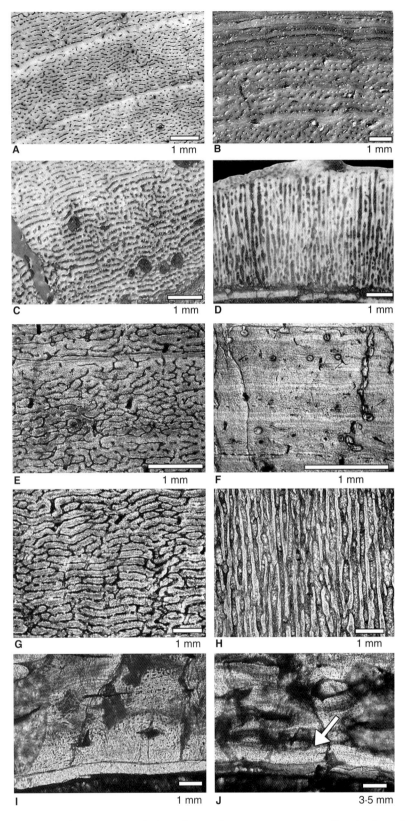

TEXT-FIG. 3. Bone tissues in *Plateosaurus engelhardti* long bones. All photomicrographs depict cross-sections, and in all the outer bone surface lies beyond the top of the image. A, polished section of laminar fibro-lamellar bone with LAGs (femur 9). B, polished section of lamellar-zonal bone (humerus 5). C, polished section of highly vascularized fibro-lamellar bone; vascular canals and laminae are much thicker than in the normal fibro-lamellar complex (femur 9). D, polished section of radial fibro-lamellar bone (RFB); fibro-lamellar complex with mainly radial vascular canal orientation and very high vascular density (vertebra 1). E, thin-section of fibro-lamellar bone with a LAG in the upper third of section (femur 9). F, thin-section of lamellar-zonal bone (tibia 1). G, thin-section of highly vascularized fibro-lamellar bone (femur 9). H, thin-section of RFB (vertebra 1). I, thin-section showing endosteal bone (femur 1). J, same sample as before, enlarged; endosteal bone marked by an arrow.

some samples, an accumulation of secondary osteons near the medullary cavity could be observed. Nevertheless, the occurrence of secondary osteons does not significantly disturb the primary bone tissue or the growth record in any sample. The primary arrangement of vascular canals was never fully destroyed by remodelling and so the secondary vascularization shows mainly longitudinal Haversian canals.

In femora, isolated secondary osteons are generally rare and are completely absent in many specimens. Only femur 2 shows numerous secondary osteons up to the exterior cortex. The tibia samples show more remodelling than those of the femora, but neither exhibits as much remodelling as the flat bones. Secondary osteons occur close to the medullary cavity and are also scattered throughout the cortex. Erosional cavities are sometimes present up to the middle of the cortex, predominately in the region with the thickest cortex. Thus, the distribution of remodelling is relatively irregular in the tibia. Fibula 2 clearly shows along its fracture planes how remodelling increases towards the proximal and distal ends of the shaft. Only the mid-shaft region is not significantly affected by remodelling. At the caudomedial part of the shaft, remodelling extends into the middle of the cortex. In other parts of the mid-shaft, remodelling occurs adjacent to the medullary cavity only. In humeri, secondary osteons and erosional cavities are most numerous close to the medullary cavities and occur in a moderately wide band around them. Only a few secondary osteons are present, scattered throughout the cortex and extending to the outer cortex or the bone surface. Although remodelling is relatively strong in the humerus, only in the inner cortex is the primary bone tissue sufficiently overprinted partially to destroy the growth record.

Endosteal bones. Endosteal bone is only present in one long bone sample (femur 1: Text-fig. 3I–J). It appears as a thin layer of lamellar bone along the boundary of the medullary cavity. The endosteal bone consists of typical lamellar bone and is separated from the cortex by a cementing line.

Growth marks. In the fibro-lamellar complex of *Plateosaurus* long bones, growth cycles can be divided into zones and annuli and clear LAGs mark the end of each growth cycle. The zones consist of the fibro-lamellar complex. The annuli often appear as bright bands to the naked eye in the polished sections. In the polarizing microscope the thin-section shows a band of dense avascular lamellar bone. The lamellar-zonal bone does not have the additional division into zones and annuli, but LAGs occur and separate the growth cycles from each other. In both bone types the annual LAGs always run parallel to the bone surface. LAGs may also appear as polish lines (Sander 2000), especially in the material from Frick.

The spacing of the LAGs is relatively wide in the inner cortex, becomes denser in the middle and increases towards the outer cortex. This is because of a slower growth rate with increasing age and consequent closer deposition of growth cycles. However, cycle distance is relatively irregular in *Plateosaurus* long bones. Narrow and poorly vascularized cycles can appear in the inner cortex and wide and highly vascularized cycles can appear in the outer cortex. Such irregularities are independent of bone tissue type. In some *Plateosaurus* long bones (Table 4), bone tissue type changes from the fibro-lamellar complex to lamellar-zonal bone, mainly in the outermost cortex. At this stage LAGs were deposited extremely close together within the lamellar bone tissue. The close spacing makes it difficult to distinguish and count growth cycles in the exterior cortex. Similar problems were encountered by Zug *et al.* (1986, p. 10) and Castanet and Smirina (1990).

Some thick growth cycles of *Plateosaurus* long bones are subdivided by second-order cycles, mainly delimited by second-order LAGs. Second-order LAGs differ from first-order LAGs in a distinctly lesser decrease of vascular density towards the LAG. In addition, they are thinner and less clear in appearance. Usually second-order LAGs cannot be followed completely around the entire cross-section. Lines hypothesized to be second-order LAGs were not used in skeletochronology. Wide cycles with them appear predominately in the inner cortex. A single major growth cycle can be divided in one, two or (occasionally) three second-order cycles. Second-order cycles are the result of short growth decreases during the normal annual growth period. They were probably caused either by seasonal changes in climate or food availability or by disease. Such second-order lines are mentioned frequently in the literature (Zug *et al.* 1986; Castanet *et al.* 1993). Sometimes, first-order LAGs are accompanied by small assemblages of second-order lines, usually consisting of two or three above and below the major LAG. This occurs more often in the outer cortex, where cycle distances are less than in the inner cortex. Such assemblages look similar to closely spaced LAGs in the outermost cortex in lamellar-zonal bone.

Flat bones

Sampling location. In contrast to long bones, the sandwich-like structure of flat bones results in only two possible coring sites. Thus, the lateral and medial surfaces of the scapula and the dorsal and ventral surfaces of the pubis are used for sampling. However, remodelling in the form of erosional cavities, cancellous tissue and secondary osteons often disturbs the primary bone tissue at these locations. The most complete and undisturbed growth record in the scapula was found in the cortex of the lateral surface. Shaft thickness in the scapula was between 2 and 3 cm, measured mediolaterally. The most complete and undisturbed growth record in the pubis was found in the dorsolateral part. The ventrolateral part also preserves a relatively good growth record, but because the inner cortex shows much more remodelling than in the dorsal part, the growth record here is usually incomplete. Combining information from the cortices of both sides therefore results in the most complete growth record. The greatest thickness of the pubis, as measured dorsoventrally, is found near the lateral edge and was between 2 and 3 cm. Often the medullary cavity in the scapula and pubis is not clearly separated from the surrounding cancellous bone and remodelling area, so the areas in the centre of the bones are summarized here under the term 'medullary region'.

Bone tissues, vascular density and remodelling. The primary bone type in the scapula is intermediate between the typical laminar fibro-lamellar complex known from the long bones and purely lamellar-zonal bone (Text-fig. 4A, C). It shows areas with relatively highly vascularized fibro-lamellar bone, which gradually pass into less vascularized lamellar and parallel-fibred bone. Longitudinal vascular canals dominate the vascular system, but their density is considerably higher than in the pubis samples. Additionally, a few reticular canals can occur. No fibrous bone tissue can be identified in the available pubis samples, and the primary bone in the pubis is of the pure lamellar-zonal type (Text-fig. 4B, D). It consists mainly of lamellar or parallel-fibred bone and is dominated by longitudinal vascular canals. The osteocytes are usually flattened. Deposition of lamellar-zonal bone demonstrates that *Plateosaurus engelhardti* pubes experience lower growth rates than long bones and scapulae.

Remodelling is generally stronger in the scapula and pubis than in the long bones. Depending on the location in the cross-sections, the primary bone tissue is largely replaced by secondary compact or cancellous bone. The cancellous bone of the medullary region is surrounded by a moderately broad remodelling zone. This contains a few erosional cavities, which are predominately rounded in shape, and secondary osteons, often representing several generations. The vascular canals of these follow the primary vascular system and are arranged longitudinally. The secondary osteons in both bones are scattered in all regions of the primary bone tissue up to the bone surface. The transition from secondary to primary bone tissue is usually irregular and is not clearly delimited.

Growth marks. Growth cycles with a clear division into zones and annuli are seldom visible in flat bones. This is because of the high amount of remodelling by secondary osteons. Only LAGs are well preserved in the samples and are unaffected by remodelling. However, this is only at the described sampling locations; it is not possible to follow the LAGs continuously around the entire cross-sec-

tion owing to the large medullary region of the flat bones. Second-order cycles are scarcer than in long bones, probably owing to the thinness of the cortex and possibly because of a generally slower growth rate in the scapula and, in particular, the pubis. However, a few less distinct, and therefore subordinate, LAGs often occur before and after the main LAGs. In addition to the general decrease in vascular density from the inner to the outer cortex, vascular density decreases within the single growth cycles. If present, zones of the scapula are generally more highly vascularized than the pubes. Often the LAGs were deposited in small bands of avascular or nearly avascular tissue of lamellar bone (i.e. annuli). Growth cycle thickness is fairly irregular in the scapula and pubes, but there is a tendency for cycle thickness to decrease from the inner to the outer cortex. Generally, cycles are thinner than in the long bones because of the thinner cortex and lower growth rates. Finally, the growth cycles disappear in the outer cortex, with only closely spaced LAGs set in very poorly vascularized or avascular lamellar bone tissue remaining. Growth cycle thickness decreases more suddenly in the outer cortex of the pubes than in the scapulae. The resorbed LAG number is generally lower than in long bones owing to a general lower growth rate in flat bones.

The pubis and scapula differ in the quality of their growth record. The pubis has high numbers of visible LAGs and very low numbers of extrapolated LAGs. Thus, it probably shows the most realistic growth record. The bone tissue of the scapula is difficult to evaluate because of its tissue type, which is intermediate between that of the fibro-lamellar complex and lamellar-zonal bone, and its strong remodelling.

Other bones: vertebrae, ribs, ischium and pedal phalanx

Sampling location. One aim of this study was to test the suitability of different bones from the *Plateosaurus* skeleton for histological research. Thus, in addition to typical long bones, scapulae and pubes, other bones were sampled, including dorsal vertebrae, ribs, ischia and a pedal phalanx. The ischium is discussed here and not in the section on flat bones because only one was available and it was studied only in a fracture surface of its shaft. Sampling locations for all of these elements were chosen on the basis of general bone morphology and the principles of bone growth. Thus, the cores from dorsal vertebrae 1 and 2 were drilled on the ventral surfaces of the vertebral centra. The cross-section of dorsal rib 1 was obtained from the middle of the element. Rib 2 was sampled at two places, one proximal to the middle of the rib and a second sample taken more distally. The core from the pedal phalanx was drilled through its

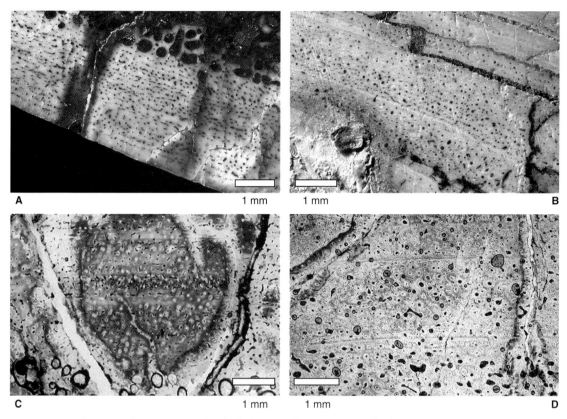

TEXT-FIG. 4. Bone tissues in *Plateosaurus engelhardti* flat bones. All photomicrographs depict cross-sections, and in all the outer bone surface lies beyond the top of the image. A, polished section of a scapula (scapula 3), lateral bone side. B, polished section of a pubis (pubis 3), dorsal bone side. C, thin-section of the primary bone of a scapula sample (scapula 4). D, thin-section of the primary bone of a pubis sample (NAA uncatalogued).

shaft from dorsal to ventral, through the entire thickness of the bone.

Histology of vertebrae. No cross-sections are available for the vertebrae. The thin- and polished sections of the cores show a thick (1–1·5 cm) cortex that consists mainly of primary bone tissue. The primary bone tissue is built up of the fibro-lamellar complex typical for *Plateosaurus*, with clear LAGs. However, in these sections only longitudinal vascular canals are visible. The upper one-third of the cortex of vertebra 1 consists of RFB. In both vertebrae, only a few secondary osteons are scattered throughout the cortex. The cortex continues inward into cancellous bone with preserved interstitial primary bone tissue. This coarse cancellous bone again grades inwards into the medullary region.

Histology of ribs. The cross-section of rib 1 is oval in outline, but in the middle, it is slightly constricted. The medullary cavity makes up 39 per cent of the dorsoventral rib diameter, with cancellous bone occupying 36 per cent, and the cortex contributing 25 per cent. The cortex percentage is slightly greater craniocaudally (32 per cent). Thus, the medullary region dominates the

rib. Only one side of the rib preserves a growth record: the other side is strongly remodelled by erosional cavities and secondary osteons. Dorsally and ventrally, the cortex is generally thin. The primary tissue of the rib is fibro-lamellar bone with clear LAGs in the inner cortex and lamellar-zonal bone in the exterior cortex. Only longitudinal vascular canals are present. Vascular density is intermediate and decreases slightly from the inner to the outer cortex.

Rib 2 shows similar features. The cancellous bone area is larger but more diffuse than in rib 1 so that quantification is not possible. Remodelling by erosional cavities and secondary osteons is very strong, and a true medullary region cannot be distinguished. The primary cortex is formed by a fibro-lamellar complex with clear LAGs and is only preserved in the exterior areas of the more proximal cross-section. The section has an elliptical outline and tapers to one side. At the other side of the section, a relatively thick cortex is preserved, but it is strongly remodelled. The LAGs in the exterior cortex are closely spaced and expressed in nearly avascular lamellar-zonal tissue. The more distal rib section has a regular elliptical cross-section, and consists of secondary compact and cancellous bone.

Histology of ischium. The fracture surface of the ischium has a nearly triangular to oval outline. A distinct medullary cavity is absent. The section contains a large area of cancellous bone and an area of erosional cavities of varying diameters with a diffuse boundary with the cortex. A few secondary osteons are scattered throughout the cortex. The primary bone tissue is fibro-lamellar with clear LAGs and predominately longitudinal vascular canals. The ischium contains RFB in its exterior cortex.

Histology of pedal phalanx. The centre of the pedal phalanx shows a large medullary cavity. The rest of the section is built up of cancellous bone containing large erosional cavities of varying diameter. Primary bone tissue is present only in the exterior cortex and in small areas between the erosional cavities. In the inner cortex it consists of fibro-lamellar bone with a very low, predominately longitudinal vascularization. The exterior cortex is built up of avascular lamellar-zonal bone.

Differences between the Frick and Trossingen localities

Although the histological samples from the two localities can easily be distinguished on the basis of the different colour and preservation of the bones, few histological differences are present. The only obvious differences are the more frequent appearance of polish lines and the larger number of secondary osteons in the Frick samples.

GROWTH RECORD

Qualitative growth record and ontogenetic status

Laminar fibro-lamellar complex with LAGs. As in other dinosaurs, the main tissue in *Plateosaurus engelhardti* bones is fibro-lamellar. Three types were observed: fibro-lamellar bone with clear LAGs (Text-fig. 3A, E); highly vascularized fibro-lamellar bone with clear LAGs (Text-fig. 3C, G); and RFB (Text-fig. 3D, H). These differ mainly in vascular density and in the arrangement and form of vascular canals. All exhibit subdivision into growth cycles, indicated by regularly appearing LAGs. Unfortunately, they do not represent an ontogenetic series, as in some other dinosaurs (Varricchio 1993; Curry 1999; Erickson and Tumanova 2000; Horner *et al.* 2000; Sander 2000). The two unusual fibro-lamellar bone types probably represent special features due to their rare occurrence and are not important for interpretation of the normal growth record.

In the long bones of *Plateosaurus engelhardti*, the laminar fibro-lamellar complex with clear LAGs represents the usual bone tissue of still-growing adults. This tissue indi-

cates that long bones of adult individuals still grew at a relatively high rate, despite cyclical interruption. Nevertheless, a general decrease in vascular density and growth cycle distance can be observed from the inner to the outer cortex in all long bones sampled. The increase in the number of deposited LAGs is accompanied by a change in the spatial organization of bone tissue and a general decrease in cycle thickness. This indicates that growth rate decreased gradually with age and size.

Juvenile growth. Juvenile *Plateosaurus* specimens are necessary for investigating early ontogenetic histology, including identification of the first primary bone tissue to be deposited and the early growth record in general, as well as for providing an exact calibration for the extrapolated part of the growth record of larger individuals. Owing to the lack of juvenile *Plateosaurus* specimens their growth history is rather speculative here. Presumably bones of juveniles grew at least at the same rate as adults, producing similar bone tissues (Text-fig. 5). Indeed, on the basis of the results of bone histological studies on other dinosaurs (Varricchio 1993; Curry 1999; Erickson and Tumanova 2000; Horner *et al.* 2000; Sander 2000; Sander and Tückmantel 2003), it is likely that the bones of juveniles grew at higher rates than those of adults (Text-fig. 5).

Lamellar-zonal bone and determinate growth. The second type of regularly appearing bone tissue in *Plateosaurus engelhardti* is lamellar-zonal bone. It occurs in the outermost cortex of some long bones and the scapulae and is the main bone tissue type in the pubes. Decreasing vascular density initiates the transition from fibro-lamellar to lamellar-zonal bone. The lamellar-zonal bone often grades into poorly vascularized or nearly avascular tissue that shows closely spaced LAGs (ESF) that record the time interval during which bone growth substantially decreased and finally ceased. Growth was therefore determinate in *Plateosaurus engelhardti*.

Bone tissue change and sexual maturity. The changes in primary bone tissue types observed from the inner to the outer cortex are hypothesized to reflect events in individual life history. However, there are three competing hypotheses (Text-fig. 5) to explain the meaning of the transition from laminar fibro-lamellar bone to lamellar-zonal bone in the long bones (Table 4).

Hypothesis I (Text-fig. 5) posits that growth was divided into three ontogenetic stages that found their expression in three different bone tissue types. The first is the unknown juvenile stage with assumed higher juvenile growth rates. This stage ended with the transition from the unknown juvenile bone tissue to laminar fibro-lamellar bone with clear LAGs, reflecting the onset of sexual maturity. Note that this transition is not preserved in any of the specimens studied, meaning that all were sexually mature. Young adults would have continued to grow for several years, during the second stage, but at a decreased rate. When bone tissue type changed from laminar fibro-lamellar to lamellar-zonal bone, this indicates that the individual was approaching its final size. The final stage in hypothesis I is documented histologically by the deposition of the EFS, which indicates termination of growth.

Hypothesis II suggests that *Plateosaurus engelhardti* growth was characterized by laminar fibro-lamellar bone with clear LAGs and a uniform growth rate that was sustained from hatching until the attainment of final body size (Text-fig. 5). Sexual maturity would correlate with the bone tissue change to lamellar-zonal bone. The juvenile stage thus would have lasted much longer under this hypothesis and growth would have terminated soon after sexual maturity was attained. Many of the sampled individuals, especially those from Trossingen, would not have been sexually mature at the time of death, although some of them had already reached large body sizes. On the basis of extrapolated data from the long bones, the bone tissue change occurred no earlier than the sixteenth year of life and later still in many specimens (see below). Such a prolonged juvenile phase appears unlikely, calling the validity of hypothesis II into question.

This leads to hypothesis III, which posits that the onset of sexual maturity is difficult to determine from the bone microstructure of *Plateosaurus engelhardti* (see also Chinsamy-Turan 2005; Erickson 2005). It could have occurred at any time during the deposition of fibro-lamellar bone or at its termination, and nothing could be said about the reproductive status of the individuals sampled in this study.

Meaning of radial fibro-lamellar bone (RFB). It is clear that RFB represents very rapid growth with high rates of

bone deposition. Erickson and Tumanova (2000, p. 559) described similar tissue as 'highly porous radially vascularized bone' in three large long bones of *Psittacosaurus mongoliensis*. As in *P. engelhardti*, this tissue is locally restricted in *Psittacosaurus* and is deposited in the exterior cortex of older individuals (Erickson and Tumanova 2000). Erickson and Tumanova (2000) regarded this special tissue as unusual for dinosaurs. In extant vertebrates, highly porous, radially vascularized bone is only formed in bony calli or is associated with substantial shape changes and is deposited very fast (Erickson and Tumanova 2000). Erickson and Tumanova (2000) discussed a pathological origin for the radial bone tissue in *Psittacosaurus*, but no calli were observed. Another possibility they considered was rapid ontogenetic migration of muscle insertions, but this could not be documented. A third hypothesis discussed was that the tissue reflected a change in loading patterns, perhaps associated with a shift from facultatively bipedal to quadrupedal locomotion.

Several hypotheses regarding the function and origin of RFB in *Plateosaurus engelhardti* are discussed below, but a satisfactory explanation remains elusive. The hypothesis that RFB is pathological appears unlikely because the normal and intact bone surface is underlain by RFB. Moreover, bones with RFB show no other conspicuous differences in morphology. We are not aware of any bone pathology that would resemble RFB. Another hypothesis is that RFB is an expression of sexual dimorphism. It is possible that one of the sexes had a higher growth rate than the other, as is common in recent tetrapods (Kabigumila 2000). Thus, the faster growing sex would have grown at a higher bone apposition rate, expressed by higher vascularity. However, RFB is only seen in selected bones and not in the complete skeleton of a single individual (e.g. SMNS F14, Tables 4–5), contrary to the predictions of this hypothesis. In addition, the occurrence of RFB in the middle and outermost cortex of nearly fully

TEXT-FIG. 5. Two alternative growth curves for *Plateosaurus engelhardti*. The black curve represents a more reptilian life history, with sexual maturity reached well before final size. The grey curve represents a more mammalian or avian life history in which sexual maturity is at approximately the same time as final adult size.

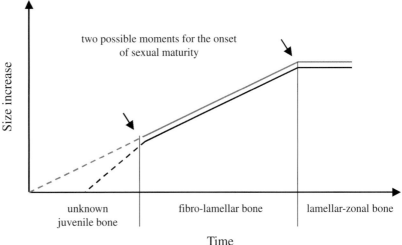

grown individuals indicates a strongly localized growth spurt in a late stage of ontogeny, when slower growth is usually initiated. Thus, a correlation between RFB and either sexual maturity or sexual dimorphism seems unlikely, unless sexual maturity was reached late in the ontogeny of *Plateosaurus* (see above). The fact that RFB is known from both Trossingen and Frick shows that although it is rare it is not restricted to a single population. In a few samples, an intermediate stage between laminar fibro-lamellar and RFB can be observed, marked by a clear increase in radial vascular canals in the laminar fibro-lamellar complex. This suggests that most (if not all) bones in *Plateosaurus* had the potential to grow with RFB, but the trigger for RFB growth remains unknown.

Meaning of highly vascularized laminar fibro-lamellar bone. Highly vascularized fibro-lamellar bone is known only from two femora (femur 9, Text-fig. 3C, G; femur 12). It is reminiscent of sauropod laminar tissue (Sander and Tückmantel 2003), suggesting a higher growth rate than in normal laminar fibro-lamellar bone of *Plateosaurus*. The isolated occurrence of this tissue cannot be explained satisfactorily. Owing to the large size of the femora and the occurrence of this tissue in the fifth and eighth or ninth total growth cycle, juvenile tissue must be excluded. Nevertheless, the probable juvenile bone tissue of *Plateosaurus* may have had an appearance similar to the tissue observed in this femur.

Meaning of secondary osteons and remodelling. The size of the remodelling zone and cancellous bone area, and the expansion of the medullary cavity, can give some information about the age of an individual. Strong remodelling, indicated by much substitution of primary bone tissue by secondary bone tissue and unclear boundaries between the cancellous bone region and the remodelling zone, suggests a greater individual age than a bone with a clearly delimited and small medullary region and little remodelling.

Meaning of endosteal bone. Castanet *et al.* (1988) described endosteal bone in young tuatara. Reid (1997*b*) noted that endosteal bone is formed in marrow-filled limb bones after active growth has ceased or is formed and resorbed during pauses in medullary expansion. This bone type can also appear when bone drift occurs to one side of the bone (Reid 1997*b*). It is clear that the endosteal layer in the *Plateosaurus* specimen (femur 1, Text-fig. 3I–J) is locally restricted to the cranial bone side, because it is not documented in either the core or cross-section from the caudal bone side of the same specimen. Based on skeletochronology, this bone belonged to a relatively young individual in which growth had not ceased.

Because of its isolated occurrence, endosteal bone is not useful for ageing *Plateosaurus* specimens.

Quantitative growth record: age estimates based on single bones

Femur. Femur samples represent a wide size range (Tables 2, 4), but no individuals with a femur length of < 50 cm were represented. Only two specimens are from Frick, and both belong to individuals with a femur length < 60 cm, but both show clear cessation of growth in the outer cortex. The number of visible LAGs in the femur samples from both localities varies between five and 14 (averaging eight), with a total femur length of 50–99 cm (Tables 2, 4; Text-fig. 1). This LAG count corresponds to 6–15 growth cycles. The resorbed LAG number is between three and 14 and also averages eight. The total LAG number varies between 12 and 24, with an average value of 16–17.

Femora with a length of 50·0 cm (femur 1, Text-fig. 6A–B), 59·5 cm (femur 4) and 77·5 cm (femur 11) were still growing rapidly, as indicated by the deposition of normally vascularized fibro-lamellar bone. Other femora (femur 9, Text-fig. 8A; femur 7) show a decrease in vascular density but continued slow growth with mainly fibro-lamellar bone. Femur 12 (Text-fig. 8B) shows normally vascularized fibro-lamellar bone, but distances between cycles have become conspicuously smaller. In these bones a decrease in growth rate is indicated, but changes in bone tissue type (indicating approaching cessation of growth) were not yet complete. Several other femora had clearly stopped growing, as indicated by deposition of nearly or completely avascular lamellar-zonal bone. Cessation of growth is recorded in femora with lengths of 56·5 cm (femur 2, Text-fig. 8D), 59·0 cm (femur 3), 63·5 cm (femur 6) and 99·0 cm (femur 13, Text-fig. 8C). An approximate general trend is that some femora show fast growth with fibro-lamellar bone to a point coinciding with total LAG 14 (femora 1, 9 and 11), after which growth decreased (this growth decrease can also contain several LAGs/growth cycles) and finally bone tissue type changes to lamellar-zonal bone (= cessation of growth) at a point coinciding with total LAG 18–20 (femora 2, 4 and 13). Exceptions to this include femur 12, which records continued growth with fibro-lamellar bone to total LAG 19/20, and another two samples (femora 3 and 6) in which growth ceased around total LAG 12 and 14/15, respectively. Thus, the femur samples indicate that growth stopped or slowed down at varied body sizes.

After growth had stopped, *Plateosaurus* individuals may have continued to live for several years, as shown by femur 13. In this individual, growth decreased after visible LAG 4, which corresponds to total LAG 18, as indicated by the deposition of low vascularized fibro-lamellar bone. It had reached most of its full body size (of approximately 10 m) after visible LAG 6 (total LAG 20). Growth then stopped, as recorded by a change in bone tissue type to avascular lamellar-zonal bone. Four closely spaced LAGs deposited in the avascular lamellar-zonal bone record the lifespan after growth had stopped.

Tibia. The total number of visible LAGs in the tibia varies between five and 11, averaging eight, with tibia lengths of 51·0–66·0 cm and corresponding femur lengths of 63·0–77·5 cm (Table 4). The visible LAG count corresponds to 6–12 growth cycles. The resorbed LAG number is between four and 15 (averaging nine). The total LAG number varies between ten and 26 with an average value of 17. Considering the relatively small size range of the tibiae sampled, they show a high range in total LAG numbers and therefore in age distribution. As shown in Table 4, the tibia samples were derived from medium-sized individuals only. The Frick tibiae all belong to the same size class, 51·0–53·0 cm, corresponding to a femur size of 63·0–65·0 cm. Tibia 1 (from an individual with a body length of 6·30 m) shows continued growth with laminar fibro-lamellar bone and 10–11 LAGs, whereas the others of the same size had strongly decreased or stopped growth. Among the Trossingen samples, growth had decreased in two relatively small tibiae but the other two tibiae show continued growth, including the largest tibia (tibia 8). Thus, there are individuals from Frick with body lengths of 6·30–6·50 m in which growth decreased greatly after a total LAG number of 23/24 (tibia 2), 22 (tibia 4: Text-fig. 8F) and 22/23 (tibia 5). After growth had stopped, in tibia 4 two more very closely spaced LAGs were deposited and tibia 5 exhibits deposition of three more LAGs in lamellar-zonal avascular bone. Tibia 2 possesses RFB on one side of the exterior cortex, so it is not clear if this bone had already stopped growth, as the avascular lamellar-zonal bone in the exterior cortex on the other bone suggests. In the Trossingen material there are tibiae of approximately the same size in which growth had continued with laminar fibro-lamellar bone, but with clearly decreased vascularization. These show a total LAG count of 11 (tibia 6, Text-fig. 8E) and 13 (tibia 3, Text-fig. 6C–D). The larger tibiae, belonging to individuals of 7·40 and 7·75 m body length, respectively, also continued to grow, as shown by the presence of laminar fibro-lamellar bone, and show total LAG numbers of 17 (tibia 7) and 14/15 (tibia 8). These examples show significantly lower LAG numbers at a greater body size. Therefore, the tibia samples also indicate that growth stopped at variable final body sizes in *Plateosaurus*.

Fibula. Fibulae were only sampled via the fracture surfaces of the Trossingen material. All sampled fibulae seem to display continuous growth with laminar fibro-lamellar bone, but some display decreased vascular density in the outer cortex, which may foreshadow a change in bone tissue. The total count of visible LAGs in the fibula samples varies between four and eight, averaging six LAGs. The size range of the sampled fibulae is 46·5–59·0 cm, corresponding to femur lengths of around 58·0–77·5 cm (Table 4). The visible LAG count corresponds to 5–9 growth cycles. The resorbed LAG number is between five and 15 (averaging eight). The total LAG number varies between 12 and 20, with an average value of 14. The fibula samples represent medium-sized individuals that also have intermediate total LAG numbers. A total LAG number of 13 occurs most frequently. This is shown by the smallest specimen (fibula 1), a medium-sized fibula (fibula 2), and the largest specimens (fibula 5, Text-fig. 6E–F; fibula 6). Two other medium-sized fibulae show somewhat greater LAG numbers (fibula 3, 15/16 LAGs; fibula 4, 19/20 LAGs) without displaying clear cessation of growth. To generalize, growth in the fibula continued with laminar fibro-lamellar bone to a point at least coincident with LAG 13. Following this, a decrease in growth rate was initiated and growth stopped at LAG 19/20 or slightly later.

Humerus. The total count of visible LAGs in humerus samples varies between four and ten (averaging seven) and the total humerus length ranges from 41·0 to 53·0 cm. The corresponding femur lengths are 63·5–85·0 cm (Table 4). The size range of the humeri is relatively great, at least based on corresponding femur size. Medium-sized individuals are represented along with one large individual. The resorbed LAG number ranges from three to 13 (averaging 7–8). The total LAG number varies between seven and 20 (averaging 15). Except for the Frick sample (humerus 5, Text-fig. 6G–H), growth continued with laminar fibro-lamellar bone in all Trossingen samples, although in most it occurred with decreasing vascular density. Only one humerus continues normal (fast) growth (humerus 6). Humeri 2–4 were only studied in fracture surfaces. Thus, it is possible that a bone tissue type change was initiated in these samples but was not detected. Humerus 5 belongs to a medium-sized individual of around 6·50 m in body length that stopped growth after total LAG 14, with deposition of closely spaced LAGs in the outermost cortex continuing to total LAG 18. The medium-sized humerus 2 exhibits a total LAG number of 19/20. In the fracture surface, vascular density is decreased, and the exterior cortex appears to be nearly avascular. A large difference in LAG count is seen in similar-sized specimens (humeri 3–4). Both represent individuals of approximately 7·0 m body length, but humerus 3 shows a total LAG count of 13 whereas humerus 4 contains 20 LAGs. In humerus 3, high vascular density and

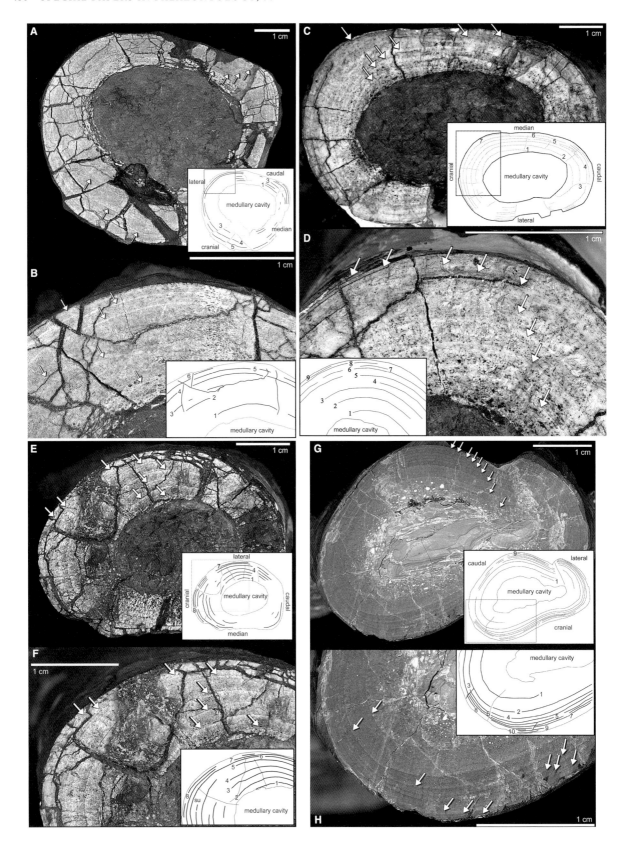

fibro-lamellar bone extends into the exterior cortex, indicating that this individual would have attained even larger body size. Humerus 4 belongs to a large medium-sized individual. The fracture surface shows that a decrease in growth rate had started. The largest humerus sampled (humerus 6) belongs to an individual of approximately 8·5 m body length and shows a low total LAG count (11/12). In its exterior cortex, this sample records continued growth with laminar fibro-lamellar bone and a high vascular density. However, in the outer part of the middle cortex two cycles of lamellar bone in nearly avascular tissue and two closely spaced LAGs were deposited, after visible LAG 3 (total number of LAGs was 7–8). After this point, growth had resumed with high vascular density and laminar fibro-lamellar bone. It is not clear if the two cycles containing lamellar-zonal bone are major cycles or subordinate cycles within visible growth cycle 4. They were counted here as major cycles based on the generally low total LAG count of this large humerus. However, this is an interesting observation as it suggests the possibility that *Plateosaurus engelhardti* could have stopped growth well before final size was reached, perhaps because of poor environmental conditions or disease, but resumed growth when conditions became better. For humerus 1 no bone length is available, but it shows decreased growth with a total LAG number of only 7/8.

The humerus samples indicate that growth rate decreased and finally stopped at variable sizes in *Plateosaurus* individuals. However, the timing of growth cessation is not similar to that of the other long bones, but this difference may be owing to errors introduced by extrapolation. Growth cessation was initiated in some samples at total LAG 19/20 and at body lengths between 6·35 and 7·0 m, whereas in the Frick specimen growth stopped at total LAG 14 and at a body length of only 6·5 m. Conversely, growth continued with fibro-lamellar bone in specimens with total LAG numbers of 7/8, 11/12 and 13 and at body lengths of 7·0–8·5 m.

Scapula. The size range of the scapulae is relatively great, representing small to medium-sized individuals. The total count of visible LAGs in the scapula samples varies between five and 11 (averaging seven). Total length of the sampled scapulae ranges from 36·5 to 49·5 cm. These scapula lengths correspond to femur lengths of 48·0–73·0 cm (Table 4). The resorbed LAG number is between three and ten, with an average value of six resorbed LAGs. The total LAG number varies between eight and 18 with an average total LAG number of 13. In comparison with the long bones and the pubis, scapula samples show an intermediate resorbed LAG number. The scapula has a lower visible LAG number, higher resorption and therefore higher total LAG numbers in contrast to the pubis samples. As in the other bones, small scapulae may show high total LAG counts and large ones lower or similar LAG counts as are present in the small scapulae. Scapula 1 contains a few closely deposited LAGs in an avascular bone tissue in the exterior cortex that indicates a complete growth stop at 4·80 m body length. The total LAG number in this sample is 16/17. Growth decreased strongly after total LAG 14/15 (scapula 1). Scapula 2 had a clearly decreased growth rate with a total LAG number of 18, but did not show avascular lamellar-zonal bone. The samples from Trossingen belong to larger individuals. However, in scapulae 4 and 6 a decrease in growth rate had been initiated at a point coincident with total LAG number of 13/14 and 11/12 and body lengths of 6·35 and 7·30 m, respectively. Scapulae 3 and 5 (Text-fig. 7A–B) would have continued to grow with a total LAG number of eight and 10/11 and inferred body lengths exceeding 6·2 and 7·1 m, respectively, at the time of death.

Pubis. The size differences and the range of LAG counts in the sampled pubes are minor. Pubis size ranges between 48·0 and 53·5 cm, corresponding to femur lengths of 59·0–66·0 cm (Table 4). The total number of visible LAGs varies between nine and 11 (averaging ten): the number of resorbed LAGs is 2–4 (averaging three). Total LAG number is 11–13, with an average value of 13. Among the bones studied, the pubis was the slowest growing bone in *Plateosaurus*. This is indicated by its purely lamellar-zonal bone tissue. The pubis has the lowest number of resorbed LAGs and the visible LAG count is the highest of all the bones. The total LAG number is

TEXT-FIG. 6. Cross-sections of long bones studied in fracture surfaces: photographs and interpretive drawings. A–B right femur 1. A, photograph of fracture surface; arrows mark the visible LAGs (NB, some LAGs are not visible at this magnification); the rectangle marks the enlargement of the lateral bone section figured in B. B, enlargement and drawing of lateral cortex; owing to the higher magnification, one more LAG is visible in the exterior cortex. C–D, right tibia 3. C, photograph of fracture surface; arrows mark the visible LAGs (NB, some LAGs are not visible at this magnification); the rectangle marks the enlargement of the lateral bone section figured in D. D, enlargement and drawing of cranial cortex. E–F, ?right fibula 5. E, photograph of fracture surface; arrows mark the visible LAGs (NB, some LAGs are not visible at this magnification); the rectangle marks the enlargement of the lateral bone section figured in F. F, enlargement and drawing of craniolateral cortex. G–H right humerus 5. G, photograph of fracture surface; arrows mark the visible LAGs (NB, some LAGs are not visible at the present magnification); the rectangle marks the enlargement of the lateral bone section figured in H. H, enlargement and drawing of cranio-medial cortex.

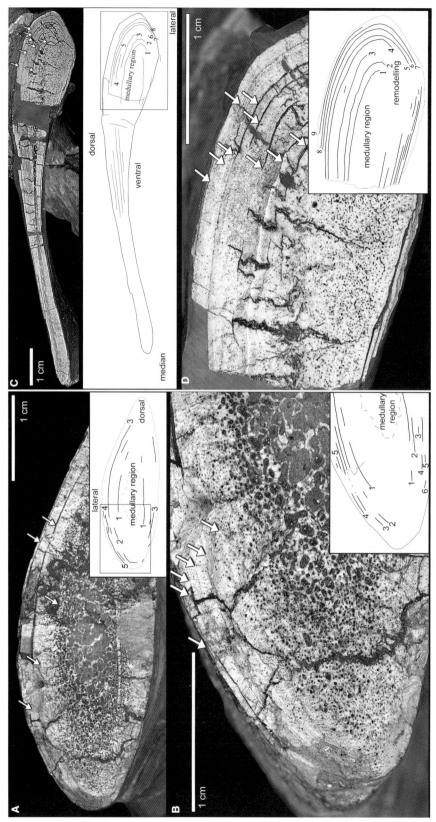

TEXT-FIG. 7. Cross-sections of flat bones studied in fracture surfaces; photographs and interpretive drawings. A–B, left scapula 5. A, photograph of fracture surface; arrows mark the visible LAGs (NB, some LAGs are not visible at the present magnification); the rectangle marks the enlargement of the lateral bone section figured in B. B, enlargement and drawing of ventral cortex. C–D, left pubis 2. C, photograph of fracture surface; arrows mark the visible LAGs (NB, some LAGs are not visible at the present magnification); the rectangle marks the enlargement of lateral bone section figured in D. D, enlargement and drawing of lateral cortex.

TABLE 4. Quantitative and qualitative growth record. Tibia, fibula, humerus, scapula, and pubis length are scaled to their corresponding femur size. FBLC, fibro-lamellar complex with LAGs; LZB, lamellar-zonal bone; NM, not measurable.

Specimen	Visible LAGs	Histology of exterior cortex	Resorbed LAGs	Total LAGs	Corresponding femur size (cm)
femur 1	6	FBLC, high vascular density	6–7	12–13	c. 50
femur 2	14	LZB in an avascular tissue, at caudal side RFB	9	23	56·5
femur 3	6	deposition of closely spaced LAGs in an avascular LZB has started	6	12	59·0
femur 4 (incomplete sample)	5	–	–	–	59·5
femur 5	7	growth continued with FBLC	7	14	62·5
femur 6	7	deposition of closely spaced LAGs in a nearly avascular LBZ has started	7–8	14–15	63·5
femur 7	8	growth continued with FBLC, but with a lower vascular density	10	18	c. 72·0
femur 8	11	growth continued with FBLC, but with a lower vascular density	5	16	74·0
femur 9	12	relative closely spaced LAGs, but still in a FBLC	3	15	74·0
femur 10 (right)	5	growth continued with FBLC	9	14	76·0
femur 11 (left)					77·5
femur 12	11	cycle distance becomes closer, but still FBLC	8–9	19–20	81·0
femur 13	10	LZB in a nearly avascular tissue	14	24	99·0
tibia 1	6	growth continued with FBLC	4–5	10–11	63·0
tibia 2	9	deposition of closely spaced LAGs in a nearly avascular LZB has started at the caudal bone side RFB	14–15	23–24	63·0
tibia 3	7	growth continued with FBLC, but with a lower vascular density	6	13	64·0
tibia 4	11	avascular LZB with closely spaced LAGs	12	23	≫ 64·0
tibia 5	11	LZB with closely spaced LAGs, but still a few longitudinally vascular canals	14–15	25–26	65·0
tibia 6	6	growth continued with FBLC, but very low vascular density	5	11	63·5
tibia 7	5	two or three cycles of RFB at the caudal bone side; the other bone sides seems to be continued with FBLC	12	17	74·0
tibia 8	7–8	growth continued with FBLC	7	14–15	77·5
fibula 1	7	very low vascularized FBLC	6	13	58·0
fibula 2	6–7	growth continued with FBLC, but vascular density starts to decrease	6	11–12	63·5
fibula 3	5–6	low vascularized FBLC	10	15–16	66·0
fibula 4	4–5	low vascularized FBLC	15	19–20	72·0
fibula 5	7–8	very low vascularized FBLC	6	13–14	74·0
fibula 6	7–8	difficult to judge, but growth seems to have continued with FBLC	5	12–13	77·5
humerus 1	4–5	continued growth with low vascularized FBLC	3	7–8	NM

TABLE 4. Continued.

Specimen	Visible LAGs	Histology of exterior cortex	Resorbed LAGs	Total LAGs	Corresponding femur size (cm)
humerus 2	8–9	continued growth with low vascularized FBLC	11	19–20	63·5
humerus 3	7	continued growth with FBLC	6	13	70·0
humerus 4	7	continued growth with low vascularized FBLC	13	20	70·0
humerus 5	10	deposition of closely spaced LAGs in a nearly avascular LZB	8	18	65·0
humerus 6	8	continued growth with FBLC	4–5	12–13	85·0
scapula 1	11	closely spaced LAGs in LZB	5–6	16–17	48·0
scapula 2	8	closely spaced LAGs in LZB with low vascularization	10	18	51·0
scapula 3	5	continued growth with FBLC	3	8	≫ 62·0
scapula 4	6	the last two LAGs were deposited more closely, still a few vascular canals	8–9	14–15	63·5
scapula 5	7	continued growth with FBLC	3–4	10–11	71·0
scapula 6	5	continued growth, but only low vascularization	5–6	11–12	73·0
pubis 1	11	avascular LZB, but no closely spaced LAGs yet	2	13	NM
pubis 2	10	closely spaced LAGs, but still with low vascularization	2–3	12–13	59·0
pubis 3	9	closely spaced LAGs in an avascular tissue	4	13	63·5
pubis 4	10	closely spaced LAGs in an avascular tissue	3	13	65·0
pubis 5	9	closely spaced LAGs in a partially avascular tissue	2	11	66·0

low to intermediate in comparison with the long bones. This is also consistent with the small medullary region in pubis samples, at least on the lateral side of the element (sampling location). Thus, it appears that the growth record in the pubis is the most realistic and complete for all *Plateosaurus* bones.

As shown in Table 4, the five pubis samples represent only medium-sized individuals 5·90–6·60 m in body length. In pubes 3–5, the outer LAGs were deposited in avascular tissue, indicating that these individuals had ceased to grow. Growth rate decrease in all pubes was initiated after total LAG 8 or 9: the total LAG counts in pubes at time of death are 11 and 13. Pubis 1 shows a clear decrease in vascular density in the exterior cortex (Table 4), indicating that cessation of growth was initiated at a total LAG count of 13. Only one sample, pubis 2 (Text-fig. 7C–D), continued with slow growth, but vascular density had already decreased at total LAG 12/13.

Thus, on the basis of the pubes, maximum size appears to lie between 6·35 and 6·50 m body length, owing to a clear cessation of growth at this point. The other pubes also show decreased vascular density at the same total LAG number. Owing to the lamellar-zonal bone tissue and a generally low vascular density in the pubis, it cannot be determined if slow growth continued or if these individuals would have stopped growing soon after deposition of this final LAG.

Vertebrae. Growth cycles are clear in the vertebrae, with each ending in a LAG. Zones and annuli are not as clearly distinguished as in other bones sampled. Vascularization is generally high, but decreases both within each growth cycle and overall from the inner to the outer cortex. Growth cycle thickness also decreases gradually from the inner to the outer cortex. In both samples, seven LAGs are visible (vertebra 1, Text-fig. 8G).

TEXT-FIG. 8. Polished sections of different kinds of bones. A, femur 9; LAGS 8–12 are not visible at this magnification. B, femur 12 (su, subordinate cycle). C, femur 13; LAGS 5–10 are not visible at this magnification. D, femur 2. E, tibia 6. F, tibia 4; LAGS 6–10 are not visible at this magnification. G, vertebra 1; note the thick layer of RFB in the outermost cortex. H, cross-section of the median part of rib 1; the arrows mark several LAGs deposited. I, foot phalanx 1; no growth cycles are visible at this magnification; numbers mark visible LAGs.

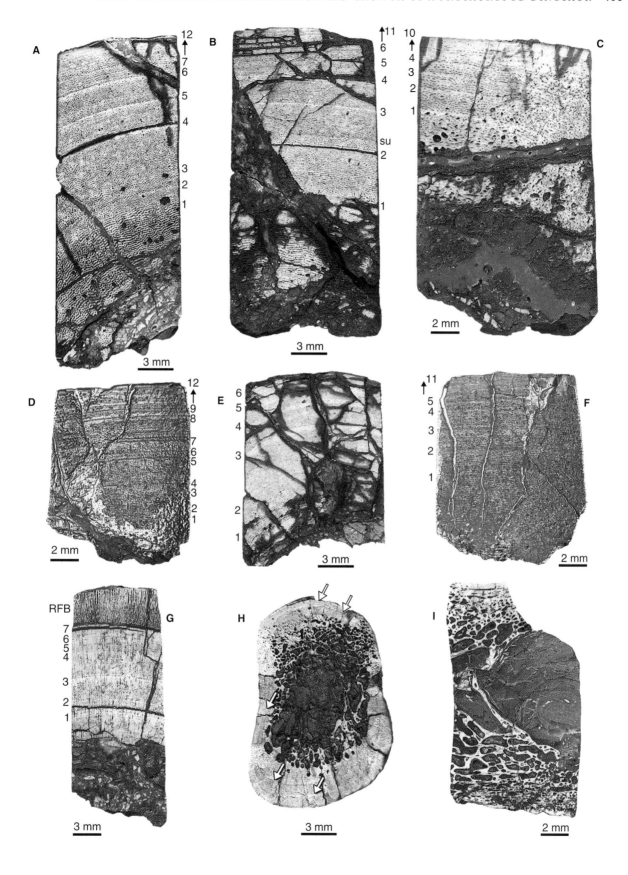

TABLE 5. LAG counts in different bones of the single individuals SMNS F 14, F 29B and F 48.

Sample	Visible LAGs	Resorbed LAGs	Total LAGs
SMNS F 14 femur length: 63·5 cm			
femur 6	7	7–8	14–15
tibia 6	6	5	11
fibula 2	6–7	6	12–13
humerus 2	8–9	11	19–20
scapula 4	6	8–9	14–15
pubis 3	9	4	13
SMNS F 48 femur length: 77·5 cm			
femur 10, 11	5	9	14
tibia 8	7–8	7	14–15
fibula 6	7–8	5	12–13
SMNS F 29 femur length: c. 65·5 cm			
pubis 4	10	3	13
pubis 5	9	2	11

Ribs. In rib 1, growth cycles are generally thin and remain of approximately constant width from the inner to the outer cortex. At least 5–6 LAGs are visible, and extrapolation suggests a total LAG count of 12–15, depending on which bone side is used in the calculation (Text-fig. 8H). Rib 2 shows similar features and only in the exterior cortex of the more proximally positioned sample can LAGs be counted (4–5 present). They are closely spaced and developed in nearly avascular lamellar-zonal tissue. The more distal rib sample preserves no growth marks because of strong remodelling.

Ischium. Only the lateral part possesses a relatively good growth record with three clear LAGs. Extrapolation suggests a total of 10–11 LAGs or 11–12 growth cycles.

Pedal phalanx. Closely spaced LAGs are only seen in the exterior cortex, where the primary bone tissue is lamellar-zonal. Approximately 5–6 LAGs can be counted (Text-fig. 8I).

Quantitative growth record: age estimates based on several bones of an individual

Sampling different bones from single individuals is always crucial for a comprehensive understanding of the growth record. Two individuals from Trossingen were sampled from several bones (Table 5). SMNS F14 (femur length 63·5 cm) was sampled from six bones: femur 6, tibia 6, fibula 2, humerus 2, scapula 4 and pubis 3. SMNS F48 is larger (femur length of 77·5 cm) and is represented by four bones: femora 10–11, tibia 8 and fibula 6. It is probable that a left and right pubis (pubes 4 and 5) of SMNS

F 29B also belong to the same individual because of their similar size. The histology of all of these elements has been described above. Below we focus on how the growth record of the separate elements can be considered together to yield the growth record of a single individual.

SMNS F14. The femur sample suggests that the animal was close to ceasing growth, as indicated by the onset of lamellar-zonal bone deposition in its exterior cortex. The other long bones of SMNS F14 also show a clear decrease of vascular density within the laminar fibro-lamellar complex of the outer cortex, but no lamellar-zonal bone was deposited, except in the pubis. The scapula shows high vascular density up to the bone surface and a slight decrease in vascular density can be observed on the lateral side only. The external cortex of the pubis shows that growth had nearly stopped, as indicated by the presence of nearly avascular bone tissue. In summary, the growth record of all of the bones indicates that this individual had undergone a clear growth decrease or was approaching cessation of growth, and had therefore nearly reached its final size. However, the differences in histology mean that bones of a single individual not only grew at different rates but also stopped growing in diameter at varying times. The visible LAG number does not vary significantly in the sampled bones of this individual (6–9). In addition, the observed growth decrease or growth stop in the exterior cortex occurs consistently in all sampled bones after visible LAG 5. The total LAG number for this individual lies between 11 and 20. Most frequent are total LAG counts of 13–15, with the tibia showing the lowest total LAG number (11) and the humerus showing the highest (19–20). The discrepancy between the humerus and the other bones is difficult to explain: either SMNS F14 is not a single individual, or this discrepancy is a result of extremely varied resorption rates in the different bones, or results from the extrapolation method used. Cessation of growth occurred from total LAG 9 to total LAG 13/14. On the basis of the average of the values, with the exception of those from the humerus, this individual is hypothesized to have grown at high rates until it was 12 years old. Afterwards, growth rates started to decrease significantly, and the individual presumably stopped growing completely after the fifteenth or sixteenth year of life (Text-fig. 9).

SMNS F48. Despite the fairly large size of the individual (7·75 m), the femur, tibia and fibula of SMNS F48 record that growth continued with deposition of laminar fibro-lamellar bone until the time of death. This individual had therefore not reached its possible maximum size. The differences between the visible LAG numbers in the different bones are relatively high. In the femur, there are only five visible LAGs, but 7–8 are visible in the

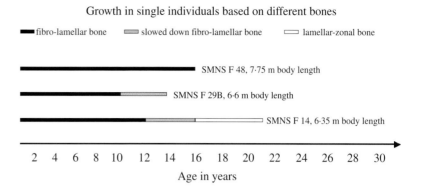

TEXT-FIG. 9. Comparison of growth curves of individuals sampled from several bones (SMNS F14, F48, F29); note the high variability in body sizes and growth patterns.

tibia and fibula. However, the total LAG number is less variable and amounts to 13–15 LAGs in all elements sampled, suggesting an age of around 16 years at the time of death (Text-fig. 9). The less variable total LAG count in comparison with SMNS F14 can possibly be explained by the relatively wide growth cycles of this fast-growing individual. This suggests that in younger individuals extrapolation does not result in as large an overestimate of total LAGs as occurs in older individuals, because of the larger average cycle thickness used for extrapolation.

SMNS F29. SMNS F29 consists mainly of left body sides from at least three individuals. Owing to their proportions and morphology, a left and a right pubis are assumed to belong to a single individual, but no other elements of this individual can be identified. The corresponding femur length for the pubes is approximately 66·0 cm, indicating a medium-sized individual. In their exterior cortex both pubes show a clear decrease in growth rates as indicated by the presence of closely spaced LAGs in either nearly (pubis 4) or partially (pubis 5) avascular bone tissue. Thus, it can be concluded that maximum size had almost been reached and that growth had nearly stopped in the pubes. Visible LAG number differs only by one between left and right pubes and growth decreased in both pubes after visible LAG 6, which corresponds to total LAG counts of eight and nine, respectively. Total LAG number differs by two cycles between the left and right pubis, but taking into account the possible error sources, this is not considered to be significant. Thus, the individual from which these two pubes originated had reached its maximum size of approximately 6·60 m after 12–14 years (Text-fig. 9).

Summary of age estimates

This section summarizes the quantitative aspects of *P. engelhardti* life history, bearing in mind the approxi-

mate nature of the values owing to the need for extrapolation. The youngest individuals have a total LAG number of eight (scapula 3; humerus 1). Scapula 3 belonged to an individual of approximately 6·20 m body size, but no data are available for humerus 1. On the basis of femur 13 and tibia 5, a maximum total LAG number of 25, representing an age of 26–27 years, is suggested for *Plateosaurus engelhardti*. Because these individuals, like all the others at Trossingen and Frick, presumably died by accident (Sander 1992) and not of old age, the maximum age of *Plateosaurus engelhardti* can be assumed to have been higher. In contrast to femur 13, which represents the maximum documented size for *P. engelhardti* (around 10·0 m), tibia 5 belongs to a medium-sized individual of only 6·50 m body length. The smallest sampled specimen had stopped growth and shows a total LAG number of 16/17. It was only 4·80 m long (scapula 1). Most individuals grew up to their fifteenth year of life with laminar fibro-lamellar bone, then growth started to decrease and finally stopped. Most individuals had stopped growing completely around their twentieth year, but lived for several more years in total (as shown by femur 13).

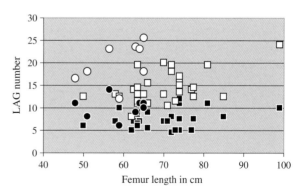

TEXT-FIG. 10. Body size expressed by femur length plotted against visible and total LAG number. Black symbols mark the visible LAG number and white symbols the total LAG number; circles represent Frick specimens, squares Trossingen specimens. All bones were scaled to their corresponding femur length. There is no correlation between LAG number and size.

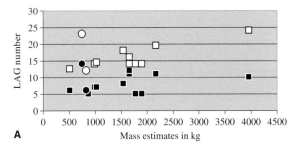

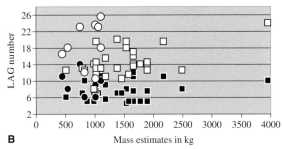

TEXT-FIG. 11. LAG count plotted against mass estimate. A, femora only. B, all sampled bones, scaled to their femur length. Mass estimates for the specimens are based on the mean value for the largest femur (femur 13: see Table 1) of mass estimates given by Seebacher (2001) and Sander (1992). Black symbols represent visible LAG number and white symbols the total LAG number; squares represent Trossingen specimens and circles Frick specimens. There is no correlation between age and body mass.

Correlation between ontogenetic status, age and body size

In Text-figures 10–11 visible and total LAG numbers are plotted against femur length and body mass estimates in order to determine whether there is a correlation between any of these variables. Text-figure 12 shows the relationship between deposited bone tissue in the exterior cortex (= growth stage) and femur length (= body size). It is striking that the longest bones do not necessarily have the highest LAG numbers (Text-figs 10–12). In fact, no correlation between bone length (= body size) and LAG number (= age) is apparent in *Plateosaurus* (Table 4; Text-figs 10–11). Text-figure 12 shows that there is no correlation between bone length and the time of growth cessation. Consequently, some individuals grew to considerably larger final sizes than others.

COMPARISON OF BONE HISTOLOGY

Other prosauropods

In her detailed histological and skeletochronological study of the prosauropod *Massospondylus carinatus*, Chinsamy

(1993a, 1994; Chinsamy-Turan 2005) described highly vascularized fibro-lamellar bone similar to that in *Plateosaurus*. The distribution of secondary osteons and cancellous bone is also similar in both prosauropods, which also share cortical stratification owing to the presence of LAGs, reflecting periods of fast and slow growth. In small, presumably young *M. carinatus* specimens, the same bone tissue occurs as in large, presumably adult specimens, but growth rates decrease during ontogeny as indicated by a diminishing spacing of the growth marks (Chinsamy 1993a). Chinsamy (1993a) calculated that in the first 5 years, 50 per cent of femoral growth was attained, with growth slowing down subsequently. The equation provided by Chinsamy (1993a: age = 0·05 femur length$^{1·52}$) indicates that in *M. carinatus*, in contrast to *Plateosaurus*, size and age were much more closely correlated and that reliable age estimates on the basis of bone length are possible. Thus, Chinsamy (1993a) was able to reconstruct a growth trajectory for *M. carinatus* on the basis of her histological data. *M. carinatus* did not have the high variability in adult size seen in *P. engelhardti*, implying a very different growth pattern and life history for *M. carinatus*. The life expectancy in *M. carinatus* is lower (maximum lifespan of 15 years has been documented) than in *Plateosaurus engelhardti* (which could live for 26–27 years). However, this difference could partially be accounted for by differences in the extrapolation methods used. The method used by Chinsamy (1993a) is likely to be more accurate because resorbed growth marks were estimated on the basis of a juvenile specimen, which was not possible in *P. engelhardti*. Although the bone tissues, vascular system, cyclical growth and irregular cycle distances are generally similar in the two prosauropods (Chinsamy 1993a, p. 325, fig. 6; Chinsamy-Turan 2005), they differ in another significant way. Namely, closely spaced LAGs in avascular lamellar tissue (the EFS or OCL) are absent from the outer cortex of *M. carinatus*, even in the largest individuals. This observation led Chinsamy (1993a) to conclude that *M. carinatus* had an indeterminate growth strategy as supposedly seen in extant ectotherms. In *Plateosaurus engelhardti*, on the other hand, growth was clearly determinate and stopped in the last third of its life. Possible reasons for these differences are: (1) adult individuals of *M. carinatus* in which growth had stopped were not sampled (Chinsamy-Turan 2005); (2) *M. carinatus* and *P. engelhardti* are not as closely related as generally believed; and (3) they are closely related and the observed differences are real.

Comparison of *P. engelhardti* growth with that of *Thecodontosaurus antiquus* (Cherry 2002; pers. obs.) shows that their bone tissues and vascular system are similar, but that LAGs are missing from the limb bones of *T. antiquus*. However, she noted that the sampled bones may all have been derived from juvenile individuals and therefore that there may not have been sufficient time for the formation of LAGs in this material. Cherry (2002) noted that larger individuals of *Thecodontosaurus* are known, but that these were not sampled in her study.

Preliminary examination of a fracture surface bisecting a *Sellosaurus gracilis* pubis (SMNS 5715 from Stuttgart-Herlach, southern Germany) revealed nearly identical bone tissue to that observed in *P. engelhardti* pubes, including the presence of RFB. RFB has also been observed in other prosauropods, including a *Thecodontosaurus* bone from England (IPB R555) and a long bone section recovered from a drill core (Hurum *et al.* 2006). The borehole is located in the Norwegian North Sea and the sediment is early Rhaetian in age. In other aspects of its histology, the bone section is also close to that observed in the *Plateosaurus* samples. Thus, this specimen probably belongs to *Plateosaurus* (Hurum *et al.* 2006).

Sauropods

De Ricqlès (1983) was the first to describe sauropod bone tissue with LAGs, in the humerus of a half grown 'Bothriospondylus' (now referred to *Lapparentosaurus*). Bone microstructure consisted of fibro-lamellar bone with typical plexiform to reticular or (in restricted areas only) longitudinal primary osteons. Erickson (2005) noted that sauropods usually lack growth lines or that they are produced only late in development. However, de Ricqlès (1983) counted LAGs in the above-mentioned *Bothriospondylus* sample and concluded that this individual had required at least 16 and at most 26 years to reach slightly less than half of its maximum adult size. De Ricqlès (1983) suggested that approximately 43 years were necessary for an individual to reach two-thirds of its maximum adult size, agreeing with Case's (1978) estimates of 40 years for the onset of sexual maturity of another sauropod, 'Hypselosaurus'. This represents a very great age for sauropods, and a very late onset of reproduction: such high ages seem unrealistic in comparison with modern tetrapods (Dunham *et al.* 1989). More recent studies on sauropods suggest substantially younger ages were needed for reaching sexual maturity and maximum size. For example, the basal titanosaur *Janenschia* reached sexual maturity at around 11 years (Sander 2000) and *Apatosaurus* needed approximately 8–10 years to reach subadult size (Curry 1998, 1999). Rimblot-Baly *et al.* (1995) concluded that *Lapparentosaurus* needed two decades to reach 'adult conditions', but they did not specify the meaning of the latter term.

Recent histological studies of sauropod bones document a uniform bone microstructure that is already present in the earliest known sauropod, *Isanosaurus* (Sander *et al.* 2004). However, varied interpretations of this

microstructure may lead to seemingly different interpretations by different authors. For example, Curry (1999) and Sander (1999, 2000) suggested determinate growth on the basis of the deposition of lamellar-zonal bone in the exterior cortex, whereas Rimblot-Baly *et al.* (1995) proposed indefinite growth for *Lapparentosaurus*, although they noted a similar 'lamellar, accretionary tissue at the cortical periphery in their largest samples identical to that observed in *Apatosaurus* bone' (Curry 1999, p. 662). Curry (1999) interpreted these features as an indication of determinate growth in *Lapparentosaurus*. Cyclical growth patterns, mainly related to differences in the vascular system, are observable in many sauropod long bones. On the basis of these and other structural characteristics, ontogenetic stages can be documented (Sander 2000). True LAGs occur in sauropods mainly in the exterior cortex at the time of lamellar-zonal bone deposition (Sander 2000; Sander and Tückmantel 2003; Erickson 2005). Exceptions to this are seen in *Janenschia* and some *Barosaurus* specimens in which LAGs also appear in fibro-lamellar bone in the inner cortex (Sander 2000). Thus, in general, sauropods started life with very rapid continuous growth, followed by a gradual decline of growth rates in adulthood.

Bone tissues and vascular systems are similar in *Plateosaurus* and sauropods, but there are major differences in bone microstructure, in the amount of fibro-lamellar bone in the cortex, the cyclical interruptions of growth by clear growth marks/LAGs through the whole cortex, lamina density and vascular density. Lamina density in normal adult laminar fibro-lamellar bone of *Plateosaurus* averages 8 lamina/mm. This relatively high number of laminae agrees with an observation by Currey (1962, p. 243), who found a greater density of laminae in a prosauropod bone than a mammal bone. Sander and Tückmantel (2003) gave an average number of 5 laminae/mm in sauropods that is more similar to the lamina density in mammals (Currey 1962; Sander and Tückmantel 2003; this study) than to prosauropods (Currey 1962; this study). Vascular density in sauropods is considerably greater than in prosauropods. All three differences indicate higher growth rates in sauropods than in prosauropods. Data from skeletochronology agree relatively well with this observation. As in some sauropods, *Plateosaurus* probably displayed an early onset of sexual maturity that was independent from the attainment of final size. Growth is generally determinate in both groups and final size was reached in *Plateosaurus* within time-spans similar to those seen in sauropods (i.e. in the second or third decade of life). Both groups seem to share the pattern of growth continuing after sexual maturity. Thus, it can be concluded that *Plateosaurus* and sauropods show similar life histories, but that sauropods reached larger body sizes during similar time intervals owing to their higher growth rates and their strategy of continuously faster, uninterrupted growth (Sander et al. 2004). An important difference between both life histories is the high variability in the adult size of *Plateosaurus*, indicating a more 'reptilian' growth pattern and life history, whereas most sauropods show little variability in adult size range.

Living reptiles

Low to intermediate vascularized lamellar-zonal bone is characteristic for living reptiles, all of which are ectotherms. Vascular canals are mainly longitudinal and bone tissue becomes avascular in adulthood. Only monitors show exceptionally extensive vascularization in their primary cortex (Francillon-Vieillot and de Buffrénil 2001). Secondary osteons are lacking, except in restricted areas in turtles and crocodiles. Throughout their whole ontogeny, the usually slow growth in living reptiles is seasonally interrupted, resulting in cyclical LAGs. These interruptions are caused by exogenous factors (seasonal climatic change) as well as by endogenous cycles of reproduction and rest. In addition, poor environmental conditions over long periods can affect the growth history of all tetrapods (Wikelski and Thom 2000; Chinsamy-Turan 2005). Growth marks are known to be formed annually in living reptiles (using mark-release-recapture methods and marking cycles with tetracycline: Castanet and Smirina 1990; Castanet et al. 1993; Grosse 1999). Living reptiles initially also possess relatively high growth rates that, nevertheless, exhibit cyclical interruptions. In contrast to mammals and birds, they continue to grow after reaching sexual maturity, sometimes for several years, although at much lower growth rates than in their first years of life. Thus, sexual maturity is often reached long before maximum size (Neill 1971; Castanet and Cheylan 1979; Chinsamy 1994; Chinsamy and Dodson 1995).

Andrews (1982) reviewed the high variability in living reptile growth rates that were dependent on environmental conditions. The advantage of such a growth strategy and the ectothermic metabolism is that low growth rates allow living reptiles to persist at times or in places of low resource availability. In many living reptiles 'maximum size' may cover a wide size range (notably in larger taxa such as crocodiles, turtles and some giant snakes, but not in lizards). The age range at maturity is as large as the range of final sizes because sexual maturity is not reached at a determined age but at a determined body size and mass. The age at which this mass/body size is reached depends strongly on the personal fitness of an individual and on the environmental conditions in which it grew (climate and food availability/quality). Therefore, size and age are not necessarily correlated in many living reptiles (e.g. Zug et al. 1986, fig. 9; Neill 1971), leading to high phenotypic diversity in terms of adult size in many living reptiles. In other living reptiles, body size variation is limited, but maximum size is attained at a variety of possible ages, strongly depending on individual history.

Birds and mammals

The growth patterns and strategies of endothermic birds and mammals are also well known. Both groups show a similar, rather simple growth pattern with very fast rates, depositing highly vascularized and continuously growing fibro-lamellar tissue (Francillon-Vieillot et al. 1990; Chinsamy 1994). Growth lines are generally rare in birds and mammals and usually occur only in the outermost cortex in mature and adult individuals in the form of closely deposited LAGs within poorly vascular to avascular lamellar-zonal bone, the EFS or OCL (Chinsamy 1994; Chinsamy and Dodson 1995; Reid 1997a; Horner et al. 2001; Chinsamy-Turan 2005; Sander and Andrassy in press), indicating determinate growth in these groups. Attainment of adult size is nearly coincident with sexual maturity in birds and mammals: onset of sexual maturity usually corresponds to a drop in growth rate (Chinsamy and Dodson 1995; Chinsamy-Turan 2005; Erickson 2005) and there is little variation in adult size. The male African elephant is the only mammal for which prolonged growth after the onset of sexual maturity has been published (Jarman 1983), but this is also typical for the males of other megaherbivores such as giraffes (J. Hummel, pers. comm. 2006) whereas the females stop growing soon after attaining sexual maturity. In all other mammals growth is determinate and age is correlated with size. Strong remodelling and Haversian bone are common in bird and mammal bones. Larger mammals need longer times for growth and to reach sexual maturity. Smaller mammals, like most rodents, need only a few weeks to attain sexual maturity and maximum size. Most birds grow to maximum size and maturity within their first year, while this process takes only a few weeks in many small songbirds (Starck and Ricklefs 1998). Thus, there are few similarities between the growth patterns of *Plateosaurus* and extant birds and mammals.

DISCUSSION

Differences between the Frick and Trossingen localities

As mentioned previously, in contrast to Trossingen, Frick has yielded only small to medium-sized individuals (Textfig. 1), with all of the sampled bones from Frick representing a size range of 4·80–6·50 m total body length.

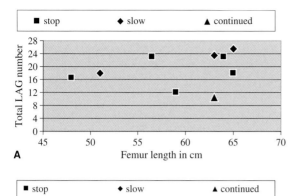

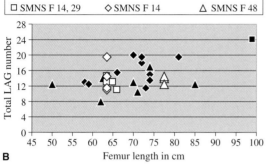

TEXT-FIG. 12. Relationship between growth stage and body size. Bones separated into the categories: 'continued growth', 'slowed down growth' and 'stopped growth'. Continued growth means that growth had continued with laminar fibro-lamellar bone without any significant growth decrease. Slowed growth means that vascular density became very low in the fibro-lamellar complex and bone tissue change was initiated resulting in deposition of highly vascularized lamellar-zonal bone. Stopped growth means that lamellar-zonal bone had been deposited with several closely spaced LAGs in avascular bone tissue. A, relationship between growth stage and body size in Frick specimens. B, relationship between growth stage and body size in the Trossingen material. This distribution does not represent a clear trend, but a broad range of data.

Most individuals from Frick had decreased growth rates or had completely stopped growing and only one shows continued growth at a body length of 6·30 m (Table 4; Text-fig. 12). Thus, the Frick animals are not young individuals, but this population consists of smaller and older individuals than those found at Trossingen. The maximum body length may have been approximately 6·50 m at Frick, with several individuals having stopped growing earlier. By contrast, maximum size at Trossingen is approximately 10·0 m, but most of the individuals are smaller (6·0–7·50 m in length). The lack of larger individuals among the Frick finds may just be a result of chance or may be a sampling artefact because the MNI at Frick is so much lower than at Trossingen. However, if the finds from Frick do record a population of smaller individuals, two explanations are possible. One may be a biological cause, such as clinal variation, or the Frick

specimens might have belonged to a different species or subspecies of *Plateosaurus*, with a different life history from the animals living at Trossingen.

Sexual dimorphism

The great variability in growth rate and maximum size in *P. engelhardti* may have been caused by sexual dimorphism, but sexual dimorphism in dinosaurs has been poorly documented, particularly in terms of histology. The histological study on *Syntarsus rhodesiensis* (Chinsamy 1990) concluded that the robust specimens that show erosional cavities that may have been formed during formation of eggshell represent a female morph (Carpenter 1999). However, Chinsamy-Turan (2005) mentioned that the sample size of *Syntarsus rhodesiensis* was probably too small to substantiate this. Schweitzer *et al.* (2005) reported the presence of an endosteally derived bone tissue in a *Tyrannosaurus rex* individual lining the interior medullary cavity. They homologized this bone tissue with the medullary bone unique to female birds (Schweitzer *et al.* 2005).

Prosauropods lacked clear display structures, but sexual dimorphism expressed in terms of body size and robustness seems plausible and has been suggested by many authors (Gow *et al.* 1990; Weishampel and Chapman 1990; Galton 1997, 2000; Benton *et al.* 2000), but unequivocal documentation has remained elusive. Galton (2001) has suggested that the less abundant, more robust and larger morph at Trossingen was male. On the basis of bone-histological work on the Tendaguru sauropods, Sander (2000) discussed possible sexual dimorphism in *Barosaurus*. He interpreted the fast growing morph (Type A) as male and the slower growing morph as female (Type B). However, he could not also exclude the possibility that the two morphs represent different taxa (Sander 2000; Remes 2004, 2005; Chinsamy-Turan 2005, p. 114).

If sample size is large enough, as in our study, sexual dimorphism expressed as variable body size and/or robustness should be reflected in bone microstructure (e.g. deposition of different bone tissues and different growth rates expressed by the degree of vascularization), because of the different growth strategies/patterns of the sexes. This is seen in living reptiles in two ways (Andrews 1982): (1) the juveniles of both sexes grow at similar rates until the individuals of the smaller sex reach their final size; individuals of the larger sex continue to increase in size; and (2) juveniles of both sexes start to grow at different growth rates; the larger sex grows faster, and sexual maturity is reached at nearly the same age but at different body sizes. On the basis of our study, no evidence in bone microstructure could be found to support sexual dimorphism in the bones of *Plateosaurus engelhardti*. The qualitative and quantitative growth record indicates that growth had stopped at strongly variable final sizes (see above and Text-fig. 12). Various parameters were checked on the basis of measurements and bone histology but a division of the sample into two groups that may represent the two sexes is not possible on the basis of these data. Histological results suggest that the morphological differences that underpinned previous divisions of these specimens into robust and gracile morphs only reflect very high variability of growth rates between individuals.

Sexual maturity

From studies on recent tetrapods we know that sexual maturity is accompanied by a clear decrease in growth rate (reptiles) or complete cessation of growth (birds and mammals). This should be visible in the bone histology of dinosaurs as well, although Erickson (2005) noted that it is difficult to detect. Thus, if sexual maturity is documented in bone microstructure, it may appear in the form of a clear decrease in growth cycle thickness and/or in a bone tissue change. At least two different histological features could represent sexual maturity in *Plateosaurus* specimens (Text-fig. 5). One is a change in bone tissue from laminar fibro-lamellar bone to lamellar-zonal bone. A similar situation is known in recent birds and mammals where, in most species, sexual maturity is reached together with maximum body size (de Ricqlès 1976, 1983; Erickson 2005). While many authors consider that this bone tissue change reflects the onset of sexual maturity in specific dinosaurs (Varricchio 1993; Chinsamy 1994; Chinsamy and Dodson 1995; Reid 1997*a*; Curry 1999; Horner *et al.* 2001; Chinsamy-Turan 2005; Erickson 2005), others (Sander 1999, 2000; Horner *et al.* 2000) concluded that sexual maturity was reached well before maximum size, as in reptiles, and well before the change from fibro-lamellar tissue to lamellar-zonal tissue. Thus, if this bone tissue change does reflect the onset of sexual maturity in *Plateosaurus engelhardti*, this would mean that sexual maturity was reached late in ontogeny (not before 16 years).

In most *Plateosaurus* samples a general decrease in growth cycle thickness occurs without a bone tissue change in the outer parts of the cortex: in some samples this decrease is relatively sudden. Such a sudden decrease is correlated with the onset of sexual maturity in modern reptiles (Chinsamy-Turan 2005) and also in some dinosaurs (e.g. *Janenschia*; Sander 2000). However, this may not be valid evidence for the onset of sexual maturity in *Plateosaurus* as this decrease in cycle thickness is not very distinct and additionally, as discussed before, cycle thickness is irregular. A general decrease in cycle thickness without a clear change of bone tissue type towards the outer cortex is not surprising because of a general decrease in growth with increasing size and age. Additionally, *Plateosaurus* growth was strongly dependent on environmental conditions, which also produces irregular cycle thickness (Chinsamy-Turan 2005).

Another possibility is that the onset of sexual maturity in *P. engelhardti* occurred earlier in ontogeny and is not preserved in the bone histology of the bones studied either owing to resorption activities or because this feature is generally not reflected in bone histology (Erickson 2005). This interpretation represents a more 'reptilian' life history, where sexual maturity is reached long before maximum body size and growth continues after the time of first reproduction for several years at a decreased rate. According to Dunham *et al.* (1989), first reproduction at an age of more than 20 years is unlikely for demographic reasons, even in very large dinosaurs. *Plateosaurus* would have been close to this age and we have excluded the possibility that this late bone tissue change represents sexual maturity. We support the hypothesis that sexual maturity was reached earlier, long before it reached final size. This is a typical pattern in most reptiles. For example, turtles (Zug *et al.* 1986), crocodiles (Neill 1971; Pooley and Gans 1976) and alligators (Magnusson *et al.* 2002) attain sexual maturity at certain body sizes/body masses, depending on climate, food availability and food quality, whereas the age at first reproduction varies within limits. For *Plateosaurus* we assume the onset of sexual maturity at a body size of approximately 4·0 m, which is approximately one-half to two-thirds of the final size of most studied specimens.

Final size

Possibly the most striking, but also the best supported, conclusion of this study is that *Plateosaurus engelhardti* had very wide range of final body sizes, ranging between 4·8 and 10·0 m (see also Sander and Klein 2005). The strength of this observation rests on the tissue change from fibro-lamellar to lamellar-zonal bone and an ESF observed in the outer cortex of the long and flat bones in some individuals, which is unequivocal evidence that these individuals were nearly or fully grown (Chinsamy-Turan 2005; Erickson 2005). The only alternative explanations for this observation would be that there is more than one species represented in the material, but all recent workers (Moser 2003; Yates 2003; Galton and Upchurch 2004) independently reached the conclusion that the material from Trossingen and Frick all belongs to the same or two closely related species, as discussed earlier.

Variability in growth rates

The observation that *Plateosaurus engelhardti* grew at variable rates is also striking. Given that only a single species is represented in the material, there are three possible explanations: (1) an extreme methodical error in recording visible LAGs and extrapolating resorbed LAGs; (2) LAGs are, in spite of experiments with recent tetrapods, not annual, but random; and (3) sexual dimorphism is hidden by the sample's variability. Because sexual dimorphism cannot be determined on the basis of current bone histological data (see above), and because the annual deposition of growth marks is the main premise of the current study, the best explanation remains very high intraspecific variability in growth rates among individuals of *Plateosaurus engelhardti* (see also Sander and Klein 2005).

SUMMARY AND CONCLUSIONS

Histological study of various cross-sections, polished sections and thin-sections of long and girdle bones of *Plateosaurus engelhardti* shows that the primary bone tissue was a laminar fibro-lamellar complex regularly interrupted by LAGs. The presence of laminar fibro-lamellar bone indicates high bone deposition rates, which are coupled with high metabolic rates in extant birds and mammals. However, deposition of LAGs is a clear indication that *Plateosaurus engelhardti* individuals did not maintain continuously high rates of bone deposition during their life history. Apart from the cycles in the fibro-lamellar bone, changes in overall growth rate during ontogeny are recorded in the exterior cortex where the fibro-lamellar complex changes into nearly avascular lamellar-zonal tissue with closely spaced LAGs in some specimens, indicating a determinate growth pattern.

Different rates of bone deposition are documented in a single skeleton. The primary cortex of the pubis consists only of lamellar-zonal bone, indicating either slower or negative allometric growth. Large medullary cavities in the long bones makes extrapolation necessary and leads to a general overestimate of total LAG numbers (= age) in the long bones. The total LAG numbers in the pubes are more realistic because of lower resorption rates.

Skeletochronology results in ages for different life history events similar to those obtained for other dinosaurs. A body length of approximately 5 m could be reached in 8 years. The maximum documented age was around 26–27 years, but longevity is assumed to be higher. Most individuals approached maximum size between 16 and 20 years. There is very high variability in maximum adult size. The sample contains specimens that stopped growth at lengths of 6·50 m and others

that continued to grow at body sizes of over 8·0 m. The largest sampled bone represents an individual of 10·0 m body length that stopped growth at around the age of 20 years and lived for several (5) years longer. Sexual maturity is assumed to be correlated with a certain size/body mass, possibly at a body size of approximately 4·0 m (half of the final size of most studied specimens). Sexual dimorphism could not be documented but remains possible.

Bone microstructure and some aspects of life history (cyclical interrupted growth, continued growth after onset of sexual maturity, determinate growth) of *Plateosaurus* were similar to other dinosaurs, but the high dependency on environmental conditions, and the wide range of adult body sizes indicate a more reptilian (plesiomorphic) growth pattern than in most other dinosaurs. Conversely, high growth rates and the deposition of fibro-lamellar bone show that *Plateosaurus* was not simply reptilian in these characters. Like many other dinosaurs, *Plateosaurus engelhardti* combines features of ectothermy and endothermy, and it cannot be easily placed into either of these categories.

Acknowledgements. First and foremost, we are very grateful to R. Wild (SMNS), R. Foelix (NAA), the Saurierkomission Frick (MSF) and H.-U. Pfretzschner (IFG), who allowed us to sample and study the *Plateosaurus* bones from Trossingen and Frick. This work would not have been possible without their confidence in the usefulness of our endeavour. The paper benefited greatly from the reviews by G. Erickson and A. Chinsamy-Turan and the hard editorial work of P. Barrett, and also of D. Batten. We thank O. Dülfer (IPB) for his untiring help in sample processing; he spent much time trying to obtain good thin-sections from the exceptionally brittle Trossingen samples. Many thanks must also go to G. Oleschinski (IPB) for the high-resolution photographs of fracture surfaces, cross-sections, polished sections and thin-sections. D. Kranz (IPB) deserves special thanks for her kind support and advice in using different graphics programs and for her constructive ideas for the layout of the figures. This research was funded by the Graduiertenförderung Nordrhein-Westfalen and the Deutsche Forschungsgemeinschaft. Our paper is a contribution of the DFG Research Unit 533 'Biology of the Sauropod Dinosaurs'.

REFERENCES

ANDREWS, R. M. 1982. Patterns of growth in reptiles. 273–320. *In* GANS, C. and POUGH, F. H. (eds). *Biology of the Reptilia. Volume 13. Physiology D.* Academic Press, London, 360 pp.

BENTON, M. J., JUUL, L., STORRS, G. W. and GALTON, P. M. 2000. Anatomy and systematics of the prosauropod dinosaur *Thecodontosaurus antiquus* from the Upper Triassic of southwest England. *Journal of Vertebrate Paleontology*, **20**, 77–108.

BRENNER, K. and VILLINGER, E. 1981. Stratigraphie und Nomenklatur des süddeutschen Sandsteinkeupers. *Jahrbuch für Geologie. Landesamt Baden-Württemberg*, **23**, 45–86.

CARPENTER, K. 1999. *Eggs, nests, and baby dinosaurs*. Indiana University Press, Bloomington, IN, 336 pp.

CASE, T. J. 1978. Speculations on the growth rate and reproduction of some dinosaurs. *Paleobiology*, **4**, 320–326.

CASTANET, J. 1994. Age estimation and longevity in reptiles. *Gerontology*, **40**, 174–192.

—— and CHEYLAN, M. 1979. Les marques de croissance des os et des ecilles comme indicateur de l'age chez *Testudo hermanni* et *Testudo graeca* (Reptilia, Chelonia, Testudinidae). *Canadian Journal of Zoology*, **57**, 1649–1665.

—— and SMIRINA, E. 1990. Introduction to the skeletochronological method in amphibians and reptiles. *Annales des Sciences Naturelles, Zoologie, Paris, Série 13*, **11**, 191–196.

—— FRANCILLON-VIEILLOT, H., MEUNIER, F. J. and RICQLES, A. DE 1993. Bone and individual aging. 245–283. *In* HALL, B. K. (ed.). *Bone. Volume 7: Bone Growth – B*. CRC Press, Boca Raton, FL, 368 pp.

—— NEWMAN, D. G. and SAINT GIRONS, H. 1988. Skeletochronological data on the growth, age, and population structure of the tuatara, *Sphenodon punctatus*, on Stephens and Lady Alice islands, New Zealand. *Herpetologica*, **44**, 25–37.

CHERRY, C. 2002. Bone histology of the primitive dinosaur, *Thecodontosaurus antiquus*. Unpublished MSc thesis, University of Bristol, 74 pp.

CHINSAMY, A. 1990. Physiological implications of the bone histology of *Syntarsus rhodesiensis* (Saurischia: Theropoda). *Palaeontologia Africana*, **27**, 77–82.

—— 1993*a*. Bone histology and growth trajectory of the prosauropod dinosaur *Massospondylus carinatus* (Owen). *Modern Geology*, **18**, 319–329.

—— 1993*b*. Image analysis and the physiological implications of the vascularisation of femora in archosaurs. *Modern Geology*, **19**, 101–108.

—— 1994. Dinosaur bone histology: implications and inferences. 213–227. *In* ROSENBERG, G. D. and WOLBERG, D. L. (eds). *Dino Fest*. Special Publication of the Paleontological Society, **7**, 504 pp.

—— and DODSON, P. 1995. Inside a dinosaur bone. *American Scientist*, **83**, 174–180.

—— and HILLENIUS, W. J. 2004. Physiology of nonavian dinosaurs. 643–659. *In* WEISHAMPEL, D. B., DODSON, P. and OSMÒSLKA, H. (eds). *The Dinosauria*. Second edition. University of California Press, Berkeley, CA, 861 pp.

CHINSAMY-TURAN, A. 2005. *The microstructure of dinosaur bone*. The Johns Hopkins University Press, Baltimore, MD, 195 pp.

CURREY, J. D. 1962. The histology of the bone of a prosauropod dinosaur. *Palaeontology*, **5**, 238–246.

CURRY, K. A. 1998. Histological quantification of growth rates in *Apatosaurus*. *Journal of Vertebrate Paleontology*, **18** (Supplement to No. 3), 36A.

—— 1999. Ontogenetic histology of *Apatosaurus* (Dinosauria: Sauropoda): new insights on growth rates and longevity. *Journal of Vertebrate Paleontology*, **19**, 654–665.

DUNHAM, A. E., OVERALL, K. L., PORTER, W. P. and FORSTER, C. A. 1989. Implications of ecological energetics and biophysical and developmental constraints for life-history variation in dinosaurs. 1–21. *In* FARLOW, J. O. E. (ed.). *Paleobiology of the dinosaurs*. Special Paper of the Geological Society of America, **238**, 100 pp.

ENLOW, D. H. and BROWN, S. O. 1957. A comparative histological study of fossil and recent bone tissues. Part II. *Texas Journal of Science*, **9**, 186–214.

ERICKSON, G. M. 2005. Assessing dinosaur growth patterns: a microscopic revolution. *Trends in Ecology and Evolution*, **20**, 677–684.

—— and TUMANOVA, T. A. 2000. Growth curve of *Psittacosaurus mongoliensis* Osborn (Ceratopsia: Psittacosauridae) inferred from long bone histology. *Zoological Journal of the Linnean Society*, **130**, 551–566.

—— CURRY-ROGERS, K. and YERBY, S. A. 2001. Dinosaurian growth patterns and rapid avian growth rates. *Nature*, **412**, 429–432.

FERGUSON, M. W. J., HONIG, L. S., BRINGAS, P. Jr and SLAVKIN, H. C. 1982. *In vivo* and *in vitro* development of first branchial arch derivates in *Alligator mississippiensis*. 275–296. *In* DIXON, A. and SARNAT, B. (eds). *Factors and mechanisms influencing bone growth*. Alan R. Liss, New York, NY, 657 pp.

FOELIX, R. 1997. Neue Saurierfunde in Frick. *Aarauer Neujahrsblätter*, **1997**, 40–46.

—— 1999. Röntgen im Museum. *Aargauische Naturforschende Gesellschaft, Mitteilungen*, **35**, 147–160.

FRANCILLON-VIEILLOT, H. and BUFFRÉNIL, V. DE 2001. Ontogenetic changes in bone compactness in male and female Nile monitors (*Varanus niloticus*). *Journal of Zoology, London*, **254**, 539–546.

—— —— CASTANET, J., GÉRAUDIE, J., MEUNIER, F. J., SIRE, J. Y., ZYLBERBERG, L. and RICQLÈS, A. DE 1990. Microstructure and mineralization of vertebrate skeletal tissues. 471–530. *In* CARTER, J. E. (ed.). *Skeletal biomineralization: patterns, processes and evolutionary trends*. Van Nostrand Reinhold, New York, NY, 832 pp.

GALTON, P. M. 1985. Cranial anatomy of the prosauropod dinosaur *Plateosaurus* from the Knollenmergel (Middle Keuper, Upper Triassic) of Germany. Part II: all the cranial material and details of soft-part anatomy. *Geologica et Palaeontologica*, **19**, 119–159.

—— 1986. Prosauropod dinosaur *Plateosaurus* (=*Gressylosaurus*) (Saurischia: Sauropodomorpha) from the Upper Triassic of Switzerland. *Geologica et Palaeontologica*, **20**, 167–183.

—— 1990. Basal Sauropodomorpha–Prosauropoda. 320–344. *In* WEISHAMPEL, D. B., DODSON, P. and OSMÒSLKA, H. (eds). *The Dinosauria*. University of California Press, Berkeley, CA, 733 pp.

—— 1997. Comments on sexual dimorphism in the prosauropod dinosaur *Plateosaurus engelhardti* (Upper Triassic, Trossingen). *Neues Jahrbuch für Geologie und Paläontologie, Monatshefte*, **1997**, 674–682.

—— 2000. The prosauropod dinosaur *Plateosaurus* Meyer, 1837 (Saurischia: Sauropodomorpha). I. The syntypes of *P. engelhardti* Meyer, 1837 (Upper Triassic, Germany), with notes on other European prosauropods with 'distally straight' femora. *Neues Jahrbuch für Geologie und Paläontologie, Abhandlungen*, **216**, 233–275.

—— 2001. The prosauropod dinosaur *Plateosaurus* Meyer, 1837 (Saurischia: Sauropodomorpha; Upper Triassic). II. Notes on the referred species. *Revue de Palèobiology*, **20**, 435–502.

—— and UPCHURCH, P. 2004. Prosauropoda. 232–258. *In* WEISHAMPEL, D. B., DODSON, P. and OSMÒSLKA, H. (eds). *The Dinosauria*. Second edition. University of California Press, Berkeley, CA, 861 pp.

GOW, C. E., KITCHING, J. W. and RAATH, M. A. 1990. Skulls of the prosauropod dinosaur *Massospondylus carinatus* Owen in the collections of the Bernard Price Institute for Palaeontological Research. *Palaeontologia Africana*, **27**, 45–58.

GROSS, W. 1934. Die Typen des mikroskopischen Knochenbaues bei fossilen Stegocephalen und Reptilien. *Zeitschrift der Anatomie und Entwicklungsgeschichte*, **203**, 731–764.

GROSSE, W. R. 1999. Altersbestimmung bei mitteleuropäischen Amphibien mittels Skelettochronologie am Beispiel der Kreuz-, Erd- und Wechselkröte (Anura, Bufonidae). *Elaphe*, **3**, 73–76.

HORNER, J. R. and PADIAN, K. 2004. Age and growth dynamics of *Tyrannosaurus rex*. *Proceedings of the Royal Society of London, Series B*, **271**, 1875–1880.

—— —— and RICQLÈS, A. DE 2001. Comparative osteohistology of some embryonic and perinatal archosaurs: developmental and behavioral implications for dinosaurs. *Paleobiology*, **27**, 39–58.

—— RICQLÈS, A. DE and PADIAN, K. 1999. Variation in dinosaur skeletochronology indicators: implications for age assessment and physiology. *Paleobiology*, **25**, 295–304.

—— —— —— 2000. Long bone histology of the hadrosaurid dinosaur *Maiasaura peeblesorum*: growth dynamics and physiology based on an ontogenetic series of skeletal elements. *Journal of Vertebrate Paleontology*, **20**, 115–129.

HUENE, F. VON 1926. Vollständige Osteologie eines Plateosauriden aus dem Schwäbischen Keuper. *Geologische und Paläontologische Abhandlungen*, **15**, 139–179.

—— 1932. Die fossile Reptil-Ordnung Saurischia, ihre Entwicklung und Geschichte. *Monographie Geologie Paläontologie*, **1**, 1–361.

HURUM, J. H., BERGAN, M., MÜLLER, R., NYSTUEN, J. P. and KLEIN, N. 2006. A Late Triassic dinosaur bone, offshore Norway. *Norwegian Journal of Geology*, **86**, 93–99.

JAEKEL, O. 1913–14. Über die Wirbeltierfunde in der oberen Trias von Halberstadt. *Paläontologische Zeitschrift*, **1**, 155–215.

JARMAN, P. 1983. Mating systems and sexual dimorphism in large, terrestrial, mammalian herbivores. *Biological Reviews*, **58**, 485–520.

KABIGUMILA, J. 2000. Growth and carapacial colour variation of the leopard tortoise, *Geochelone pardalis babcocki*, in northern Tanzania. *African Journal of Ecology*, **38**, 217–223.

KLEVEZAL, G. A. 1996. *Recording structures of mammals: determination of age and reconstruction of life history*. Balkema, Rotterdam, 274 pp.

MAGNUSSON, E. W., VLIET, K., POOLEY, A. C. and WHITAKER, R. 2002. Fortpflanzung. 118–135. *In* ROSS, C. A. (ed.). *Krokodile und Alligatoren*. Orbis Verlag, Verlagsgruppe Falken/Mosaik, Niedernhausen, 239 pp.

MEYER, H. H. VON 1837. Mitteilung an Prof. Bronn (*Plateosaurus engelhardti*). *Neues Jahrbuch für Geologie und Paläontologie*, **1837**, 316.

MOSER, M. 2003. *Plateosaurus engelhardti* Meyer, 1837 (Dinosauria: Sauropodomorpha) aus dem Feuerletten (Mittelkeuper, Obertrias) von Bayern. *Zitteliana*, **B24**, 1–188.

NEILL, W. T. 1971. *The last of the ruling reptiles: alligators, crocodiles, and their kin*. Columbia University Press, New York, NY, and London, 486 pp.

POOLEY, A. C. and GANS, G. 1976. The Nile crocodile. *Scientific American*, **234**, 114–124.

REID, R. E. H. 1990. Zonal 'growth rings' in dinosaurs. *Modern Geology*, **15**, 19–48.

—— 1997*a*. How dinosaurs grew. 403–413. *In* FARLOW, J. O. and BRETT-SURMAN, M. K. (eds). *The complete dinosaur*. Indiana University Press, Bloomington, IN, 752 pp.

—— 1997*b*. Histology of bones and teeth. 329–339. *In* CURRIE, P. J. and PADIAN, K. (eds). *Encyclopedia of dinosaurs*. Academic Press, London, 869 pp.

REISZ, R. R., SCOTT, D., SUES, H.-D., EVANS, D. C. and RAATH, M. A. 2005. Embryos of an Early Jurassic prosauropod dinosaur and their evolutionary significance. *Science*, **309**, 761–764.

REMES, K. 2004. Revision von '*Barosaurus*' *africanus* (Sauropoda, Diplodocinae) aus den Tendaguru-Schichten Tansanias. Unpublished Master's thesis, University of Berlin, 121 pp.

—— 2005. New insights into the origin and evolution of diplodocoid sauropods. *Journal of Vertebrate Paleontology*, **25** (Supplement to No. 3), 104A.

RICQLÈS, A. DE 1968. Recherches paléohistologiques sur les os longs des tetrapodes. I.- Origine du tissu osseux plexiforme des dinosaurians sauropodes. *Annales de Paléontologie*, **54**, 133–145.

—— 1976. Recherches paléohistologiques sur les os longs des tetrapodes. VII. Sur le classification, la signification fonctionelle et l'histoire des tissus osseux des tetrapodes. *Annales de Paléontologie*, **62**, 71–126.

—— 1983. Cyclical growth in the long limb bones of a sauropod dinosaur. *Acta Palaeontologica Polonica*, **28**, 225–232.

—— MEUNIER, F. J., CASTANET, J. and FRANCILLON-VIEILLOT, H. 1991. Comparative microstructure of bone. 1–78. *In* HALL, B. K. (ed.). *Bone. Volume 3: bone matrix and bone specific products*. CRC Press, Boca Raton, FL, 500 pp.

RIMBLOT-BALY, F., RICQLES, A. DE and ZYLBERBERG, L. 1995. Analyse paléohistologique d'une série de croissance partille chez *Lapparentosaurus madagascariensis* (Jurassique Moyen): essai sur la dynamique de croissance d'un dinosaure sauropode. *Annales de Paléontologie (Invert – Vert.)*, **81**, 49–86.

SANDER, P. M. 1992. The Norian *Plateosaurus* bonebeds of central Europe and their taphonomy. *Palaeogeography, Palaeoclimatology, Palaeoecology*, **93**, 255–299.

—— 1999. Life history of the Tendaguru sauropods as inferred from long bone histology. *Mitteilungen aus dem Museum für Naturkunde der Humboldt-Universität Berlin, Geowissenschaftliche Reihe*, **2**, 103–112.

—— 2000. Long bone histology of the Tendaguru sauropods: implications for growth and biology. *Paleobiology*, **26**, 466–488.

—— and ANDRASSY, P. in press. Lines of arrested growth and long bone histology in Pleistocene large mammals from Germany: What do they tell us about dinosaur physiology? *Palaeontographica, Abteilung A.*

—— and KLEIN, N. 2005. Unexpected developmental-plasticity in the life history of an early dinosaur. *Science*, **310**, 1800–1802.

—— and TÜCKMANTEL, C. 2003. Bone lamina thickness, bone apposition rates, and age estimates in sauropod humeri and femora. *Paläontologische Zeitschrift*, **77**, 161–172.

—— —— BUFFETAUT, E., CUNY, G., SUTEETHORN, V. and LOEUFF, J. 2004. Adaptive radiation in sauropod dinosaurs: bone histology indicates rapid evolution of giant body size through acceleration. *Organisms, Diversity and Evolution*, **4**, 165–173.

SCHWEITZER, M. H., WITTMEYER, J. L. and HORNER, J. R. 2005. Gender-specific reproductive tissue in ratites and *Tyrannosaurus rex. Nature*, **308**, 1456–1460.

SEEBACHER, F. 2001. A new method to calculate allometric length-mass relationships of dinosaurs. *Journal of Vertebrate Paleontology*, **21**, 51–60.

SEEMANN, R. 1932. Verlauf und Ergebnis der Trosssinger Sauriergrabung. *Jahreshefte des Vereins für Vaterländische Naturkunde, Wuerttemberg*, **88**, 52–54.

—— 1933. Das Saurischierlager in den Keupermergeln bei Trossingen. *Jahreshefte des Vereins für Vaterländische Naturkunde, Wuerttemberg*, **89**, 129–160.

SEITZ, A. L. 1907. Vergleichende Studien über den makroskopischen Knochenbau fossiler und rezenter Reptilien. *Nova Acta Academiae Caesareae Leopoldino-Carolinae Germanicae Naturae Curiosorum*, **87**, 230–370.

STARCK, J. M. and CHINSAMY, A. 2002. Bone microstructure and developmental plasticity in birds and other dinosaurs. *Journal of Morphology*, **254**, 232–246.

—— and RICKLEFS, R. 1998. *Avian growth and development. Evolution within the altricial-precocial spectrum.* Oxford University Press, Oxford, 456 pp.

TÜTKEN, T. 2004. Paleobiology and skeletochronology of Jurassic dinosaurs: implications from the histology and oxygen isotope composition of bones. *Palaeogeography, Palaeoclimatology, Palaeoecology*, **206**, 217–238.

VAN HEERDEN, J. 1997. Prosauropods. 242–263. *In* FARLOW, J. O. and BRETT-SURMAN, M. K. (eds). *The complete dinosaur.* Indiana University Press, Bloomington, IN, 752 pp.

VARRICCHIO, D. J. 1993. Bone microstructure of the Upper Cretaceous theropod dinosaur *Troodon formosus. Journal of Vertebrate Paleontology*, **13**, 99–104.

WEISHAMPEL, D. B. 1984. Trossingen: E. Fraas, F. v. Huene, R. Seemann and the 'Schwäbische Lindwurm' *Plateosaurus.* 249–253. *In* REIF, W.-E. and WESTPHAL, F. (eds). *Third Symposium on Mesozoic Terrestrial Ecosystems. Short papers.* Attempto Verlag, Tübingen, 259 pp.

—— and CHAPMAN, R. E. 1990. Morphometric study of *Plateosaurus* from Trossingen (Baden-Württemberg, BRD/Germany). 43–51. *In* CARPENTER, K. and CURRIE, P. J. (eds). *Dinosaur systematics: perspectives and approaches.* Cambridge University Press, Cambridge, 318 pp.

—— and WESTPHAL, F. 1986. *Die Plateosaurier von Trossingen.* Attempto Verlag, Tübingen, 27 pp.

WELLNHOFER, P. 1993. Prosauropod dinosaurs from the Feuerletten (Middle Norian) of Ellingen near Weissenburg in Bavaria. *Revue de Paléobiologie, Volume Special*, **7**, 263–271.

WIKELSKI, M. and THOM, C. 2000. Marine iguanas shrink to survive 'El Nino'. *Nature*, **403**, 37.

YATES, A. 2003. The species taxonomy of the sauropodomorph dinosaurs from the Löwenstein Formation (Norian, Late Triassic) of Germany. *Palaeontology*, **46**, 317–337.

—— 2004. *Anchisaurus polyzelus* (Hitchcock): the smallest known sauropod dinosaur and the evolution of gigantism among sauropodomorph dinosaur. *Postilla*, **230**, 1–58.

ZIEGLER, B. 1986. *Der Schwäbische Lindwurm. Funde aus der Urzeit im Museum am Löwentor.* Konrad Theiss Verlag, Stuttgart, 171 pp.

—— 1988. Führer durch das Museum am Löwentor. *Stuttgarter Beiträge zur Naturkunde, Serie C*, **27**, 1–100.

ZUG, G. R. P., WYNN, A. H. and RUCKDESCHEL, C. 1986. Age determination of loggerhead sea turtle, *Caretta caretta*, by incremental growth marks in the skeleton. *Smithsonian Contributions to Zoology*, **427**, 1–34.

[Special Papers in Palaeontology 77, 2007, pp. 207–222]

WHAT PNEUMATICITY TELLS US ABOUT 'PROSAUROPODS', AND VICE VERSA

by MATHEW WEDEL

University of California Museum of Paleontology and Department of Integrative Biology, 1101 Valley Life Sciences Building, Berkeley, CA 94720-4780, USA; e-mail: sauropod@berkeley.edu

Typescript received 15 February 2006; accepted in revised form 24 October 2006

Abstract: Pneumatic (air-filled) bones are an important feature of the postcranial skeleton in pterosaurs, theropods and sauropods. However, there is no unambiguous evidence for postcranial pneumaticity in basal sauropodomorphs and even the ambiguous evidence is scant. Patterns of skeletal pneumatization in early sauropods and theropods suggest that basal saurischians had cervical air sacs like those of birds. Furthermore, patterns of pneumaticity in most pterosaurs, theropods and sauropods are diagnostic for abdominal air sacs. The air sacs necessary for flow-through lung ventilation like that of birds may have evolved once (at the base of Ornithodira), twice (independently in pterosaurs and saurischians) or three times (independently in pterosaurs, theropods and sauropods). Skeletal pneumaticity appears to be more evolutionarily malleable than the air sacs and diverticula that produce it. The evolution of air sacs probably pre-dated the appearance of skeletal pneumaticity in ornithodirans.

Key words: Prosauropoda, Sauropodomorpha, Saurischia, Ornithodira, pneumaticity, air sacs, diverticula.

PNEUMATICITY is a prominent feature of the postcranial skeleton in theropod and sauropod dinosaurs. In contrast, there is little evidence for postcranial pneumaticity in basal sauropodomorphs (informally referred to as 'prosauropods' in this paper), although from time to time some aspects of prosauropod osteology have been posited as evidence of pneumaticity (Britt 1997) or compared with unequivocal pneumatic structures in sauropods (Yates 2003; Galton and Upchurch 2004). My goals in this paper are to review the evidence for postcranial skeletal pneumaticity (PSP) in prosauropods and to discuss the origin of air sacs and pneumaticity in early dinosaurs and their relatives.

Prosauropod taxonomy is currently in a state of flux, as other papers in this volume attest (Sereno 2007; Upchurch *et al.* 2007; Yates 2007). Prosauropods were traditionally considered a paraphyletic assemblage that gave rise to sauropods. Sereno (1998) recovered a monophyletic Prosauropoda, defined this clade (anchored upon *Plateosaurus*) as a monophyletic sister taxon to Sauropoda and united the two in a node-based Sauropodomorpha. A similar phylogenetic hypothesis was described by Galton and Upchurch (2004). However, other recent phylogenetic analyses (Yates 2003, 2004; Yates and Kitching 2003) have found that some prosauropods are closer to *Saltasaurus* than to *Plateosaurus*; thus, under current phylogenetic definitions they should be regarded as basal sauropods. Some other taxa (e.g. *Saturnalia*) lie outside Sauropodomorpha as defined by Sereno (1998) altogether (Yates 2003, 2004; Yates and Kitching 2003; Langer 2004).

However, the monophyly or paraphyly of the group of taxa traditionally called prosauropods is not critical to the purposes of this paper. What is important is that all traditional 'prosauropods' have two things in common: they lack unequivocal evidence of pneumatic cavities in their vertebrae and ribs, and they are phylogenetically bracketed by sauropods and theropods (Text-fig. 1).

Institutional abbreviations. BMNH, The Natural History Museum, London; CM, Carnegie Museum of Natural History, Pittsburgh, USA; FMNH, Field Museum of Natural History, Chicago, USA; MSM, Mesa Southwest Museum, Mesa, USA; OMNH, Oklahoma Museum of Natural History, Norman, USA; SMNS, Staatliches Museum fur Natürkunde, Stuttgart, Germany.

Anatomical abbreviations. ACDL, anterior centrodiapophyseal lamina; AL, accessory lamina; AVF, anteroventral fossa; NAF, neural arch fossa; PCDL, posterior centrodiapophyseal lamina; PDF, posterodorsal fossa; PODL, postzygodiapophyseal lamina; PPDL, paradiapophyseal lamina; PRDL, prezygodiapophyseal lamina; SPOL, spinopostzygapophyseal lamina; SPRL, spinoprezygapophyseal lamina (lamina abbreviations after Wilson 1999).

POSTCRANIAL PNEUMATICITY IN THEROPOD AND SAUROPOD DINOSAURS

Before examining the evidence for PSP in 'prosauropods', I will review the conditions present in other saurischian

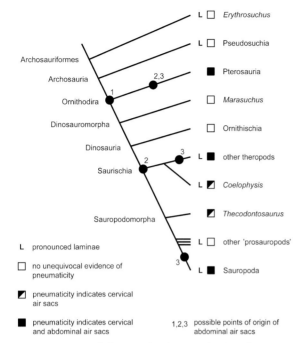

TEXT-FIG. 1. A phylogeny of archosaurs showing the evolution of postcranial skeletal pneumaticity and air sacs. *Thecodontosaurus* is shown as having limited postcranial pneumaticity. The evidence for this is ambiguous; see text for discussion. Based on the phylogenetic framework of Brochu (2001) and Yates (2003).

skin. Where a diverticulum comes into contact with a bone, it may (but does not always) induce bone resorption, which can produce pneumatic tracks, fossae or foramina. If resorption of the cortex produces a foramen, the diverticulum may enter the medullary space and replace the existing internal structure with a series of air-filled chambers of varying complexity. The best description of this process is provided by Bremer (1940).

The extent of PSP varies in different avian clades. Almost any postcranial bones can become pneumatized; in large soaring birds such as pelicans, almost the entire skeleton is pneumatic, including the distal limb elements (O'Connor 2004). Although many large volant and flightless birds have highly pneumatic skeletons, the correlation between body size and the extent of PSP in birds is weak (O'Connor 2004). PSP tends to be reduced or absent in diving birds (Gier 1952; O'Connor 2004). Different parts of the skeleton become pneumatized by diverticula of different air sacs in extant birds (Table 1); this is important because it allows us to make inferences regarding the evolution of air sacs in fossil taxa. PSP in non-avian theropods generally follows the avian model (Britt 1993, 1997; O'Connor and Claessens 2005; O'Connor 2006). Patterns of pneumatization along the vertebral column indicate that both anterior and posterior air sacs (presumably cervical and abdominal) had evolved by the time of the ceratosaur-tetanuran divergence (O'Connor and Claessens 2005).

Fossae are present in the presacral vertebrae of basal sauropods such as *Shunosaurus* and *Barapasaurus* (Britt 1993; Wilson and Sereno 1998). These fossae are similar to the unequivocally pneumatic foramina and camerae of more derived sauropods, both in their position on individual vertebrae and in their distribution along the vertebral column, and because of these similarities they have usually been regarded as pneumatic in origin (Britt 1993, 1997; Wedel 2003a). However, similar fossae are present in other tetrapods that lack PSP, so the presence of fossae alone is at best equivocal evidence for PSP (O'Connor 2006; see below). The vertebrae of more derived sauropods have foramina that communicate with large internal

dinosaurs. The sister taxon of Sauropodomorpha is Theropoda; consequently, 'prosauropods' are phylogenetically bracketed in part by birds, the only clade of extant vertebrates with extensive PSP. The relationship between the respiratory system and pneumatic postcranial bones in birds has been described many times (e.g. Müller 1908; King 1966; Duncker 1971; O'Connor 2004), and is briefly summarized here. The relatively small, constant-volume, unidirectional flow-through lungs of birds are ventilated by the attached air sacs, which are large, flexible and devoid of parenchymal tissue. The lungs and air sacs also produce air-filled tubes called diverticula that pass between the viscera, between the muscles, and under the

TABLE 1. Parts of the postcranial skeleton that are pneumatized by diverticula of different parts of the respiratory system in extant birds. Pneumaticity varies widely within populations and clades, and not all elements are pneumatized in all taxa (based on Duncker 1971 and O'Connor 2004).

Respiratory structure	Skeletal elements
Lung (parenchymal portion)	Adjacent thoracic vertebrae and ribs
Clavicular air sac	Sternum, sternal ribs, pectoral girdle and humerus
Cervical air sac	Cervical and anterior thoracic vertebrae and associated ribs
Anterior thoracic air sac	Sternal ribs
Posterior thoracic air sac	(none reported)
Abdominal air sac	Posterior thoracic, synsacral and caudal vertebrae, pelvic girdle and femur
Subcutaneous diverticula	Distal limb elements

chambers; the combination of foramina and large internal chambers is an unambiguous indicator of PSP (O'Connor 2006). There is a general trend in sauropod evolution for PSP to spread posteriorly along the vertebral column, albeit to different extents in different clades and with a few reversals (Wedel 2003*b*; Text-fig. 2). In both sauropods and theropods, fossae in basal forms were replaced by large-chambered (camerate) vertebrae and eventually small-chambered (camellate) vertebrae in more derived taxa (Britt 1993, 1997; Wedel 2003*a*).

The evolution of PSP in sauropods mirrors in detail that of non-avian theropods. At the level of individual elements (e.g. vertebrae and ribs), pneumatic features in sauropods compare very closely with those of both avian and non-avian theropods (Text-fig. 3). In terms of the ratio of bony tissue to air space within a pneumatic element, sauropod vertebrae are, on average, comparable with the limb bones of many extant birds: about 60 per cent air by volume (Wedel 2004, 2005; Woodward 2005; Schwarz and Fritsch 2006). At the level of the skeleton, osteological indicators of pneumaticity spread as far back as the mid-caudal vertebrae in at least two groups of sauropods, the diplodocines and saltasaurines (Osborn 1899; Powell 1992). Among non-avian theropods, extensive pneumatization of the caudal series evolved only in oviraptorosaurs (Osmólska *et al.* 2004). Finally, limited appendicular pneumaticity was probably present in both sauropods and non-avian theropods. The dromaeosaur *Buitreraptor* has a pneumatic furcula (Makovicky *et al.* 2005), and a large foramen in the proximal femur of the oviraptorid *Shixinggia* is probably also pneumatic in origin (Lü and Zhang 2005). Large chambers have been reported in the ilia of the basal diplodocoid *Amazonsaurus* (Carvalho *et al.* 2003) and in several titanosaurs (Powell 1992; Sanz *et al.* 1999; Xu *et al.* 2006). Although these chambers are similar to unequivocally pneumatic spaces in the other saurischians, it has not yet been

shown that the ilial chambers are connected to foramina, which are necessary for pneumatization to occur (see O'Connor 2006).

EVIDENCE OF PNEUMATICITY IN 'PROSAUROPODS'

Historically, postcranial pneumaticity in 'prosauropods' has received little attention, which is to be expected given the paucity of available evidence. Janensch (1947) posited that a foramen in a dorsal vertebra of *Plateosaurus* might have been pneumatic, but he attached no great weight to this hypothesis. Britt (1997) considered vertebral laminae evidence of pneumaticity in 'prosauropods'. Most recently, Yates (2003, p. 14, fig. 12) identified 'pleurocoel-like pits' in the mid-cervical vertebrae of *Thecodontosaurus caducus*, and Galton and Upchurch (2004, p. 245) referred to fossae in the posterior dorsals of some prosauropods as 'pleurocoelar indentations'. The 'pleurocoel-like' structures were not explicitly described as pneumatic in either work. Although fossae are not unambiguous indicators of pneumaticity (O'Connor 2006), vertebral fossae seem to be an early step toward full pneumatization, both ontogenetically and phylogenetically (Wedel 2003*a*). Putative pneumatic characters in 'prosauropods' can be divided into three categories: vertebral laminae, foramina and fossae, which will be discussed in this order, below.

Vertebral laminae. Vertebral laminae are struts or plates of bone that connect the various apophyses of a vertebra to each other and to the centrum. The landmarks that are usually connected in this way are the pre- and postzygapophyses, the diapophyses and parapophyses, and the neurapophysis. The form and occurrence of the major laminae in saurischian dinosaurs were reviewed by Wilson

TEXT-FIG. 2. A diagram showing the distribution of fossae and pneumatic chambers (black boxes) along the vertebral column in sauropods. Only the lineage leading to diplodocines is shown here. The same caudal extension of pneumatic features also occurred independently in macronarian sauropods, culminating in saltasaurines, and several times in theropods. The format of the diagram is based on Wilson and Sereno (1998, fig. 47). Phylogeny based on Wilson (2002), Yates (2003) and Upchurch *et al.* (2004).

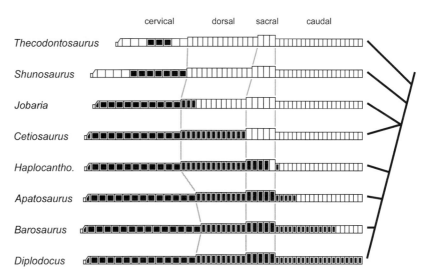

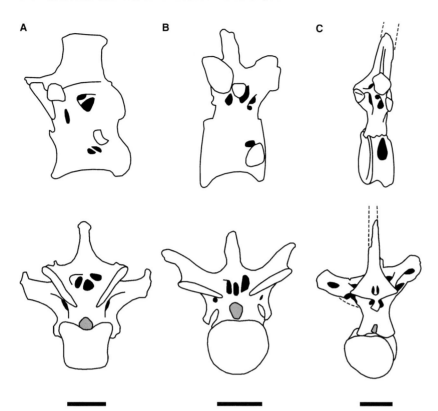

TEXT-FIG. 3. Pneumatic foramina (black) in thoracic or dorsal vertebrae of an extant bird, a non-avian theropod and a sauropod, in right lateral view (above) and posterior view (below). A, a crane, *Grus*. B, an abelisaurid, *Majungatholus*. C, a diplodocid, *Apatosaurus*. A and B traced from O'Connor and Claessens (2005, fig. 3); C traced from a photograph of OMNH 1382. Scale bars represent 1 cm in A, 3 cm in B and 20 cm in C.

(1999). In addition to a basic set of laminae common to all saurischians, many sauropods and theropods have other irregularly developed laminae that are usually not named but are collectively called accessory laminae. Laminae tend to be more numerous and more sharply defined in camerate than camellate vertebrae (Wilson and Sereno 1998; Wedel 2003*a*). Camellate vertebrae evolved relatively early in the radiation of non-avian theropods (Britt 1993, 1997), and most derived theropods have less elaborate systems of laminae than neosauropods. This may explain why the literature on laminae has tended to focus on sauropods (e.g. Osborn 1899; Osborn and Mook 1921; Janensch 1929, 1950; Wilson 1999).

Two problems with the identification of laminae that are relevant to the question of pneumaticity are how well developed a ridge of bone must be before we call it a lamina, and whether laminae are primarily additive structures formed by the deposition of new bone, or are simply bone that is left over following the formation of fossae. The first problem is important because, as shown below, incipient laminae are broadly distributed among archosaurs. To what extent are the distinct laminae of saurischian dinosaurs new (= apomorphic) structures, rather than modifications of pre-existing ones? This question has ramifications for the evolution of laminae and for coding of laminae as characters in phylogenetic analyses.

The second question can be stated: do laminae grow out from the corpus of the vertebra to define the fossae that they bound, or do we only recognize laminae as distinct structures because the bone between them has been removed? For example, the cervical vertebra of *Nigersaurus* illustrated by Sereno and Wilson (2005, fig. 5.8) has on the lateral face of the neural spine two fossae divided by an accessory lamina (Text-fig. 4). At its edges, the anteroventral fossa approaches both the prezygapophysis and the diapophysis. This region is flat or convex in most other neosauropods, which have a lateral fossa in roughly the same position as the posterodorsal fossa in *Nigersaurus*. It seems likely therefore that the anteroventral fossa in *Nigersaurus* is a new morphological feature, and that the accessory lamina can only be recognized as a lamina because a fossa has been excavated below it. Conversely, the vertebrae of most tetrapods do not have straight bars of bone that connect the zygapophyses to the neurapophysis, but this is exactly what the spinopre- and spinopostzygapophyseal laminae of some sauropods do (Text-fig. 4). In comparison with the condition in other tetrapods, including prosauropods, these laminae appear to be additive structures. These potentially opposing processes of lamina formation should be kept in mind while reading the following descriptions.

The laminae of sauropods often form the boundaries of fossae that have been interpreted as pneumatic, either

TEXT-FIG. 4. Laminae, fossae and foramina in cervical vertebrae of *Nigersaurus* and *Apatosaurus*. A, fifth cervical vertebra of *Nigersaurus*, traced from Sereno and Wilson (2005, fig. 5.8). B, tenth cervical vertebra of *Apatosaurus*, traced from Gilmore (1936, pl. 24). Scale bars represent 5 cm in A and 20 cm in B.

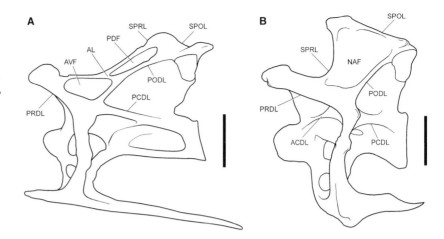

because they contain foramina that lead to internal chambers or because they are heavily sculpted, with numerous subfossae (*sensu* Wilson 1999) and a distinct bony texture (although texture alone is not necessarily a good indicator of pneumaticity; see O'Connor 2006). Wilson (1999) considered whether sauropod laminae existed to provide mechanical support or to subdivide pneumatic diverticula, and concluded that they probably served both functions simultaneously. Following from the aforementioned discussion, we might also ask if sauropod laminae exist because the pneumatic diverticula are subdivided, as they often are in birds (e.g. Wedel 2003*b*, fig. 2), and these subdivisions are impressed into the bone, leaving laminae between them. Rather than try to determine which structure has morphogenetic precedence, it may be more useful to view sauropod vertebrae in light of Witmer's (1997) hypothesis that the form of a pneumatic bone can be viewed as the outcome of a struggle between bone tissue, which grows partly in response to biomechanical stress, and pneumatic diverticula, which are opportunistic and invasive and spread wherever possible (see Sadler *et al.* 1996 and Anorbe *et al.* 2000 for examples of proliferating diverticula).

The laminae of 'prosauropods' differ from those of sauropods in three important ways. The first is that prosauropods have fewer laminae. The laminae that connect the diapophysis to the centrum, parapophysis and zygapophyses are usually present (Wilson 1999), but those that connect the neurapophysis to other landmarks are absent (Text-fig. 5; but see Bonaparte 1999, figs 13–16 on *Lessemsaurus*). The second is that laminae are confined to the presacral vertebrae in 'prosauropods', whereas the sacral vertebrae of neosauropods and the caudal vertebrae of diplodocids also bear laminae.

The third and most important difference between the laminae of sauropods and 'prosauropods' is that the fossae bounded by the latter are blind. These fossae do not contain foramina or subfossae and they do not have a distinctive bone texture. Consequently, there is no strong reason to suspect that they contained pneumatic diverticula. O'Connor (2006) found that similar fossae in extant crocodilians and birds may contain cartilage or adipose tissue. Considering whether the laminae are additive structures or remnants of fossa formation sheds little light on the problem. Some laminae, such as the PRDLs of *Plateosaurus* cervicals, are straight-line structures that

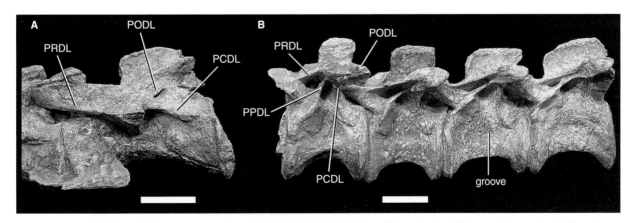

TEXT-FIG. 5. Vertebrae of *Plateosaurus trossingensis* (SMNS 13200) in left lateral view. A, the eighth cervical vertebra. B, dorsal vertebrae 1–4. Scale bars represent 5 cm.

appear to have been added, compared with the condition in vertebrae that lack laminae (Text-fig. 5). Others, such as the PODLs in the same vertebrae, are only detectable because they have been undercut by a fossa. The form of the fossae themselves provides no obvious clues to their contents *in vivo*.

Laminae like those of 'prosauropods' occur in many other archosaurs. Desojo *et al.* (2002) and Parker (2003) recognized that many of the laminae described by Wilson (1999) for saurischian dinosaurs are also present in basal archosaurs and pseudosuchians. The full complement of diapophyseal laminae is present in dorsal vertebrae of the basal archosauriform *Erythrosuchus* and in those of poposaurs such as *Sillosuchus* and *Arizonasaurus*, including the PCDL, PODL, PPDL and PRDL (Text-fig. 6; see Alcober and Parrish 1997; Nesbitt 2005). At least in *Erythrosuchus*, the fossae bounded by these laminae contain aggregates of vascular foramina; obvious foramina like these are not present or at least not common in the interlaminar fossae of 'prosauropods' (pers. obs.). Incipient laminae are also present in some neosuchian crocodyliforms. Most dorsal vertebrae of *Goniopholis stovalli* have rudimentary PCDLs and PODLs (Text-fig. 7). The PODL is bounded dorsally by a shallow fossa on the lateral face of the neural spine and ventrally by a deep infrapostzygapophyseal fossa. In at least some of the vertebrae, the fossa on the side of the neural spine has a distinct margin (Text-fig. 7B). Although most neosuchian crocodyliforms have extensive skull pneumatization (Witmer 1997; Tykoski *et al.* 2002), PSP is absent in the clade (O'Connor 2006).

Vertebral laminae also occur in non-amniotes. The best example is probably the plethodontid salamander *Aneides lugubris*, in which plate-like shelves of bone connect the parapophyses of dorsal vertebrae to the ventrolateral margins of the centra (Wake 1963). These shelves of bone are absent in other species of *Aneides* and in other plethodontid genera, and they are thus additive structures that are apomorphic for *A. lugubris* (compare Wake 1963, fig. 9 with Wake and Lawson 1973, fig. 6). *A. lugubris* has the most prolonged ontogeny of any plethodontid, and it is peramorphic relative to other species in the genus, with a more extensively ossified skeleton (Wake 1963; Wake *et al.* 1983). The development of laminae in the species is probably an epiphenomenon of the extensive ossification of the skeleton, which in turn is related to adaptations for arboreality and feeding (Larson *et al.* 1981). As such, the laminae of *A. lugubris* are not homologous with those of archosaurs in a taxic sense, and they are probably produced by different developmental processes. Still, *A. lugubris* demonstrates that laminae can evolve in vertebrates that are far removed from basal dinosaurs in both genealogy and body size, and it provides a potential system in which to investigate the development of laminae in an extant tetrapod.

On one hand, it is possible that the laminae and fossae of basal archosauriform and pseudosuchian vertebrae appear pneumatic because they are pneumatic (Gower 2001), but the morphology is not compelling. Like the fossae of 'prosauropod' neural spines, those of *Erythrosuchus* and *Arizonasaurus* lack subfossae, foramina that lead to large internal chambers, or altered texture. The presence of similar features in crocodyliforms and salamanders is strong evidence that the morphologies in question can be produced in the absence of pneumaticity. Unlike 'prosauropods', basal archosauriforms and pseudosuchians are not bracketed by taxa with unequivocal evidence of pneumaticity, so inferring that they had pneumatic vertebrae would require pulmonary diverticula and possibly also air sacs to have evolved much earlier than otherwise supposed (see 'Palaeobiological implications' below).

Regardless of when the capacity for PSP evolved, the laminae of 'prosauropods' bound fossae that are not unequivocally pneumatic. Similar laminae are present in crocodilians, a group in which postcranial pneumaticity is entirely absent. Some of the osteological traces of diverticula are subtle, and the possibility that the neural arch fossae of 'prosauropods' accommodated pneumatic diverticula cannot be ruled out, but there is no strong evidence for it.

Foramina. The only putative pneumatic foramen in a 'prosauropod' is that described by Janensch (1947) in a cervical vertebra of *Plateosaurus*. Janensch argued that the size (11×4 mm) and form of the foramen were more consistent with a pneumatic than a vascular interpretation. The identity of the foramen can only be settled by first-hand observation, preferably with a computed tomographic (CT) scan to determine if the foramen leads to any large internal chambers. Unfortunately, such an examination has yet to be conducted. However, the caudal vertebrae of some whales are similar in size to *Plateosaurus* dorsal vertebrae (*c.* 20 cm in maximum linear dimension) and have vascular foramina up to 30 mm in diameter (pers. obs.), so large foramina do not necessarily indicate the presence of pneumaticity. In general, the prominent foramina and internal chambers that are typical of sauropod vertebrae are absent in the vertebrae of 'prosauropods'.

Fossae. The first step in recognizing pneumatic fossae is to distinguish between vertebrae that have distinct fossae and those that are merely waisted (narrower in the middle than at the ends). The vertebrae of most vertebrates are waisted to some extent. In humans the effect is barely noticeable, but in some archosaurs the 'waist' of the vertebra may be only half the diameter of the ends of the centrum (e.g. Nesbitt 2005, fig. 16). Some degree of waisting is to be expected based on the early development

TEXT-FIG. 6. Dorsal vertebrae of *Erythrosuchus africanus* (BMNH R533). A, right lateral and B, ventrolateral views. Scale bar represents 5 cm.

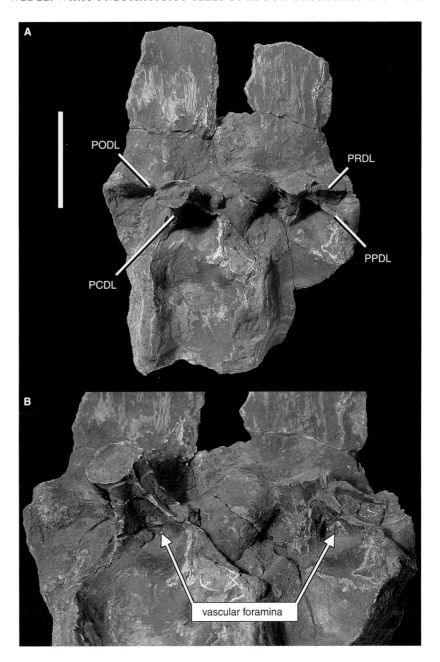

of vertebrae. Cell-dense regions of the embryonic axial column become intervertebral discs, and lower-density regions become vertebral bodies. This produces centra that are inherently waisted (Verbout 1985, pl. 10; Wake 1992, figs 6.5, 6.7). The degree of waisting has occasionally been used as a taxonomic character (Case 1907), but to date there is no clear explanation of why some vertebrae are more waisted than others. Regardless, vertebral waisting is widespread in vertebrates and is not evidence for pneumaticity.

Waisting aside, fossae still suffer from a problem of definition. Consider a spectrum of morphological possibilities (Text-fig. 8). At one end is a vertebra that is waisted but lacks distinct fossae: for example, a thoracic vertebra of an artiodactyl. At the other end is a vertebra with large foramina that open into internal chambers, such as a dorsal vertebra of *Saltasaurus*. The 'chamber morphospace' between these endpoints is filled with a continuum of deeper and more distinct fossae and camerae. Adjacent to the artiodactyl vertebra we might put a vertebra that has fossae with a distinct margin on one side but not the other, like those in the cervical centra of *Arizonasaurus*, which are bounded dorsally by the PCDL; next, a fossa that has a distinct bony rim on all sides, but that is not enclosed by a bony lip, like those in dorsal centra of adult *Barapasaurus* or juvenile *Apatosaurus*. The

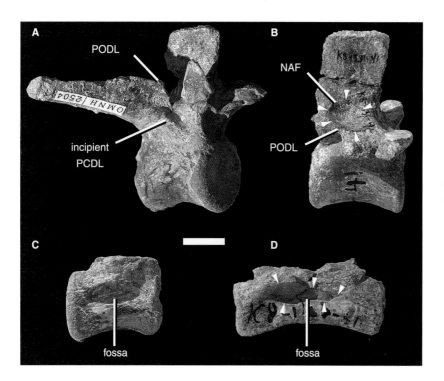

TEXT-FIG. 7. Dorsal and caudal vertebrae of *Goniopholis stovalli*. These vertebrae are part of an associated collection of several individuals from the type locality of the species. A, a dorsal vertebra (OMNH 2504) in left posterolateral view. B, a dorsal vertebra (OMNH 2470) in right lateral view. C, a middle caudal centrum (OMNH 2448) in right lateral view. D, a distal caudal centrum (OMNH 2454) in left lateral view. White arrows in B and D highlight the margins of fossae. Scale bar represents 1 cm.

penultimate example is a fossa that is enclosed by a bony lip, but that is little expanded beyond the boundaries of the opening, such as the fossae in presacral centra of *Haplocanthosaurus* [Britt (1993) referred to these chambers as camerae, whereas Wedel termed them fossae (Wedel *et al.* 2000; Wedel 2003*a*). The morphology of these features is intermediate between that of fossae and camerae, and either term could reasonably be applied]. Finally, in neosauropods such as *Camarasaurus* and *Saltasaurus* the space beyond the bony lip is greatly expanded, so that the result is a foramen that leads to camerae or camellae.

The fossae along this spectrum vary in geometry and they are not all pneumatic. Although *Goniopholis* is extinct and not part of the crown-group Crocodylia, it is highly unlikely that the caudal vertebrae of this semi-aquatic neosuchian were pneumatic. Nevertheless, they bear lateral fossae with distinct margins that are very similar to structures that are sometimes interpreted as pneumatic in dinosaurs, such as the sacral 'pleurocoels' of ornithomimosaurs. However, distinct margins alone are not compelling evidence of pneumaticity. Conversely, truly pneumatic fossae need not have distinct margins. For example, the fossae behind the prezygapophyses of ratites lack clear margins, but CT scans show that they house pneumatic diverticula, and they sometimes contain pneumatic foramina (Text-fig. 9). In extant birds, the pneumatic canalis intertransversarius lies alongside the centrum (Müller 1908), but many birds have cervical centra that are laterally convex and lack any fossae (the

pneumatic foramina are usually located inside the cervical rib loop or ansa costotransversaria).

The foregoing discussion implies that where chambers lack a distinct lip of bone, geometry alone is a poor clue to whether or not a given fossa has a pneumatic origin. Other lines of evidence must be used, such as position in the body, the presence or absence of adjacent pneumatic foramina, subfossae, or textural differences (and even the last two may be misleading; see O'Connor 2006).

Vertebral centra of 'prosauropods' can be quite narrow-waisted, and some have lateral grooves or fossae that are bounded on one side by a lamina. As with neural arch laminae, these features are sometimes associated with pneumaticity but they are not diagnostic for it. The 'pleurocoelar indentations' mentioned by Galton and Upchurch (2004) do not have a distinct boundary or lip in any of the specimens that I have examined (e.g. Moser 2003, pl. 16). The only known 'prosauropod' with distinctly emarginated lateral fossae is *Thecodontosaurus caducus* (Yates 2003). Cervical vertebrae 6–8 of BMNH P24, the holotype of *T. caducus*, have small, distinct fossae just behind the diapophyses (Text-fig. 10). The fossae are high on the centra and may have crossed the neurocentral sutures, which are open. The fossa on the eighth cervical looks darker than it should because it is coated with glue. The ninth cervical has a very shallow, teardrop-shaped hollow in the same region of the centrum. The bone texture in this hollow is noticeably smoother than on the rest of the centrum (this is especially apparent under low-angle lighting). That the fossa on the ninth

TEXT-FIG. 8. Diagram showing the evolution of fossae and pneumatic chambers in sauropodomorphs and their outgroups. Vertebrae are shown in left lateral view with lines marking the position of the cross-sections, and are not to scale. The omission of 'prosauropods' from the figure is deliberate; they have no relevant apomorphic characters and their vertebrae tend to resemble those of many non-dinosaurian archosaurs. Cross-sections are based on first-hand observation (*Giraffa* and *Arizonasaurus*), published sections (*Barapasaurus*, *Camarasaurus* and *Saltasaurus*) or CT scans (*Apatosaurus* and *Haplocanthosaurus*). *Giraffa* based on FMNH 34426. *Arizonasaurus* based on MSM 4590 and Nesbitt (2005, fig. 17). *Barapasaurus* based on Jain *et al.* (1979, pls 101–102). *Apatosaurus* based on CM 11339. *Haplocanthosaurus* based on CM 572. *Camarasaurus* based on Ostrom and McIntosh (1966, pl. 24). *Saltasaurus* modified from Powell (1992, fig. 16).

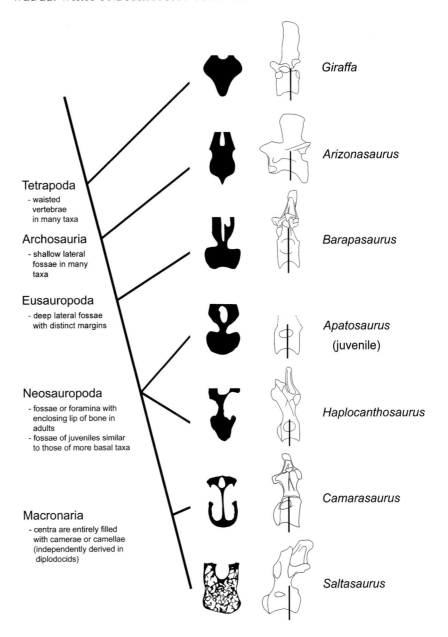

Tetrapoda
- waisted vertebrae in many taxa

Archosauria
- shallow lateral fossae in many taxa

Eusauropoda
- deep lateral fossae with distinct margins

Neosauropoda
- fossae or foramina with enclosing lip of bone in adults
- fossae of juveniles similar to those of more basal taxa

Macronaria
- centra are entirely filled with camerae or camellae (independently derived in diplodocids)

Giraffa

Arizonasaurus

Barapasaurus

Apatosaurus (juvenile)

Haplocanthosaurus

Camarasaurus

Saltasaurus

cervical is shallower and less distinct than those on cervicals 6–8 is reminiscent of the diminution of pneumatic features observed at the transition from pneumatic to apneumatic vertebrae, as seen in the anterior dorsal vertebrae of *Jobaria* (Sereno *et al.* 1999, fig. 3) and the middle caudal vertebrae of *Diplodocus* (Osborn 1899, fig. 13). The holotype specimen of *T. caducus* represents an immature individual (Yates 2003), however, so the shallow fossa on the ninth cervical may be incompletely developed.

Are the fossae of *T. caducus* pneumatic? If so, they are the only good evidence for invasive pneumatic features in the postcrania of 'prosauropods'. Previously, I have assumed that they were pneumatic, based in part on the

distinct margins of the fossae in cervicals 6–8, and also on the fact that the fossae only occur on cervicals 6–9 (Wedel 2006). The first line of evidence is inadequate to diagnose pneumaticity unequivocally. The second is also problematic. Cervical vertebrae 5–9 are the only ones that are always pneumatized in the chicken (Hogg 1984a), and the cervical and anterior thoracic vertebrae are the first parts of the axial skeleton to be pneumatized during the ontogeny of birds (Cover 1953; Hogg 1984b). The spread of pneumaticity posteriorly along the vertebral column in the ontogeny of birds appears faithfully to recapitulate the evolution of pneumaticity in theropods and sauropods (Wedel 2003b, 2005). The presence of fossae on the midcervical vertebrae of *T. caducus* is easily explained if

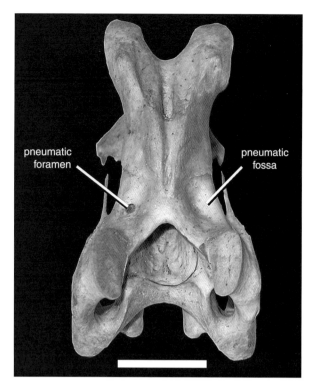

TEXT-FIG. 9. An uncatalogued cervical vertebra of an emu (*Dromaius novaehollandiae*) from the OMNH comparative collection. Scale bar represents 2 cm.

the fossae are pneumatic; their appearance in that part of the skeleton mirrors early ontogeny in birds and is also consistent with later trends in the evolution of PSP in sauropodomorphs (Text-fig. 2). On all four vertebrae, the fossae are not closely associated with laminae and cannot be dismissed as epiphenomena of lamina formation (see O'Connor 2006); a specific soft-tissue influence was causally related to the formation of the fossae. The geometry of the fossae is not sufficient to specify that soft-tissue influence because adipose, muscular and pulmonary tissues have all been found to occupy similar fossae in other tetrapods (O'Connor 2006). On the other hand, the presence of the fossae only on the midcervical vertebrae is difficult to explain if they were not produced by pneumatic diverticula like those of more derived sauropods.

Summary. Vertebral laminae and shallow depressions on the centra are widespread in archosauriforms and not diagnostic of pneumaticity, although it is difficult to rule out the possibility that they may have been associated with pneumatic diverticula. 'Prosauropods' have fewer laminae than most sauropods, fewer vertebrae with laminae, and the fossae adjacent to the laminae are almost always blind (with no large foramina or chambers). A foramen in a vertebra of *Plateosaurus* and distinct fossae in the cervical vertebrae of *Thecodontosaurus caducus* are

the best evidence for potential pneumaticity in 'prosauropods', but neither is an unambiguous indicator of PSP and both would benefit from further study. In any case, the diagnostic osteological correlatives of pneumaticity that are common in sauropods and theropods are absent or extremely rare in 'prosauropods', and the putative pneumatic features that are widespread in 'prosauropods' (laminae and shallow fossae) are not compelling evidence of pneumaticity. To leave aside for a moment the question of 'prosauropod' monophyly, 'prosauropods' are unusual as the only sizeable group (or grade) of saurischian dinosaurs that lack extensive PSP.

PALAEOBIOLOGICAL IMPLICATIONS

Pneumatic bones are of palaeobiological interest in two ways. We may be interested in the bones themselves: in their external and internal morphology, in the ratio of bone to air space, and in the ways that they develop. They are also important, arguably more important, as osteological markers of the pulmonary system. In this section I discuss the origins of pneumaticity and of air sacs, and the implications for the respiratory physiology of sauropodomorphs.

Origin of the diverticular lung and PSP. The first part of the postcranial skeleton to be pneumatized in any saurischian dinosaur is the cervical column. The fossae in the mid-cervical vertebrae of *Thecodontosaurus caducus* are not definitely pneumatic on the basis of geometry alone. However, their placement in the skeleton is suspiciously similar to the early stages of pneumatization in birds. The same is true of fossae in the cervical column of the basal sauropod *Shunosaurus* (Wilson and Sereno 1998). Among basal theropods, *Coelophysis bauri* is the earliest well-represented taxon with evidence of pneumaticity. The postaxial cervical vertebrae of *C. bauri* have pneumatic cavities that occupy most of the neural spine and that communicate with the outside through several large foramina (Colbert 1989).

The pattern of pneumatization in these early diverging saurischians indicates the presence of cervical air sacs like those of birds. It is true that in sauropsids diverticula may develop from practically any portion of the respiratory system. However, it does not follow that the diverticula that pneumatize the skeleton can come from anywhere (contra Hillenius and Ruben 2004), for two reasons. First, in extant birds the cervical vertebrae are only pneumatized by diverticula of cervical air sacs. Diverticula of the cranial air spaces, larynx and trachea are never known to pneumatize the postcranial skeleton (King 1966), and diverticula of the parenchymal portion of the lung only pneumatize the vertebrae and ribs adjacent to

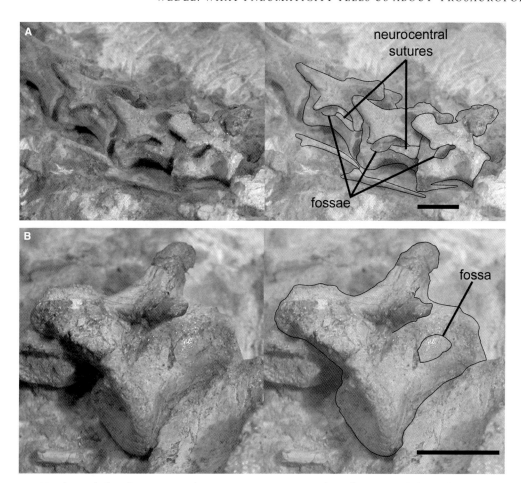

TEXT-FIG. 10. Vertebrae of *Thecodontosaurus caducus*, BMNH P24. A, cervical vertebrae 6–8 in left lateral view. B, cervical vertebra 9 in right lateral view. Scale bars represent 1 cm.

the lungs (O'Connor 2004). Second, as discussed below, pneumatization of the posterior half of the body is accomplished only by diverticula of abdominal air sacs (O'Connor and Claessens 2005). These observations of extant taxa provide valuable guidelines for interpreting patterns of skeletal pneumatization in fossil taxa. Pneumatization by diverticula of cervical air sacs is the only mechanism for pneumatizing the neck that is (1) known to occur in extant taxa and (2) consistent with the pattern of pneumatization found in basal saurischians (Wedel 2006).

Pneumaticity in basal saurischians is extremely limited. The bone removed by pneumatization of the postcranial skeleton (or fossa formation, if the fossae of *Thecodontosaurus* are not pneumatic) accounted for much less than 1 per cent of the total body volume in both *Coelophysis* and *Thecodontosaurus* (see Appendix), compared with several per cent for more derived sauropods and theropods (Wedel 2004, 2005). PSP probably did not evolve as an adaptation for lightening the skeleton, although it seems to have been exapted for that purpose later in saurischian evolution (Wedel 2003*b*).

Furthermore, diverticula did not evolve to pneumatize the skeleton. In the first place, many of the diverticula of birds are visceral, subcutaneous or intermuscular, and do not pneumatize any bones (Duncker 1971). Skeletal pneumatization cannot be invoked to explain the presence of these diverticula. In the second place, the presence of diverticula is a prerequisite for pneumatization of the skeleton. The immediate ancestors of *Coelophysis* and *Thecodontosaurus* must have already had cervical diverticula (assuming that the fossae of the latter are pneumatic in origin). Pneumatization of the cervical series could not happen until these diverticula were already in place, so the diverticula must have evolved for some other reason.

Alternatively, the origins of paravertebral diverticula and of PSP may have been coincident. The first step may have been a developmental accident that allowed the diverticula to push beyond the coelom and these 'unleashed' diverticula may have pneumatized the vertebral column immediately. This sort of morphogenetic behaviour on the part of diverticula is plausible on the basis of cases in the human clinical literature (e.g. Sadler *et al.* 1996; Anorbe *et al.* 2000). The main argument against this near-saltational scenario is that the first vertebrae to be pneumatized in both sauropodomorphs (*Thecodontosaurus*, *Shunosaurus*) and theropods (*Coelophysis*) are cervicals that are not adjacent to the lungs (Wedel 2006).

Origin of flow-through ventilation. Flow-through ventilation requires that air sacs be present both anterior and posterior to the parenchymal portion of the lung. Given the pattern of pneumatization found in pterosaurs, sauropods and theropods, we may infer that cervical air sacs were present in the ancestral ornithodiran (or evolved independently in pterosaurs and saurischians). The next problem is to determine when abdominal air sacs originated and how many times.

In extant birds, the posterior thoracic, synsacral and caudal vertebrae, pelvic girdle and hindlimb are only pneumatized by diverticula of abdominal air sacs (O'Connor and Claessens 2005; contra Ruben *et al.* 2003; Chinsamy and Hillenius 2004; Hillenius and Ruben 2004). So if a fossil archosaur is discovered with pneumatic vertebrae posterior to the mid-thorax, we have a compelling case for inferring that the animal had abdominal air sacs. Pneumatic vertebrae in the 'posterior compartment' are present in pterosaurs, diplodocid and macronarian sauropods, and in most clades of neotheropods, but are absent in non-dinosaurian dinosauromorphs, ornithischians, herrerasaurids, 'prosauropods', basal sauropods, dicraeosaurids, and in basal members of most neotheropod clades (e.g. *Baryonyx*, *Ceratosaurus* and *Allosaurus*; pers. obs.).

How many times did abdominal air sacs evolve? Possibly just once, before the ornithodiran divergence; possibly twice, in pterosaurs and saurischians; or possibly three times, in pterosaurs, sauropods and theropods (Text-fig. 1). We could take this to its logical conclusion and assume that abdominal air sacs evolved afresh in every group with posterior compartment pneumaticity; this would require the independent origin of abdominal air sacs in ceratosaurs, allosauroids and coelurosaurs, for example (not to mention several independent derivations within coelurosaurs).

The alternative is that some or all of the groups listed above had abdominal air sacs but failed to pneumatize any elements in the posterior compartment. The same condition pertains in many extant birds (O'Connor 2004, table 2). O'Connor and Claessens (2005) posited an origin of abdominal air sacs by the time of the ceratosaur-tetanuran divergence, based on the presence of posterior compartment pneumatization in *Majungatholus*, and despite its absence in basal ceratosaurs and basal tetanurans.

In terms of evolutionary change, ventilation mechanisms are highly conserved, PSP is highly labile and diverticula seem to lie between these extremes. All birds have essentially the same lung architecture; the biggest difference among living forms is the presence or absence of a neopulmo (Duncker 1971). On the other hand, PSP varies widely within small clades and even within populations (King 1966; Hogg 1984*a*; O'Connor 2004). Diverticula appear to be more conserved than PSP, although a dedicated study comparing the evolution of the two is needed. For example, most birds have femoral and perirenal diverticula, but the femur and pelvis are only pneumatized in a subset of these taxa (Müller 1908; King 1966; Duncker 1971). These observations are necessarily tentative, given the paucity of phylogenetically based comparative studies of pneumatic diverticula and PSP (but see O'Connor 2004). Furthermore, our knowledge of variation in the pulmonary system and its diverticula is based entirely on extant birds, and may not be applicable to other saurischians.

Nevertheless, the evolutionary malleability of lungs, diverticula and PSP in birds should not be ignored in reconstructing the pulmonary systems of fossil archosaurs. The absence of unequivocal PSP in most 'prosauropods' does not mean that they lacked air sacs. Depending on the preferred phylogenetic hypothesis, Sauropodomorpha is only one or two nodes away from Neotheropoda. Most neosauropods have pneumatic vertebrae in the posterior compartment. If these sauropods found some way to pneumatize the posterior compartment without abdominal air sacs, then surely the same could be true of some or all non-avian theropods. Likewise, if posterior compartment pneumaticity is *prima facie* evidence of abdominal air sacs in theropods, then abdominal air sacs must also have been present in sauropods (and, by extension, pterosaurs). What is good for the goose is good for *Gongxianosaurus*. It is more parsimonious to infer that cervical and abdominal air sacs were present in the ancestral saurischian, but did not pneumatize the skeleton in 'prosauropods', than to infer independent origins of air sacs in sauropods and theropods.

Most pterosaurs have extensively pneumatized skeletons, although it is not clear whether pneumaticity is present in any of the Triassic forms (Bonde and Christiansen 2003). The presence of PSP in pterosaurs, sauropodomorphs and theropods suggests that air sacs may have been present in the ancestral ornithodiran. An apparent problem with pushing the origin of air-sac-driven breathing back before the origin of Saurischia is the utter absence of PSP in ornithischians. PSP appeared in pterosaurs, sauropodomorphs and theropods relatively quickly after the divergence of each clade: by the Norian in theropods (Colbert 1989) and no later than the Early Jurassic in pterosaurs and sauropodomorphs (Bonde and Christiansen 2003; Wedel 2005). If ornithischians had air sacs and diverticula then it is odd that they never evolved PSP during the 160 million years of their existence. However, this problem may be more illusory than real. The invasion of bone by pneumatic epithelium is essentially opportunistic (Witmer 1997). Although pneumatic diverticula may radically remodel both the exterior and the interior of an affected bone, this remodelling cannot occur if the diverticula never come into contact with the bone, and may not occur even if they do. Furthermore, for all of the potential advantages it conveys, PSP is still an exaptation of a pre-existing system: in an adaptive sense, lineages that lack PSP do not know what they are missing. Recall that PSP in basal saurischians did little to lighten the skeleton (see above). Ornithischians may have had air sacs without diverticula, or diverticula without PSP. It is pointless to consider the advantages that ornithischians 'lost' by never evolving PSP, because that evolution would have hinged on the incidental contact of bone and air sac and could not have been anticipated or sought by natural selection.

The problem of determining when abdominal air sacs evolved is challenging because it forces us to decide between events of unknown probability: the possibility that ornithischians had an air sac system and never 'discovered' PSP (if abdominal air sacs are primitive for Ornithodira), vs. the possibility that a system of cervical and abdominal air sacs evolved independently in pterosaurs and saurischians. Currently, available evidence is insufficient to falsify either hypothesis.

Sauropodomorph palaeobiology. It is likely that 'prosauropods' had cervical and abdominal air sacs, given the strong evidence for both in sauropods and theropods. We may not be able to determine for certain whether 'prosauropods' had a bird-like flow-through lung, but the requisite air sacs were almost certainly present. Our null hypothesis for the respiratory physiology of 'prosauropods' should take into account some form of air-sac-driven ventilation.

The air sacs of birds mitigate the problem of tracheal dead space (Schmidt-Nielsen 1972), and some birds have improbably long tracheae (i.e. longer than the entire body of the bird; see McClelland 1989). In addition, birds can ventilate their air sacs without blowing air through the lungs, which allows them to avoid alkalosis during thermoregulatory panting (Schmidt-Nielsen *et al.* 1969). Finally, flow-through breathing allows birds to extract much more oxygen from the air than mammals can (Bernstein 1976). In general, sauropods were larger and longer-necked than 'prosauropods', and the aforementioned capabilities of a bird-like ventilation system may have helped sauropods overcome the physiological challenges imposed by long necks and large

bodies, including tracheal dead space, heat retention and oxygen uptake.

The one obvious advantage that 'prosauropods' did not share with sauropods is the very lightweight skeletal construction afforded by pneumaticity. In life, the average pneumatic sauropod vertebra was approximately 60 per cent air by volume (Wedel 2005; Woodward 2005; Schwarz and Fritsch 2006). All else being equal, a sauropod could have a neck two-thirds longer than that of a prosauropod for the same skeletal mass. Pneumaticity helped sauropods overcome constraints on neck length, and thereby opened feeding opportunities that were not available to 'prosauropods'. How important that difference was is unknown, but it is worth considering in reconstructions of sauropodomorph evolution and palaeobiology.

Acknowledgements. This work was completed as part of a doctoral dissertation in the Department of Integrative Biology, University of California, Berkeley. I am grateful to my advisors, K. Padian and W. A. Clemens, and to the members of my dissertation committee, F. C. Howell, D. Wake and M. Wake, for advice and encouragement. Many thanks to P. M. Barrett and T. Fedak for organizing the symposium that gave rise to this manuscript. I am grateful for the hospitality and patience of curators and collections managers everywhere, especially P. Barrett, S. Chapman, R. Schoch, M. Moser, A. Henrici, M. Lamanna, R. Cifelli, N. Czaplewski and J. Person. L. Claessens, R. Irmis, S. Nesbitt and P. O'Connor provided many inspiring discussions and gracious access to their unpublished work. R. Irmis and M. Taylor read early drafts of this paper and made many helpful suggestions, as did my dissertation committee members. L. Claessens and J. Harris provided thoughtful review comments that greatly improved this paper, and I thank them for their time and effort. A translation of Janensch (1947) was made by G. Maier, whose effort is gratefully acknowledged. Funding for this project was provided by the Jurassic Foundation, Sigma Xi, the Department of Integrative Biology at the University of California, Berkeley, the University of California, Museum of Paleontology, and the UCMP Doris and Samuel P. Welles Fund. This is UCMP Contribution no. 1919.

REFERENCES

ALCOBER, O. and PARRISH, J. M. 1997. A new poposaurid from the Upper Triassic of Argentina. *Journal of Vertebrate Paleontology*, **17**, 548–556.

ANORBE, E., AISA, P. and SAENZ DE ORMIJANA, J. 2000. Spontaneous pneumatocele and pneumocephalus associated with mastoid hyperpneumatization. *European Journal of Radiology*, **36**, 158–160.

BENTON, M. J., JUUL, L., STORRS, G. W. and GALTON, P. M. 2000. Anatomy and systematics of the prosauropod dinosaur *Thecondontosaurus antiquus* from the Upper Triassic of southwest England. *Journal of Vertebrate Paleontology*, **20**, 77–108.

BERNSTEIN, M. H. 1976. Ventilation and respiratory evaporation in the flying crow, *Corvus ossifragus*. *Respiration Physiology*, **26**, 371–382.

BONAPARTE, J. F. 1999. Evolución de las vertebras presacras en Sauropodomorpha. *Ameghiniana*, **36**, 115–187.

BONDE, N. and CHRISTIANSEN, P. 2003. The detailed anatomy of *Rhamphorhynchus*: axial pneumaticity and its implications. *Geological Society, London, Special Publication*, **217**, 217–232.

BREMER, J. L. 1940. The pneumatization of the humerus in the common fowl and the associated activity of theelin. *Anatomical Record*, **77**, 197–211.

BRITT, B. B. 1993. Pneumatic postcranial bones in dinosaurs and other archosaurs. Unpublished PhD thesis, University of Calgary, 383 pp.

—— 1997. Postcranial pneumaticity. 590–593. *In* CURRIE, P. J. and PADIAN, K. (eds). *The encyclopedia of dinosaurs*. Academic Press, San Diego, CA, 869 pp.

BROCHU, C. A. 2001. Progress and future directions in archosaur phylogenetics. *Journal of Paleontology*, **75**, 1185–1201.

CARVALHO, I. S., AVILLA, L. S. and SALGADO, L. 2003. *Amazonsaurus maranhensis* gen. et sp. nov. (Sauropoda, Diplodocoidea) from the Lower Cretaceous (Aptian–Albian) of Brazil. *Cretaceous Research*, **24**, 697–713.

CASE, E. C. 1907. Revision of the Pelycosauria of North America. *Carnegie Institution of Washington Publication*, **55**, 1–175.

CHINSAMY, A. and HILLENIUS, W. J. 2004. Physiology of nonavian dinosaurs. 643–659. *In* WEISHAMPEL, D. B., DODSON, P. and OSMÓLSKA, H. (eds). *The Dinosauria*. Second edition. University of California Press, Berkeley, CA, 861 pp.

COLBERT, E. H. 1989. The Triassic dinosaur *Coelophysis*. *Bulletin of the Museum of Northern Arizona*, **57**, 1–160.

COVER, M. S. 1953. Gross and microscopic anatomy of the respiratory system of the turkey. III. The air sacs. *American Journal of Veterinary Research*, **14**, 239–245.

DESOJO, J. B., ARCUCCI, A. B. and MARSICANO, C. A. 2002. Reassessment of *Cuyosuchus huenei*, a Middle–Late Triassic archosauriform from the Cuyo Basin, west-central Argentina. *Bulletin of the New Mexico Museum of Natural History and Science*, **21**, 143–148.

DUNCKER, H.-R. 1971. The lung air sac system of birds. *Advances in Anatomy, Embryology, and Cell Biology*, **45**, 1–171.

GALTON, P. M. and UPCHURCH, P. 2004. Prosauropoda. 232–258. *In* WEISHAMPEL, D. B., DODSON, P. and OSMÓLSKA, H. (eds). *The Dinosauria*. Second edition. University of California Press, Berkeley, CA, 861 pp.

GIER, H. T. 1952. The air sacs of the loon. *Auk*, **69**, 40–49.

GILMORE, C. W. 1936. Osteology of *Apatosaurus* with special reference to specimens in the Carnegie Museum. *Memoirs of the Carnegie Museum*, **11**, 175–300.

GOWER, D. J. 2001. Possible postcranial pneumaticity in the last common ancestor of birds and crocodilians: evidence from *Erythrosuchus* and other Mesozoic archosaurs. *Naturwissenschaften*, **88**, 119–122.

HILLENIUS, W. J. and RUBEN, J. A. 2004. The evolution of endothermy in terrestrial vertebrates: Who? When? Why? *Physiological and Biochemical Zoology*, **77**, 1019–1042.

HOGG, D. A. 1984*a*. The distribution of pneumatisation in the skeleton of the adult domestic fowl. *Journal of Anatomy*, **138**, 617–629.

—— 1984*b*. The development of pneumatisation in the postcranial skeleton of the domestic fowl. *Journal of Anatomy*, **139**, 105–113.

HURLBURT, G. 1999. Comparison of body mass estimation techniques, using Recent reptiles and the pelycosaur *Edaphosaurus boanerges*. *Journal of Vertebrate Paleontology*, **19**, 338–350.

JAIN, S. L., KUTTY, T. S., ROY-CHOWDHURY, T. K. and CHATTERJEE, S. 1979. Some characteristics of *Barapasaurus tagorei*, a sauropod dinosaur from the Lower Jurassic of Deccan, India. *Proceedings of the IV International Gondwana Symposium*, **1**, 204–216.

JANENSCH, W. 1929. Material und Formengehalt der Sauropoden in der Ausbeute der Tendaguru-expedition. *Palaeontographica (Supplement 7)*, **2** (1), 1–34.

—— 1947. Pneumatizitat bei Wirbeln von Sauropoden und anderen Saurischien. *Palaeontographica (Supplement 7)*, **3** (1), 1–25.

—— 1950. Die Wirbelsaule von Brachiosaurus brancai. *Palaeontographica (Supplement 7)*, **3** (2), 27–93.

JERISON, H. J. 1973. *Evolution of the brain and intelligence*. Academic Press, New York, NY, 482 pp.

KING, A. S. 1966. Structural and functional aspects of the avian lungs and air sacs. *International Review of General and Experimental Zoology*, **2**, 171–267.

LANGER, M. C. 2004. Basal Saurischia. 25–46. *In* WEISHAMPEL, D. B., DODSON, P. and OSMÓLSKA, H. (eds). *The Dinosauria*. Second edition. University of California Press, Berkeley, CA, 861 pp.

LARSON, A., WAKE, D. B., MAXSON, L. R. and HIGHTON, R. 1981. A molecular phylogenetic perspective on the origins of morphological novelties in the salamanders of the tribe Plethodontini (Amphibia, Plethodontidae). *Evolution*, **35**, 405–422.

LÜ JUN-CHANG and ZHANG BAO-KUN 2005. A new oviraptorid (Theropod: Oviraptorosauria) from the Upper Cretaceous of the Nanxiong Basin, Guangdong Province of southern China. *Acta Palaeontologica Sinica*, **44**, 412–422.

MAKOVICKY, P. J., APESTEGUÍA, S. and AGNOLÍN, F. L. 2005. The earliest dromaeosaurid theropod from South America. *Nature*, **437**, 1007–1011.

McCLELLAND, J. 1989. Larynx and trachea. 69–103. *In* KING, A. S. and McCLELLAND, J. (eds). *Form and function in birds*. Volume 4. Academic Press, London, 608 pp.

MOSER, M. 2003. *Plateosaurus engelhardti* Meyer, 1837 (Dinosauria: Sauropodomorpha) from the Feuerletten (Mittelkeuper; Obertrias) of Bavaria. *Zitteliana*, **B24**, 1–188.

MÜLLER, B. 1908. The air-sacs of the pigeon. *Smithsonian Miscellaneous Collections*, **50**, 365–420.

MURRAY, P. F. and VICKERS-RICH, P. 2004. *Magnificent mihirungs*. Indiana University Press, Bloomington, IN, 410 pp.

NESBITT, S. J. 2005. Osteology of the Middle Triassic pseudosuchian archosaur *Arizonasaurus babbitti*. *Historical Biology*, **17**, 19–47.

O'CONNOR, P. M. 2004. Pulmonary pneumaticity in the postcranial skeleton of extant Aves: a case study examining Anseriformes. *Journal of Morphology*, **261**, 141–161.

—— 2006. Postcranial pneumaticity: an evaluation of soft-tissue influences on the postcranial axial skeleton and the reconstruction of pulmonary anatomy in archosaurs. *Journal of Morphology*, **267**, 1199–1226.

—— and CLAESSENS, L. P. A. M. 2005. Basic avian pulmonary design and flow-through ventilation in non-avian theropod dinosaurs. *Nature*, **436**, 253–256.

OSBORN, H. F. 1899. A skeleton of *Diplodocus*. *Memoirs of the American Museum of Natural History*, **1**, 191–214.

—— and MOOK, C. C. 1921. *Camarasaurus, Amphicoelias*, and other sauropods of Cope. *Memoirs of the American Museum of Natural History*, **3**, 247–287.

OSMÓLSKA, H., CURRIE, P. J. and BARSBOLD, R. 2004. Oviraptorosauria. 165–183. *In* WEISHAMPEL, D. B., DODSON, P. and OSMÓLSKA, H. (eds). *The Dinosauria*. Second edition. University of California Press, Berkeley, CA, 861 pp.

OSTROM, J. H. and McINTOSH, J. S. 1966. *Marsh's dinosaurs: the collections from Como Bluff*. Yale University Press, New Haven, CT, xxiv + 388 pp.

PARKER, W. G. 2003. Description of a new specimen of *Desmatosuchus haplocerus* from the Late Triassic of northern Arizona. Unpublished MS thesis, Northern Arizona University, Flagstaff, AZ, 312 pp.

PAUL, G. S. 1997. Dinosaur models: the good, the bad, and using them to estimate the mass of dinosaurs. 129–154. *In* WOLBERG, D. L., STUMP, E. and ROSENBERG, G. (eds). *Dinofest International: proceedings of a symposium sponsored by Arizona State University*. Academy of Natural Sciences, Philadelphia, PA, 587 pp.

PECZKIS, J. 1994. Implications of body mass estimates for dinosaurs. *Journal of Vertebrate Paleontology*, **14**, 520–533.

POWELL, J. E. 1992. Osteología de *Saltasaurus loricatus* (Sauropoda – Titanosauridae) del Cretácico Superior del noroeste Argentino. 165–230. *In* SANZ, J. L. and BUSCALIONI, A. D. (eds). *Los dinosaurios y su entorno biotico: Actas del Segundo Curso de Paleontología en Cuenca*. Instituto Juan de Valdes, Cuenca, Argentina, 397 pp.

RUBEN, J. A., JONES, T. D. and GEIST, N. R. 2003. Respiratory and reproductive paleophysiology of dinosaurs and early birds. *Physiological and Biochemical Zoology*, **76**, 141–164.

SADLER, D. J., DOYLE, G. J., HALL, K. and CRAWFORD, P. J. 1996. Craniocervical bone pneumatisation. *Neuroradiology*, **38**, 330–332.

SANZ, J. L., POWELL, J. E., LE LOEUFF, J., MARTINEZ, R. and PEREDA SUPERBIOLA, X. 1999. Sauropod remains from the Upper Cretaceous of Laño (north central Spain). Titanosaur phylogenetic relationships. *Estudios del Museo de Ciencias Naturales de Alava*, **14** (Numero Especial 1), 235–255.

SCHMIDT-NIELSEN, K. 1972. *How animals work*. Cambridge University Press, Cambridge, vi + 114 pp.

—— KANWISHER, J., LASIEWSKI, R. C., COHN, J. E. and BRETZ, W. L. 1969. Temperature regulation and respiration in the ostrich. *Condor*, **71**, 341–352.

SCHWARZ, D. and FRITSCH, G. 2006. Pneumatic structures in the cervical vertebrae of the Late Jurassic Tendaguru sauropods *Brachiosaurus brancai* and *Dicraeosaurus*. *Eclogae Geologicae Helvetiae*, **99**, 65–78.

SERENO, P. C. 1998. A rationale for phylogenetic definitions, with application to the higher-level taxonomy of Dinosauria. *Neues Jahrbuch für Geologie und Paläontologie, Abhandlungen*, **20**, 41–83.

—— 2007. Basal Sauropodomorpha: historical and recent phylogenetic hypotheses, with comments on *Ammosaurus major* (Marsh, 1891). 261–289. *In* BARRETT, P. M. and BATTEN, D. J. (eds). *Evolution and palaeobiology of early sauropodomorph dinosaurs*. Special Papers in Palaeontology, **77**, 289 pp.

—— and WILSON, J. A. 2005. Structure and evolution of a sauropod tooth battery. 157–177. *In* WILSON, J. A. and CURRY-ROGERS, K. (eds). *The sauropods: evolution and paleobiology*. University of California Press, Berkeley, CA, 349 pp.

—— BECK, A. L., DUTHEIL, D. B., LARSSON, H. C. E., LYON, G. H., MOUSSA, B., SADLEIR, R. W., SIDOR, C. A., VARRICCHIO, D. J., WILSON, G. P. and WILSON, J. A. 1999. Cretaceous sauropods and the uneven rate of skeletal evolution among dinosaurs. *Science*, **286**, 1342–1347.

TYKOSKI, R. S., ROWE, T. B., KETCHAM, R. A. and COLBERT, M. W. 2002. *Calsoyasuchus valliceps*, a new crocodyliform from the Early Jurassic Kayenta Formation of Arizona. *Journal of Vertebrate Paleontology*, **22**, 593–611.

UPCHURCH, P., BARRETT, P. M. and DODSON, P. 2004. Sauropoda. 259–324. *In* WEISHAMPEL, D. B., DODSON, P. and OSMÓLSKA, H. (eds). *The Dinosauria*. Second edition. University of California Press, Berkeley, CA, 861 pp.

—— —— and GALTON, P. M. 2007. The phylogenetic relationships of basal sauropodomorphs: implications for the origin of sauropods. 57–90. *In* BARRETT, P. M. and BATTEN, D. J. (eds). *Evolution and palaeobiology of early sauropodomorph dinosaurs*. Special Papers in Palaeontology, **77**, 289 pp.

VERBOUT, A. J. 1985. The development of the vertebral column. *Advances in Anatomy, Embryology and Cell Biology*, **90**, 1–122.

WAKE, D. B. 1963. Comparative osteology of the plethodontid salamander genus *Aneides*. *Journal of Morphology*, **113**, 77–118.

—— 1992. The endoskeleton: the comparative anatomy of the vertebral column and ribs. 192–237. *In* WAKE, M. H. (ed.). *Hyman's comparative vertebrate anatomy*. Third edition. University of Chicago Press, Chicago, IL, 788 pp.

—— and LAWSON, R. 1973. Developmental and adult morphology of the vertebral column in the plethodontid salamander *Eurycea bislineata*, with comments on vertebral evolution in the Amphibia. *Journal of Morphology*, **139**, 251–300.

—— WAKE, D. B. and WAKE, M. H. 1983. The ossification sequence of *Aneides lugubris*, with comments on heterochrony. *Journal of Herpetology*, **17**, 10–22.

WEDEL, M. J. 2003*a*. The evolution of vertebral pneumaticity in sauropod dinosaurs. *Journal of Vertebrate Paleontology*, **23**, 344–357.

—— 2003*b*. Vertebral pneumaticity, air sacs, and the physiology of sauropod dinosaurs. *Paleobiology*, **29**, 243–255.

—— 2004. Skeletal pneumaticity in saurischian dinosaurs and its implications for mass estimates. *Journal of Vertebrate Paleontology*, **24** (Supplement to No. 3), 127A.

—— 2005. Postcranial skeletal pneumaticity in sauropods and its implications for mass estimates. 201–228. *In* WILSON, J. A. and CURRY-ROGERS, K., (eds). *The sauropods: evolution and paleobiology*. University of California Press, Berkeley, CA, 349 pp.

—— 2006. The origins of postcranial skeletal pneumaticity in dinosaurs. *Integrative Zoology*, **2**, 80–85.

—— CIFELLI, R. L. and SANDERS, R. K. 2000. Osteology, paleobiology, and relationships of the sauropod dinosaur *Sauroposeidon*. *Acta Palaeontologica Polonica*, **45**, 343–388.

WILSON, J. A. 1999. A nomenclature for vertebral laminae in sauropods and other saurischian dinosaurs. *Journal of Vertebrate Paleonotology*, **19**, 639–653.

—— 2002. Sauropod dinosaur phylogeny: critique and cladistic analysis. *Zoological Journal of the Linnean Society*, **136**, 217–276.

—— and SERENO, P. C. 1998. Early evolution and higher-level phylogeny of sauropod dinosaurs. *Memoir of the Society of Vertebrate Paleontology*, **5**, 1–68.

WITMER, L. M. 1997. The evolution of the antorbital cavity of archosaurs: a study in soft-tissue reconstruction in the fossil record with an analysis of the function of pneumaticity. *Memoir of the Society of Vertebrate Paleontology*, **3**, 1–73.

WOODWARD, H. 2005. Bone histology of the titanosaurid sauropod *Alamosaurus sanjuanensis* from the Javelina Formation, Texas. *Journal of Vertebrate Paleontology*, **25** (Supplement to No. 3), 132A.

XU XING, ZHANG XIAO-HONG, TAN QING-WEI, ZHAO XI-JIN and TAN LIN 2006. A new titanosaurian sauropod from Late Cretaceous of Nei Mongol, China. *Acta Geologica Sinica*, **80**, 20–26.

YATES, A. M. 2003. A new species of the primitive dinosaur *Thecodontosaurus* (Saurischia: Sauropodomorpha) and its implications for the systematics of basal dinosaurs. *Journal of Systematic Paleontology*, **1**, 1–42.

—— 2004. *Anchisaurus polyzelus* (Hitchcock): the smallest known sauropod dinosaur and the evolution of gigantism among sauropodomorph dinosaurs. *Postilla*, **230**, 1–58.

—— 2007. The first complete skull of the Triassic dinosaur *Melanorosaurus* Haughton (Sauropodomorpha: Anchisauria). 9–55. *In* BARRETT, P. M. and BATTEN, D. J. (eds). *Evolution and palaeobiology of early sauropodomorph dinosaurs*. Special Papers in Palaeontology, **77**, 289 pp.

—— and KITCHING, J. W. 2003. The earliest known sauropod dinosaur and the first steps toward sauropod locomotion. *Proceedings of the Royal Society of London, Series B*, **270**, 1753–1758.

APPENDIX

The method of calculating the volumes of bone removed by pneumatization in *Coelophysis* and *Thecondontosaurus* (see 'Palaeobiological implications' above) is provided here. To estimate the whole body volumes of the dinosaurs I used graphic double integration (GDI: Jerison 1973; Hurlburt 1999; Murray and Vickers-Rich 2004). I traced over the skeletal reconstructions of Colbert (1989, fig. 103) and Benton *et al.* (2000, fig. 19) to make lateral view body outlines. Dorsal view body outlines were drawn by hand based on those of Paul (1997) and digitally manipulated to match the dimensions of the skeletal reconstructions. Using GDI, I obtained whole body volumes of 23·5 L for *Coelophysis* and 3·3 L for the holotypic individual of *Thecodontosaurus caducus*; the latter animal is a small juvenile. Adjusted for scale, these results are consistent with previous mass estimates for both taxa (Peczkis 1994).

Pneumaticity is present throughout the cervical series of *Coelophysis*. The total length of the cervical series is *c.* 50 cm, and the vertebral centra have a mean diameter of 1 cm, based on measurements of uncatalogued CM specimens. The neural spines are roughly the same size as the centra. The combined cervical centra are treated as a simple cylinder 50 cm long with a diameter of 1 cm, which yields a volume of 40 cm^3. If the neural spines are assumed to be equal in volume to the centra, the combined volume of the cervical vertebrae is 80 cm^3. The cervical vertebrae of *Coelophysis* are probably not more than 50 per cent air by volume, based on observations of broken specimens, so the volume of bone removed during pneumatization of the cervical vertebrae was *c.* 40 cm^3, or 0·17 per cent of the volume of the body.

For *Thecondontosaurus caducus* it is simpler to calculate the volumes of the individual fossae. The fossae on cervicals 6–8 are each *c.* 5 mm long, 2·5 mm tall and 1·25 mm deep. The paired fossae on each vertebra can be thought of as forming the two halves of an oblate spheroid with *x*, *y* and *z* diameters of 5, 2·5 and 2·5 mm, respectively. The volume of this spheroid, and thus the volume of the paired fossae, is 0·016 cm^3. The fossae on cervicals 6–8 are all roughly the same size, and the visible fossa on the ninth cervical is only about half as deep. The volume of bone removed during fossa formation is therefore 0·057 cm^3, or 0·0017 per cent of the volume of the body.

These calculations are all approximate, but they are sufficient to demonstrate that PSP did not have a noticeable effect on the skeletal mass of basal saurischians. In the case that I have underestimated the volume of the pneumatic chambers in *Coelophysis* relative to the body volume by a factor of six: the volume of these chambers would still only be 1 per cent of the volume of the body. In contrast, the volume of air in the pneumatic vertebrae of *Tyrannosaurus* and *Diplodocus* accounted for 4–6 per cent of the volume of the animals. These air spaces replaced bone, a relatively dense tissue, and lightened the animals by 7–10 per cent (Wedel 2004, 2005).

[Special Papers in Palaeontology 77, 2007, pp. 223–243]

NEW INFORMATION ON *LESSEMSAURUS SAUROPOIDES* (DINOSAURIA: SAUROPODOMORPHA) FROM THE UPPER TRIASSIC OF ARGENTINA

by DIEGO POL* *and* JAIME E. POWELL†

*CONICET, Museo Palaeontológico Egidio Feruglio, Av. Fontana 140, Trelew CP 9100, Chubut, Argentina; e-mail: dpol@mef.org.ar
†CONICET, Instituto Miguel Lillo, Miguel Lillo 205, San Miguel de Tucumán CP 4000, Tucumán, Argentina; e-mail: jpowell@csnat.unt.edu.ar

Typescript received 9 February 2006; accepted in revised form 26 October 2006

Abstract: Postcranial remains of *Lessemsaurus sauropoides* are described herein, including elements of the vertebral column, pectoral girdle, forelimb, pelvis and hindlimb. These remains were closely associated with the cervicodorsal neural arches previously described from this taxon. This assemblage of bones shows numerous derived characters, including some derived similarities exclusively shared with *Antetonitrus ingenipes* from the Upper Triassic of South Africa. Additionally, this material reveals an unusual combination of plesiomorphic character states present in many non-eusauropod sauropodomorphs together with derived characters that suggest affinities with eusauropods and related taxa.

Key words: *Lessemsaurus*, Sauropodomorpha, Prosauropoda, Sauropoda, Triassic.

BASAL sauropodomorphs from the Villa Unión-Ischigualasto Basin (north-west Argentina; Stipanicic and Bonaparte 1972) are so far exclusively known from the upper section of the Los Colorados Formation (Groeber and Stipanicic 1953). Three different taxa have been described from this unit: *Riojasaurus incertus* (Bonaparte 1972; Bonaparte and Pumares 1995), *Coloradisaurus brevis* (Bonaparte 1978) and *Lessemsaurus sauropoides* (Bonaparte 1999). The original description of the last taxon focused on eight cervicodorsal vertebrae (PVL 4822–1), and several similarities with both non-eusauropod sauropodomorphs ('prosauropods') and basal sauropods ('cetiosaurids') were noted (Bonaparte 1986, 1999). Additional undescribed material (also catalogued under PVL 4822) consists of other vertebral elements and remains of the pectoral girdle, forelimb, pelvis and hindlimb. These elements are described and figured herein, providing a more complete understanding of the anatomy of *Lessemsaurus sauropoides*.

The undescribed remains are referred to *Lessemsaurus sauropoides* on the basis of their close association with the type material described by Bonaparte (1999) and the presence of numerous characters that distinguish this material from that of other non-eusauropod sauropodomorphs from the Los Colorados Formation. This assemblage of bones (PVL 4822) includes material from more than one individual (based on the recovery of several duplicate elements in the collection). *Lessemsaurus sauropoides* is of interest because of the presence of a unique combination of apomorphic and plesiomorphic character states. These remains are compared with other non-eusauropod sauropodomorphs and their significance and phylogenetic relevance is discussed.

The comparisons made in this contribution are based on the examination of specimens of different taxa and relevant literature detailed in Table 1. Unless noted explicitly, all references to other taxa are based on the sources of data listed in this table.

Institutional abbreviations. AMNH, American Museum of Natural History, New York, USA; BMNH, The Natural History Museum, London, UK; BPI, Bernard Price Institute, Johannesburg, South Africa; IVPP, Institute of Vertebrate Palaeontology and Palaeoanthropology, Beijing, People's Republic of China; NGMJ, Nanjing Geological Museum, Nanjing, China; MB, Institut für Palaontologie, Museum fur Naturkunde, Humbolt-Universitat, Berlin, Germany; MCP, Museu Pontificia Universidade Catolica, Porto Alegre, Brazil; MPEF, Museo Palaeontológico Egidio Feruglio, Trelew, Argentina; NM QR, National Museum, Bloemfontein, South Africa; PVL, Instituto Miguel Lillo, Tucumán, Argentina; SAM, Iziko – South African Museum,

Cape Town, South Africa; SMNS, Staatliches Museum für Naturkunde Stuttgart, Germany; ULR, Museo Ciencias Naturales, Universidad La Rioja, La Rioja, Argentina; YPM, Yale Peabody Museum, New Haven, USA.

Anatomical abbreviations. 4t, fourth trochanter; I.1, phalanx 1 of first digit; I.u, ungual of first digit; a-I, concave articular surface for metatarsal I; a-II, articular surface for metacarpal II; acp, acromion process; agr, groove for ascending process of astragalus; alp, anterolateral descending process; ap, ascending process; apr, pubic apron; bcr, brevis crest; cn, cnemial crest; dac, hemispherical distal condyle; d.f., flexor fossa of distal humerus; dle, dorsolateral edge of asymmetrical ungual; dno, dorsal notch of ischial symphysis; dpc, deltopectoral crest; fdp, facet for descending process; fhe, femoral head; fpo, popliteal fossa; gl, glenoid cavity; ibl, iliac blade; ilp, iliac peduncle; isp, ischial peduncle; it, humeral internal tuberosity; lc, lateral condyle; lco, lateral concavity of pubic apron; ld, lateral depression; ldc, lateral distal condyle; lgr, unbifurcated lateral groove; lp, lunar lateral process; mb, astragalar main body; mc, medial condyle; mdc, medial distal condyle; mlp, medial collateral ligament pit; mtc I-II, metacarpals I through II; mtt I-V, metatarsals I through V; mwa, medial wall of acetabulum; ncs, neurocentral suture; of, obturator foramen; ol, olecranon; pa, parapophysis; paf, proximal articular facet; pap, preacetabular process; pdl, proximodorsal lip; pdp, posterior descending process of tibia; plp, proximolateral process; pmc, subrectangular posteromedial corner; pmf, proximomedial flange; pop, postacetabular process; ppl, proximal pubic plate; ppr, parapophyseal ridge; pup, pubic peduncle; pvh, proximoventral heel; rfo, radial fossa; rp, rounded proximal surface of humerus; sac, supracetabular crest; scb, scapular blade; stm, steep distal margin of 4th trochanter; tub, flexor tubercle; uap, anterior process of ulnar proximal end; umc, medial concavity on ulnar proximal end; vk, ventral keel of ischial shaft; vme, ventromedial edge of asymmetrical ungual.

TABLE 1. Source of comparative data used in this study. All comparative references to the following taxa have been observed in the specimens listed or taken from the respective bibliographic reference. Comparisons based on other specimens or taken from additional references are explicitly indicated in the text.

Taxon and source

Anchisaurus polyzelus, YPM 1883
Antetonitrus ingenipes, BPI/1/4952
Blikanasaurus cromptoni, SAM-PK-K403
Coloradisaurus brevis, PVL 5904
Efraasia minor, SMNS 12667
Ferganasaurus verzilini, Alifanov and Averianov (2003)
Kotasaurus yamanpalliensis, Yadagiri (2001)
Lufengosaurus huenei, IVPP V15
Massospondylus carinatus, BPI/1/4934
Melanorosaurus readi, NM QR1551
Mussaurus patagonicus, PVL 4587
Omeisaurus maoianus, Tang *et al.* (2001)
Patagosaurus fariasi, PVL 4170
Plateosaurus engelhardti, SMNS 13200
Riojasaurus incertus, PVL 3808
Saturnalia tupiniquim, MCP 3844-PV
Tazoudasaurus naimi, Allain *et al.* (2004)
Tehuelchesaurus benitezi, MPEF-PV 1125
Thecodontosaurus antiquus, YPM 2195
Thecodontosaurus caducus, Yates (2003)
Vulcanodon karibaensis, Cooper (1984)
Yunnanosaurus huangi, NGMJ V116 [V20]

SYSTEMATIC PALAEONTOLOGY

SAURISCHIA Seeley, 1887
SAUROPODOMORPHA von Huene, 1932

Genus LESSEMSAURUS Bonaparte, 1999

Type species. *Lessemsaurus sauropoides* Bonaparte, 1999

Diagnosis. As for type and only species.

Lessemsaurus sauropoides Bonaparte, 1999
Text-figures 1–13

1986 unnamed advanced prosauropod Bonaparte, p. 248, fig. 19.1–2.
1999 *Lessemsaurus sauropoides* Bonaparte, p. 133.
2000 *Lessemsaurus*; Buffetaut *et al.*, p. 73.
2003 *Lessemsaurus sauropoides*; Yates and Kitching, p. 1753.
2004 *Lessemsaurus sauropoides*; Galton and Upchurch, p. 234, fig. 12.6F–I.
2005 *Lessemsaurus*; Wilson, p. 403.

Holotype. PVL 4822–1. Bonaparte (1999) described and figured eight presacral neural arches. He mentioned additional presacral vertebrae and some appendicular elements as probably associated with this specimen. Owing to the lack of articulated remains, it cannot be determined which of the PVL 4822 elements belong to the same individual. Therefore, the holotype is now restricted to the eight presacral neural arches originally described and figured by Bonaparte (1999) and catalogued as PVL 4822-1. These eight neural arches are individually identified by the collection numbers (PVL 4822-1/1–4822-1/7 and PVL 4822-1/10).

Referred material. The rest of the elements in the assemblage originally catalogued as PVL 4822 have been given additional numbers to allow identification of each individual element (PVL 4822/8–4822/9 and 4822/11–4822/79). These elements include dorsal and sacral vertebrae, scapulae, coracoid, humerus, ulna and radius, metacarpals I and II, manual digit I, ilium, ischium, pubes, femur, tibia fragmentary metacarpals and pedal phalanges. All of these remains were found closely associated with each other by Bonaparte and others during the Lillo Palaeontologic Expedition of 1971. Bonaparte (1999) interpreted these remains as belonging to three different individuals.

Locality and horizon. La Esquina (8 km west of Provincial Road 26, at the 142 km mark), La Rioja Province, Argentina. Upper section of the Los Colorados Formation (Groeber and Stipanicic 1953). The age of this unit has been considered as Norian–Rhaetian (Bonaparte 1972) or the tetrapod-based biochron late Coloradian (Bonaparte 1973). The horizon where these remains were found is located approximately 150 m below the upper limit of this unit (Bonaparte 1999).

Revised diagnosis. A large non-eusauropod sauropodomorph with the following unique combination of characters (autapomorphies indicated with an asterisk): dorsal and middle to posterior cervicals with high neural arches; strong neural arch constriction below the postzygapophyses; deep postspinal fossa; dorsoventrally high infrapostzygapophyseal depression; middle and posterior dorsals with neural spines higher than long (with a height/length ratio of 1·5–2·0); robust scapula, with its blade markedly expanded; metacarpal I extremely short, with a proximal end lateromedially wider than metacarpal length; acute lateral process on proximolateral corner of metacarpal II*; pubic peduncle of ilium with a medial flange, forming a narrow and marginal medial wall of the acetabulum*; brevis crest extending from the base of the ischial peduncle to the posterior tip of the postacetabular process*; and cross-section of the distal tibia subrectangular with its major axis orientated lateromedially and being twice as long as its anteroposterior extension.

DESCRIPTION

The neural arches that compose the holotype of *Lessemsaurus sauropoides* (PVL 4822–1) were described in detail by Bonaparte (1999) and will not be described further herein. Instead, we focus on the undescribed material, including the additional vertebral remains, the shoulder girdle, forelimb, pelvis and hindlimb.

Vertebral column

Cervical vertebrae. PVL 4822 includes three cervical vertebral centra (PVL 4822/20–4822/22; Text-fig. 1). The element 4822/20 is the most anterior centrum and probably belongs to a middle cervical. The anterior articular surface is dorsoventrally shorter than the posterior end of the centrum, the former being approximately 0·78 times as high as the latter (Text-fig. 1A). The anteroposterior length of the middle cervical centrum of *Lessemsaurus sauropoides* is approximately twice the dorsoventral height of its posterior articular surface and 2·27 times the height of its anterior articular surface. The middle cervicals of some non-eusauropod sauropodomorphs have similar proportions (e.g. *Riojasaurus incertus*, ULR 56; *Kotasaurus yamanpalliensis*). However, most non-eusauropod sauropodomorphs [e.g. *Massospondylus*

carinatus, BPI/1/5241; *Coloradisaurus brevis*, PVL 3967; *Lufengosaurus huenei* (Young 1941); *Yunnanosaurus huangi* (Young 1942); *Plateosaurus engelhardti*, MB skelett 25] and derived sauropods have middle cervical vertebrae proportionately longer than those of *Lessemsaurus sauropoides*. The centrum is acamerate (*sensu* Wedel 2003) and amphicoelous, as in all non-eusauropod sauropodomorphs. The lateral surface of the centrum is only slightly concave and bears an elongated crest-like parapophysis close to its anterior edge (Text-fig. 1). The vertebral centrum PVL 4822/20 is notably constricted lateromedially with respect to its anterior and posterior ends and is approximately half the width of the articular surfaces. The ventral surface of this centrum is rounded and lacks the sagittal crest present in several non-eusauropod sauropodomorphs (e.g. *Massospondylus carinatus*, SAM-PK-K391; *Anchisaurus polyzelus*). A flat ventral surface is present in other non-eusauropod sauropodomorphs, however (e.g. *Plateosaurus engelhardti*, MB skelett 25; *Thecodontosaurus caducus*).

The centrum PVL 4822/21 probably belongs to a posterior cervical vertebra. In contrast to the middle cervical (PVL 4822/20), the anterior and posterior articular facets of this element are sub-equal in dorsoventral height (Text-fig. 1B). As in the middle cervical, this centrum is acamerate (*sensu* Wedel 2003) and amphicoelous. The lateral surface of the centrum is slightly more concave in this vertebra and the parapophysis is similarly located at the anterior edge of the centrum. The parapophysis of PVL 4822/21 is subcircular in shape, rather than an elongated crest. The lateromedial constriction between the anterior and posterior ends is extremely marked in this vertebra, with its width approximately 0·4 times that of the articular surfaces. As with the middle cervical, the ventral surface of this centrum lacks a sagittal crest.

The centrum PVL 4822/22 is either the last cervical or the first dorsal, based on its proportions, degree of lateromedial constriction and position of the parapophysis on the centrum. As in PVL 4822/21, the centrum is acamerate and amphicoelous, and its anterior and posterior articular facets are sub-equal in dorsoventral height (Text-fig. 1C). The parapophysis is located slightly more dorsally than in the preceding elements and is subtriangular in lateral view. The lateromedial constriction at centrum mid-length is even more developed than in the preceding vertebrae, its width being approximately 0·3 times the breadth of the anterior articular surface. In contrast to the rounded ventral surface of the preceding elements, the ventral surface of PVL 4822/22 has a sharp edge. This morphology is also present in several taxa that lack a sagittal crest in the anterior and middle cervical vertebrae (e.g. *Plateosaurus engelhardti*, MB skelett 25). Both articular surfaces are dorsoventrally deeper than lateromedially wide, in contrast to the subcircular shape seen in most other non-eusauropod sauropodomorphs (e.g. *Plateosaurus engelhardti*, MB skelett 25; *Riojasaurus incertus*, PVL 3844; *Lufengosaurus huenei*). In ventral view, the anterior articular surface is more lateromedially expanded than the posterior one, mainly because of the lateral projection of the cranially located parapophyses.

Dorsal vertebrae. Bonaparte (1999) described five dorsal neural arches belonging to PVL 4822–1. The material catalogued as

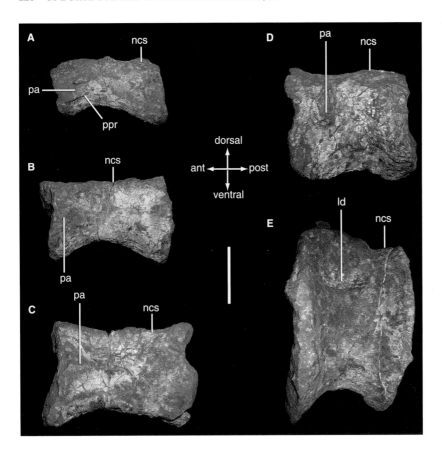

TEXT-FIG. 1. Cervicodorsal vertebral centra of *Lessemsaurus sauropoides* in lateral view. A, middle cervical PVL 4822/20. B, posterior cervical PVL 4822/21. C, posterior cervical-anterior dorsal PVL 4822/22. D, anterior dorsal PVL 4822/23. E, posterior dorsal PVL 4822/24. Scale bar represents 5 cm.

PVL 4822 also includes several dorsal centra (PVL 4822/23–4822/25). The centrum PVL 4822/23 probably belongs to the first or second dorsal vertebra (Text-fig. 1D). The height of the anterior articular surface is sub-equal to that of the posterior articular surface. However, in contrast to the cervical centra, the anterior articular surface is lateromedially narrower than the posterior articular surface. The anteroposterior length of PVL 4822/23 is only slightly greater that the dorsoventral height of its articular surfaces. The anterior dorsals of most non-eusauropod sauropodomorphs are proportionately longer than those of *Lessemsaurus sauropoides* (e.g. *Plateosaurus engelhardti*, MB skelett 25; *Massospondylus carinatus*, SAM-PK-K391). Other sauropodomorphs, however, have proportionately short and high centra, as in *Lessemsaurus sauropoides* (e.g. *Riojasaurus incertus*; *Kotasaurus yamanpalliensis*; *Patagosaurus fariasi*). The lateral surface of the centrum is concave and the centrum is acamerate and amphicoelous. The parapophysis is dorsoventrally taller than it is long anteroposteriorly and is located close to the dorsal corner of the vertebral centrum, but more posteriorly than in the preceding elements. Vertebral centrum PVL 4822/23 is notably constricted lateromedially, but its narrowest point is located on the anterior third of the vertebra, rather than at centrum mid-length, as in the cervical vertebrae. At this point, the vertebral centrum is approximately 0·5 times as wide as the anterior articular surface. As in PVL 4822/22, a sharp crest forms the ventral surface of the centrum.

The isolated dorsal centrum PVL 4822/24 and the three articulated centra PVL 4822/25 probably belong to posterior and middle dorsal vertebrae, respectively. These vertebral centra are more symmetrical than the anterior dorsals, with an anterior articular surface that is equal in depth and width to the posterior articular surface. The centra become less constricted lateromedially towards the posterior part of the dorsal series. This constriction is located at their midpoint, as in most non-eusauropod sauropodomorphs. The ventral surfaces of the centra are rounded and lack the sharp ventral edge seen in the centra around the cervicodorsal transition (except for some vertebrae that have a sharper ventral edge that may have resulted from post-mortem lateromedial crushing; PVL 4822/24). The lateral surfaces of the centra bear a central depression. This depression is deep and well delimited by sharp ridges on PVL 4822/24 (Text-fig. 1E).

Sacral vertebrae. Only two fragmentary sacral vertebrae are present in PVL 4822. These probably belong to the first and second primordial sacrals. Their centra are remarkably large and robust, with articular surfaces 108 mm wide. These vertebrae probably belong to an individual different from that pertaining to the majority of the cervicodorsal vertebrae (PVL 4822-1/1–25) owing to their significant size difference. The ventral surface is rather flattened and the lateral surface bears a notably deep small depression. Unfortunately, the transverse processes and sacral ribs are not preserved.

Pectoral girdle

Scapula. Two scapulae are included in PVL 4822: a left (PVL 4822/50) and a right (4822/51) element. These elements probably belong to different individuals as the left scapula is significantly smaller than the right and its dorsoventral extension is approximately 85 per cent of that of the right element. The small scapula (PVL 4822/50) probably belongs to the same individual as the cervicodorsal series (or a similarly-sized specimen).

The scapula of *Lessemsaurus sauropoides* is remarkably different from that of non-eusauropod sauropodomorphs. The following description is based on a vertical orientation of the scapula (Text-fig. 2). This element is strongly expanded ventrally and dorsally (Text-fig. 2). The dorsal end is approximately 0·59 times the dorsoventral extension of the scapular in the small element (PVL 4822/50), while this ratio is 0·54 in the larger left scapula (PVL 4822/51). The ventral expansion is only slightly smaller, its anteroposterior extension being approximately 0·52 times the dorsoventral length of the scapula (PVL 4822/51). PVL 4822/50 has an incomplete ventral end and cannot be precisely measured. Most non-eusauropod sauropodomorphs (e.g. *Plateosaurus engelhardti*, AMNH 6810; *Massospondylus carinatus*; *Riojasaurus incertus*, PVL 3663; *Lufengosaurus huenei*; *Yunnanosaurus huangi*; *Coloradisaurus brevis*; *Anchisaurus polyzelus*) have a much more elongate scapula with poorly expanded dorsal blades (with ratios varying between 0·22 and 0·45) and moderately expanded ventral ends (ratios varying between 0·45 and 0·5). Interestingly, the scapula of *Antetonitrus ingenipes* from the Upper Triassic of South Africa (Yates and Kitching 2003) has similarly expanded ventral and dorsal ends. Derived sauropodomorphs (e.g. *Isanosaurus attavipachi*: Buffetaut *et al.* 2000; *Kotasaurus yamanpalliensis*; and *Cetiosaurus oxoniensis*: Upchurch and Martin 2003) lack the extreme dorsal expansion present in *Lessemsaurus sauropoides* and *Antetonitrus ingenipes*, but some have similarly large ventral expansions.

The ventral end of the scapula of *Lessemsaurus sauropoides* expands anteriorly because of the acromion process, which is as reduced in the anteroposterior dimension as in most non-eusauropod sauropodomorphs. This condition contrasts with the markedly enlarged acromion process of derived sauropodomorphs (e.g. Neosauropoda; Wilson and Sereno 1998). The acromion process of *Lessemsaurus sauropoides* forms an angle of approximately 40 degrees with the dorsoventral axis of the scapula shaft (Text-fig. 2). This angle lies among those seen in non-eusauropod sauropodomorphs, which range between 40 and 50 degrees (e.g. *Riojasaurus incertus*, PVL 3663; *Plateosaurus engelhardti*, AMNH 6810; *Melanorosaurus readi*; *Antetonitrus ingenipes*). Other non-eusauropod sauropodomorphs (e.g. *Saturnalia tupiniquim*, *Coloradisaurus brevis*. *Lufengosaurus huenei*, *Massospondylus carinatus*) have angles varying between 65–80 degrees. This condition is also present in eusauropods (Yates and Kitching 2003). The ventral margin of the scapula has not been perfectly preserved in either of the scapulae referred to *Lessemsaurus sauropoides*. The lateral surface of the ventral expansion of the scapula, between the acromion process and the glenoid surface, is markedly concave, as in most sauropodomorphs. The glenoid region occupies the posterior half of the ventral expansion and is much thicker lateromedially than the rest of the scapula. The

scapula forms the dorsal half of the articular glenoid facet, which faces posteroventrally.

The scapular shaft is notably broad in comparison with other sauropodomorphs. The minimum anteroposterior width of the scapulae of *Lessemsaurus sauropoides* is approximately 0·24 times its dorsoventral extension (PVL 4822/51). In the smaller specimen (PVL 4822/50) this ratio is even larger (0·29). Most sauropodomorphs have narrower scapular shafts with ratios varying between 0·15–0·17 [e.g. *Saturnalia tupiniquim*; *Plateosaurus engelhardti*, AMNH 6810; *Massospondylus carinatus*, BPI/1/5241; *Yunnanosaurus huangi*; *Riojasaurus incertus*, PVL 3663; *Isanosaurus attavipachi* (Buffetaut *et al.* 2000)]. However, a few sauropodomorphs have a broadened scapular shaft, similar to that of *Lessemsaurus sauropoides* (*Melanorosaurus readi*; *Antetonitrus ingenipes*) and probably *Vulcanodon karibaensis*, as noted previously (Gauthier 1986; Yates and Kitching 2003). The scapula shaft of *Lessemsaurus sauropoides*, however, is notably restricted in its dorsoventral extent (Text-fig. 2), a character shared exclusively with *Antetonitrus ingenipes*. Due to its short extension and the broad ventral and dorsal ends, the anterior and posterior edges of the scapula shaft are markedly concave.

The dorsal blade of *Lessemsaurus sauropoides* is thin, laminar and has a convex dorsal margin. The posterodorsal corner of the scapula extends posteriorly as an acute process, exceeding the level of the caudal margin of the ventral expansion. The anterodorsal corner is less developed, not as acute, and is level with the anteroventral margin of the scapula.

Coracoid. Only one incomplete coracoid is preserved (PVL 4822/52). Although, the distal end of the posteroventral process is missing, the coracoid of *Lessemsaurus sauropoides* is rather small and subcircular in lateral view (as preserved), as in all other sauropodomorphs. The glenoid region is lateromedially thicker than the preserved region of the posteroventral process of the coracoid. Unfortunately, the coracoid tubercle has not been preserved, and therefore it cannot be determined if this tubercle was markedly reduced or lost as in derived sauropodomorphs (Wilson and Sereno 1998).

Forelimb

An almost complete forelimb is included in PVL 4822. These elements may belong to the smaller specimen, although the lack of comparative material precludes corroboration of this hypothesis.

Humerus. The humerus of *Lessemsaurus sauropoides* (PVL 4822/53) is particularly interesting owing to the unique combination of character states present. This element has symmetrically expanded proximal and distal ends separated by a short and narrow shaft, giving the element an hourglass shape in anterior and posterior views (Text-fig. 3). The proximal end of the humerus of *Lessemsaurus sauropoides* differs from that of non-eusauropod sauropodomorphs. The relative width of the proximal expansion with respect to humeral length is 0·4; similar to that of most non-eusauropod sauropodomorphs except for a few taxa with highly expanded proximal humeri

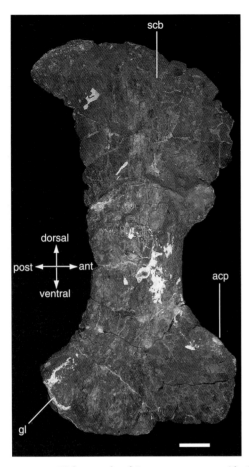

TEXT-FIG. 2. Right scapula of *Lessemsaurus sauropoides* PVL 4822/51 in lateral view. Scale bar represents 5 cm.

(e.g. *Massospondylus carinatus, Yunnanosaurus huangi, Coloradisaurus brevis*). The proximal surface of the humerus of *Lessemsaurus sauropoides* is markedly convex, with a moderately well-developed rounded articular head. A similar condition is also present in *Melanorosaurus readi, Antetonitrus ingenipes, Kotasaurus yamanpalliensis* and basal eusauropods (e.g. *Ferganasaurus verzilini*). In contrast, other non-eusauropod sauropodomorphs have a straight and transversely orientated proximal edge of the humerus (e.g. *Saturnalia tupiniquim; Plateosaurus engelhardti*, MB skelett 25; *Riojasaurus incertus; Yunnanosaurus huangi*). The medial region of the proximal end of the humerus bears a poorly developed internal tuberosity (Text-fig. 3). In contrast, some non-eusauropod sauropodomorphs have an extremely enlarged internal tuberosity (e.g. *Massospondylus carinatus, Coloradisaurus brevis, Lufengosaurus huenei, Yunnanosaurus huangi*). A less-developed internal tuberosity is a generalized feature among sauropodomorphs (e.g. *Saturnalia tupiniquim; Plateosaurus engelhardti*, MB skelett 25; *Melanorosaurus readi;* eusauropods).

The deltopectoral crest rises gradually from the rounded proximolateral corner of the humerus (Text-fig. 3A). This crest occupies 40 per cent of the length of the humerus. This proportion is slightly lower than that in most non-eusauropod sauropodomorphs, which have deltopectoral crests occupying 45–55

per cent of their humeral lengths (e.g. *Massospondylus carinatus; Lufengosaurus huenei; Plateosaurus engelhardti*, MB skelett 25; *Coloradisaurus brevis; Riojasaurus incertus*). The most basal sauropodomorph (i.e. *Saturnalia tupiniquim*), *Vulcanodon karibaensis, Kotasaurus yamanpalliensis*, and neosauropods have similarly reduced deltopectoral crests.

The deltopectoral crest is low, contrasting with the condition of most non-eusauropod sauropodomorphs in which this crest is remarkably high and sharp-edged. The condition of *Lessemsaurus sauropoides* is slightly less developed than in *Melanorosaurus readi* (SAM-PK-K3450) and *Antetonitrus ingenipes*. However, the deltopectoral crests of these taxa (including *Lessemsaurus sauropoides*) are not as low and reduced as those of eusauropod taxa (e.g. *Tehuelchesaurus benitezi, Ferganasaurus verzilini*), in which the crest is an extremely low ridge. In lateral view, the profile of the deltopectoral crest of *Lessemsaurus sauropoides* and *Antetonitrus ingenipes* is rounded and anteriorly convex. In contrast, other non-eusauropod sauropodomorphs have a subrectangular profile with a straight and vertically orientated anterior margin. In anterior view, the deltopectoral crest of *Lessemsaurus sauropoides* is slightly sinuous, being directed mediodistally along its proximal half and laterodistally along its distal half (Text-fig. 3A). The condition seen in *Lessemsaurus sauropoides*, however, is not as marked as in other sauropodomorph taxa (e.g. *Riojasaurus incertus, Coloradisaurus brevis, Lufengosaurus huenei*). The deltopectoral crest of *Lessemsaurus sauropoides* lacks the proximodistal sulcus located lateral to the crest in other non-eusauropod sauropodomorphs (e.g. *Massospondylus carinatus, Lufengosaurus huenei*).

The humeral shaft is remarkably short, occupying less than 20 per cent of the total length of the humerus (Text-fig. 3). The shaft has a circular cross-section. The distal end of the humerus is lateromedially wide and anteroposteriorly narrow. The width of the distal expansion is approximately 0·43 times the total length of the humerus. This ratio indicates a notably enlarged ventral expansion, similar to those of a few other non-eusauropod sauropodomorphs (e.g. *Yunnanosaurus huangi, Coloradisaurus brevis*). Most other sauropodomorph humeri have less expanded ventral ends, with their width/length ratios varying between 0·30 and 0·36 [e.g. *Saturnalia tupiniquim; Plateosaurus engelhardti*, MB skelett 25; *Melanorosaurus readi* (Galton *et al.* 2005); *Kotasaurus yamanpalliensis; Tehuelchesaurus benitezi; Ferganasaurus verzilini*]. The anterior surface of the distal end has a deep fossa that broadens distally (Text-fig. 3A). This fossa is deep, as in some non-eusauropod sauropodomorphs (e.g. *Plateosaurus engelhardti*, MB skelett 25; *Massospondylus carinatus*, SAM-PK-K391), although it is not circular in outline and not sharply delimited.

Because of the diversity of sizes and the lack of clear association among the assemblage of elements catalogued as PVL 4822, the relative length of the hindlimb and forelimb elements cannot be determined for *Lessemsaurus sauropoides*. However, all of the humeral remains are notably long in comparison with the known femora.

Ulna. If the only known ulna (PVL 4822/54) belongs to the same individual as the humerus PVL 4822/53, the ulna would be approximately 0·58 times the length of the humerus (Text-

TEXT-FIG. 3. Right humerus of *Lessemsaurus sauropoides* PVL 4822/53. A, anterior view. B, posterior view. Scale bar represents 5 cm.

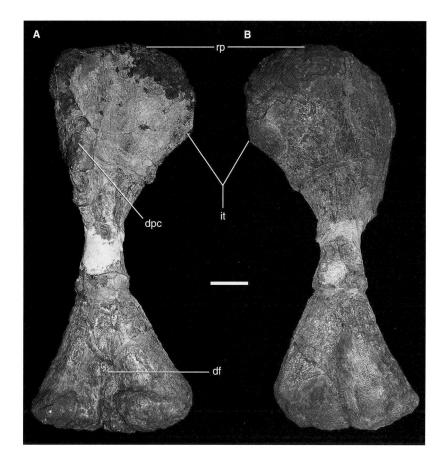

fig. 4). This ratio is similar to the condition in most non-eusauropod sauropodomorphs, with the exception of *Yunnanosaurus huangi* and *Massospondylus carinatus*, which have proportionately longer ulnae. The ulna is expanded lateromedially both proximally and distally (Text-fig. 4B). The lateromedial width of these expansions is approximately 0·35 times the length of the ulna, a proportion similar to that seen in *Yunnanosaurus huangi*, *Lufengosaurus huenei* and *Melanorosaurus readi* (SAM-PK-K3449). This metric is proportionately larger than that of *Antetonitrus ingenipes* and *Vulcanodon karibaensis*. The ulnar shaft, however, is not as lateromedially narrow as in non-eusauropod sauropodomorphs (e.g. *Yunnanosaurus huangi*, *Lufengosaurus huenei*, *Plateosaurus engelhardti*). The anteroposterior extension of the ulna shaft decreases gradually towards the distal end, where the ulna becomes anteroposteriorly flattened and lateromedially wide.

The proximal end of the ulna is expanded both lateromedially and anteroposteriorly (Text-fig. 4A). Thus, in proximal view the ulna is triradiate, as in most non-eusauropod sauropodomorphs. The anterior process of the proximal end is high and lateromedially broad, being slightly deflected laterally. This process delimits a relatively large radial fossa. In *Lessemsaurus sauropoides* this fossa is more concave than in most non-eusauropod sauropodomorphs. Some sauropodomorph taxa, however, have a similar degree of development (e.g. *Antetonitrus ingenipes*, *Yunnanosaurus huangi*). *Melanorosaurus readi* (Bonnan and Yates 2007), *Vulcanodon karibaensis*, and more derived sauropodomorphs have an even more strongly developed radial fossa (Wil-

son and Sereno 1998; Bonnan 2003; Yates and Kitching 2003). Ventrolateral to the radial fossa, the ulna of *Lessemsaurus sauropoides* has an incipiently developed lateral process that would articulate with the posterior edge of the proximal end of the radius (homologous to the 'anterolateral' process *sensu* Bonnan 2003). The posterior margin of the proximal end of the ulna of *Lessemsaurus sauropoides* is flat and extensive. The lateromedial length of this edge is approximately 0·85 times the anteroposterior length of the proximal articular surface of the ulna. The medial margin of the proximal ulna is also concave (Text-fig. 4A), although less so than the lateral edge (i.e. the radial fossa). This condition contrasts with that of most sauropodomorphs (e.g. *Yunnanosaurus huangi*, *Plateosaurus engelhardti*, *Vulcanodon karibaensis*, *Ferganasaurus verzilini*, *Tehuelchesaurus benitezi*), but is present in some of the ulnar material of *Antetonitrus ingenipes* and *Melanorosaurus readi* (SAM-PK-K3449 and NM QR 3314). The proximal end of the ulna has a poorly developed olecranon process, resembling the condition in eusauropods. The distal end of the ulna of *Lessemsaurus sauropoides* is markedly expanded lateromedially but flattened anteroposteriorly (Text-fig. 4C). The distal articular surface is subovoid in shape.

Radius. The only radius (PVL 4822/55) was probably associated with the ulna (PVL 4822/54). It is a stout and robust element (Text-fig. 4C) in comparison with most non-eusauropod sauropodomorphs. The proximal end is lateromedially narrow and anteroposteriorly elongated, as in most non-eusauropod

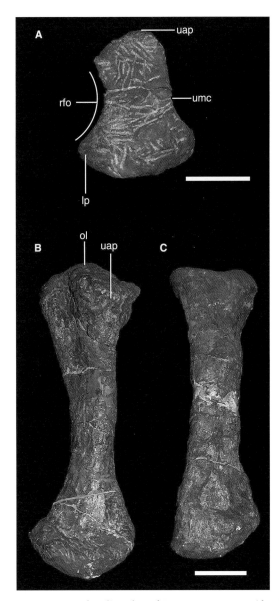

TEXT-FIG. 4. Left radius-ulna of *Lessemsaurus sauropoides*. A, ulna PVL 4822/54 in proximal view. B, ulna PVL 4822/54 in anterior view. C, radius PVL 4822/55 in lateral view. Scale bars represent 5 cm.

sauropodomorphs (including forms such as *Antetonitrus ingenipes*). Derived sauropodomorphs have a much broader proximal end of the radius, which is subtriangular in some taxa (e.g. *Camarasaurus*).

The shaft of the radius is ovoid in cross-section, with the major axis orientated anteroposteriorly. Thus, in lateral view, the shaft of the radius is poorly constricted with respect to the proximal and distal ends. This morphology contrasts with the slender, elongated, and rod-like shape of the radial shaft in most non-eusauropod sauropodomorphs (e.g. *Thecodontosaurus antiquus*, *Plateosaurus engelhardti*, *Lufengosaurus huenei*, *Riojasaurus incertus*). Some sauropodomorphs, however, have similar proportions to *Lessemsaurus sauropoides* [e.g. *Antetonitrus ingenipes*,

Jingshanosaurus xinwaensis (Zhang and Yang 1994)]. Interestingly, *Vulcanodon karibaensis* and eusauropods also have the radial shaft poorly constricted and subovoid in cross-section (e.g. *Ferganasaurus verzilini*, *Camarasaurus*). The distal end of the radius of *Lessemsaurus sauropoides* is slightly flattened lateromedially and moderately expanded anteroposteriorly (Text-fig. 4C), as in all basal saurischians. In contrast, neosauropods have a broad posterior surface that articulates with the ulna.

Manus. The manus of *Lessemsaurus sauropoides* is poorly represented in PVL 4822 and is restricted to a left metacarpal I (PVL 4822/56), a right metacarpal II (PVL 4822/57), a left phalanx I.1 (PVL 4822/58) and a left ungual of digit I (PVL 4822/59). Despite the fragmentary nature of the known remains, the manus is particularly interesting owing to its unique combination of plesiomorphic and apomorphic character states.

Although the first metacarpal is not articulated with a second metacarpal, its proximal end would probably have been proximally inset into the carpus, as in all non-eusauropod sauropodomorphs (Sereno 1999). This inference is based on the presence of a flat articular surface on the proximolateral corner of metacarpal I (Text-fig. 5A), which abuts the medial surface of distal carpal II in non-eusauropod sauropodomorphs (e.g. *Plateosaurus engelhardti*; *Riojasaurus incertus*, PVL 3662).

The first metacarpal bears many of the unique characters present in non-eusauropod sauropodomorphs, although it is remarkably broad lateromedially and short proximodistally (Text-fig. 5A). The lateromedial width of the proximal end is approximately 1·18 times the maximum proximodistal length of metacarpal I (i.e. the distance between the proximolateral end and the distal lateral condyle). A similarly short and broad metacarpal I is also present in *Antetonitrus ingenipes* and a specimen referred to *Melanorosaurus readi* (NM QR 3314; Galton *et al.* 2005). A few non-eusauropod sauropodomorphs have a slightly more elongate metacarpal I, which is sub-equal in breadth and length (e.g. *Massospondylus carinatus*, *Lufengosaurus huenei*, *Yunnanosaurus huangi*). Most non-eusauropod sauropodomorphs, however, have significantly more gracile first metacarpals, with a maximum width to maximum length ratio ranging between 0·60 and 0·76 (e.g. *Thecodontosaurus antiquus*; *Plateosaurus engelhardti*; *Riojasaurus incertus*, PVL 3662; *Anchisaurus polyzelus*). Derived sauropods also have elongated first metacarpals, which are sub-equal in length to all other metacarpals. In this derived condition, metacarpal I is part of the characteristic semi-tubular arrangement of the sauropod manus, which is interpreted as being mechanically advantageous for graviportal locomotion, as shearing and tensile forces would be reduced and redistributed (Bonnan 2003). This condition has been considered to be a synapomorphy of either Neosauropoda (Wilson and Sereno 1998) or Eusauropoda (Upchurch 1998; Bonnan 2003). However, the outgroup condition for these clades is unknown [e.g. *Barapasaurus tagorei* (Jain *et al.* 1975); *Kotasaurus yamanpalliensis*; *Vulcanodon karibaensis*].

In proximal view, metacarpal I is subtriangular, as in all non-eusauropod sauropodomorphs. The lateral surface of the proximal end is flat and extensive. The medial end is much shallower, forming the tip of the triangular proximal surface. Because of the reduced proximodistal extension of metacarpal I, the shaft is extremely short and lateromedially wide.

The distal articular condyles of metacarpal I are highly asymmetrical (Text-fig. 5A), as in all non-eusauropod sauropodomorphs. The lateral and medial condyles are large and have articular surfaces that extend along an angle of approximately 180 degrees (Text-fig. 5). The distal end of the lateral condyle only slightly exceeds that of its medial counterpart. The lateral condyle is lateromedially narrow and much higher than the medial condyle. This condition is also present in *Antetonitrus ingenipes*, but contrasts with the condition of other non-eusauropod sauropodomorphs (e.g. *Plateosaurus engelhardti*). Some derived neosauropods (e.g. *Shunosaurus*, *Brachiosaurus* and titanosauriforms) also lack this strong asymmetry (Wilson 2002). However, basal eusauropods have an asymmetrical distal end of metacarpal I (e.g. *Ferganasaurus verzilini*). The medial condyle is rather low, being approximately half of the lateral condyle's depth. Its lateromedial width is approximately twice the breadth of the lateral condyle. The ventral excursion of the articular surface of the medial distal condyle is larger than its dorsal counterpart. Thus, the entire articular surface is orientated ventrodistally rather than distally (the axis that joins the dorsal and ventral ends of the articular surface forms an angle of 50 degrees with the proximodistal axis of metacarpal I). The lateral surface of the medial condyle of metacarpal I bears a moderately well-developed ligament pit.

The second metacarpal (PVL 4822/57) is also a short and broad element (Text-fig. 5C). This metacarpal is much longer than metacarpal I, with its maximum proximodistal length approximately 1·78 times the length of the first metacarpal (PVL 4822/56). If these two elements belong to the same individual (or at least to specimens of similar size), their relative lengths would be remarkably dissimilar, a unique condition for *Lessemsaurus sauropoides*. All non-eusauropod sauropodomorphs have a metacarpal II that is only slightly longer than metacarpal I, with metacarpal II : metacarpal I length ratios varying between 1·2 and 1·4 (e.g. *Thecodontosaurus antiquus*; *Anchisaurus polyselus*; *Plateosaurus engelhardti*; *Riojasaurus incertus*, PVL 3662; *Massospondylus carinatus*; *Lufengosaurus huenei*; *Yunnanosaurus huangi*; *Antetonitrus ingenipes*). Derived sauropodomorphs (e.g. eusauropods) have a relatively longer metacarpal I than basal forms.

The proximal end of metacarpal II is rather flattened and lateromedially broad (Text-fig. 5C). The dorsal surface bears a low crest that extends proximodistally along the proximal half of metacarpal II, being slightly displaced medially as in most non-eusauropod sauropodomorphs (e.g. *Massospondylus carinatus*; *Plateosaurus engelhardti*; *Riojasaurus incertus*, PVL 3662). The medial margin of the proximal end of metacarpal II is not perfectly preserved, making it impossible to determine the presence of the flat surface that abuts the proximolateral end of metacarpal I in some non-eusauropod sauropodomorphs (e.g. *Plateosaurus engelhardti*, *Massospondylus carinatus*, *Antetonitrus ingenipes*). The proximolateral corner of the second metacarpal of *Lessemsaurus sauropoides* has a short and acute process extending laterally. This process is either absent or only incipiently present in other sauropodomorphs (Text-fig. 5C).

The shaft of metacarpal II is poorly constricted, dorsoventrally flattened, and extremely short proximodistally. The narrowest

point is located distal to the metacarpal mid-length. The subovoid distal end is poorly preserved. Despite the poor preservation, it seems to lack both well-developed condyles and an intercondylar groove.

Remains of digit I include the first phalanx of manual digit I (PVL 4822/58), which was probably associated with the first metacarpal (PVL 4822/56). This element has the characteristic torsion along the proximodistal axis that is present in all non-eusauropod sauropodomorphs. In *Lessemsaurus sauropoides*, the torsion of phalanx I.1 is approximately 29 degrees (Text-fig. 5B). This relatively low degree of torsion is similar to that of most non-eusauropod sauropodomorphs (e.g. *Thecodontosaurus antiquus*; *Plateosaurus engelhardti*; *Riojasaurus incertus*, PVL 3362). This condition contrasts with the highly twisted phalanx I.1 of a few non-eusauropod sauropodomorphs, in which the torsion is approximately 45 degrees (e.g. *Massospondylus carinatus*, *Yunnanosaurus huangi*, *Lufengosaurus huenei*), as well as with the lack of torsion in derived sauropodomorphs (e.g. eusauropods).

Manual phalanx I.1 of *Lessemsaurus sauropoides* is remarkably short and broad, being slightly wider at its proximal end than proximodistally long (Text-fig. 5B). This character is also present in *Antetonitrus ingenipes*. All other non-eusauropod sauropodomorphs have either a subquadrangular phalanx I.1 (e.g. *Massospondylus carinatus*, *Lufengosaurus huenei*, *Yunnanosaurus huangi*), or a phalanx I.1 that is longer than wide (e.g. *Thecodontosaurus antiquus*; *Anchisaurus polyzelus*; *Efraasia minor*; *Plateosaurus engelhardti*; *Riojasaurus incertus*, PVL 3662). The proximal articular surface of phalanx I.1 has two concave facets divided by a well-developed ridge. This ridge ends dorsally at the well-developed proximodorsal lip and ventrally at the tip of the proximal heel of phalanx I.1 (Text-fig. 5B). The lateral articular facet is higher and faces proximolaterally, while the reduced medial facet faces proximomedially. The medial facet extends proximoventrally much more than the lateral facet, contributing to the development of the proximal heel of phalanx I.1. Thus, in medial view, this heel seems remarkably extensive while in lateral view it is as developed as in other basal saurischians. The differential development and extension of the proximal articular surfaces of manual phalanx I.1 is clearly correlated with the asymmetrical morphology of the distal condyles of metacarpal I.

Due to the short proximodistal length of phalanx I.1, its shaft is short and rather wide lateromedially, being approximately 0·6 times the width of the proximal end. The shaft of phalanx I.1 is high dorsoventrally, being sub-equal in height with respect to the distal and proximal ends (excluding the proximal heel of the latter). The distal end of phalanx I.1 bears an extensive articular surface, extending along an angle of 180 degrees. The lateral and medial condyles diverge ventrally, so that the dorsal surface of the distal end of phalanx I.1 is much narrower lateromedially than its ventral counterpart. This articular surface is ginglymoidal and the intercondylar groove is deep along the ventral half of this articular surface. The distal articular surface extends more ventrally than dorsally, as in other non-eusauropod sauropodomorphs. The axis that joins the dorsal and ventral ends of the medial articular surface forms an angle of 60 degrees with the proximodistal axis of phalanx I.1. The lateral pit for the collateral ligament is extensive, but shallow and poorly delimited. The

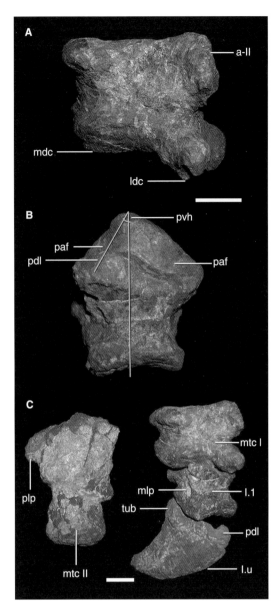

TEXT-FIG. 5. Manual remains of *Lessemsaurus sauropoides*. A, first left metacarpal PVL 4822/56 in dorsal view. B, left manual phalanx I.1 PVL 4822/58 in dorsal view. C, second right metacarpal PVL 4822/57 and left manual digit I (including the ungual PVL 4822/59) in dorsal view. Scale bars represent 2 cm.

ventral end of the first manual ungual of *Lessemsaurus sauropoides* has a weak flexor tubercle (Text-fig. 5C). This contrasts with the large flexor tubercle found in all basal saurischians, including all non-eusauropod sauropodomorphs (e.g. *Plateosaurus engelhardti*, *Yunnanosaurus huangi*, *Massospondylus carinatus*). Furthermore, some non-eusauropod sauropodomorphs have an extremely well-developed flexor tubercle (e.g. *Anchisaurus polyzelus*; *Thecodontosaurus antiquus*; *Efraasia minor*; *Melanorosaurus readi*, NM QR 3314). Basal eusauropods, in contrast, lack a well-developed flexor tubercle (e.g. *Ferganasaurus verzilini*). Ungual I also lacks the extensive proximodorsal lip present in most non-eusauropod sauropodomorphs (Text-fig. 5C).

The lateral and medial surfaces of ungual I are slightly convex and lack the characteristic groove that bifurcates proximally in all non-eusauropod sauropodomorphs (and sauropodomorph outgroups). Interestingly, the condition seen in *Lessemsaurus sauropoides* is also present in basal eusauropods (e.g. *Ferganasaurus verzilini*).

The distal half of ungual I is strongly recurved in most non-eusauropod sauropodomorphs (e.g. *Thecodontosaurus antiquus*, *Anchisaurus polyzelus*, *Plateosaurus engelhardti*, *Massospondylus carinatus*, *Lufengosaurus huenei*). In these forms, the tangents of the proximal and distal ends of the dorsal margin form an angle of approximately 90 degrees. In contrast, ungual I of *Lessemsaurus sauropoides* is significantly less recurved (Text-fig. 5C). The distribution of this character among basal eusauropods is poorly known, but all known eusauropod manual unguals from digit I are less recurved than those of non-eusauropod sauropodomorphs. The dorsoventral height of the first manual ungual of *Lessemsaurus sauropoides* tapers constantly and gradually along its length (Text-fig. 5C). In contrast, in some non-eusauropod sauropodomorphs, there is a steep reduction in the dorsoventral height at a point close to the mid-length of the first manual ungual (e.g. *Thecodontosaurus antiquus*, *Anchisaurus polyzelus*, *Efraasia minor*, *Yunnanosaurus huangi*; *Lufengosaurus huenei*).

Pelvic girdle

An almost complete pelvis is present in PVL 4822, including a right ilium (PVL 4822/60), a left and right pubis (PVL 4822/61–4822/62), and fragmentary remains of a distal ischium (PVL 4822/63).

Ilium. The ilium (PVL 4822/60) of *Lessemsaurus sauropoides* is similar to that of most non-eusauropod sauropodomorphs (Text-fig. 6). The iliac blade is dorsoventrally low and anteroposteriorly elongated. As in all non-eusauropod sauropodomorphs, the preacetabular process is short and subtriangular, with its apex directed anteriorly. A similar condition is also present in the basal sauropod *Kotasaurus yamanpalliensis*, although all eusauropods have an extensive, dorsoventrally high preacetabular process that extends anterior to the pubic peduncle (Sereno 1999; Yates and Kitching 2003). The anterior end of the preacetabular process of *Lessemsaurus sauropoides* differs from that of other non-eusauropod sauropodomorphs. Firstly, its anterior end is straight, rather than ventrally deflected as in some taxa (e.g. *Yunnanosaurus huangi*). Secondly, the dorsal margin of the

medial pit is reduced in extent but is notably deeper and well-delimited (Text-fig. 5C). This depression is located close to the dorsal end of the distal articular condyle.

The ungual of digit one (PVL 4822/59) is robust, lateromedially flattened, and well developed (Text-fig. 5C), being sub-equal in length to the combined proximodistal lengths of metacarpal I and phalanx I.1. The proximal end of ungual I bears two concave articular facets divided by a sharp ridge. In lateral and medial views, these articular facets are not as concave as in most non-eusauropod sauropodomorphs (e.g. *Plateosaurus engelhardti*, *Yunnanosaurus huangi*, *Massospondylus carinatus*). The proximo-

preacetabular process is continuous with the dorsal margin of the iliac blade (Text-fig. 6). In other non-eusauropod sauropodomorphs, these two margins are separated by a moderately developed step (e.g. *Riojasaurus incertus*; *Lufengosaurus huenei*; *Yunnanosaurus huangi*; *Massospondylus carinatus*, BPI/1/4693).

As mentioned above, the iliac blade of *Lessemsaurus sauropoides* is dorsoventrally low above the acetabulum (Text-fig. 6). The dorsal margin of the iliac blade is orientated subparallel to the longitudinal axis of the skeleton and is slightly convex. These characters represent the plesiomorphic condition for Sauropodomorpha and are strongly modified in eusauropods, which possess a high iliac blade with a strongly convex dorsal margin (McIntosh 1990).

The acetabular region is enlarged anteroposteriorly relative to the condition in most non-eusauropod sauropodomorphs. The acetabulum is rather high dorsoventrally, occupying half of the maximum dorsoventral height of the ilium (measured at the ischial peduncle). The supracetabular crest is slightly widened at the base of the pubic peduncle (Text-fig. 6). This crest forms a narrow shelf that extends anteroventrally along the dorsal half of the pubic peduncle (although the ventral end of the crest seems to be broken). The supracetabular crest represents an intermediate morphology between the condition of most non-eusauropod sauropodomorphs and eusauropods. All non-eusauropod sauropodomorphs (except *Saturnalia tupiniquim*) have a similarly located supracetabular crest, but it is significantly more developed than in *Lessemsaurus*. In ventral view, the supracetabular crest of non-eusauropod sauropodomorphs is a broad shelf with laterally convex margins. Derived sauropodomorphs (e.g. eusauropods) lack this large supracetabular crest. The acetabulum is completely open, as in all sauropodomorphs (except for *Saturnalia tupiniquim*; Langer 2003). However, the acetabulum of *Lessemsaurus sauropoides* bears a unique condition among sauropodomorphs. The anterior articular surface of the acetabulum has a narrow medial flange that faces laterally, forming a narrow and marginal medial wall of the acetabulum (Text-fig. 6). This narrow medial wall extends along most of the dorsal margin of the acetabulum. The posterior articular surface of acetabulum, extending along the ischial peduncle, lacks this

structure. The acetabular articular surface of the ischial peduncle is flat or slightly convex, while the anterior and dorsal surfaces are slightly concave.

The pubic peduncle is well developed and extends anteroventrally at an angle of approximately 50 degrees with respect to the longitudinal axis of the ilium (Text-fig. 6). In contrast to the condition in most non-eusauropod sauropodomorphs, the pubic peduncle expands anteroposteriorly towards its ventral end. Owing to the extension of the medial acetabular wall, the cross-section at the mid-point of the pubic peduncle is unique among non-eusauropod sauropodomorphs. Its anterior surface is convex, its extensive medial edge is flat and orientated anteroposteriorly, and its posterior surface (i.e. the acetabular surface) is posteriorly concave. The anterior and posterior surfaces meet at the sharp lateral edge of the pubic peduncle formed by the supracetabular crest. At the distal end of the pubic peduncle, this crest disappears and the cross-section becomes teardrop-shaped, tapering posteriorly from the convex anterior margin.

The ischial peduncle is only slightly shorter than the pubic peduncle as in several other taxa (e.g. *Thecodontosaurus caducus*, BMNH P77/1; *Efraasia minor*, SMNS 12354; *Plateosaurus engelhardti*; *Yunnanosaurus huangi*; *Riojasaurus incertus*). Other non-eusauropod sauropodomorphs have an ischial peduncle that is significantly shorter than the pubic peduncle (e.g. *Massospondylus carinatus*, BPI/1/4693; *Lufengosaurus huenei*). The latter condition is accentuated in *Kotasaurus yamanpalliensis*, *Vulcanodon karibaensis* and eusauropods, where the ischial peduncle is almost absent. The ischial peduncle has a subtriangular cross-section, with its base located anteriorly (on the acetabular surface). The posteroventral edge of the ischial peduncle lacks the distinct heel present in some non-eusauropod sauropodomorphs (e.g. *Plateosaurus engelhardti*, *Riojasaurus incertus*).

The postacetabular blade of the ilium of *Lessemsaurus sauropoides* is slightly reduced anteroposteriorly in comparison with the elongated condition of most non-eusauropod sauropodomorphs (e.g. *Plateosaurus engelhardti*, *Riojasaurus incertus*). However, the condition present in *Lessemsaurus* (Text-fig. 6) does not reach the degree of reduction seen in the postacetabular blade of *Kotasaurus*. Derived eusauropods have an extremely

TEXT-FIG. 6. Right ilium of *Lessemsaurus sauropoides* PVL 4822/60 in lateral view. Scale bar represents 5 cm.

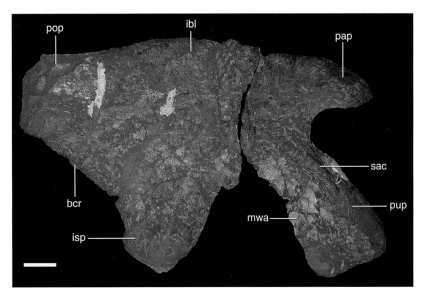

reduced postacetabular blade (Yates and Kitching 2003). The dorsal margin of the postacetabular blade is slightly concave. The posterior end is blunt and dorsoventrally extensive, as in the basal sauropodomorph *Thecodontosaurus caducus*. The posterior margin of the postacetabular process of *Lessemsaurus* is not well preserved, however, and its blunt condition could be a result of preservational causes.

The ventral margin of the postacetabular process is entirely occupied by the posterior extension of the 'brevis crest' (i.e. the sharp ridge that connects the ventral surface of the brevis fossa with the posterior edge of the ischial peduncle). The degree of development of this crest is unique among sauropodomorphs (Text-fig. 6). It extends from the base of the ischial peduncle to a point that lies almost at the posterior tip of the postacetabular process. A similar crest is present in some non-eusauropod sauropodomorphs (e.g. *Riojasaurus incertus*), although in this taxon the crest is significantly reduced, originating at the dorsoventral midpoint of the ischial peduncle and disappearing well before the posterior end of the postacetabular process (close to the anterior end of the brevis fossa). Due to the posterior extension of the brevis crest of *Lessemsaurus*, the brevis fossa is short anteroposteriorly.

Pubis. A pair of articulated pubes (PVL 4822/62) is interesting owing to its unique combination of plesiomorphic and apomorphic sauropodomorph character states. The pubes are moderately elongated and narrow, and their maximum lateromedial width would have been approximately 0·65 times their maximum proximodistal length (Text-fig. 7). Several non-eusauropod sauropodomorphs have similar proportions [e.g. *Plateosaurus engelhardti*, *Massospondylus carinatus* (Cooper 1981), *Vulcanodon karibaensis*, *Tazoudasaurus naimi*]. Eusauropods, in contrast, have much wider and shorter pubes, as noted by Cooper (1984).

The pubic plate is extensive, occupying approximately 40 per cent of the entire length (Text-fig. 7). The proximal plate is relatively large, resembling the condition of *Vulcanodon karibaensis*, *Tazoudasaurus naimi* and basal eusauropods [*Kotasaurus yamanpalliensis*, *Shunosaurus lii* (Zhang 1988), *Omeisaurus maoianus*], in which the proximal plate occupies 40–56 per cent of pubis length. Basal sauropodomorphs have a proportionately more restricted pubic plate that usually occupies less than 33 per cent of total pubic length. As in most sauropodomorphs, the lateral margin of the pubic plate lacks lateral pubic tubercles. These are present in basal saurischians (e.g. *Herrerasaurus ischigualastensis*; Novas 1993) and a few non-eusauropod sauropodomorphs (e.g. *Saturnalia tupiniquim*; *Efraasia minor*, SMNS 12354; *Plateosaurus engelhardti*, SMNS 12950). The margins of the obturator foramen of *Lessemsaurus sauropoides* are not well preserved, although this opening seems to be rather reduced, as in derived sauropodomorphs.

Distal to the pubic plate, the pubis is flat, lateromedially wide, and orientated transversely (Text-fig. 7). This orientation and morphology of the pubic apron is the plesiomorphic condition for Sauropodomorpha as it is present in sauropodomorph outgroups (*Herrerasaurus ischigualastensis*: Novas 1993) and all non-eusauropod sauropodomorphs (including *Vulcanodon karibaensis* and *Tazoudasaurus naimi*). The lateral margins of the pubic apron are slightly concave, as in some non-eusauropod

sauropodomorphs (e.g. *Coloradisaurus brevis*, *Massospondylus carinatus*, *Lufengosaurus huenei*, *Tazoudasaurus naimi*). However, in contrast to these forms, the pubic apron of *Lessemsaurus* tapers gradually along its distal end. The lateromedial width of the pubic apron is slightly more developed with respect to the pubic apron's length than in non-eusauropod sauropodomorphs. The minimum lateromedial width of the pubic apron of *Lessemsaurus* is approximately 80 per cent of its proximodistal length, while the maximum lateromedial width is sub-equal to its proximodistal length. The blade of the pubic apron is thick at its lateral margin and remarkably thin along its medial edges. The distal end of the pubes (Text-fig. 7) is less expanded than in most non-eusauropod sauropodomorphs.

Ischium. The ischia are represented by two conjoined distal ends (PVL 4822/63; Text-fig. 8). As in all non-neosauropod sauropodomorphs, the preserved portion of the ischial shaft is subtriangular in cross-section. The distal end of the conjoined ischia is subtriangular, and its dorsoventral height is approximately 0·8 times its lateromedial width. A similar condition is present in *Vulcanodon karibaensis* and some non-eusauropod sauropodomorphs (e.g. *Plateosaurus engelhardti*).

Hindlimb

Femur. Two femora are present: one is poorly preserved (PVL 4822/64), while a second (right) is complete (PVL 4822/65). The femur has a posteriorly bent distal end and is sigmoid in

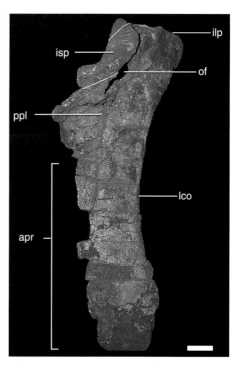

TEXT-FIG. 7. Right pubis of *Lessemsaurus sauropoides* PVL 4822/61 in ventral view. Scale bar represents 5 cm.

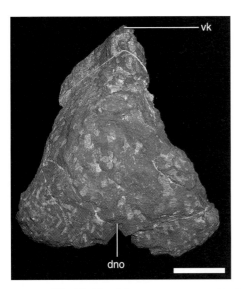

TEXT-FIG. 8. Distal ischia of *Lessemsaurus sauropoides* PVL 4822/63 in ventrodistal view. Scale bar represents 5 cm.

The fourth trochanter is a well-developed crest located at femoral mid-length (Text-fig. 9), as in *Antetonitrus ingenipes*, *Vulcanodon karibaensis* and eusauropods. In contrast, non-eusauropod sauropodomorphs have a fourth trochanter located on the proximal half of the femoral shaft. In medial view, the fourth trochanter has a straight profile. As in most non-eusauropod sauropodomorphs (Langer 2003), the proximal end of this ridge merges into the femoral shaft gradually while its distal end is remarkably steep (Text-fig. 9B). The fourth trochanter is located close to the medial margin of the femoral shaft, a character present in some non-eusauropod sauropodomorphs [*Riojasaurus incertus*, *Coloradisaurus brevis*, *Anchisaurus polyzelus*, *Melanorosaurus readi* (Galton *et al.* 2005), *Antetonitrus ingenipes*] and eusauropods. In posterior view, the fourth trochanter has a sigmoid profile. The medial surface of the fourth trochanter bears a pronounced depression for the insertion of the m. caudofemoralis longus, as in *Melanorosaurus readi* and *Antetonitrus ingenipes*. Unfortunately, the lesser trochanter is poorly preserved in both preserved femora (PVL 4822/64–4822/65) and it cannot be determined if it was as well developed as in *Riojasaurus incertus* or *Melanorosaurus readi* or if it was reduced, as in other sauropodomorphs.

The femoral shaft is subovoid, being slightly wider lateromedially and anteroposteriorly (Text-fig. 9). This condition is intermediate between the subcircular femoral shaft of most non-eusauropod sauropodomorphs and the derived elliptical cross-section of *Antetonitrus ingenipes* and eusauropods. The distal end of the femur markedly expands lateromedially at the level of the condyles with respect to the lateromedial width of the femoral shaft. The popliteal fossa on the posterior surface of the distal femur is remarkably deep owing to the large posterior extension of the tibial and fibular condyles.

lateral view (Text-fig. 9B) as in most non-eusauropod sauropodomorphs (Galton 1990). In posterior view, however, the femur of is straight (Text-fig. 9A) and lacks the lateral curvature of the femoral distal end that is present in some non-eusauropod sauropodomorphs (e.g. *Plateosaurus engelhardti*). The femoral head is well developed, having its major axis perpendicular to the proximodistal axis of the femur and to the longitudinal axis of the skeleton.

TEXT-FIG. 9. Right femur of *Lessemsaurus sauropoides* PVL 4822/65. A, posterior view. B, posteromedial view. Scale bar represents 10 cm.

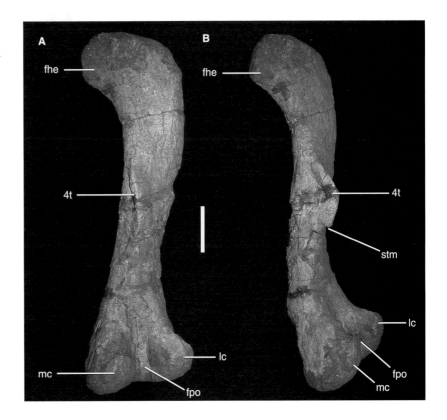

Tibia. Two tibiae are present (Text-fig. 10). The right tibia (PVL 4822/66) is slightly larger than the left one (PVL 4822/67): they probably belonged to different individuals. In comparison with the only complete femur of *Lessemsaurus sauropoides* (PVL 4822/65), the largest tibia is approximately 0·60 times the proximodistal length of the femur. However, these comparisons must be taken cautiously as these elements may belong to different-sized individuals and the femur/tibia length ratio is subject to strong ontogenetic variation in other non-eusauropod sauropodomorphs (e.g. *Mussaurus patagonicus*; Bonaparte and Vince 1979). It is interesting to note, however, that this ratio is similar to that of *Vulcanodon karibaensis* and eusauropods.

The proximal surface is subtriangular, having a remarkably large lateromedial extension along its posterior edge. The cnemial crest is lateromedially broad in the right tibia (PVL 4822/67) and lateromedially flattened in the larger left element (PVL 4822/66). These differences are probably the result of lateromedial crushing of PVL 4822/66. In both specimens, the cnemial crest does not project dorsally (Text-fig. 10), as in all non-eusauropod sauropodomorphs. The proximal half of the tibial shaft is subcircular in cross-section. Towards the distal end, however, the tibial shaft expands lateromedially. This expansion is present in all sauropodomorphs, except for basal forms (e.g. *Saturnalia*, *Thecodontosaurus caducus*). The distal portion of the tibial shaft of *Lessemsaurus* is uniquely flattened anteroposteriorly (Text-fig. 10). Thus, the cross-section of the distal tibia is subrectangular with its major axis orientated lateromedially and is twice as long transversely as anteroposteriorly.

The distal tibia has the articular socket for the ascending astragalar process characteristic of sauropodomorphs (Text-fig. 10B). This articular concavity extends proximodistally between the posterolateral and anterolateral processes of the distal tibia. In lateral view, the anterolateral flange exceeds distally the limit of the posterolateral process. Interestingly, the anterolateral descending process of the tibia is more laterally extensive than the posterolateral process (Text-fig. 10B). Thus, the articular socket of the tibia is visible in posterior view but is hidden in anterior view by the extensive anterolateral process. As noted by Yates (2004), this morphology only occurs in neosauropods and the non-eusauropod sauropodomorphs *Anchisaurus polyzelus* and *Antetonitrus ingenipes*. The medial surface of the distal tibia is markedly convex, whereas the anterior and posterior surfaces of the distal tibia are flat.

Astragalus. The astragalus of *Lessemsaurus* is only known from a single right element (PVL 4822/68; Text-fig. 11). In most respects, this element resembles the morphology seen in non-eusauropod sauropodomorphs. The main body of the astragalus is subrectangular, having its major axis orientated lateromedially. The lateral end of the astragalar body is slightly shorter anteroposteriorly (approximately 80 per cent) than the medial end (Text-fig. 11B). This condition is also present in some non-eusauropod sauropodomorphs (e.g. *Melanorosaurus readi*, *Mussaurus patagonicus*, *Coloradisaurus brevis*) and most basal eusauropods (Upchurch 1995, 1998; Wilson 2002). The posteromedial corner of the astragalus of *Lessemsaurus* is formed by straight medial and posterior edges that meet at a right angle (Text-fig. 11B), as in most non-eusauropod sauropodomorphs.

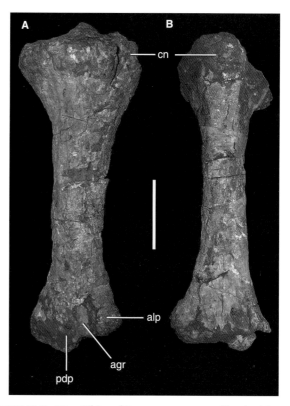

TEXT-FIG. 10. Tibiae of *Lessemsaurus sauropoides*. A, right tibia PVL 4822/66 in posterolateral view. B, left tibia 4822/67 in anterior view. Scale bar represents 10 cm.

In contrast, neosauropods have a medially tapering subtriangular astragalar body with an anteroposteriorly long lateral edge and an extremely short medial end (Upchurch 1995; Wilson and Sereno 1998). The condition of this character in the Early Jurassic *Vulcanodon karibaensis* and *Tazoudasaurus naimi* appears to be intermediate between the subrectangular astragalar body of non-eusauropod sauropodomorphs and the triangular-shaped astragalus of neosauropods.

The proximodistal depth of the astragalar main body is approximately constant, as in other non-eusauropod sauropodomorphs. The distal surface of the astragalus of *Lessemsaurus* is slightly convex, similar to all non-eusauropod sauropodomorphs and basal eusauropods. Neosauropods, in contrast, have a strongly convex ventral surface of the astragalar body (Upchurch 1995; Wilson 2002).

The astragalar ascending process is remarkably extensive. It occupies approximately 70 per cent of the lateromedial extension of the astragalus and 83 per cent of its anteroposterior length. Thus, the posterior and medial basins of the astragalar dorsal surface are reduced in *Lessemsaurus* (Text-fig. 11B). Most non-eusauropod sauropodomorphs have smaller astragalar ascending processes and extensive posterior and medial concave surfaces (*Plateosaurus engelhardti*; *Riojasaurus incertus*, PVL 3663; *Coloradisaurus brevis*). Despite its reduction, the posterior concave facet of the astragalar dorsal surface still separates the ascending process form the astragalar posterior margin in *Lessemsaurus*, in contrast to the condition of *Mamenchisaurus* and neosauropods

(Wilson 2002). The posterior concave facet of *Lessemsaurus* bears small foramina but lacks the vertical crest present in *Barapasaurus* and more derived sauropods (Wilson 2002).

The proximal articular surface of the ascending process of *Lessemsaurus* faces proximomedially and is slightly deflected anteriorly. As in most non-eusauropod sauropodomorphs, the lateral and anterior surfaces of the astragalar ascending process are vertically orientated. The anterior surface of the ascending process has well-developed fossa (present in most non-eusauropod sauropodomorphs). This structure is present in *Blikanasaurus cromptoni*, but is lost in *Vulcanodon karibaensis* and Eusauropoda (Wilson and Sereno 1998).

Pes. Pedal remains are fragmentary, but include: a complete, poorly preserved, metatarsal I (PVL 4822/69); a proximal end of metatarsal II (PVL 4822/70); a complete metatarsal III (PVL 4822/71); proximal and distal ends of metatarsal IV (PVL 4822/72–4822/73); and a complete metatarsal V (PVL 4822/74). The metatarsus of *Lessemsaurus* is relatively broad and robust (Text-fig. 12A) in comparison with those of non-eusauropod sauropodomorphs, but it is more slender and elongate than in *Blikanasaurus cromptoni* or eusauropods.

TEXT-FIG. 11. Left astragalus of *Lessemsaurus sauropoides* PVL 4822/68. A, posterior view. B, dorsal view. C, ventral view. Scale bar represents 5 cm.

Metatarsal I (PVL 4822/69) is a long and robust element that is poorly constricted at its midpoint (Text-fig. 12A). The proximodistal length of this metatarsal is approximately 0·79 times the length of the third metatarsal. In most non-eusauropod sauropodomorphs, the first metatarsal is significantly shorter, being approximately 0·60–0·65 times the length of metatarsal III. The condition in *Blikanasaurus cromptoni* and *Vulcanodon karibaensis* also falls within this range. However, more derived forms (e.g. *Omeisaurus maoianus*) have a metatarsal I that is enlarged and similar to that of *Lessemsaurus*. The proximal end of metatarsal I is slightly expanded lateromedially, but its preservation is extremely poor. The expanded proximal region of metatarsal I overlaps the proximal end of metatarsal II. The shaft of metatarsal I is only slightly constricted (Text-fig. 12A), but its lateromedial width is similar to that of metatarsal III (as in non-eusauropod sauropodomorphs). Derived sauropodomorphs generally have a first metatarsal that is much wider than the other elements (Wilson and Sereno 1998). The distal end of metatarsal I is also poorly preserved, although it can be noted that the articular condyles are asymmetrically developed. The surface of the distal end of metatarsal I is rather expanded and bears a shallow ligament pit. The medial articular surface is dorsoplantarly short as in most non-neosauropod sauropodomorphs.

The second metatarsal is known only from its proximal end (PVL 4822/70). As in all non-eusauropod sauropodomorphs (including *Blikanasaurus cromptoni*), the proximal articular surface has an hourglass shape in proximal view, with strongly concave lateral and medial margins for the articulation of metatarsals I and III (Sereno 1999). The lateral concavity is secondarily lost in *Vulcanodon karibaensis* and more derived sauropods. The proximal articular surface is flat and dorsoventrally elongated with straight dorsal and plantar edges (Text-fig. 12A).

The third metatarsal is complete but imperfectly preserved (PVL 4822/71). The proximal articular facet is subtriangular, having a broad plantar surface and a narrow dorsal apex. The metatarsal III of *Lessemsaurus sauropoides* tapers gradually along its shaft, reaching its minimum width close to its distal end (Text-fig. 12A). At this point, the third metatarsal of *Lessemsaurus sauropoides* is slightly flattened dorsoventrally. The distal end of the metatarsal III is slightly deflected medially with respect to its proximodistal axis. Although this could be accentuated by its preservation, a similar deflection is present in other sauropodomorphs (e.g. *Coloradisaurus brevis*; *Massospondylus carinatus*, BPI/1/4377; *Blikanasaurus cromptoni*; *Vulcanodon karibaensis*). Unfortunately, the distal articular surface of PVL 4822/71 is broken. The lateral and medial surfaces of the distal end of metatarsal III have shallow but clearly defined collateral ligament pits. The dorsal depression for the extensor ligament seems to be absent from the third metatarsal of *Lessemsaurus*, although as noted above, this region is poorly preserved so this could be a preservational artefact.

The fourth metatarsal is only represented by proximal (PVL 4822/72) and distal ends (PVL 4822/73). These two fragments might belong to the same element, as they have comparable sizes and preservational attributes (Text-fig. 12A). As in all non-neosauropod sauropodomorphs, metatarsal IV is remarkably broad at its proximal end but narrow distally. The dorsoventral height of the proximal end is low, as in other sauropodomorphs

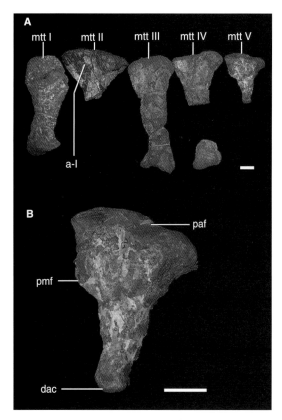

TEXT-FIG. 12. Metatarsal elements of *Lessemsaurus sauropoides*. A, metatarsal I (PVL 4822/69), metatarsal II (PVL 4822/70), metatarsal III (PVL 4822/71), proximal and distal ends of metatarsal IV (PVL 4822/72–4822/73) and metatarsal V (PVL 4822/74) in dorsal view. B, detail of reduced metatarsal V PVL 4822/74. Scale bars represent 3 cm.

[e.g. *Plateosaurus engelhardti*; *Massospondylus carinatus*, BPI/1/ 4377; *Blikanasaurus cromptoni*; *Vulcanodon karibaensis*; *Omeisaurus tinafuensis* (Tang *et al.* 2001)]. The dorsal surface of the proximal end of metatarsal IV lacks a proximodistally orientated crest. The poor preservation of this region, however, precludes determining if this crest was truly absent in *Lessemsaurus* or if it is just missing in PVL 4822/72 because of incomplete preservation. The determination of this character with more material would be interesting as the crest of eusauropods is reduced in comparison with that of most non-eusauropod sauropodomorphs (e.g. *Plateosaurus engelhardti*; *Massospondylus carinatus*, BPI/1/4377; *Coloradisaurus brevis*; *Riojasaurus incertus*, PVL 3526; *Blikanasaurus cromptoni*). The distal end of metatarsal IV (PVL 4822/73) is poorly expanded with respect to the proximal end (PVL 4822/72). Its articular surface is subtrapezoidal in cross-section, being much narrower than in metatarsal III (Text-fig. 12A). The dorsal half of the articular surface is markedly convex and lacks an intercondylar groove, as in *Riojasaurus* (PVL 3526). The ventral surface of the distal articular surface bears a broad groove between the articular condyles. The lateral condyle extends ventrolaterally in a well-developed flange, as in most non-eusauropod sauropodomorphs (including *Blikanasau-*

rus cromptoni). Derived sauropodomorphs (e.g. eusauropods), have a less prominent ventrolateral flange on the lateral distal condyle of metatarsal IV. As in most non-eusauropod sauropodomorphs, metatarsal IV of *Lessemsaurus* has a well-developed lateral ligament pit. The dorsal fossa for the extensor ligament is absent from metatarsal IV.

Metatarsal V (PVL 4822/74) is flat and triangular as in all non-eusauropod sauropodomorphs (Text-fig. 12A–B). It is remarkably short, being 0·44 times the length of metatarsal III. The reduced condition in *Lessemsaurus* is also present in all other non-eusauropod sauropodomorphs (including *Blikanasaurus cromptoni*). Derived sauropodomorphs (e.g. *Vulcanodon karibaensis*, *Omeisaurus maoianus*) have a much more elongated metatarsal V, which is approximately 0·7 times the length of metatarsal III. The proximal end is dorsoventrally low and lateromedially wide. Metatarsal V of *Lessemsaurus* tapers distally along its entire length. The medial margin of this element has a pronounced flange that extends medially and would have been overlapped by metatarsal IV (Text-fig. 12B) The distal end of metatarsal V bears a hemispherical terminal surface (Text-fig. 12B). It is unclear if this surface is indicative of the presence of an ossified pedal phalanx V.1. Several other non-eusauropod sauropodomorphs have similar distal ends of metatarsal V but were found without a phalanx in this digit (e.g. *Thecodontosaurus caducus*; *Anchisaurus polyzelus*; *Riojasaurus incertus*, PVL 3526; *Blikanasaurus cromptoni*).

Remains of the pedal digits of *Lessemsaurus* are also incomplete. These are restricted to three non-terminal phalanges (PVL 4822/75–4822/77) and two unguals (PVL 4822/78–4822/79). The three non-terminal pedal phalanges are probably proximal elements based on their size (relative to the metatarsals) and proportions (Text-fig. 13A). Two of these phalanges are subquadrangular in dorsal view, with marked lateromedial constriction in their shafts (PVL 4822/75–4822/76). These phalanges probably belong to pedal digits I and II based on the development of their ginglymoidal articular surface. The third non-terminal phalanx (PVL 4822/77) is slightly longer than wide and is less constricted at its proximodistal midpoint. Its distal end has two well-developed articular condyles, although the intercondylar groove is much shallower than in the two other phalanges. The morphology of this phalanx is most congruent with the first pedal phalanx of digit III of other non-eusauropod sauropodomorphs. If these interpretations are correct, the pedal digits of *Lessemsaurus* would resemble the short and broad condition present in some non-eusauropod sauropodomorphs (e.g. *Melanorosaurus readi*, NM QR 3314; *Blikanasaurus cromptoni*). Other non-eusauropod sauropodomorphs have a much more elongated pes (e.g. *Plateosaurus engelhardti*; *Coloradisaurus brevis*).

One of the pedal unguals (PVL 4822/78) is significantly taller dorsoventrally than the other (PVL 4822/79). These unguals are large, with their proximodistal height approximately 0·5 times the length of metatarsal III and 0·64 times the length of metatarsal I (Text-fig. 13B). The larger pedal ungual is lateromedially flattened with a narrow ventral surface. The two concave proximal articular facets are slightly asymmetrical and are divided by a sharp longitudinal ridge. The proximal flexor tubercle is present but small (Text-fig. 13). The flattened lateral and medial

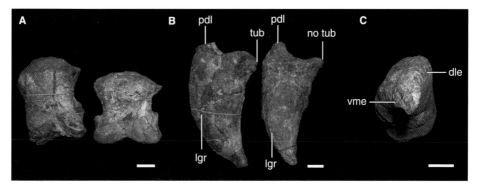

TEXT-FIG. 13. Pedal phalanges of *Lessemsaurus sauropoides*. A, proximal pedal phalanges PVL 4822/75–4822/76 in dorsal view (ordered by number from right to left). B, pedal unguals PVL 4822/78–4822/79 in lateral view (ordered by number from left to right). C, pedal ungual phalanx II? (PVL 4822/79) in distal view. Scale bars represent 2 cm.

surfaces bear a shallow proximodistal groove. The ventromedial edge of this ungual phalanx is distinctly sharp. All of these characters are exclusively present in pedal ungual I of non-eusauropod sauropodomorphs (e.g. *Plateosaurus engelhardti*, MB skelett 25; *Massospondylus carinatus*, BPI/1/4377; *Blikanasaurus cromptoni*). However, in contrast to the pedal ungual I of these forms, the ungual PVL 4822/78 lacks a proximal bifurcation of the lateral and medial grooves. The second ungual (PVL 4822/79) is markedly asymmetrical, having a dorsomedially facing surface and a ventrolaterally facing surface separated by sharp ridges (Text-fig. 13B–C). The ventrolateral surface bears a shallow groove while the dorsomedial surface seems to be smooth. Its proximal articular surface is rather narrow lateromedially and lacks a flexor tubercle on its ventral surface. This combination of characters is present in the second digit ungual of some sauropodomorphs (e.g. *Antetonitrus ingenipes*). In contrast, most non-eusauropod sauropodomorphs have a rather symmetrical and ventrally flattened second pedal ungual. Interestingly, the condition in *Lessemsaurus* resembles the sickle shaped, asymmetrical unguals of some derived eusauropods in lacking a ventrally flattened surface.

DISCUSSION

Taxonomic identity of PVL 4822

The assemblage PVL 4822 was found in close association, but not in articulation. It includes elements from several individuals as demonstrated by the duplication of elements and size differences among the material. Nevertheless, we regard all of this material as referable to *Lessemsaurus* for the reasons given below.

Firstly, the assemblage of bones seems to be monospecific as none of the duplicate elements is distinguishable from each other (and some elements, such as the scapulae, bear apomorphic features). Secondly, many of the elements found in this association can be distinguished from those of other sauropodomorphs known from the Los Colorados Formation (*Riojasaurus incertus* and

Coloradisaurus brevis) owing to the presence of the numerous characters (characters observed in PVL 4822-1 were originally noted by Bonaparte 1999):

1. Dorsoventrally elongated cervicodorsal neural arches (PVL 4822-1).
2. Strong neural arch constriction below the postzygapophyses (PVL 4822-1).
3. Deep postspinal fossa (PVL 4822-1).
4. Dorsoventrally high infrapostzygapophyseal depression (PVL 4822-1).
5. Articular surface of cervicodorsal centrum higher than wide (PVL 4822/22).
6. Middle and posterior dorsals with neural spines higher than broad (PVL 4822-1).
7. Scapular dorsal blade and shaft expanded with respect to scapula dorsoventral height (PVL 4822/50).
8. Rounded proximal edge of humeral surface (PVL 4822/53).
9. Well-developed radial fossa on ulna (PVL 4822/54).
10. Radial shaft poorly constricted (PVL 4822/55); unknown in *C. brevis*.
11. Proximodistal length of metacarpal I sub-equal to its lateromedial width, and remarkably smaller than metacarpal II (PVL 4822/56); unknown in *C. brevis*.
12. Acute lateral process on proximolateral corner of metacarpal II (PVL 4822/57); unknown in *C. brevis*.
13. Manual phalanx I.1 wider than long (PVL 4822/58); unknown in *C. brevis*.
14. Manual ungual I poorly curved and lacking lateral grooves (PVL 4822/59); unknown in *C. brevis*.
15. Preacetabular process of ilium straight and with its dorsal margin continuous with the iliac blade (PVL 4822/60); unknown in *C. brevis*.
16. Pubic peduncle of ilium flaring distally, lacking a large supracetabular crest and forming a thin and marginal medial wall of the acetabulum (PVL 4822/60).
17. Brevis crest extending from the base of the ischial peduncle to the posterior tip of the reduced postacetabular process (PVL 4822/60); unknown in *C. brevis*.
18. Flat pubic apron with concave lateral margins (PVL 4822/62).

19. Distal end of femur straight in posterior view with fourth trochanter located at the midpoint of the diaphysis (PVL 4822/65).
20. Cross-section of the distal tibia subrectangular with its major axis orientated lateromedially and twice as long as its anteroposterior extension (PVL 4822/66–4822/67).
21. Subquadrangular pedal phalanges (PVL 4822/75–4822/76).
22. Narrow ventral surface of pedal unguals lacking a flattened surface (PVL 4822/78–4822/79).

Finally, the diagnostic postcranial characters of *Riojasaurus* (Bonaparte 1972) and *Coloradisaurus* (D. Pol, pers. obs.) are absent from the material catalogued under PVL 4822. Consequently, we refer all of this material to *Lessemsaurus sauropoides* on the basis of current data. However, future discoveries of articulated remains are necessary to test the taxonomic identity of the material described herein.

Affinities of Lessemsaurus

Lessemsaurus sauropoides was originally referred to the Melanorosauridae (Bonaparte 1999), a group of large, robust non-eusauropod sauropodomorphs of disputed monophyly (Galton 1985; Van Heerden and Galton 1997; Yates 2003, 2004; Galton and Upchurch 2004; Barrett *et al.* 2005). Bonaparte (1999) considered this form to be closer to sauropod origins than other non-eusauropod sauropodomorphs, such as *Plateosaurus* or *Riojasaurus*. More recently, Yates and Kitching (2003) noted the similarity in the distally flared dorsal neural spine present in the *Lessemsaurus* material described by Bonaparte (1999) and that of *Antetonitrus ingenipes*, a new taxon from the lower Elliot Formation (Norian) of South Africa.

Galton and Upchurch (2004) included *Lessemsaurus* in a cladistic analysis, which recovered it as the sister taxon of *Camelotia* on the basis of the shared presence of anteroposteriorly short posterior dorsal centra. This clade clustered with *Melanorosaurus* and *Riojasaurus* to form a monophyletic Melanorosauridae, but none of the synapomorphies of this more inclusive clade could be scored for *Lessemsaurus* (based on the information published at that time: PVL 4822-1). Owing to the large amount of missing data, the position of *Lessemsaurus* was weakly supported in the context of that dataset: only one extra step was necessary for it to become the sister-taxon of *Euskelosaurus*, *Massospondylus* or *Blikanasaurus*, or for it to become the most basal sauropodomorph.

The remains described herein offer new information on the anatomy of *Lessemsaurus*. Interestingly, they provide a suite of derived characters shared exclusively with *Antetonitrus*. These include: the presence of a scapula with a broad dorsal blade and shaft (with respect to scapula dorsoventral height); a short scapular shaft; the distal lateral condyle of metacarpal I taller dorsoventrally than the medial condyle; and a manual phalanx I.1 that is slightly wider (at its proximodistal end) than proximodistally long. The absence of these derived conditions in all other sauropodomorphs suggests that these characters may be synapomorphies of a clade including these two taxa. This suggestion must be tested within the context of an inclusive phylogenetic analysis: however, the latter lies outside the scope of this contribution.

Lessemsaurus *and the origin of eusauropod morphology*

Bonaparte (1999) interpreted the anatomy of the cervicodorsal neural arches of *Lessemsaurus* (PVL 4822-1) as displaying a morphology that was approaching the condition present in eusauropods (e.g. posterior cervical neural spines transversally wide; well-delimited depressions on anterior surface of neural arch pedicels of the cervicals; infrapostzygapophyseal constriction in cervicals; dorsal neural spine high, with the neural arch more than half the total height of the vertebra; and sharply delimited pneumatic fossa on posterior dorsal centra). The combination of these derived characters with the retained presence of sauropodomorph plesiomorphies (e.g. the absence of pleurocoels in cervical vertebrae; dorsal transverse process not deflected dorsally; dorsolateral surface of the diapophyseal lamina lacking a depression; absence of spinodiapophyseal lamina on middle and posterior dorsals) suggested that this taxon was potentially relevant for understanding the origin of the morphological features that traditionally characterized Eusauropoda. The new information provided herein identifies an additional set of plesiomorphic and derived characters in the rest of the skeleton that supports this view, placing *Lessemsaurus* as a particularly interesting form owing to the unique combination of character states.

The additional set of derived characters shared with eusauropods provides potential synapomorphies for a clade formed by *Lessemsaurus*, eusauropods and related forms, such as *Vulcanodon*. Several of the characters discussed below are also present in *Antetonitrus ingenipes*, but are absent in all 'prosauropods' (exceptions noted below):

1. Proportionately short and high dorsal centra (also present in *Camelotia borealis*: Galton 1998; Galton and Upchurch 2004).
2. Well-developed acromion process and ventral expansion of scapula.
3. Humerus with a low deltopectoral crest and markedly convex proximal surface.
4. Radial shaft that is poorly constricted and subovoid in cross-section.
5. Poorly curved manual ungual I with a reduced flexor tubercle and proximodorsal lip, and lacking a collateral groove.

6. Pubic plate occupying more than 40 per cent the total length of pubis.
7. Distal end of femur straight in posterior view.
8. Fourth trochanter located at femoral mid-length.
9. Anterolateral descending process of the tibia exceeds laterally the posterolateral descending process (also present in *Anchisaurus polyzelus*; Yates 2004).
10. Relatively long metatarsal I (being approximately 80 per cent of the length of metatarsal III).
11. Subquadrangular non-terminal pedal phalanges [also present in *Melanorosaurus readi* (NM QR 3314) and *Blikanasaurus cromptoni*].
12. Pedal ungual with narrow (instead of flattened) ventral surface and lacking a proximal bifurcation of the lateral and medial grooves.

As mentioned above, the new material described herein reveals that *Lessemsaurus sauropoides* lacks numerous derived characters present in basal eusauropods (some of which are also recorded in *Vulcanodon*). For these characters, *Lessemsaurus* (and in many cases *Antetonitrus ingenipes*) possesses the plesiomorphic condition present in other non-eusauropod sauropodomorphs (or 'prosauropods'). Among these we can note the following:

1. Acamerate and amphicoelous centra.
2. Distal end of the radius lacking a broad posterior surface for articulation with the ulna.
3. Metacarpal I with a subtriangular proximal surface and which is reduced in length with respect to other elements of the metacarpus.
4. Iliac blade that is low and slightly convex.
5. Well-developed ischial peduncle of the ilium.
6. Moderately elongated pubes with a flattened and transversely orientated pubic apron.
7. Steep distal end of fourth trochanter.
8. Distal end of the femur bent posteriorly.
9. Subrectangular astragalar main body, with posterior and medial margins meeting at a right angle.
10. Distal surface of astragalus flattened or slightly convex.
11. Width of metatarsal I sub-equal to that of the other metatarsals.
12. Reduced triangular metatarsal V that is less than half the length of the other metatarsals.

Other characters, such as the degree of development of the radial fossa of the ulna, the supracetabular crest, the postacetabular process of ilium, and the width/length ratio of the metatarsus, seem to show an intermediate condition between the morphology present in non-eusauropod sauropodomorphs ('prosauropods') and that of eusauropods.

Lessemsaurus and the monophyly of Prosauropoda

The new material described herein also has a bearing on the debated monophyly of Prosauropoda, because of the presence of characters that have been considered as pro-

sauropod synapomorphies. The simultaneous presence in *Lessemsaurus* of such characters, together with those that suggest affinities of this taxon with eusauropods, is of particular interest because it provides relevant information for testing prosauropod monophyly. Among the proposed prosauropod synapomorphies of recent studies (Sereno 1999; Galton and Upchurch 2004), *Lessemsaurus* can be scored for the following characters:

1. Absence of prezygadiapophyseal lamina on caudal dorsals (Galton and Upchurch 2004).
2. Metacarpal I inset into the carpus, evidenced by the flat articular surface on the proximolateral corner of this element for the distal carpal II (Sereno 1999).
3. Metacarpal I basal width more than 65 per cent of the maximum length (Sereno 1999).
4. Twisted manual phalanx I.1 (Sereno 1999; Galton and Upchurch 2004).
5. Proximal heel in manual phalanx I.1 (Galton and Upchurch 2004).
6. Subtriangular preacetabular process of ilium (Sereno 1999).
7. Subtriangular ischial distal shaft in cross-section (Sereno 1999).
8. Metatarsal II proximal articular surface hourglass-shaped (Sereno 1999).

Four other characters proposed as prosauropod synapomorphies (Sereno 1999; Galton and Upchurch 2004) can be scored in the present material of *Lessemsaurus* but show a similar morphology to eusauropods: anteroposteriorly short caudal dorsal centra (length/height ratio < 1·0); deltopectoral crest orientated obliquely to the long axis through the distal humeral condyles; deltopectoral crest occupying less than 50 per cent of the humeral length; and reduced obturator foramen in pubis (less than 50 per cent of the acetabulum).

All of these sets of characters provide evidence for understanding the evolution of the characteristic morphology of eusauropods and testing prosauropod monophyly, owing to the presence of previously unrecorded combinations of morphologies present in 'prosauropods' and eusauropods. This will be discussed elsewhere, as integrating and testing this phylogenetic information lies outside the scope of this paper.

CONCLUSIONS

The remains of *Lessemsaurus sauropoides* described herein increase our knowledge of sauropodomorph diversity in the Late Triassic Los Colorados Formation (north-west Argentina). The similarities noted between *Lessemsaurus* and *Antetonitrus* suggest a possible close relationship between the South African and South American tetrapod faunas during the Late Triassic, which needs to be thoroughly tested following taxonomic revision of the abundant material collected from the lower Elliot and Los Colorados formations.

This material adds information that will be useful in understanding several outstanding problems in the early evolutionary history of Sauropodomorpha. Some of these problems, such as understanding the evolutionary processes behind the origin of sauropod *bauplan*, the changes in diversity patterns (e.g. extinctions, radiations), and the evolution of character complexes, cannot be approached without an inclusive phylogenetic analysis of all relevant taxa. In particular, the outcome of these studies will be strongly dependent upon whether a monophyletic (Gauffre 1995; Sereno 1999; Galton and Upchurch 2004) or paraphyletic (Gauthier 1986; Yates 2003, 2004) Prosauropoda is supported. As discussed above, these remains, as well as those of the large sauropodomorphs from the lower Elliot Formation (Galton and Van Heerden 1998; Yates and Kitching 2003; Galton *et al.* 2005), will play a critical role in these analyses as they possess a particularly interesting combination of plesiomorphic and apomorphic sauropodomorph characters.

Acknowledgements. We thank J. F. Bonaparte for his support and comments on this manuscript. The remains described herein were found during the course of the palaeontological expedition of the Lillo Institute in 1971, conducted by J. F. Bonaparte with the support of CONICET and the Fundación Miguel Lillo. Part of the present study was developed with the financial support of the Division of Palaeontology (AMNH), the Theodore Roosevelt Fund, and the Annette Kade Fund (to DP). We thank M. Norell, J. Clark and P. Olsen for critically reading earlier versions of this paper. A. Yates, P. Upchurch and P. Barrett provided critical comments that enhanced the quality and clarity of this contribution. Discussions on these topics with J. Bonaparte, J. Wilson, P. Sereno and P. Galton are also greatly appreciated. Comparisons with relevant material were possible thanks to collections access provided by: M. Norell (AMNH); X. Xu (IVPP); R. Wild (SMNS); S. Kaal (SAM); J. Gauthier (YPM); D. Unwin (MB); M. Feng (NGMJ); M. Langer; B. Rubidge and A. Yates (BPI); and R. Nuttal and E. Butler (NM QR).

REFERENCES

ALIFANOV, V. R. and AVERIANOV, A. O. 2003. *Ferganasaurus verzilini*, gen. et sp. nov. A new neosauropod (Dinosauria, Saurischia, Sauropoda) from the Middle Jurassic of Fergana Valley, Kirghizia. *Journal of Vertebrate Paleontology*, **23**, 358–372.

ALLAIN, R., AQUESBI, N., DEJAX, J., MEYER, C., MONBARON, M., MONTENAT, C., RICHIR, P., ROCHDY, M., RUSSELL, D. and TAQUET, P. 2004. A basal sauropod dinosaur from the Early Jurassic of Morocco. *Comptes Rendus Palevol*, **3**, 199–208.

BARRETT, P. M., UPCHURCH, P. and WANG XIAO-LIN 2005. Cranial osteology of *Lufengosaurus huenei* (Dinosauria: Prosauropoda) from the Lower Jurassic of Yunnan, People's Republic of China. *Journal of Vertebrate Paleontology*, **25**, 806–822.

BONAPARTE, J. F. 1972. Los tetrápodos del sector superior de la Formacion Los Colorados, La Rioja, Argentina (Triássico superior). *Opera Lilloana*, **22**, 1–183.

—— 1973. Edades Reptil para el Triásico de Argentina y Brasil. *Actas V Congreso Geológico Argentino*, **3**, 93–129.

—— 1978. *Coloradia brevis* n. g. et n. sp. (Saurischia Prosauropoda), dinosaurio Plateosauridae de la Formaciòn Los Colorados, Triàsico superior de La Rioja, Argentina. *Ameghiniana*, **15**, 327–332.

—— 1986. The early radiation and phylogenetic relationships of the Jurassic sauropod dinosaurs, based on vertebral anatomy. 247–258. *In* PADIAN, K. (ed.). *The beginning of the Age of Dinosaurs.* Cambridge University Press, Cambridge, 390 pp.

—— 1999. Evolución de las vértebras presacras en Sauropodomorpha. *Ameghiniana*, **36**, 115–187.

—— and PUMARES, J. A. 1995. Notas sobre el primer craneo de *Riojasaurus incertus* (Dinosauria, Prosauropoda, Melanorosauridae) del Triásico Superior de La Rioja, Argentina. *Ameghiniana*, **32**, 341–349.

—— and VINCE, M. 1979. El hallazgo del primer nido de Dinosaurios Triásicos (Saurischia, Prosauropoda), Triásico Superior de Patagonia, Argentina. *Ameghiniana*, **16**, 173–182.

BONNAN, M. F. 2003. The evolution of manus shape in sauropod dinosaurs: implications for functional morphology, forelimb orientation and sauropod phylogeny. *Journal of Vertebrate Paleontology*, **23**, 595–613.

—— and YATES, A. M. 2007. A new description of the forelimb of the basal sauropodomorph *Melanorosaurus*: implications for the evolution of pronation, manus shape, and quadrupedalism in sauropod dinosaurs. 157–168. *In* BARRETT, P. M. and BATTEN, D. J. (eds). *Evolution and palaeobiology of early sauropodomorph dinosaurs.* Special Papers in Palaeontology, **77**, 289 pp.

BUFFETAUT, E., SUTEETHORN, V., CUNY, G., TONG, HAI-YAN, LE LOEUFF, J., KHANSUBHA, S. and JONGAUTCHARLYAKUL, S. 2000. The earliest known sauropod dinosaur. *Nature*, **407**, 72–74.

COOPER, M. R. 1981. The prosauropod dinosaur *Massospondylus carinatus* Owen from Zimbabwe: its biology, mode of life and phylogenetic significance. *Occasional Papers of the National Museums and Monuments of Rhodesia, Series B*, **6**, 689–840.

—— 1984. A reassessment of *Vulcanodon karibaensis* Raath (Dinosauria: Saurischia) and the origin of the Sauropoda. *Palaeontologia Africana*, **25**, 203–231.

GALTON, P. M. 1985. Notes on the Melanorosauridae, a family of large prosauropod dinosaurs (Saurischia: Sauropodomorpha). *Geobios*, **18**, 671–676.

—— 1990. Basal Sauropodomorpha–Prosauropoda. 320–344. *In* WEISHAMPEL, D. B., DODSON, P. and OSMÓLSKA, H. (eds). *The Dinosauria.* University of California Press, Berkeley, CA, 733 pp.

—— 1998. Saurischian dinosaurs from the Upper Triassic of England: *Camelotia* (Prosauropoda, Melanorosauridae) and *Avalonianus* (Theropoda, ?Carnosauria). *Palaeontographica Abteilung A*, **250**, 155–172.

—— and UPCHURCH, P. 2004. Prosauropoda. 232–258. *In* WEISHAMPEL, D. B., DODSON, P. and OSMÓLSKA, H. (eds). *The Dinosauria*. Second edition. University of California Press, Berkeley, CA, 861 pp.

—— and VAN HEERDEN, J. 1998. Anatomy of the prosauropod dinosaur *Blikanasaurus cromptoni* (Upper Triassic, South Africa), with notes on the other tetrapods from the lower Elliot Formation. *Paläontologische Zeitschrift*, **72**, 163–177.

—— —— and YATES, A. 2005. Postcranial anatomy of referred specimens of the sauropodomorph dinosaur *Melanorosaurus* from the Upper Triassic of South Africa. 1–37. *In* TIDWELL, V. and CARPENTER, K. (eds). *Thunderlizards: the sauropodomorph dinosaurs*. Indiana University Press, Bloomington, IN, 495 pp.

GAUFFRE, F.-X. 1995. Phylogeny of prosauropod dinosaurs. *Journal of Vertebrate Paleontology*, **15** (Supplement to No. 3), 31A.

GAUTHIER, J. A. 1986. Saurischian monophyly and the origin of birds. *Memoirs of the California Academy of Sciences*, **8**, 1–55.

GROEBER, P. F. C. and STIPANICIC, P. N. 1953. Triásico. 13–141. *In* GROEBER, P. F. C. (ed.). *Mesozoico Geografía de la Republica Argentina*. Sociedad Argentina de Estudios Geográficos, Buenos Aires, 541 pp.

HUENE, F. von 1932. Die fossile Reptil-Ordnung Saurischia, ihre Entwicklung und Geschichte. *Monographien zur Geologie und Paläontologie, Series 1*, **4**, 1–361.

JAIN, S. L., KUTTY, T. S., ROY-CHOWDHURY, T. and CHATTERJEE, S. 1975. The sauropod dinosaur from the Lower Jurassic Koto Formation of India. *Proceedings of the Royal Society of London, Series B*, **188**, 221–228.

LANGER, M. C. 2003. The pelvic and hind limb anatomy of the stem-sauropodomorph *Saturnalia tupiniquim* (Late Triassic, Brazil). *Paleobios*, **23**, 1–40.

McINTOSH, J. S. 1990. Sauropoda. 345–401. *In* WEISHAMPEL, D. B., DODSON, P. and OSMÓLSKA, H. (eds). *The Dinosauria*. University of California Press, Berkeley, CA, 733 pp.

NOVAS, F. E. 1993. New information on the systematics and postcranial skeleton of *Herrerasaurus ischigualastensis* (Theropoda: Herrerasauridae) from the Ischigualasto Formation (Upper Triassic) of Argentina. *Journal of Vertebrate Paleontology*, **13**, 400–423.

SEELEY, H. G. 1887. On the classification of the fossil animals commonly named Dinosauria. *Proceedings of the Royal Society of London*, **43**, 165–171.

SERENO, P. C. 1999. The evolution of dinosaurs. *Science*, **284**, 2137–2147.

STIPANICIC, P. N. and BONAPARTE, J. F. 1972. Cuenca triásica de Ischigualasto-Villa Unión (Provincias de San Juan y La Rioja). 507–536. *In* LEANZA, A. F. (ed.). *Geología regional Argentina*. Academia Nacional de Ciencias, Córdoba, 869 pp.

TANG FENG, JIN XING-SHENG, KANG XI-MIN and ZHANG GOU-JUN 2001. Omeisaurus maoianus, *a complete Sauropoda from Jingyan, Sichuan*. China Ocean Press, Beijing, 128 pp. [In Chinese, English summary].

UPCHURCH, P. 1995. The evolutionary history of sauropod dinosaurs. *Philosophical Transactions of the Royal Society of London, Series B*, **349**, 365–390.

—— 1998. The phylogenetic relationships of sauropod dinosaurs. *Zoological Journal of the Linnean Society*, **124**, 43–103.

—— and MARTIN, J. 2003. The anatomy and taxonomy of *Cetiosaurus* (Saurischia, Sauropoda) from the Middle Jurassic of England. *Journal of Vertebrate Paleontology*, **23**, 208–231.

VAN HEERDEN, J. and GALTON, P. M. 1997. The affinities of *Melanorosaurus*, a Late Triassic prosauropod dinosaur from South Africa. *Neues Jahrbuch für Geologie und Paläontologie, Monatshefte*, **1997**, 39–55.

WEDEL, M. J. 2003. The evolution of vertebral pneumaticity in sauropods dinosaurs. *Journal of Vertebrate Paleontology*, **23**, 344–357.

WILSON, J. A. 2002. Sauropod dinosaur phylogeny: critique and cladistic analysis. *Zoological Journal of the Linnean Society*, **136**, 217–276.

—— 2005. Integrating ichnofossil and body fossil records to estimate locomotor posture and spatiotemporal distribution of early sauropod dinosaurs: a stratocladistic approach. *Paleobiology*, **31**, 400–423.

—— and SERENO, P. C. 1998. Early evolution and higher-level phylogeny of sauropod dinosaurs. *Memoir of the Society of Vertebrate Paleontology*, **5**, 1–68.

YADAGIRI, P. 2001. The osteology of *Kotasaurus yamanpalliensis*, a sauropod dinosaur from the Early Jurassic Kota Formation of India. *Journal of Vertebrate Paleontology*, **21**, 242–252.

YATES, A. M. 2003. A new species of the primitive dinosaur, *Thecodontosaurus* (Saurischia: Sauropodomorpha) and its implications for the systematics of early dinosaurs. *Journal of Systematic Palaeontology*, **1**, 1–42.

—— 2004. *Anchisaurus polyzelus* Hitchcock: the smallest known sauropod dinosaur and the evolution of gigantism amongst sauropodomorph dinosaurs. *Postilla*, **230**, 1–58.

—— and KITCHING, J. W. 2003. The earliest known sauropod dinosaur and the first steps towards sauropod locomotion. *Proceedings of the Royal Society of London, Series B*, **270**, 1753–1758.

YOUNG CHUNG-CHIEN 1941. A complete osteology of *Lufengosaurus huenei* Young (gen. et sp. nov.). *Palaeontologica Sinica, Series C*, **7**, 1–53.

—— 1942. *Yunnanosaurus huangi* (gen. et sp. nov.), a new Prosauropoda from the Red Beds at Lufeng, Yunnan. *Bulletin of the Geological Society of China*, **22**, 63–104.

ZHANG YI-HONG 1988. *The Middle Jurassic dinosaur fauna from Dashanpu, Zigong, Sichuan. Sauropod dinosaurs. Part 1. Shunosaurus*. Sichuan Science and Technology Publishing House, Chengdu, 89 pp. [In Chinese, English summary].

—— and YANG ZHAO-LONG 1994. *A complete osteology of Prosauropoda in Lufeng Basin, Yunnan, China, Jingshanosaurus*. Yunnan Science and Technology Publishing House, Kunming, 100 pp. [In Chinese, English summary].

[Special Papers in Palaeontology 77, 2007, pp. 245–260]

NEW INFORMATION ON THE BRAINCASE AND SKULL OF *ANCHISAURUS POLYZELUS* (LOWER JURASSIC, CONNECTICUT, USA; SAURISCHIA: SAUROPODOMORPHA): IMPLICATIONS FOR SAUROPODOMORPH SYSTEMATICS

by TIM J. FEDAK* *and* PETER M. GALTON†

*Department of Biology, Dalhousie University, Life Sciences Centre, Halifax, Nova Scotia, B3H 4J1, Canada; e-mail: tfedak@dal.ca

†Professor Emeritus, University of Bridgeport, Bridgeport, CT 06604, USA; current address, 315 Southern Hills Drive, Rio Vista, CA 94571, USA; e-mail: pgalton@bridgeport.edu

Typescript received 6 February 2006; accepted in revised form 21 October 2006

Abstract: A skull of *Anchisaurus polyzelus* (YPM 1883) has been misinterpreted for over 120 years as a result of deformation during preservation and loss of a small piece of the skull block shortly after collection in 1884. The only other skull of this taxon (YPM 209), from a smaller individual, has been largely ignored in previous studies owing to distortion and incomplete preparation. Additional preparation of the latter specimen has exposed several new elements, including the nearly complete parabasisphenoid, a region that is damaged and incomplete in YPM 1883. Based on this new infor-

mation, a phylogenetic analysis supports the recent hypothesis that *Anchisaurus* was a basal sauropod. However, the strength of this hypothesis has been greatly reduced, and is also undermined further by the possibility that the specimens of *Anchisaurus* are skeletally immature. In general, the skull and braincase of *Anchisaurus* resembles those of 'prosauropods' more closely than those of derived sauropods.

Key words: Sauropodomorpha, Lower Jurassic, skull, braincase, *Anchisaurus*, *Ammosaurus*, enamel.

HISTORICALLY the basal sauropodomorph specimens from the Portland Formation (Lower Jurassic, Pliensbachian) of the Connecticut Valley, eastern USA, have been attributed to one of two genera, *Anchisaurus* Marsh, 1885 or *Ammosaurus* Marsh, 1891 (see historical review in Galton 1976). However, Yates (2004) has argued convincingly that *Ammosaurus major* (Marsh, 1891) should be regarded as a junior synonym of *Anchisaurus polyzelus* (Hitchcock, 1865), demonstrating that the purported differences between these taxa could be accounted for by ontogenetic variation (as in the case of the width of pes) or specimen deformation during preservation (with respect to the length of laterodistal groove on the tibia and the absence of a rim to the obturator foramen of pubis, for example).

Only two specimens with cranial material are referable to *Anchisaurus polyzelus* (YPM 1883 and YPM 209). The specimens were collected at the same time (in 1884) and were separated by less than 5 m horizontally (with YPM 209 about 60 cm higher in the section) within Wolcott's sandstone quarry (Lull 1953; Galton 1976), in the Lower Jurassic Portland Formation

(Olsen and Galton 1977), in Manchester, Connecticut. Although the skull of YPM 1883 has been described in several studies (see below), the braincase has been misinterpreted for nearly 120 years (Fedak 2003) owing to a missing portion (Text-fig. 1A–B). Furthermore, very little has been published on YPM 209 as the specimen has remained largely unprepared until now, with only a few elements visible that were exposed some time prior to 1906.

The skull of YPM 1883 was first described by Marsh (1891), who provided only brief comments and offered no specific details about the condition of the braincase. Marsh (1892, p. 543) published reconstructions of the 'somewhat crushed and distorted' skull, along with another short description that includes a brief comparison with the braincase of *Thecodontosaurus* (YPM 2192: Upper Triassic, England). Marsh (1892, p. 546) suggested that *Thecodontosaurus* differs from YPM 1883 (*Anchisaurus*) in its 'extended parasphenoid, and the very long basipterygoid processes', and later (1896, p. 148) amended his observations on the braincase, stating that 'the basipterygoid processes are unusually short'.

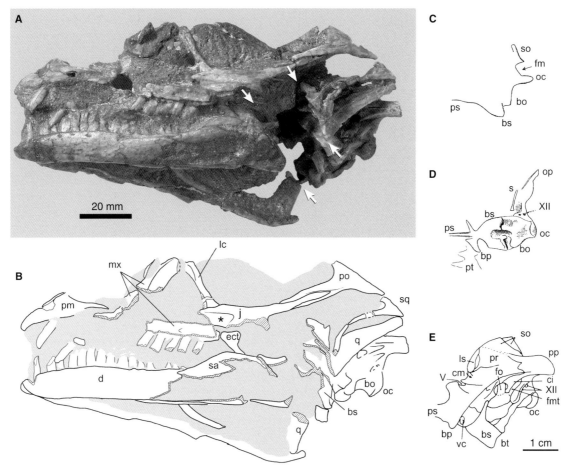

TEXT-FIG. 1. Photograph and illustrations of the *Anchisaurus* (YPM 1883) skull demonstrate previous interpretations of the braincase region. A, a similar view to the first published photograph (von Huene 1906) of the skull in oblique left lateral view; white arrows have been added to demarcate a piece missing since the specimen was collected in 1884. B, the labelled schematic of the same view as the photograph, which shows that the posterior end of the maxilla has become disarticulated from its contact (*) with the jugal. Interpretive line drawings of the braincase region from von Huene (1914a) in left lateral (C) and ventral view (D) and, modified from Galton and Bakker (1985), in left lateral view (E). Portions of the broken bone surface of the basisphenoid (between the posteriormost arrows in A) were previously misinterpreted as the basipterygoid processes (bp) in D–E.

The first photographs of the YPM 1883 skull, published by von Huene (1906), clearly demonstrate that a portion of the braincase (Text-fig. 1A) has been missing since the time of these early descriptions. Von Huene (1914a) redescribed the skull, providing line drawings of the braincase (Text-fig. 1C–E). Contrary to Marsh's opinion, von Huene (1914a, p. 70) contended that the basipterygoid processes were not short: 'Near the long basipterygoid processes there is still to be seen on the right the attachment of the pterygoid' (translated from the German original). In these studies, von Huene misidentified a cross-section of the braincase (Text-fig. 1D) as the basipterygoid processes.

Lull (1915, 1953) made the important observation that, as a consequence of crushing, the basisphenoid appears to have been bent upward and separated from its natural articulation with the basioccipital tubera. Galton (1976) provided a detailed description of the skull of YPM 1883 and, in this and subsequent publications (Galton 1985a, 1990; Galton and Bakker 1985; Galton and Upchurch 2004), followed Marsh's interpretation, suggesting that, in comparison with *Thecodontosaurus* and *Plateosaurus*, YPM 1883 has relatively small basipterygoid processes (Text-fig. 1E). The presence of small/short basipterygoid processes continues to be cited as an apomorphic character for *Anchisaurus* (Yates 2002, 2004).

After re-examination of the braincase and a consideration of the historical documentation of the specimen, it is now apparent that a portion of the skull block (Text-fig. 1A), containing parts of the braincase and basipterygoid processes, was lost when the block was broken shortly after the specimen was collected (Fedak 2003). Previous comments on the size and orientation of the basipterygoid processes are based on misidentification of the ventral and bisected surfaces of the basisphenoid. The matrix between the braincase and the stapes, noted in the descriptions and illustrations of von Huene (1914*a*) and Galton (1976), was removed by W. W. Amaral during preparation of the skull in the early 1980s, revealing many additional details of the braincase anatomy.

Further preparation of the second, smaller skull (YPM 209) has exposed new elements, including the nearly complete parabasisphenoid (Fedak 2005), details of which are provided herein. The new data provide the first opportunity to consider the proposed synonymy of *Ammosaurus* (YMP 209, holotype of *Anchisaurus solus* Marsh, 1892; referred to *Ammosaurus* by von Huene 1932; and referred to *A. major* by Galton 1976), with *Anchisaurus* YPM 1883 based on cranial characters. A well-preserved braincase of *Thecodontosaurus* (YPM 2192) has been described and figured recently by Benton *et al.* (2000, p. 85, fig. 3), and we provide brief descriptive comments and an illustration of this specimen here (Text-fig. 2A) for comparative purposes and in order to note minor differences in interpretation from previous studies. These new data were added to a phylogenetic analysis to re-evaluate the relationships of *Anchisaurus polyzelus*, which has recently been identified by Yates (2002, 2004) as a basal (and also the smallest) sauropod dinosaur (a result confirmed by Upchurch *et al.* 2005).

Institutional abbreviations. ACM, Amherst College Museum, Amherst, Massachusetts; AMNH, American Museum of Natural History, New York; SMNS, Staatliches Museum für Naturkunde in Stuttgart; YPM, Yale Peabody Museum, New Haven, Connecticut.

Abbreviations on text-figures. ar, articular; bo, basioccipital; bp, basipterygoid process; bs, basisphenoid; bt, basal tubera; c, carotid artery foramen; cb, ceratobranchial; cc, cavum cochlear; ci, crista interfenestra; cm, vena cerebalis media; cp, crista prootica; cs, crista sellaris; d, dentary; ect, ectopterygoid; f, frontal; fj, foramen jugularis; flp, foramen lacerum posterior; fm, foramen magnum; fmt, fissure metotica; fo, fenestra ovalis; j, jugal; lc, lachrymal; ls, laterosphenoid; lw, lateral walls of sella turcica; mf, metotic fissure; mx, maxilla; oc, occipital condyle; op, opisthotic; pa, parietal; pra, prearticular; pm, premaxilla; po, postorbital; pp, paroccipital process; pr, prootic; ps, parasphenoid rostrum; pt, pterygoid; q, quadrate; qj, quadratojugal; s, stapes; sa, surangular; so, supraoccipital; sp, splenial; sq, squamosal; t, teeth; vc, vidian canal for internal carotid artery; V, VII and XII, foramina for cranial nerves.

MATERIAL AND METHODS

Anatomical terminology follows that used by Bellairs and Kamal (1981) and Baumel and Witmer (1993). The braincases of '*Thecodontosaurus antiquus*' (YPM 2192: taxon

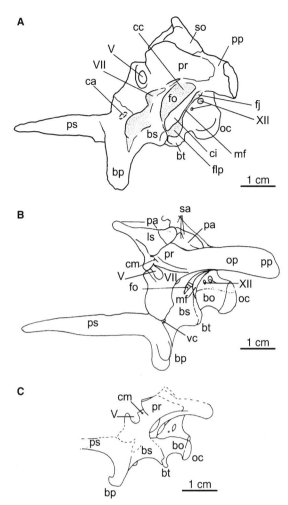

TEXT-FIG. 2. Schematic illustrations of basal sauropodomorph braincases in left lateral view. A, '*Thecodontosaurus*' (YPM 2192); the metotic fissure is a continuous structure, composed of the foramen lacerum posterior located anteroventrally and the foramen jugularis posterodorsally; the Vidian canal (internal carotid artery) enters the posterior portion of the sella turcica. B, *Plateosaurus* (AMNH 6810). C, *Efraasia* (SMNS 12667). All from Galton and Bakker (1985), with A modified slightly based on direct observations of the specimen.

restricted to the holotype dentary by Galton 2005) and *Anchisaurus polyzelus* (YPM 1883 and YPM 209) were examined directly. During mechanical preparation of the smaller skull (YPM 209) (by TF), a long-wave (365 nm) ultraviolet (UV) lamp (UVP Inc. 'Mineralight' lamp, Model UVGL-25) was used in an otherwise dark room, which allowed improved discrimination between the thin fossil bone and the hard calcite matrix. Under long-wave UV light, the skeletal elements luminesce bright white against the dark background of the surrounding matrix, allowing even the smallest of elements to be recognized immediately on exposure. Without UV light it was nearly impossible to discriminate between the small bone elements and large calcite crystals in the matrix. Digital photographs of YPM 209 were taken under similar lighting conditions and using a Tiffen (Haze-1) UV filter mounted on the camera lens.

In order to examine the surface texture of the tooth enamel in YPM 209, epoxy resin casts of a newly exposed left dentary tooth were produced from a high-resolution rubber peel. Several attempts were required in order to capture an accurate surface impression of the tooth, as the initial peels captured emulsion artefacts from the consolidant present on the tooth surface; a similar artefact appears to have occurred during the analysis of the tooth enamel in YPM 1883 (Yates 2004; see below). A suitable mould was produced only after thoroughly rinsing the tooth with acetone to remove all consolidant. The epoxy cast was examined in a scanning electron microscope (SEM).

SYSTEMATIC PALAEONTOLOGY

DINOSAURIA Owen, 1842
SAURISCHIA Seeley, 1887
SAUROPODOMORPHA von Huene, 1932

Genus ANCHISAURUS Marsh, 1885

Type species. *Megadactylus polyzelus* Hitchcock, 1865.

Diagnosis. As for type species.

Anchisaurus polyzelus (Hitchcock, 1865)
Text-figures 1, 3–8

    1858   No Name Hitchcock, p. 186.
v * 1865   *Megadactylus polyzelus* Hitchcock, p. 39, pl. 9, fig. 6.
  v 1870   *M. polyzelus* (Hitchcock); Cope, 122A–G, pl. 13.
  v 1882   *Amphisaurus* Marsh, p. 84 (*Megadactylus* preoccupied).
  v 1885   *Anchisaurus* Marsh, p. 169 (*Amphisaurus* preoccupied).
  v 1889   *Anchisaurus major* Marsh, p. 331, fig. 1.
  v 1891   *Ammosaurus major* (Marsh); Marsh, p. 267.
  v 1891   *Anchisaurus colurus* Marsh, p. 267.
v * 1891   *Anchisaurus polyzelus* (Hitchcock); Marsh, p. 169.
  v 1892   *Anchisaurus colurus* (Marsh); Marsh, p. 543, pls 15, 16 (figs 1–2).
  v 1892   *Anchisaurus solus* (Marsh); Marsh, p. 545.
  v 1896   *M. polyzelus* (Hitchcock); Marsh, p. 147.
  v 1906   *Thecodontosaurus polyzelus* (Hitchcock); von Huene, p. 19, fig. 10.
vp 1912   *Anchisaurus colurus* (Marsh); Lull, p. 414, figs 2–3.
v 1914a   *T. polyzelus* (Hitchcock); von Huene, p. 75, figs 23–24.

vp 1914b   *Anchisaurus colurus* (Marsh); von Huene, p. 3.
vp 1915   *Anchisaurus colurus* (Marsh); Lull, p. 130, figs [copied from Marsh 1892] 18–21, pls 4, 10.
  v 1932   *Ammosaurus solus* (Marsh); von Huene, p. 27, pl. 49, fig. 1.
  v 1932   *Thecodontosaurus polyzelus* (Hitchcock); von Huene, p. 116.
  v 1932   *Yaleosaurus colurus* (Marsh); von Huene, p. 119, pl. 14, fig. 1; pl. 54, fig. 3.
vp 1953   *Yaleosaurus colurus* (Marsh); Lull, p. 107, pl. 4, figs 15–18 [copied from Marsh 1892].
  v 1976   *Ammosaurus major* (Marsh); Galton and Cluver 1976, p. 143, fig. 8A, J [see also for earlier synonymy].
vp 1976   *Ammosaurus major* (Marsh); Galton, p. 3, figs 2A, 23–31.
  v 1976   *Anchisaurus colurus* (Marsh); Galton and Cluver, p. 132 (= *Anchisaurus polyzelus*, Hitchcock) [see also for earlier synonymy].
  v 1976   *Anchisaurus polyzelus* (Hitchcock); Galton and Cluver, p. 132, figs 7A, C, 8E, 9E, H–K, 10A, G, K, 11L, 12B [see also for earlier synonymy].
  v 1976   *Anchisaurus polyzelus* (Hitchcock); Galton, p. 2, figs 1B, 3, 5–24.
  v 1976   *Anchisaurus solus* (Marsh); Galton and Cluver, p. 143 (= *Ammosaurus major*, Marsh) [see also for earlier synonymy].
vp 1990   *Ammosaurus major* (Marsh); Galton, p. 335.
  v 1990   *Anchisaurus polyzelus* (Hitchcock); Galton, p. 335, figs [copied from Galton 1976; Galton and Bakker 1985] 15.3A, 15.4A, 15.6N, 15.8G.
  v 2004   *Ammosaurus major* (Marsh); Galton and Upchurch, p. 234.
  v 2004   *Anchisaurus polyzelus* (Hitchcock); Galton and Upchurch, p. 234, figs [copied from Galton 1976; Galton and Bakker 1985] 12.3A, 12.5A, 12.8N, 12.10G.
  v 2004   *Anchisaurus polyzelus* (Hitchcock); Yates, p. 3, figs 1–8, 9D, 10–11, 12F.

Holotype. ACM 41109, including several vertebrae, part of the left scapula, distal ends of the left radius and ulna, left manus, articulated distal ischial blades, the left femur (estimated length of 180 mm), fibula, proximal tibia and metatarsal IV, and the proximal part of right metatarsal II.

Type locality. Early Jurassic (Pliensbachian) Portland Formation, Springfield Massachusetts.

Referred material. YPM 209, a very small (femur *c.* 110 mm in length) and nearly complete specimen, which lacks only the distal portion of the tail and right lower fore limb; YPM 1883, a well-preserved specimen, femur length 210 mm, which lacks only cervicals 4–10, the tail and left lower fore limb, left ilium and hind limb, and the majority of both ischia; and YPM 208, the largest referred specimen (1·35 times larger than YPM 1883 based on length of metatarsal 3), which includes only the last six dorsal vertebrae, three dorsal ribs and sacrum, proximal part of right scapula and nearly complete pelvis, left hind limb, partial right femur and right pes. These specimens are all from the Wolcott sandstone quarry that was located 1·6 km north of Buckland train station, Manchester, Connecticut; Early Jurassic (Pliensbachian) Portland Formation. See Galton (1976) and Yates (2004) for further details.

Emended diagnosis. Of the autapomorphies of *Anchisaurus polyzelus* listed by Yates (2004), the following remain valid: lateral pit on the distal end of the quadrate above the articular condyle; ventrally opening foramen at the base of the second sacral rib; a large fenestra that pierces the third sacral rib; a long and narrow preacetabular blade of the ilium (twice as long as high at its base); ischium with ventrally emarginated obturator plate and flat co-planar blades; and large obturator foramen that occupies most of the obturator plate of the pubis. In addition, *Anchisaurus polyzelus* possesses a uniquely wide prootic trough.

Additional information on YPM 1883

Deformation and damage. During or shortly after the collection of YPM 1883, the skull block was broken. Lull (1915, 1953, p. 61), who summarized Marsh's notes, stated that 'Part of the large block was split off at New Haven, and this smaller piece contained the head and part of the neck.' Based on the specimen's current condition, the skull block was apparently broken into at least three pieces. A repaired fracture is still visible, traversing the frontals at mid-length, the sutural contact between the jugal and the postorbital on the right side, and truncating the basisphenoid, left quadrate and left mandibular ramus (Text-figs 1A, 3B–D). Although the skull is presently composed of two re-attached fragments, at least one other piece of the skull block was not re-attached and has apparently been lost. The (missing) third piece was at least 1 cm wide in lateral view; the posterior edge demarcated by the fractured surface of the basisphenoid and the left quadrate, while the anterior surface contacted the broken surface of the left mandibular ramus (Text-fig. 1A). The missing portion would thus have included the basipterygoid processes and the ventral end of the left quadrate, including the articular surface for the quadratojugal.

The distal portion of the posteroventral process of the left jugal has also been truncated by the missing area (Text-fig. 1A–B). Because of the loss of this area and displacement of the left postorbital (and attached jugal) as a consequence of the transverse crushing of the skull, it is very difficult to estimate the anterior extent of the lower temporal fenestra under the orbit.

The right posterior portion of the skull is deformed, with many of the elements broken, lost or displaced posteromedially (Text-fig. 3). The right paroccipital process was bent posteriorly resulting in its fracture. A previously unidentified element, located medial to the right quadrate and anterior to the squamosal (Text-fig. 3B), is interpreted as a fragment of the paroccipital process. The morphology of the fragment matches the broken surface of the remainder of the process: when joined these two pieces frame a semicircular canal foramen (Text-fig. 3D).

The right lateral wall of the braincase has been rotated clockwise in dorsal view, resulting in the medial surface now being partially visible from the right side, whereas most of the right prootic and the entire exoccipital were apparently destroyed (Text-fig. 3A–B, D) during preburial deformation. The exposure of the medial surface of the right side of the braincase appears to have been incorporated into the previous reconstruction of the braincase (compare 'cm' in Text-fig. 1E with '**' of Text-fig. 3B).

The right and left postorbitals are separated from their articulations with the frontal (Text-fig. 4) and are displaced posteromedially. The parietals and postorbital did not contact each other and the frontals do contribute to the supratemporal fenestra (contra Yates 2004). The right quadrate is fractured and the supraoccipital is twisted clockwise in dorsal view. The region anterior to the right orbit was also badly distorted: all but the most ventral part of the right lachrymal is missing; the posterior maxilla is displaced ventrally (Text-fig. 1A–B); and the anterior edge of the frontals is displaced towards the left side. The remains of three right maxillary teeth (Text-fig. 3B) are orientated perpendicular (83–92 degrees) to the long axis of the maxilla, while the posterior teeth preserved in the left maxilla (Text-fig. 1B) are directed more anteriorly (59–77 degrees). However, the left maxilla has been rotated anticlockwise in lateral view, with its posterior edge disarticulated from the jugal (Text-fig. 1B), so the apparent procumbent orientation of the posterior maxillary teeth is likely to be a deformational artefact (contra Yates 2004).

The posterior region of the skull is largely intact on the left side although, as Lull (1915, 1953) pointed out, the anterior braincase has been displaced dorsally, separating the basisphenoid from the basioccipital. In left lateral view it appears that the anterior portion of the braincase has been rotated clockwise. This deformation accounts for the separation of the supraoccipital from its sutural contact with the parietal, the fracturing of the right quadrate, and separation of the basisphenoid from the basioccipital at the basal tubera.

The fracture of the right paroccipital process into several pieces, outward rotation of the right side of the braincase and displacement of the other skull elements suggest that a force was applied to the left lateral area of the skull while it was orientated with its right side on the ground.

Braincase anatomy. Only the anteriormost 25 mm of the parasphenoid rostrum is preserved, which is visible on the right side. The cultriform process of the parasphenoid is emarginated dorsally, with a U-shaped trough and only a slight extension of the lateral walls, making the cultriform process as a whole as wide as it is tall, at least anteriorly. A small piece (c. 12 mm) of the parabasisphenoid is missing immediately posterior to the cultriform process. The anterior end of the parasphenoid extends 33 mm from the base of the basioccipital tubera: the distance is 41 mm in YPM 2192 ('*Thecodontosaurus*').

The posterior part of the basisphenoid is bisected by an oblique break that is inclined anterodorsally from a point on the ventral surface of the basisphenoid, slightly anterior to the fenestra ovalis. The break passes anterior to the sella turcica of the left side. The resulting gap is wedge-shaped, being widest on the left side, narrowing towards the middle of the skull and not affecting the dermatocranial elements of the right side.

The right and left Vidian canals are visible on the fractured ventral surface of the basisphenoid. A transverse buttress that is visible on the dorsalmost part of the broken section is interpreted as the crista sellaris (sensu Bellairs and Kamal 1981), the posterior wall of the sella turcica, but deformation has repositioned the crista sellaris slightly anteroventrally. Therefore, the internal carotid arteries entered the anteriormost end of the sella turcica, as in 'Thecodontosaurus' (Text-fig. 2B). The left, and part of the right, lateral wall of the sella turcica can also be seen in section, between the crista sellaris and the Vidian canals (Text-fig. 3C).

Only the posteriormost parts of the ventral basisphenoid remain undamaged. The posterior tubera of the basisphenoid are cup like, with a short lateral component (Text-fig. 3C), although they have been displaced anteroventrally, so there is now a significant gap between the basisphenoid and the basioccipital. Evidence to suggest that this gap is the result of displacement rather than a true morphological feature (as suggested by Yates 2004) is that the crista prootica (Text-fig. 3C), located immediately anterior to the fenestra ovalis, is fractured and displaced anteroventrally from the prootic on the left side. There is an even more pronounced fracture of the right crista prootica. These fractures and displacements demonstrate that the basisphenoid was rotated clockwise in left lateral view, separating the basisphenoid from the basioccipital posteroventrally and the prootic dorsally.

Anterior to the crista prootica, the lateral surface of the basisphenoid forms the widest part and the anterior limit of a deep trough, referred to here as the prootic trough. The trough narrows posterodorsally, terminating at the prootic-opisthotic suture; a similarly located but narrower trough is also found in 'Thecodontosaurus' (Text-fig. 2A). Two foramina are located within this trough in YPM 1883, the facial foramen (for cranial nerve VII) and a foramen for VIII_2, the posterior or acoustic ramus of the auditory nerve, that opens posteriorly into the cochlear cavum (as in Plateosaurus: Galton 1985b). Based on the strong deformation of the skull, the open nature of the prootic trough may possibly represent a distortional artefact. However, on the right side the trough is partially visible anteriorly and also appears to be wide, so this may represent a genuine, and autapomorphic, feature.

Dorsal to the prootic trough, the edge of the left prootic is broken and missing, but the trigeminal foramen (for cranial nerve V) appears to be located along this break (Text-fig. 3C). The prootic extends posterodorsally and slightly laterally, form-

ing the dorsal margin of the fenestra ovalis, and overlaps the opisthotic posteriorly.

The anteroventral region of the left opisthotic currently exhibits a broken surface where the crista interfenestralis originated (Text-fig. 3C). This crest was apparently lost during preservation or preparation because its distal (anteroventral) tip is still preserved dorsal to the suture between the basisphenoid and the basioccipital. The foramen jugularis (exit of cranial nerves X and XI and the vena jugularis), located within the posterior region of the metotic fissure, is immediately ventral to the broken surface of the opisthotic that would have borne the crista interfenestralis. As most of the crista interfenestralis is missing, the cochlear cavum is clearly visible within the left side of the fenestra ovalis. The cochlear cavum exhibits a foramen (for cranial nerve VIII) on its dorsal surface and an anteriorly positioned foramen for VIII_2 that communicates with the prootic trough mentioned above.

The basioccipital has a concave ventral surface, is narrow at mid-length and is widest posteriorly. The exoccipital-basioccipital suture is still visible (Text-fig. 3C) and demonstrates that the exoccipital contributes to the lateral corner of the occipital condyle as in all 'prosauropods' with the exception of Coloradisaurus (Galton 1990). The two hypoglossal foramina (for cranial nerve XII) are located dorsal to this suture.

The left paroccipital process is nearly complete, missing only a small fragment distally. The opisthotic merges indistinguishably into the exoccipital posteroventrally to form the paroccipital process, which is overlapped by the prootic anterolaterally and the supraoccipital dorsomedially (Text-figs 3–4).

There are two bones located anterior and ventral to the basisphenoid (Text-fig. 3C), which represent broken pieces of the right opisthotic or possibly a posterior fragment of the right pterygoid. Regardless, the fragments further demonstrate that considerable damage to the right side of the braincase occurred prior to, or during, burial.

The skull of YPM 209 ('Ammosaurus major')

Deformation and damage. The anterior portion of the skull and mandibular rami remain in articulation, but the posterior portion of the skull is extensively crushed dorsoventrally. Based on the splayed limbs of the fully articulated skeleton (Galton 1976, fig. 30), the skull was buried with the lateral surface of the right mandible resting on the ground and the dorsal edge of the left lateral surface facing upwards, an orientation that also accounts for the oblique nature of the dorsoventral crushing of the posterior skull elements.

Marsh (1892, p. 545) provided the first description of the skull of YPM 209 in which he briefly noted that the specimen

TEXT-FIG. 3. Photographs and line drawings of the *Anchisaurus* (YPM 1883) skull (A–B) and braincase (C–D) in occipital (A), right lateral (B), left ventrolateral (C) and right ventrolateral (D) views. A, the missing ventral portion of the left quadrate (arrowhead) is truncated by the missing piece of the skull block. The right paroccipital process and posterior portion of the right side of the braincase (two black arrows) were fractured prior to preservation. B, the deformation to the right side of the skull exposes the medial surface of the braincase (**), and a fragment of the paroccipital process (not shown: location marked with *) was preserved at the ventral tip of the right squamosal. The white arrows demarcate the repaired oblique crack in the skull block (B–C), and the missing area of the left quadrate and basisphenoid (D), which represents the posterior edge of the missing portion of the skull block. Broken bone surfaces are identified with diagonal lines; matrix and irrelevant elements are shaded grey.

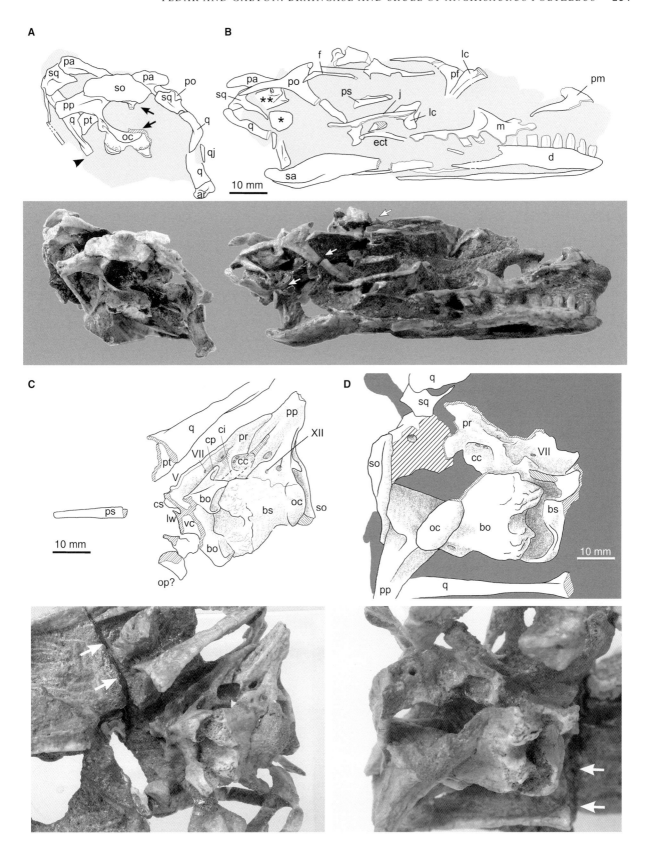

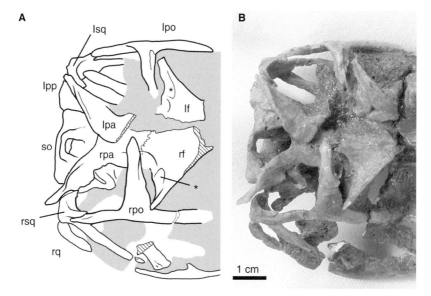

TEXT-FIG. 4. A, line drawing, and B, photograph, of the posterior dorsal region of the *Anchisaurus* skull (YPM 1883). The parietal-frontal contact has been lost on the left side, but the open sutural contact remains visible on the right. The postorbitals are displaced posteromedially from their sutural contacts with the frontals (*); the postorbitals did not contact the parietals and the supratemporal fossa on the frontal remains visible, particularly on the right side. The gap between the parietals and supraoccipital is a consequence of deformation and displacement of the braincase. Abbreviations as listed in text, with the addition of prefixes for left (l) and right (r) sides.

exhibited large orbits, a quadrate and numerous teeth that were inclined forward, and that 'the skeleton is embedded in a very coarse matrix, so difficult to remove that preparation is only in part complete'. Von Huene (1906) was the first to publish photographs and illustrations (von Huene 1914a) of the skull block, which demonstrate that no significant preparation of the specimen was carried out between 1906 and 2003 (when this project started). Based on the appearance of the rounded and smooth surfaces of the matrix, preparation prior to 1906 was carried out with a hand-held grinder, and these early preparation techniques, matched with the hard calcitic matrix and very thin cranial elements, unfortunately resulted in tremendous damage to the specimen. Additional preparation has revealed numerous new anatomical details.

Skull anatomy

Braincase elements. The parabasisphenoid is now exposed from the tip of the parasphenoid rostrum to the basisphenoid portion of the basal tubera. The posterior edge of the basisphenoid was discovered directly under the surface where preparation had stopped in 1906, so it is likely that the posteriormost part of the braincase was destroyed during this earlier preparation.

The long axis of the parabasisphenoid rostrum is aligned with the ventral surface of the basisphenoid. The straight line that joins the ventral surface between the basal tubera and the ridge between the basipterygoid processes is parallel to and 1 mm ventral to the ventral edge of the parasphenoid rostrum. The lateral sides of the cultriform process are well developed, but the process is as wide as it is dorsoventrally tall; a condition that is shared with YPM 1883 (see above),

Massospondylus (Gow 1990) and *Coloradisaurus* (Bonaparte 1978). The anterior tip of the parabasisphenoid is located 22 mm from the posterior edge of the basisphenoid portion of the basal tubera.

The left and most of the right basipterygoid processes are preserved. A transverse ridge connects the proximal ends of the processes, extending along the ventral surface of the basisphenoid (Text-fig. 5A, C), similar to the situation in *Plateosaurus* (Galton 1984). The left basipterygoid process measures 10 mm from its distal tip to the dorsal margin of the parabasisphenoid rostrum. The process is directed anterolaterally, and in lateral view the distal end is located 8 mm below the dorsal margin. The tip of the left basipterygoid process is located 7 mm from the sagittal midline; therefore, the span between the basipterygoid processes is 14 mm, a span that is greater than the width of the basisphenoid portion of the basal tuberae (c. 10 mm). The right basipterygoid process appears to have been fractured at its proximal base (Text-fig. 5B), and the distal edge was lost during previous preparation of the skull.

Premaxilla. The right premaxilla is now partially exposed in medial and lateral view. Most of the dorsal process was lost during earlier preparation, but a small portion of its base remains on the anterior margin of the bone. The roots of the right teeth are exposed because of the loss of the lateral surface of the premaxilla. The long posterior process contacted the medial surface of the maxilla, extending 16 mm posteriorly from the anterior tip of the premaxilla to a point at the base of the ascending process of the maxilla. Four premaxillary teeth are preserved, representing the total tooth count (Text-fig. 5E). The teeth are orientated slightly posteroventrally in relation to the posterior

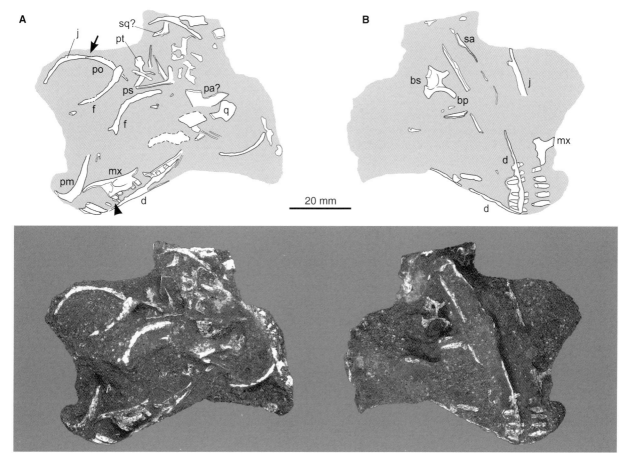

TEXT-FIG. 5. Line drawings and long-wave UV light photographs of the small *Anchisaurus polyzelus* skull (YPM 209), in A, left dorsolateral, and B, right ventrolateral views. Anterior is toward the bottom left (A) or bottom right (B). The black arrow (A) identifies the boundary between the right postorbital ventral process and right jugal, whereas the arrowhead identifies the location of the dentary tooth examined by SEM.

process of the premaxilla, and are mildly recurved. The left premaxilla contains only three partial teeth (Text-fig. 5F), which were badly damaged during preparation that took place prior to 1906.

Maxilla. As in other basal sauropodomorphs (Galton 1984; Sues *et al.* 2004), the maxilla is triradiate, with a subnarial ramus, an ascending process that separates the external naris and antorbital fenestra, and a posterior ramus that contacts the jugal posteriorly. The ascending process and subnarial ramus of the right maxilla are now exposed in lateral view (Text-figs 5E, 6B). Although the ascending process of the left maxilla was damaged prior to 1906, preparation has exposed the delicate features at the base of this process, as well as the anterior edge of the antorbital fenestra and the medial surface of the anterior ramus (Text-fig. 6A).

The preserved portion of the ascending process is orientated perpendicular to the ventral surface of the maxilla, and extends 5 mm from the dorsal surface of the subnarial ramus (Text-fig. 5E), but the anterior edge of this process was damaged prior to 1906, so the full length is uncertain. Three empty

alveoli and one partial tooth are visible in the preserved portion of the right maxilla; two of these tooth positions are located anterior to the base of the dorsal process. The dorsolateral margin of the subnarial ramus is partially depressed, which is consistent with the narial fossa being present on the anterior ramus of the maxilla.

Additional preparation of the left maxilla demonstrates that the medial sheet of the ascending process did not extend posteriorly, but rather follows the lateral edge of the process (antorbital fossa), so that the antorbital fenestra is strongly embayed anteriorly. The medial edge of the subnarial ramus is also exposed on the left side, although the lateral surface was apparently destroyed during earlier preparation. A pronounced longitudinal groove is visible on the medial edge that would have accommodated the long posterior process of the premaxilla described above, as in *Plateosaurus* (Galton 1984). Based on the length of the exposed subnarial ramus of the left maxilla, the anteriormost portion probably represents the 'anterior process' (Galton 1984), a small knob-like process on the maxilla that wrapped medially around the posterior process of the premaxilla. The more lateral, tooth-bearing portion of the anterior ramus would not therefore have extended as far anteriorly as the currently exposed medial edge.

Four incomplete teeth are preserved within the left maxilla but additional teeth were present prior to 1906; at least one anteriorly and several posteriorly. The preserved maxillary teeth are all orientated perpendicular to the ventral surface of the maxilla and are not inclined anteriorly (contra Marsh 1892). As the anterior maxillary teeth of both YPM 209 and YPM 1883 are not procumbent, the orientation of the posterior maxillary teeth of YPM 1883 is likely to represent a deformational artefact (contra Yates 2004).

Frontal. The orbital margins of the right and left frontals were first identified by von Huene (1914*a*). Galton and Cluver (1976) admitted that these fragments may represent portions of the frontals; he also suggested that the elements may represent parietal fragments. However, von Huene's identification was correct, an interpretation that is consistent with the orientation and dimensions of the other skull elements. The central portions of the frontals were almost certainly ground away by preparation prior to 1906. Alternatively, however, the more medial portion of the frontals may have been minimally ossified due to the immature status of YPM 209. The ossification of the frontals commences along the orbital margins in many extant taxa (e.g. *Alligator mississippiensis*: Rieppel 1993) and the same pattern is apparently present in embryos of *Massospondylus* (Reisz *et al.* 2005, fig. 2B): only later in ontogeny does the more medial region become fully ossified.

Postorbital. The right postorbital is missing the posterior process and the posterior margin of the ventral process. The ventral process remains in articulation with the jugal (see below); Galton (1976) erroneously identified the latter as a single element (right jugal). As a consequence of the broken surface along the posterior margin of the ventral process, the lateromedial width cannot be determined. The postorbital medial process is disarticulated from, and overlapped, by the frontal.

Jugal. A fragment of the right jugal remains in articulation with the postorbital, and the division between these two elements can still be observed (Text-fig. 6A). The posterodorsal process of the jugal contacts the posterior surface of the ventral process of the postorbital, and the orbital margin of the jugal is represented by the external bone surface. However, the remainder of the jugal fragment is truncated by broken bone surfaces. A newly exposed fragment on the right side of the skull appears to represent the ventral portion of the jugal, an interpretation that would require the jugal to have been broken longitudinally during burial.

Quadrate. Prior to further preparation, no element could be identified as a quadrate (contra Marsh 1892). However, the left quadrate is now exposed, having been displaced and rotated prior to burial, so that it is now exposed in medial view. The distal end was lost during previous preparation, but no additional morphological features can be determined from this partially exposed fragment.

Pterygoid. The right pterygoid is now exposed in medial view, located immediately lateral to the parasphenoid rostrum. The rectangular quadrate ramus lacks the thin dorsal process that articulates with the anterior edge of the medial quadrate in *Plateosaurus* (Galton 1984), but this process would probably have been destroyed when this area was removed during earlier preparation. An unidentified thin fragment of bone that overlaid the medial portion of the pterygoid (Text-fig. 6A) was removed, exposing the basipterygoid flange. The flange is short, blunt and directed medially. The palatal ramus is also partially visible in medial view, the anterior tip being located underneath the left frontal fragment.

Mandible. The lateral surfaces of both mandibular rami were obliterated prior to 1906, so little can be said about the morphology of the individual elements. The mandibular rami diverge from the symphysis at an angle of 45 degrees.

The right dentary is 5 mm deep at its anterior end and reaches a height of 7 mm ventral to the tooth row. The greatest depth of the mandibular ramus (at the surangular) is 11 mm. As in YPM 1883 (Yates 2004), there is no evidence to suggest that the dorsoventral height of the dentary below the tooth row was greater than 20 per cent of dentary length. The dentary length is at least 35 mm based on the tooth-bearing portion visible on the left mandibular ramus, but might have approached 40 mm if the posterior termination occurred at the deepest part of the surangular as in other basal sauropodomorphs (Galton 1984; Sues *et al.* 2004).

Only five teeth are visible in the right dentary. The small tooth crowns are 3 mm in length apicobasally and orientated perpendicular to the dorsal margin of the dentary. A space is located adjacent to the symphyseal edge but this represents a missing tooth. The overlying premaxillary teeth obscure the symphyseal edge of the left dentary, but the root and the distal edge of a crown are partially visible, demonstrating that a tooth is located immediately adjacent to the symphysis; consequently, the first tooth is not inset in YPM 209 or YPM 1883 (Sereno 1999; Yates 2004).

A single empty alveolus is located posterior to the first tooth of the left dentary, followed by four teeth, another empty alveolus, an erupted partial crown, an empty alveolus and one containing a tooth root (the crown was lost during previous preparation). The last preserved tooth is located 25 mm from the symphysis. Two additional tooth positions may have been present more posteriorly, yielding a dentary tooth count of 12. However, the low number of dentary teeth is of uncertain phylogenetic importance due to the immaturity of the specimen (see 'Discussion').

The recently exposed third left dentary tooth has the best preserved crown. However, like the other teeth of YPM 209 and YPM 1883, it has a centrally located longitudinal crack, suggestive of at least a short period of subaerial exposure prior to burial (Behrensmeyer 1978). A transverse crack near the tooth apex is also taphonomic, but several delicate striations appear to be visible in the area near the mesial edge of the tooth (Text-fig. 7A–B).

The enamel of this tooth is very lightly textured, and this texture varies over different regions of the tooth crown (Text-fig. 7). This texture is much finer than that described for YPM 1883 by Yates (2004), and the size, shape and distribution of the

TEXT-FIG. 6. *Anchisaurus polyzelus* (YPM 209), line drawings and long-wave UV photographs of the posterior basisphenoid in ventral view (A) and parasphenoid rostrum in right lateral view (B). The area demarcated by dashed lines (A) is not a foramen but a preparation artefact. The reconstructions of the YPM 209 parabasisphenoid in ventral (C) and left lateral (D) views demonstrate that the ventral edge of the parasphenoid rostrum is parallel to, and 1 mm above, the ventral sagittal surface of the basisphenoid (dashed lines in D). E, the right, and F, left maxilla, premaxillary teeth and dentary in lateral views; the posterior process of the premaxilla is not visible in lateral view (dashed line) but can be seen on the opposite side of the block. Cross-hatched area represents preparation-damaged bone surface, and the black arrow (F) is the dentary tooth that was examined by SEM and is shown in Text-figure 7.

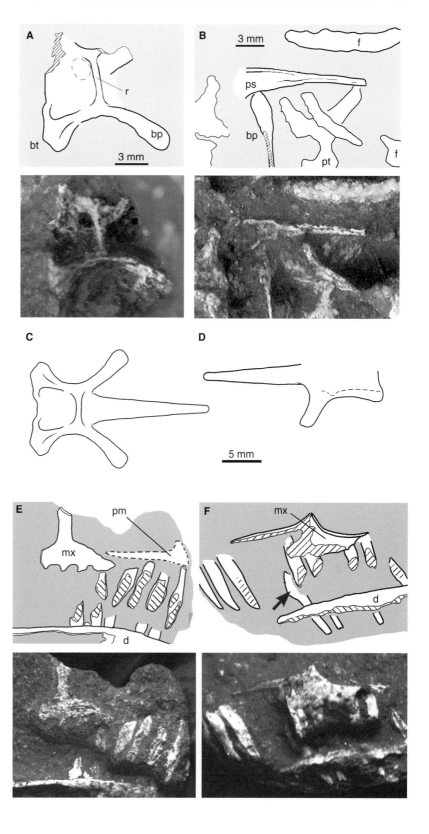

texture present on the teeth of both YPM 209 and YPM 1883 are distinct from the wrinkled enamel of derived sauropods (Wilson and Sereno 1998).

The variability of the enamel textures noted for *Anchisaurus* may represent either variation within the dentition, ontogenetic variation, trauma or disease (Franz-Odendaal 2003, 2004).

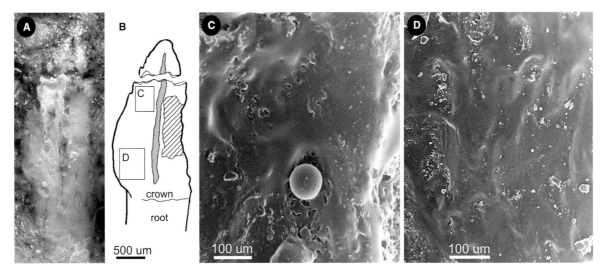

TEXT-FIG. 7. A, photograph, B, schematic drawing, and C–D, SEM photomicrographs of a recently exposed *Anchisaurus* dentary tooth (YPM 209). The tooth was damaged by transverse and longitudinal (solid grey) taphonomic cracks and a small portion of enamel is missing (diagonal lines). The preserved tooth enamel exhibits a very fine textured pattern (C–D). The sphere in the middle of C is an air bubble present in the mould.

Also, the textured enamel noted in YPM 1883 (Yates 2004, fig. 6) includes conspicuous fine lines running obliquely to 'the short longitudinal wrinkles'; therefore, the recorded surface may represent an emulsion (consolidant?) artefact. The significance of the enamel textures on the teeth of *Anchisaurus* is difficult to determine owing to an absence of systematic analyses of surface structure among sauropodomorph enamels (Hwang 2005) and the limited quality and number of samples available for *Anchisaurus*.

Other skull elements. A fragment located near the posterolateral edge of the skull may represent the right squamosal, and the parietals may also be exposed in dorsal view (Text-fig. 6A), but little can be said about the morphology of these fragments.

Specimen maturity
Galton (1976) was the first to propose that the skeletons of *Anchisaurus* (and '*Ammosaurus*') represent juvenile or subadult

specimens. Conversely, Yates (2005) proposed that two of the specimens (YPM 1883 and YPM 208) were mature, or nearly mature, as the smaller specimen (YPM 1883) has closed sutures on the posterior dorsal vertebrae. Thus, *Anchisaurus* was identified as a very small, but morphologically mature basal sauropod dinosaur (Yates 2002, 2004).

The prootic-opisthotic, basisphenoid-basioccipital, parietal-parietal and parietal-frontal sutures remain open, as do the neurocentral sutures of the preserved (anterior) cervical vertebrae, which contradict the hypothesis that YPM 1883 was nearly mature. The small number of teeth also argues for immaturity, as tooth count increases ontogenetically in other 'prosauropods' (e.g. Sues *et al.* 2004). A future histological study (by TF) of the *Anchisaurus* material will, it is hoped, provide further evidence with which to evaluate the maturity of the specimens. If all specimens of *Anchisaurus* do prove to be immature, phylogenetic studies incorporating these skeletons should be cautious when

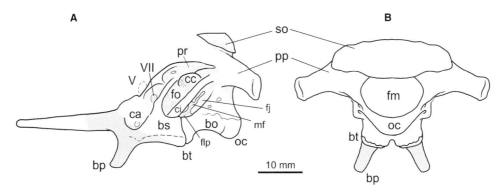

TEXT-FIG. 8. Reconstructions of the braincase of *Anchisaurus polyzelus* based on YPM 1883 scaled up parabasisphenoid from YPM 209 in A, left lateral, and B, occipital views. The solid grey area identifies the portion of the YPM 1883 parabasisphenoid that is missing, and the dashed line represents the sagittal ventral surface of the basisphenoid.

scoring characters in order to avoid potential problems with ontogenetic variability.

PHYLOGENETIC ANALYSIS

No significant differences in skull anatomy can be found between YPM 209 and YPM 1883. Thus, all of the specimens from the Portland Formation are considered to represent a single taxon, *Anchisaurus polyzelus* (Hitchcock, 1865). Minor differences between the two skulls can be attributed to ontogenetic variation. However, the smaller skull (YPM 209) is badly deformed and damaged, limiting the comparisons that could be made between the two skulls.

Yates (2004) presented a phylogenetic analysis of basal sauropodomorph relationships on the basis of 205 characters scored for 17 sauropodomorph ingroup taxa, with two different pairs of outgroup taxa. This analysis included the novel result that *Anchisaurus* was the smallest sauropod dinosaur and was positioned at the base of the sauropod clade (following the stem-based definition of Wilson and Sereno 1998). This was a robust result, with the node uniting *Anchisaurus* and other sauropods supported with a bootstrap value of 97 per cent. In addition, five additional steps were required to remove *Anchisaurus* from this group. Here, we re-examine the strength of this hypothesis by including new information on the skulls and braincases of YPM 209 and YPM 1883 in a phylogenetic analysis.

As a result of the new preparation work described above, several character scorings for *Anchisaurus* used in the Yates (2004) analysis now require revision (referred to in the following discussion by Yates' original character numbers). As a result of preservational or preparation damage, the anterior portion of the left lachrymal dorsal process is missing (22.1 now 22.?); this has the additional consequence that the anteroventral contribution of the lachrymal to the antorbital fossa is uncertain (23.1 now 23.?). The missing piece of the YPM 1883 skull block truncates the posterior portion of the right jugal, and transverse crushing of the postorbitals confounds assessment of the position of the anterior limit of the lower temporal fenestra (29.2 now 22.?). Also, the preserved ventral edge of the left quadrate is a broken surface and does not represent the articular surface for the quadratojugal, so the position of the quadrate foramen cannot be determined in either YPM 1883 or YPM 209 (38.1 now 38.?). Further scoring modifications are required as a consequence of misinterpretations of the distorted skull elements of YPM 1883: the frontals did contribute to the supratemporal fenestra (30.1 now 30.0); and there is a lateral component to the cup-like posterior basisphenoid portion of the basal tubera (46.0 now 46.1).

New information from the skull of YPM 209 also provides additional data to update several character scorings. The region adjacent to the interbasipterygoid space is not preserved in YPM 1883, but there is a well-developed interbasipterygoid septum in YPM 209 (48.0 now 48.1). Moreover, the floor of the braincase has been misinterpreted in YPM 1883 as a result of displacement and the missing portion of the skull block. A new reconstruction, based on the new parabasisphenoid from YPM 209, demonstrates that the ventral edge of the braincase was nearly straight (49.2 now 49.0). Also, the basipterygoid processes of YPM 209 are long and, when scaled to the YPM 1883 braincase, are equal to the dorsal height of the braincase (50.0 now 50.1).

In YPM 1883 and YPM 209, there is no anterior inclination of the maxillary tooth crowns (64.1 now 64.0) or of the dentary tooth crowns (65.? now 65.0) (contra Yates 2004). While the tooth enamel of YPM 209 is textured, this pattern and the 'wrinkled enamel' of YPM 1883 are not necessarily comparable with the wrinkled enamel of derived sauropods (71.1 now 71.?). Finally, the length of the manus of YPM 1883 is 38 per cent of that of the humerus length plus radius length (see Galton 1976; Yates 2004, p. 15), so an apparent miscoding was corrected (132.2 now 132.1).

A reanalysis of the Yates (2004) matrix was conducted with the incorporation of the above-mentioned modified character scores. Whereas 14 characters were listed as ordered (Yates 2004, p. 23), two additional characters (90 and 94) were also listed as ordered in the character list and were therefore treated as such in the reanalysis of the matrix. The revised data set was analysed using PAUP 4·0 beta version (Swofford 2002), with branch and bound searches and multistate characters treated as polymorphisms.

A single tree (length 441 steps, CI = 0·58, RI = 0·72) was recovered from this analysis with the same topology as that proposed by Yates (2004), including the position of *Anchisaurus* at the base of the sauropod clade. However, the robustness of this relationship has been weakened. In the new analysis this position has a bootstrap value of 70 per cent and a decay index of 2. Only two additional steps are required to place *Anchisaurus* among the 'prosauropod' taxa. A similar result was also found when the modified matrix was further altered to include new information on the skull of *Massospondylus* (Sues et al. 2004).

As mentioned above, it is likely that the specimens of *Anchisaurus* are immature, which would affect the scoring of at least two characters that currently support its inclusion within the sauropod clade. If the scores for two characters are reversed as would be likely with skeletal maturity, characters 66.0–66.1 (the number of dentary teeth) and 123.0–123.1 (relating to the proportions of the appendicular skeleton), then two equally parsimonious trees are recovered. The two trees differ only in the placement of *Anchisaurus*; one tree proposes a topology identical to that of Yates (2004) while the other places *Anchisaurus* as the sister taxa of 'prosauropods' more derived than *Efraasia minor* [*Anchisaurus* + (*Riojasaurus* + (*Plateosaurus* + (*Coloradisaurus* + (*Massospondylus* + *Lufengosaurus*))))].

The classification of *Anchisaurus* also depends on the definition of Sauropoda adopted. Based on a phylogenetic analysis that included an almost complete specimen of *Melanorosaurus* (Upper Triassic, South Africa), Yates (2005) noted that many taxa, such as *Riojasaurus*, *Anchisaurus*, *Jingshanosaurus*, *Yunnanosaurus*, *Massospondylus*, *Coloradisaurus* and *Lufengosaurus*, are 'captured' by the current diagnosis of Sauropoda and, because of this, a new definition of Sauropoda may be required. Yates (2007, in press) proposes a stem-based definition of Sauropoda that excludes *Anchisaurus*, and the other taxa listed above. Although there are differences in the analyses, *Anchisaurus* is also excluded from the stem-based definition of the Sauropoda used by Upchurch *et al.* (2007).

DISCUSSION

A new reconstruction of the braincase of *Anchisaurus* (Text-fig. 8) is based upon the reinterpretation of the braincase elements in YPM 1883 and YPM 209. The estimated basipterygoid lengths for YPM 1883 and the reconstruction are based upon the relative lengths of the basipterygoid processes and parabasisphenoid of YPM 209 (22 mm: 8 mm = 36 per cent). The estimated basipterygoid length (12 mm), from distal tip of basipterygoid process to dorsal edge of parasphenoid rostrum, suggests that *Anchisaurus* had long basipterygoid processes, as von Huene (1914*a*) stated, although his statement was based upon misinterpretation of the YPM 1883 skull.

This new reconstruction differs significantly from that presented in previous studies (Text-fig. 1E) and closely resembles those of other 'prosauropods' such as *Thecodontosaurus*, *Efraasia minor* (Galton and Bakker 1985; Yates 2003) and *Plateosaurus* (Galton 1984, 1985*b*) (Text-fig. 2). The braincase of *Anchisaurus* differs significantly from that of *Camarasaurus* (Chatterjee and Zheng 2005) and other sauropods in that the cultriform process of the parasphenoid is elongate, the ventral surface of the braincase is straight (with the cultriform process being parallel to the ventral surface of the basisphenoid: see reconstruction, Text-fig. 8), and the metotic fenestra and foramen ovalis are visible in lateral view (not obscured by a large crista prootica). Interestingly, however, the junction between the basisphenoid and basioccipital does appear to be a deep U-shaped

ventral depression in both YPM 209 and YPM 1883, a feature otherwise only found in *Lufengosaurus* (Barrett *et al.* 2005) and derived sauropods.

For over 100 years the skull of YPM 1883 has been misinterpreted because the skull block had fractured and one of the pieces was lost. Both of the skulls of *Anchisaurus* have suffered deformation owing to dorsoventral and lateral crushing, which is important to the interpretation of these small and delicate specimens. While commercial developments currently cover the original quarry where the specimens of *Anchisaurus* were recovered, other nearby sites have potential for producing new specimens, which are necessary to further our understanding of this taxon.

Acknowledgements. We are grateful to L. Murray, M. A. Turner, M. Fox and J. Gauthier (YPM) for their help while examining the specimens in their care, M. Benton (University of Bristol) for providing a translation of von Huene (1906), and A. Yates (University of the Witwatersrand) for providing information from his unpublished papers. M. Vickaryous and B. Hall (Dalhousie University) provided useful comments on an earlier version of this manuscript, as did reviewers D. Pol (Ohio State University) and M. Benton and P. Barrett. TF is grateful to Ping Li (Dalhousie University) for assistance with the operation of the SEM, the Fundy Geological Museum for preparation equipment, and financial support provided by the Jurassic Foundation, an NSERC postgraduate scholarship, and an NSERC research grant awarded to B. Hall. PMG is grateful to W. W. Amaral (Museum of Comparative Zoology, Harvard University) for his delicate preparation work on YPM 1883 that was funded by USA Grant No. DEB 81-01969.

REFERENCES

BARRETT, P. M., UPCHURCH, P. and WANG XIAO-LIN 2005. Cranial osteology of *Lufengosaurus huenei* Young (Dinosauria: Prosauropoda) from the Lower Jurassic of Yunnan, People's Republic of China. *Journal of Vertebrate Paleontology*, **25**, 806–822.

BAUMEL, J. J. and WITMER, L. M. 1993. Osteologia. 45–132. *In* BAUMEL, J. J., KING, A. S., BRAEAZILE, J. E., EVANS, H. E. and VANDEN BERGE, J. C. (eds). *Handbook of avian anatomy: Nomina anatomica avium*. The Nuttall Ornithological Club, Cambridge, MA, 779 pp.

BEHRENSMEYER, A. K. 1978. Taphonomic and ecological information from bone weathering. *Paleobiology*, **4**, 150–162.

BELLAIRS, A. d'A. and KAMAL, A. M. 1981. The chondrocranium and the development of the skull in recent reptiles. 1–263. *In* GANS, C. and PARSONS, T. S. (eds). *Biology of the Reptilia. Volume 11. Morphology F.* Academic Press, Toronto, 475 pp.

BENTON, M. J., JUUL, L., STORRS, G. W. and GALTON, P. M. 2000. Anatomy and systematics of the prosauropod dinosaur *Thecodontosaurus antiquus* from the Upper Triassic of southwest England. *Journal of Vertebrate Paleontology*, **20**, 77–108.

BONAPARTE, J. F. 1978. *Coloradia brevis* n. g. et n. sp. (Saurischia Prosauropoda), dinosaurio Plateosauridae de la Formación Los Colorados, Triásico Superior de La Rioja, Argentina. *Ameghiniana*, **15**, 327–332.

CHATTERJEE, S. and ZHENG ZHONG 2005. Neuroanatomy and dentition of *Camarasaurus lentus*. 199–211. *In* TIDWELL, V. and CARPENTER, K. (eds). *Thunder-lizards: the sauropodomorph dinosaurs*. Indiana University Press, Indianapolis, IN, 495 pp.

COPE, E. D. 1870. Synopsis of the extinct Batrachia, Reptilia and Aves of North America. *Transactions of the American Philosophical Society*, **14**, 1–252.

FEDAK, T. J. 2003. A new interpretation and description of the *Anchisaurus polyzelus* (Saurischia: Sauropodomorpha) braincase and its implications for prosauropod systematics. *Journal of Vertebrate Paleontology*, **23** (Supplement to No. 3), 49A.

—— 2005. Two heads are better than one: considering *Anchisaurus* as a small sauropod. *Journal of Vertebrate Paleontology*, **25** (Supplement to No. 3), 56A.

FRANZ-ODENDAAL, T. 2003. A fresh look at a developmental enamel defect in humans, mutant mice, and fossil giraffes: a contribution to evo devo. *Developmental Biology*, **259**, 452–452.

—— 2004. Enamel hypoplasia provides insights into early systemic stress in wild and captive giraffes (*Giraffa camelopardalis*). *Journal of Zoology*, **263**, 197–206.

GALTON, P. M. 1976. Prosauropod dinosaurs (Reptilia: Saurischia) of North America. *Postilla*, **169**, 1–98.

—— 1984. Cranial anatomy of the prosauropod dinosaur *Plateosaurus* from the Knollenmergel (Middle Keuper, Upper Triassic) of Germany. I – Two complete skulls from Trossingen/Württ. with comments on the diet. *Geologica et Palaeontologica*, **18**, 139–172.

—— 1985*a*. Cranial anatomy of the prosauropod dinosaur *Sellosaurus gracilis* from the Middle Stubensandstein (Upper Triassic) of Nordwürttemberg, West Germany. *Stuttgarter Beiträge zur Naturkunde, Serie B*, **118**, 1–39.

—— 1985*b*. Cranial anatomy of the prosauropod dinosaur *Plateosaurus* from the Knollenmergel (Middle Keuper, Upper Triassic) of Germany II – All the cranial material and details of soft-part anatomy. *Geologica et Palaeontologica*, **19**, 119–159.

—— 1990. Basal Sauropodomorpha-Prosauropoda. 320–344. *In* WEISHAMPEL, D. B., DODSON, P. and OSMÓLSKA, H. (eds). *The Dinosauria*. University of California Press, Berkeley, CA, 733 pp.

—— 2005. Basal sauropodomorph dinosaur taxa *Thecodontosaurus* Riley & Stutchbury, 1836, *T. antiquus* Morris, 1843 and *T. caducus* Yates, 2003: their status re. humeral morphs from the 1834 fissure fill (Upper Triassic) in Clifton, Bristol, UK. *Journal of Vertebrate Paleontology*, **25** (Supplement to No. 3), 61A.

—— and BAKKER, R. T. 1985. The cranial anatomy of the prosauropod dinosaur *Efraasia diagnostica*, a juvenile individual of *Sellosaurus gracilis* from the Upper Triassic of Nordwürttemberg, West Germany. *Stuttgarter Beiträge zur Naturkunde, Serie B*, **117**, 1–15.

—— and CLUVER, M. A. 1976. *Anchisaurus capensis* (Broom) and a revision of the Anchisauridae (Reptilia, Saurischia). *Annals of the South African Museum*, **69**, 121–159.

—— and UPCHURCH, P. 2004. Prosauropoda. 232–258. *In* WEISHAMPEL, D. B., DODSON, P. and OSMÓLSKA, H. (eds). *The Dinosauria*. Second edition. University of California Press, Berkeley, CA, 861 pp.

GOW, C. E. 1990. Morphology and growth of the *Massospondylus* braincase (Dinosauria Prosauropoda). *Palaeontologia Africana*, **27**, 59–75.

HITCHCOCK, E. 1858. *Ichnology of New England. A report on the sandstone of the Connecticut Valley, especially its fossil footprints.* Commonwealth of Massachusetts, 220 pp.

HITCHCOCK, E. Jr 1865. *A supplement to the ichnology of New England*. Wright and Potter, Boston, 90 pp.

HUENE, F. VON 1906. Über die Dinosaurier der ausser-europäischen Trias. *Geologische und Paläontologische Abhandlungen*, **8**, 97–156.

—— 1914*a*. Nachträge zu meinen früheren Beschreibungen triassischer Saurischia. *Geologische und Paläontologische Abhandlungen*, **13**, 69–82.

—— 1914*b*. Saurischia et Ornithischia Triadica ('Dinosauria' Triadica). *Fossilium Catalogus I: Animalia*, **4**, 1–21.

—— 1932. Die fossile Reptil-Ordnung Saurischia, ihre Entwicklung und Geschichte. *Monographien zur Geologie und Paläontologie, Serie 1*, **4**, 1–361.

HWANG, S. H. 2005. Phylogenetic patterns of enamel microstructure in dinosaur teeth. *Journal of Morphology*, **266**, 208–240.

LULL, R. S. 1912. The life of the Connecticut Triassic. *American Journal of Science*, **4**, 397–422.

—— 1915. Triassic life of the Connecticut Valley. *Connecticut State Geological and Natural History Survey, Bulletin*, **24**, 1–285.

—— 1953. Triassic life of the Connecticut Valley. Revised. *Connecticut Geological and Natural History Survey, Bulletin*, **81**, 1–331.

MARSH, O. C. 1882. Classification of the Dinosauria. *American Journal of Science, Series 3*, **23**, 81–86.

—— 1885. Names of extinct reptiles. *American Journal of Science, Series 3*, **29**, 169.

—— 1889. Notice of new American dinosaurs. *American Journal of Science, Series 3*, **37**, 331–336.

—— 1891. Notice of new vertebrate fossils. *American Journal of Science, Series 3*, **42**, 265–269.

—— 1892. Notes on Triassic Dinosauria. *American Journal of Science, Series 3*, **43**, 543–546.

—— 1896. The dinosaurs of North America. *United States Geological Survey, 16th Annual Report*, **1894–96**, 133–244.

OLSEN, P. E. and GALTON, P. M. 1977. Triassic–Jurassic tetrapod extinctions: are they real? *Science*, **197**, 983–986.

OWEN, R. 1842. Report on British fossil reptiles. *Reports of the British Association for the Advancement of Science*, **11**, 60–294.

REISZ, R. R., SCOTT, D., SUES, H.-D., EVANS, D. C. and RAATH, M. A. 2005. Embryos of an Early Jurassic prosauropod dinosaur and their evolutionary significance. *Science*, **309**, 761–764.

RIEPPEL, O. 1993. Studies on skeletal formation in reptiles. V. Patterns of ossification in the skeleton of *Alligator mississippiensis* Daudin (Reptilia, Crocodylia). *Zoological Journal of the Linnean Society*, **109**, 301–325.

SEELEY, H. G. 1887. On the classification of the fossil animals commonly named Dinosauria. *Proceedings of the Royal Society of London*, **43**, 165–171.

SERENO, P. C. 1999. The evolution of the dinosaurs. *Science*, **284**, 2137–2147.

SUES, H.-D., REISZ, R. R., HINIC, S. and RAATH, M. A. 2004. On the skull of *Massospondylus carinatus* Owen, 1854 (Dinosauria: Sauropodomorpha) from the Elliot and Clarens formations (Lower Jurassic) of South Africa. *Annals of the Carnegie Museum*, **73**, 239–257.

SWOFFORD, D. L. 2002. PAUP * Phylogenetic Analysis Using Parsimony (*and other methods). Sinauer Associates, Sunderland, MA.

UPCHURCH, P., BARRETT, P. M. and GALTON, P. M. 2005. The phylogenetic relationships of basal sauropodomorphs: implications for the origin of sauropods. *Journal of Vertebrate Paleontology*, **25** (Supplement to No. 3), 126A.

—— —— —— 2007. A phylogenetic analysis of basal sauropodomorph relationships: implications for the origin of sauropod dinosaurs. 57–90. *In* BARRETT, P. M. and BATTEN, D. J. (eds). *Evolution and palaeobiology of early sauropodomorph dinosaurs*. Special Papers in Palaeontology, **77**, 289 pp.

WILSON, J. A. and SERENO, P. C. 1998. Early evolution and higher-level phylogeny of sauropod dinosaurs. *Memoir of the Society of Vertebrate Paleontology*, **5**, 1–68.

YATES, A. M. 2002. A re-examination of the phylogenetic position of the unusual sauropodomorph *Anchisaurus*. *The Palaeontological Association, Newsletter*, **50**, 55.

—— 2003. A new species of the primitive dinosaur *Thecodontosaurus* (Saurischia: Sauropodomorpha) and its implications for the systematics of early dinosaurs. *Journal of Systematic Palaeontology*, **1**, 1–42.

—— 2004. *Anchisaurus polyzelus* (Hitchcock): the smallest known sauropod dinosaur and the evolution of gigantism amongst sauropodomorph dinosaurs. *Postilla*, **230**, 1–58.

—— 2005. The skull of the Triassic sauropodomorph, *Melanorosaurus readi*, from South Africa and the definition of Sauropoda. *Journal of Vertebrate Paleontology*, **25** (Supplement to No. 3), 132A.

—— 2007. The first complete skull of the Triassic dinosaur *Melanorosaurus* Haughton (Sauropodomorpha: Anchisauria) 9–55. *In* BARRETT, P. M. and BATTEN, D. J. (eds). *Evolution and palaeobiology of early sauropodomorph dinosaurs*. Special Papers in Palaeontology, **77**, 289 pp.

—— in press. Solving a dinosaurian puzzle: the identity of *Aliwalia rex* Galton. *Historical Geology*.

[Special Papers in Palaeontology 77, 2007, pp. 261–289]

BASAL SAUROPODOMORPHA: HISTORICAL AND RECENT PHYLOGENETIC HYPOTHESES, WITH COMMENTS ON *AMMOSAURUS MAJOR* (MARSH, 1889)

by PAUL C. SERENO

University of Chicago, Organismal Biology and Anatomy, 1027 East 57th St, Chicago, Illinois 60637, USA; e-mail: dinosaur@uchicago.edu

Typescript received 14 June 2006; accepted in revised form 30 October 2006

Abstract: Basal sauropodomorphs, historically referred to as 'prosauropods', include approximately 20 genera of Late Triassic–Early Jurassic age. Recent discoveries on several continents and taxonomic review of important species have brought to bear significant new taxon and character data. After review of the taxonomic status of *Anchisaurus polyzelus*, *Ammosaurus major* is recommended as the appropriate genus and species for basal sauropodomorph material from the Portland Formation of eastern North America. Traditional (precladistic) and cladistic interpretation of basal sauropodomorph phylogeny has varied between two extremes: a monophyletic clade of 'prosauropods' or a sequence of basal sauropodomorphs that increasingly approach the sauropod condition. Given new species that exhibit a range of derived features, future resolution will probably lie somewhere between these polar viewpoints. Conflicting results from recent analyses suggest that greater resolution of basal sauropodomorph phylogeny will come with continued clarification of anatomical details and a comparative methodology that focuses on character data rather than simply the most parsimonious tree.

Key words: Sauropodomorpha, prosauropod, dinosaur, *Anchisaurus*, *Ammosaurus*, phylogeny.

SAUROPODOMORPHS comprise the longest-lived, most speciose radiation of large-bodied, terrestrial vertebrate herbivores. The focus here is on basal sauropodomorphs, the non-sauropod portion of the radiation traditionally referred to as 'prosauropods'. This is an anatomically conservative assemblage of approximately 20 genera that flourished during a brief interval of approximately 30 million years from the Late Triassic (Late Carnian, *c.* 220 Ma) through the Early Jurassic (Hettangian–Sinemurian, *c.* 195 Ma). By the close of the Triassic, they had achieved a global distribution, extending their range east to west across Pangaea and toward each pole. They constitute the first morphologically closely knit, global radiation of dinosaurian herbivores (Sereno 1997; Galton and Upchurch 2004; Pol 2004).

In 1976, Peter Galton revived interest in basal sauropodomorphs with a seminal review of North American material, which includes some of the earliest dinosaur remains recovered on that continent. In the 30 years since, many discoveries and taxonomic studies have dramatically transformed the basis for understanding basal sauropodomorph phylogeny. The oldest basal sauropodomorph recorded to date, *Saturnalia tupiniquim*, was discovered recently in the Santa Maria Formation in southern Brazil (Langer *et al.* 1999, 2007; Langer 2003). Probably close to the Middle/Late Triassic boundary in age, the Santa Maria fauna is roughly coeval with the Ischigualasto fauna from Argentina, radiometrically dated to approximately 228 Ma (Rogers *et al.* 1993). *Saturnalia* provides tangible fossil evidence of the roots of the sauropodomorph radiation, which previously existed only as a ghost lineage generated by older members of their saurischian sister group, Theropoda (*Eoraptor*, Herrerasauridae).

By the close of the Triassic (*c.* 200 Ma), basal sauropodomorphs had diversified in southern Africa (*Antetonitrus*, *Melanorosaurus*, '*Euskelosaurus*', *Blikanasaurus*: Van Heerden 1979; Cooper 1981; Galton and Van Heerden 1998; Yates and Kitching 2003; Yates 2005), South America (*Coloradisaurus*, *Riojasaurus*, *Lessemsaurus*, *Mussaurus*, *Unayasaurus*: Bonaparte 1978, 1999; Casamiquela 1980; Leal *et al.* 2004; Pol 2004), Europe (*Efraasia*, *Plateosaurus*, *Ruehleia*, *Thecodontosaurus*: von Huene 1926; Galton 1984, 2001; Benton *et al.* 2000; Yates 2003*a*, *b*) and Greenland (cf. *Plateosaurus*: Jenkins *et al.* 1995).

Ten million years later, during the Early Jurassic (*c.* 190 Ma), basal sauropodomorphs had diversified in North America (*Ammosaurus*, *Anchisaurus*: Galton 1976; Yates 2004) and Asia (*Lufengosaurus*, *Jingshanosaurus*, *Yimenosaurus*, *Yunnanosaurus*: Young 1941, 1942; Bai *et al.* 1990; Zhang and Yang 1994; Barrett *et al.* 2005).

Well-preserved material from India is now known that is also probably Early Jurassic in age (Kutty 1969), and postcranial bones of similar age have been reported from Antarctica (Hammer and Hickerson 1994). Australia, a landmass with little exposure of fossiliferous terrestrial deposits of Mesozoic age, remains the only continent with no evidence of basal sauropodomorphs, the only record (*Agrosaurus*) now shown to be based on fossil material from England (Vickers-Rich *et al.* 1999).

This paper first briefly reviews the historical emergence of our current palaeontological record for basal sauropodomorphs. Second, a vexing taxonomic question is addressed: the proper assignment of material from eastern North America to either *Anchisaurus* or *Ammosaurus*. Third, traditional (precladistic) and cladistic hypotheses for basal sauropodomorphs are summarized and compared and several notable characters are discussed. Finally, a phylogenetic taxonomy is proposed that would accommodate alternative phylogenetic interpretations.

Institutional abbreviations: AM, Amherst College Museum, Amherst; AMNH, American Museum of Natural History, New York; IVPP, Institute for Vertebrate Paleontology and Paleoanthropology, Beijing; PVL, Paleontological Collection of the Fundación-Instituto Miguel Lillo, Tucumán; SMNS, Staatlichen Museums für Naturkunde, Stuttgart; UCR, University College of Rhodesia (Zimbabwe); YPM, Yale Peabody Museum, New Haven.

FOSSIL DISCOVERY

Early period (1836–1900)

Relatively complete skulls or skeletons of basal sauropodomorphs first came to light at the close of the 19th century, and so most of the early taxonomy was based on incomplete, disarticulated material. The first described genera, *Thecodontosaurus* from England and Wales (Riley and Stutchbury 1836), *Plateosaurus* from Germany (von Meyer 1837) and *Massospondylus* (Owen 1854; Seeley 1895) from southern Africa, survive to the present as valid taxa, although now represented by much more complete material (Cooper 1981; Galton 1984, 2001; Kermack 1984; Benton *et al.* 2000; Yates 2003*a*; Sues *et al.* 2004). The first reasonably complete skeleton pertained to the genus *Anchisaurus* and was discovered in a rock quarry near Manchester, Connecticut (Marsh 1891). Although most of the skull was preserved, preservational factors had complicated its interpretation. Marsh, for example, described *Anchisaurus* as a carnivorous theropod in a paper on its skeletal restoration (Marsh 1893). Just before the close of the 19th century, nonetheless, the general form of the skull and skeleton of a representative basal sauropodomorph had been realized.

Middle period (1900–1950)

In the first two decades of the 20th century, multiple articulated skeletons with skulls were discovered in rock quarries in Trossingen, Germany. Now collectively attributed to the genus *Plateosaurus* (Text-figs 4–8), this material formed the basis of von Huene's influential descriptive account of the skull and skeleton (von Huene 1926).

Trained in Germany, C. C. Young returned to China in the 1930s and directed the recovery of well-preserved remains of basal sauropodomorphs from the Lower Lufeng Formation in Yunnan Province. The majority of these pertain to the genera *Lufengosaurus* and *Yunnanosaurus*, which were described in a series of papers (Young 1941, 1942, 1947, 1951). Until recently, Young's reports constituted the only information available for basal sauropodomorphs from Asia.

Recent work (1970–present)

During the recent period, many new genera have been described, including: *Saturnalia* (Langer *et al.* 1999; Langer 2003) and *Unaysaurus* (Leal *et al.* 2004) from southern Brazil; *Coloradisaurus, Riojasaurus, Lessemsaurus* and *Mussaurus* from Argentina (Bonaparte 1969, 1978, 1999; Bonaparte and Vince 1979); *Yimenosaurus* and *Jingshanosaurus* from China (Bai *et al.* 1990; Zhang and Yang 1994); and *Efraasia* and *Ruehleia* from Germany (Galton 1973, 2001; Yates 2003*b*), and two unnamed genera from India (Kutty 1969).

Detailed taxonomic and descriptive studies have been undertaken, the most comprehensive of these involving reviews of *Anchisaurus* and *Ammosaurus* (Galton 1976; Yates 2004), *Efraasia* (Galton 1973; Yates 2003*b*), *Plateosaurus* (Galton 1984, 2000, 2001; Yates 2003*b*), *Massospondylus* (Cooper 1981; Gow *et al.* 1990; Sues *et al.* 2004), *Lufengosaurus* (Barrett *et al.* 2005), *Riojasaurus* (Bonaparte and Pumares 1995), *Mussaurus* and *Lessemsaurus* (Pol 2004), and *Thecodontosaurus* (Kermack 1984; Yates 2003*a*).

Finally, new material currently under study will better document *Saturnalia*, the Argentine genera *Mussaurus* and *Lessemsaurus* (Pol 2004; Pol and Powell 2005), and new remains from northern Argentina (R. Martinez, pers. comm. 2006). More information is needed for Asian genera, especially '*Gyposaurus*' *sinensis* (Young 1941, 1948), *Jingshanosaurus, Yunnanosaurus* and *Yimenosaurus*. New remains from North America and Greenland now under study will also impact future phylogenetic hypotheses. Well-preserved skull and postcranial material from the Kayenta Formation and Navajo Sandstone of western

North America, for example, represent new taxa (Attridge *et al.* 1985; Irmis 2005; Loewen *et al.* 2005), and specimens from the Newark Supergroup in Nova Scotia (Olsen *et al.* 1987) and from the Fleming Fjord Formation of eastern Greenland (Jenkins *et al.* 1995) are under study. Finally, a well-preserved skull and skeleton has recently been referred to the southern African genus *Melanorosaurus* (Yates 2005, 2007), which will strongly impact its phylogenetic interpretation.

AMMOSAURUS OR *ANCHISAURUS*?

Two genera, *Ammosaurus* and *Anchisaurus*, and their various species were described from the Lower Jurassic Portland Formation of Connecticut (Galton 1976). Sereno (1999*a*) regarded the Portland material as representing the single species *Ammosaurus major*, although no explanation was provided. Yates (2004), on the other hand, regarded *Anchisaurus polyzelus* as the appropriate name for this taxon. Galton and Upchurch (2004, p. 251), by contrast, maintained the generic separation of *Ammosaurus* and *Anchisaurus*, arguing that there are observable differences among the specimens. More recently, Fedak (2005) re-examined the same specimens. He regarded them as a single taxon and followed Yates by using the genus *Anchisaurus*. As detailed below, *Ammosaurus major* is regarded herein as the appropriate genus and species for diagnostic material from the Portland Formation. Referral of specimens to *Ammosaurus* from western North America (Navajo Formation of Arizona and southern Utah: Galton 1971, 1976) is not supported here, in agreement with Yates (2004) and Irmis (2005). *Ammosaurus major* will continue to play an important role in phylogenetic analysis, and so its taxonomic status should be carefully considered.

A single geographically disparate species from southern Africa, *Anchisaurus capensis*, was based on a small partial postcranial skeleton lacking much of the anterior half. Originally described by Broom (1906) and later named as a new genus and species, *Gyposaurus capensis* (Broom 1911), the taxon was initially placed within the Anchisauridae (Broom 1911). More recently, the genus was considered a junior synonym of *Anchisaurus*, based principally on the shared presence of an elongate iliac preacetabular process, the supposedly enlarged and ventrally incomplete obturator foramen, and a proportionately narrow foot (Galton and Cluver 1976). The larger size of the ungual of pedal digit I was the only diagnostic feature listed for *Anchisaurus capensis*; this relative proportion, however, is widespread among basal sauropodomorphs and may well be primitive within the clade.

Cooper (1981) listed *Gyposaurus capensis* as a junior synonym of *Massospondylus carinatus*, a conclusion

followed by Galton (1990), and most recently the species was listed as a *nomen dubium* (Galton and Upchurch 2004). The elongate proportions of the iliac preacetabular process have long been recognized as distinctive of the North American material. Unfortunately, the dorsal margin of this process is incomplete in the holotype specimen of *Gyposaurus capensis*; it was reconstructed with deeper proportions (Galton and Cluver 1976, fig. 3). Although review of the holotype material in South Africa is necessary, the status of *Gyposaurus* and its type species (*G. capensis*) as doubtful names appears to be the most reasonable option at present.

Anchisaurus polyzelus

Hitchcock (1855) reported the discovery of a skeleton (AM 41/109) during a blasting operation at an armory in Springfield, Massachusetts. Initially described as *Megadactylus polyzelus* (Hitchcock 1865), preoccupation forced transfer first to the genus *Amphisaurus* (Marsh 1882) and then *Anchisaurus* (Marsh 1885). The holotype specimen consists of vertebral fragments, the proximal portion of a scapula, a partial right forelimb and manus, conjoined distal ischial shafts, and portions of the left hind limb (Galton 1976, figs 3, 5–10). The enlarged manual digit I, triangular cross-section of the conjoined ischial shafts at mid length and other features in the preserved material of the holotype clearly allow placement among basal sauropodomorphs. There are, however, no diagnostic features at generic or specific levels. None of the features listed by Galton (1976, p. 88) or Galton and Upchurch (2004, p. 251) is observable in the holotype. Galton and Upchurch, for example, cited the 'emarginated proximal portion of the pubis' when there is little, if any, of either pubis preserved in the holotype. These authors are surely referring to the much more complete referred specimen YPM 1883.

Galton (1976, p. 82) stated succinctly his reasoning, which attempts to salvage the taxon *Anchisaurus polyzelus* on the basis of the fragmentary holotype: 'It is apparent that the differences between AM 41/109 and YPM 1883 are minimal and can be attributed either to differences in preservation or to individual variation, a factor that tends to be overlooked. I conclude that YPM 1883 should be referred to *Anchisaurus polyzelus* because, on the basis of the available material, YPM 1883 cannot be distinguished from AM 41/109 by any characters of taxonomic significance'. The holotype specimen (AM 41/109), on which the validity of the generic and species name rests, must exhibit at least one diagnostic feature for referral of other specimens. Galton, by contrast, bases his referral of YPM 1883 only on the absence of differences.

Yates (2004, p. 5), aware of the need for the holotype to exhibit diagnostic traits, cited one feature in AM 41/109 as fulfilling this role: 'flattened coplanar ischial shafts'. Galton (1976, fig. 6I–J, p. 17) figured the ischial shafts of the holotype, and they have substantial dorso-ventral depth in lateral and distal views. Galton described these shafts as 'subtriangular ischial rods'. Sereno (1999a) used the inverted subtriangular cross-section of conjoined ischial shafts as a prosauropod synapomorphy. This is the condition toward the distal end of the ischia in AM 41/109 (Galton 1976, fig. 6J), *Massospondylus carinatus* (Cooper 1981, fig. 55) and many other basal sauropodomorphs. There seems little basis for arguing that the form of the ischial shafts in AM 41/109 is diagnostic.

Yates (2004) also argued that the form of the ischial shafts is derived in other specimens, such as YPM 209. In this specimen, however, the proximal portion of the ischial shafts is angled at about 30 degrees to the horizontal (Galton 1976, figs 30A, 31D) as in *Massospondylus carinatus* (Cooper 1981, fig. 55). Galton (1976, p. 66) described the shafts as 'ischial rods', not as flattened and coplanar. Although he also remarked that they were 'horizontal' and 'oval' in cross-section, the critical distal portion of the shaft is broken away on both sides. The ischial shafts are not preserved in YPM 1883 and are broken at mid-length in YPM 208. The latter specimen comes closest to the cited condition of having coplanar ischial shafts (Yates 2004, fig. 2). However, only the proximal ends are preserved, and post-mortem dorsoventral compression cannot be ruled out. Should YPM 208 prove diagnostic in this regard, in any case, it is the holotype of *Ammosaurus major* not *Anchisaurus polyzelus*.

Other features listed by Yates (2004) as diagnostic for *Anchisaurus polyzelus* include a foramen at the base of the second sacral rib and an enlarged obturator fenestra on the pubis. The 'foramen' does not pass through the sacral rib but rather appears to be developed as a shallow depression, and the obturator opening appears to be artificially enlarged by loss of the thin bone margin around the foramen. In any case, these features are not observable on the holotype specimen (AM 41/109). The elongate proportions of the iliac preacetabular process, the final feature listed by Yates, has been recognized as diagnostic, but again is not preserved in the holotype specimen of *Anchisaurus polyzelus*.

In summary, evidence for diagnostic features in the holotypic specimen of *Anchisaurus polyzelus* is weak at best. AM 41/109, found in a Massachusetts quarry, may well be the same species as the sauropodomorph material recovered in Connecticut quarries to the south in the same river valley (East Windsor, Manchester) and perhaps the same as material collected recently in Nova Scotia

(Olsen *et al.* 1987). However, it is imprudent to assume so without justification on the basis of diagnostic features, which, first and foremost, must be present in the holotype specimen. As such features are not apparent, *Anchisaurus polyzelus* is here regarded as a *nomen dubium*. Anchisauridae (Marsh 1885), furthermore, should not be employed as a higher taxon for species from the Portland Formation, because it is not clear that it would include the fragmentary holotype specimen on which the familial name is based. Although *Anchisaurus* has gained wider usage in the recent literature than *Ammosaurus*, both were listed as valid genera in recent taxonomic compilations (Galton 1990; Galton and Upchurch 2004). A special appeal to save the genus *Anchisaurus* and its type species *A. polyzelus* over *Ammosaurus major* does not seem warranted.

Ammosaurus major

The first fragmentary skeleton (YPM 2125) from the Connecticut River valley in Connecticut was discovered in 1818 during quarrying operations in the Upper Portland Formation near East Windsor. Three better-preserved specimens (YPM 208, 209, 1883) were collected in the 1880s from a single sandstone quarry some 20 km to the south near Manchester, Connecticut (Lull 1915, 1953; Galton 1976; Olsen *et al.* 1987). Sadly, the skull and anterior half of one of these specimens was incorporated into the abutments of a bridge, and only small fragments were later recovered (YPM 208: Ostrom 1969; Galton 1976).

These three specimens were initially allocated by Marsh (1889a, 1891, 1892) to three species of *Anchisaurus*, *A. solus* (YPM 209), *A. colurus* (YPM 1883), and *A. major* (YPM 208), the last of which he referred to a new genus as *Ammosaurus major* (Marsh 1891). Von Huene (1932) later assigned the most complete specimen (YPM 1883) to a new genus, *Yaleosaurus*. Galton (1976, pp. 82–83) subdivided these three specimens, assigning YPM 1883 to *Anchisaurus polyzelus* and referring YPM 208 and 209 to *Ammosaurus major*. Finally, Yates (2004) regarded *Ammosaurus major* as a junior synonym of *Anchisaurus polyzelus* as discussed above.

Galton's subdivision was based primarily on the proportions of the metatarsus. Because the metatarsus is not preserved in the holotype specimen (AM 41/109), *Anchisaurus polyzelus* was identified as 'narrow-footed' on the basis of a referred specimen (YPM 1883) and compared to the 'broad-footed' metatarsus in *Ammosaurus major* (YPM 208, 209). Cooper (1981), Yates (2004) and others have criticized this distinction as ontogenetic or as an artefact of preservation. The case is re-examined in the taxonomic revision presented below.

SYSTEMATIC PALAEONTOLOGY

SAUROPODOMORPHA von Huene, 1932

Genus AMMOSAURUS Marsh, 1891

1891 *Ammosaurus* Marsh, p. 267
1932 *Yaleosaurus* von Huene, p. 119.

Type species. Ammosaurus major (Marsh, 1889a).

Diagnosis. As for type and only species.

Ammosaurus major (Marsh, 1889a)

1889a *Anchisaurus major* Marsh, p. 331, text-fig. 1.
1891 *Ammosaurus major* (Marsh); Marsh, p. 267.
1891 *Anchisaurus colurus* Marsh, p. 267.
1892 *Anchisaurus solus* Marsh, p. 545.
1932 *Ammosaurus solus* (Marsh); von Huene, p. 27, pl. 49, fig. 1.
1932 *Yaleosaurus colurus* (Marsh); von Huene, p. 119, pl. 54, fig. 3.

Holotype. YPM 208, partial articulated skeleton in two blocks including several middle and posterior dorsal vertebrae and ribs, three sacral vertebrae, partial right scapula, pelvic girdle, and partial left and right hindlimbs.

Type locality and horizon. Manchester, Connecticut; Upper Portland Formation (Lower Jurassic).

Referred material. YPM 209, immature, partially articulated skeleton lacking the tail (originally described as *Anchisaurus solus* Marsh 1892); YPM 1883, well-preserved articulated adult skeleton lacking most of the cervical vertebrae and all caudal vertebrae (originally described as *Anchisaurus colurus* Marsh, 1891).

Diagnosis. Basal sauropodomorph with spool-shaped dorsal vertebrae with length approximately twice the dorsoventral diameter of the centrum face and an elongate preacetabular process on the ilium (length twice basal depth) that extends as far anteriorly as the pubic peduncle.

Remarks. Both of the diagnostic features are present in the holotype and referred specimens. The spool-shaped dorsal vertebrae, present in the holotype but better exposed in YPM 209 and 1883, have unusually elongate proportions (Galton 1976, figs 15I–J, 30). The preacetabular process is unusually long compared with that in other basal sauropodomorphs, as measured with ischial and pubic peduncles positioned along a horizontal (Galton and Cluver 1976; Yates 2004). The process, however, has been figured as slightly longer than preserved relative to the remainder of

the ilium (Galton 1976, figs 19A, 26E). The preacetabular process does not extend beyond the distal end of the pubic peduncle. The process in YPM 208, which is identical to that in YPM 1883, was also shown as more pointed than preserved.

Galton (1976) did not figure the carpus and manus accurately in *Ammosaurus major*. Misinformation has arisen, as a result, regarding the inset of metacarpal 1 into the carpus (Yates 2004; Irmis 2005). In YPM 1883 the enlarged distal carpal 1, the only carpal preserved, is not positioned directly over metacarpal 1 (Galton 1976, fig. 17C) but rather is inset from the medial edge of metacarpal 1 so that its lateral margin would overlap metacarpal 2. The proximal end of metacarpal 1 is not aligned with the bases of the other metacarpals (Galton 1976, fig. 17C) but rather is inset approximately 4 mm into the carpus relative to metacarpal 2 (Galton 1976, fig. 18). The strong overlap of metacarpals 2 and 3 and metacarpals 4 and 5 as preserved is lost in the flattened reconstruction of the metacarpus. The specimens clearly suggest that *Ammosaurus major* had a carpus and manus of similar design to that found in other basal sauropodomorphs in which these parts have been preserved in natural articulation (Cooper 1981, fig. 35); distal carpal 1 overlaps distal carpal 2, metacarpal 1 is inset into the carpus relative to metacarpal 2, and metacarpal 1 articulates laterally with distal carpal 2.

The shallow dorsoventral height of the ischium had been used to distinguish YPM 208 (Galton 1976, pp. 56, 82), but a comparable portion of the ischium is not preserved in any other specimen from the Manchester quarry. In YPM 208 the ventral margins of both ischia are eroded and have broken edges. The shallow subacetabular depth of the ischia and the supposed presence of an ischial obturator process appear to be artefacts of preservation (Galton 1976, fig. 26E–F). Reconstructions of the pubis differ markedly in YPM 208 and 1883. The former was shown with a broad margin under the acetabulum and an unusually large obturator foramen (Galton 1976, figs 19A, 20A, 26E–F). The margins of the foramen, however, are poorly preserved, bringing into question the true size of the opening. The latter specimen was reconstructed with a narrow margin under the acetabulum with an open obturator notch. Although Galton remarked that the margin of the foramen is preserved, it appears broken like adjacent areas of the pubic blade (Galton 1976, fig. 12).

Galton's description of the pes in YPM 1883 as 'narrow-footed' and that in YPM 208 as 'broad-footed' is not supported by re-examination of the specimens. First, the metatarsals in both specimens have been distorted postmortem, severely so in YPM 208. In both specimens the broad medial side of the shaft of metatarsal 1 faces dorsally, rather than medially, as a result of dorsoventral flattening. Second, it is difficult to compare the degree of metatarsal overlap in each specimen, because the metatarsus is exposed only in dorsal view in YPM 208 and ventral view in YPM 1883. Third, YPM 1883 is approximately 25 per cent smaller than YPM 208, leaving open the possibility that proportional differences may reflect this differential in body size.

Despite these caveats, it is possible to measure the metatarsals in these specimens to assess the qualitative claim long held in the literature that one (YPM 1883) is discordantly more slender than the other (YPM 208: Galton 1976, 1990; Galton and Cluver 1976; Galton and Upchurch 2004). The maximum width across the metatarsus (metatarsals 1–5) and the maximum length of metatarsal 3 are, respectively, 60 mm and 98 mm in YPM 1883 (right side) and 81 mm and 135 mm in YPM 208 (left side). The width of the metatarsus relative to the length of metatarsal 3, thus, is 61 per cent in YPM 1883 and 60 per cent in YPM 208. They have virtually identical metatarsal proportions as noted previously by Yates (2004).

An identical pair of comparative measurements taken on the reconstruction of the metatarsus of these two specimens (Galton 1976, figs 22E, 29D), however, yields different results. Using the associated scale bar for determining the length and width of the metatarsus, YPM 1883 (37, 49 mm) and YPM 208 (23, 41 mm) yield relative widths of 56 per cent (YPM 1883) and 76 per cent (YPM 208). Thus, a 20 per cent difference in width was generated in the process of reconstructing specimens that appear very similar and were found in the same quarry. Post-mortem distortion has also played a role, generating an 11 per cent difference in length between right and left third metatarsals in a single individual (Galton 1976, table 4; YPM 208, 120 vs. 135 mm).

PHYLOGENY: TRADITIONAL INTERPRETATIONS

Prosauropods as 'carnosaurs', ornithischians or more distant archosauromorphs

When the first partial skeletons of basal sauropodomorphs came to light, they were placed among theropods (Marsh 1884). Genera now regarded as basal sauropodomorphs have since been linked to, or placed within, the two remaining dinosaurian clades, Ornithischia and Sauropoda (von Huene 1914a, 1932; Romer 1956; Colbert 1964; Charig et al. 1965; Bakker and Galton 1974; Paul 1984; Gauthier 1986). The early history of these disparate taxonomic assignments can be traced to an erroneous association with jaw fragments and teeth of rauisuchian archosaurs. The mix-up first occurred among fossils collected in the Middle Stubensandstein (Norian, Late Triassic) of Germany. Von Huene (1908, 1914a, 1932) referred

articulated and disarticulated remains now identified as *Plateosaurus gracilis* (=*Sellosaurus*) to *Teratosaurus suevicus*, now regarded as a poposaurid (Galton 1985; Benton 1986). Young (1951) repeated von Huene's spurious association in his work with fossils from the Lower Lufeng Formation, assigning basal sauropodomorph material to a taxon (*Sinosaurus triassicus*) based on jaw fragments with carnivorous teeth. Finally, Raath (1972) described isolated carnivorous teeth found near the pelvis of the holotype specimen of the basal sauropod *Vulcanodon karibaensis*, an association that continued to haunt the more recent detailed revision of this important taxon (Cooper 1984).

These apparently carnivorous forms had to be reconciled with other basal sauropodomorphs characterized by lightweight skulls with herbivorous teeth, a group von Huene (1920) collectively called Prosauropoda. At the time, all of these forms were regarded as Late Triassic in age. The view that emerged and remained dominant until the mid 1960s regarded 'prosauropods' as the central stock of saurischian phylogeny, with a herbivorous subgroup (Plateosauria) more closely related to sauropods and a carnivorous subgroup (Palaeosauria) more closely related to theropods (Text-fig. 1A; von Huene 1914a, 1932, 1956; Romer 1956; Colbert 1964).

Several basal sauropodomorphs, often very immature specimens, were initially attributed to other clades, such as the 'ornithischian' *Tawasaurus* (Young 1982) and the 'lepidosaurian' *Fulengia* (Carroll and Galton 1977) from the Lower Lufeng Formation of China (Evans and Milner 1989; Sereno 1991). On the other hand, jaws of disparate archosauromorphs have been regarded as basal sauropodomorphs. 'Two new prosauropods' were briefly described from rocks of probable Late Triassic age in Madagascar (Flynn et al. 1999, p. 763), although only one feature was cited that is currently regarded as a basal sauropodomorph synapomorphy (ventral deflection of the anterior alveolar margin of the dentary). The presence of low triangular crowns, some apparently with wear facets (Flynn et al. 1999, fig. 1), raises questions about the affinity of this material as well as similar jaw material described previously under the genus *Azendohsaurus* (Dutuit 1972; Gauffre 1993). More complete material recently recovered from Madagascar is expected to clarify the non-dinosaurian status of these specimens, which continue to lurk around basal nodes within Sauropodomorpha even in the most recent analyses (Pol 2004).

Prosauropods as ancestors or side-branch?

Ancestors. As the name implies, Prosauropoda has long been viewed as the ancestral group that gave rise to sauropods. Prosauropoda, in other words, is paraphyletic with some basal sauropodomorphs, in particular the 'melanorosaurids', more closely related to Sauropoda than others. Colbert (1964) crystallized this view (Text-fig. 1A), versions of which were presented long before and after his timely review (von Huene 1929; Romer 1968; Cooper 1981; Bonaparte 1986; Bonaparte and Pumares 1995).

Supporting evidence to link 'melanorosaurids' and sauropods has always been meagre at best and typically based on femoral morphology or assumptions about quadrupedal posture. As summarized by Romer (1966, p. 150): 'Even as regards *Plateosaurus* and its relatives, most, at least, seem to have left no descendants. But in the case of such a form as *Melanorosaurus* of South Africa, in which little of a trend toward bipedalism is present, we may be dealing with forms close to the ancestry of later sauropods.'

Stocky long bones, a more columnar femoral shaft and fourth trochanter positioned closer to mid shaft are features that were used to group together forms such as South American *Riojasaurus incertus* (Bonaparte 1972), European *Camelotia borealis* (Galton 1998), and South African *Melanorosaurus readi* as Melanorosauridae (Galton 1985; Van Heerden and Galton 1997) in close proximity to Sauropoda. A more complete skeleton including a skull with a surprising combination of features has recently been referred to *Melanorosaurus readi* (Yates 2005, 2007). When described it will doubtless shed much light on the affinities of at least the nominotypical genus.

Side-branch. Another traditional interpretation is that Prosauropoda constitutes a side-branch, or monophyletic sister taxon, to Sauropoda. Cruickshank (1975) was the first to clearly depict this alternative hypothesis in graphical form (Text-fig. 1B). There were two principal lines of evidence driving this view, both of which are now regarded as unsatisfactory: (1) the overlapping temporal range of prosauropods and early sauropods suggesting their independent evolution (Charig *et al.* 1965); (2) reduction patterns in metatarsals or phalanges that preclude bipedal precursors or are irreversible (Charig *et al.* 1965; Cruickshank 1975). Surprisingly, traditional arguments for the monophyly of prosauropods were not based on shared derived features, such as the twisted pollex.

Basal sauropodomorphs and sauropods were initially regarded as Late Triassic and Jurassic–Cretaceous in age, respectively. It soon became clear from associations in the field as well as re-dating of various formations that these two groups broadly overlapped each other during the Early Jurassic (Charig *et al.* 1965). That begged the question as to which basal sauropodomorphs were more closely related to sauropods, as a simple linear progression was clearly an inadequate hypothesis.

Temporal relations alone, however, cannot provide the basis for phylogenetic relationships, and so this argument ultimately fails to clarify relations at the base of Sauropodomorpha.

The morphology of the manus and pes in sauropodomorphs has been used to posit separate origins for prosauropods. Charig *et al.* (1965) argued that prosauropods and sauropods evolved along separate lineages because sauropods showed no evidence in their manus or pes that their forebears were bipedal. They argued further that facultative bipedalism in prosauropods was probably a derived condition. Cruickshank (1975) and Van Heerden (1978) focused on the fifth pedal digit, contrasting the reduced paddle-shaped fifth metatarsal in prosauropods as compared with the strong weight-bearing fifth metatarsal in sauropods. Cruickshank (1975, p. 90) posited that 'This character must indicate that the lineage leading to *Vulcanodon* separated from the typical prosauropod at a time, when their common ancestor had a "normal" [unreduced] fifth metatarsal.' Reduction of the fifth pedal digit, in other words, was a condition that seemed irreversible to Cruickshank. Van Heerden (1978) provided a hypothesis to suit this argument; he set prosauropods aside, linking the early dinosaur *Herrerasaurus*, with its proportionately longer fifth metatarsal, to later sauropods.

Although it is true that strong digital reduction or loss of phalanges most often does not reverse in descendants in dinosaurs and other tetrapods, such reversal cannot be excluded *a priori*. Character reversal must be entertained if the preponderance of phylogenetic data favours that interpretation. Furthermore, Cooper (1984) and others have noted the transitional form of the fifth metatarsal in the basal sauropod *Vulcanodon*. In the time since these arguments were forwarded, a similar remarkable reversal in the reduction of pedal digit I has been shown to have occurred within therizinosauroid theropods (Kirkland *et al.* 2005) that involves proximal relocation and renewed articular contact of metatarsal 1 with the astragalus.

Narrow-footed vs. broad-footed prosauropods

Galton (1971, 1973, 1976) and Galton and Cluver (1976) subdivided basal sauropodomorphs into 'narrow-footed' species placed in Anchisauridae and 'broad-footed' species placed in Plateosauridae and Melanorosauridae. The narrow condition was alleged to be present in both the manus and the pes. Furthermore, it was defended as being unrelated to size (Galton and Cluver 1976, p. 132). Galton and Upchurch (2004, p. 251) recently have reiterated the importance of metatarsal proportions, citing Galton (1976) and referring to the original distinctions that were made among materials from the Manchester quarry in Connecticut: 'Ammosaurus major has a broad metatarsus, while it is slender in the larger Anchisaurus.'

Cooper (1981, p. 696) was first to strongly criticize the distinction between narrow- and broad-footed sauropodomorphs. He argued that all of the differences were size related and could be found in a growth series for the species Massospondylus carinatus. Yates (2004) concurred and provided a critique using measurements. I have extended this critique to the influential pedal reconstructions for Anchisaurus and Ammosaurus (see above). The distinction clearly lacks justification and is based on specimens here referred to the same species (Ammosaurus major). The classificatory scheme that separated narrow- from broad-footed genera has not been supported by any subsequent cladistic analysis.

PHYLOGENY: CLADISTIC INTERPRETATIONS

Cladistic interpretation of basal sauropodomorph phylogeny, similar to traditional interpretation, has varied between two extremes: a sequence of basal sauropodomorphs that increasingly approach the sauropod condition or a monophyletic clade of 'prosauropods' (Text-figs 2–3). The first analyses were qualitative. Subsequent quantitative analyses incorporated new characters, specimens and taxa (Table 1). The large increase to more than 100 characters in analyses after 2000 represents, in large part, an increase in the taxonomic scope of the phylogenetic problem under consideration. Whereas the character data in Sereno (1999a) focused on non-sauropod sauropodomorphs (traditional 'prosauropods'), later analyses include character data relevant to adjacent portions of the tree (Sauropodomorpha and or more basal nodes; nodes within Sauropoda). Although this has advantages regarding character sampling, one of the disadvantages is that the data relevant to a particular portion of the tree (basal Sauropodomorpha) are admixed with data relevant only to distant nodes.

Prosauropod paraphyly

Prosauropod paraphyly (Text-fig. 2) was first framed in a cladistic context by Gauthier (1986) and later argued in more detailed analyses by Yates (2003a, 2004), Yates and Kitching (2003) and Pol (2004).

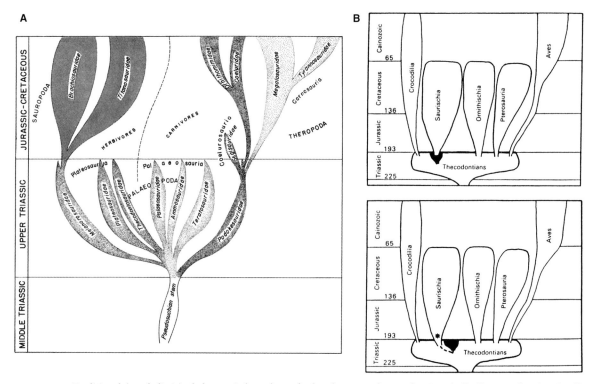

TEXT-FIG. 1. Traditional (precladistic) phylogenetic hypotheses for basal sauropodomorphs. A, spindle diagram showing families of herbivorous and carnivorous basal sauropodomorphs collected together in the paraphyletic taxon Palaeopoda (from Colbert 1964). B, earliest clear distinction of paraphyletic (upper) vs. monophyletic (lower) hypotheses for basal sauropodomorph dinosaurs (from Cruickshank 1975). Asterisk indicates the position of the basal sauropod Vulcanodon karibaensis.

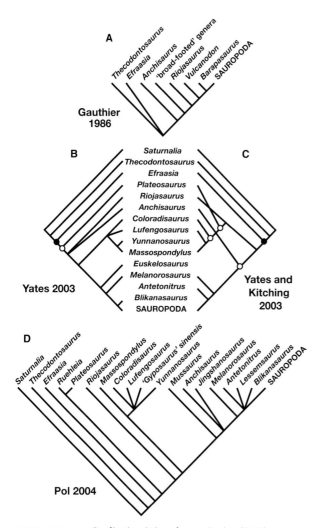

TEXT-FIG. 2. Qualitative (A) and quantitative (B–D) phylogenetic hypotheses for basal Sauropodomorpha that support 'prosauropod' paraphyly. For ease of comparison, terminal taxa outside Sauropodomorpha are excluded and sauropods are collapsed to a single taxon. A, cladogram as indicated by text and character lists in Gauthier (1986). B, reduced consensus cladogram (after Yates 2003*a*, fig. 23). C, consensus cladogram from analysis of 212 characters (after Yates and Kitching 2003). D, consensus hypothesis from analysis of 212 characters (Pol 2004). Unmarked branch points = 1–4 unambiguous synapomorphies; circled branch points = 5–9 unambiguous synapomorphies; solid nodes = 10 or more unambiguous synapomorphies.

Gauthier (1986). Gauthier (1986) was first to outline a cladistic hypothesis for basal sauropodomorphs, citing several genera and listing associated synapomorphies. I have translated his text into a cladogram (Text-fig. 2A). The 'narrow-footed' *Thecodontosaurus* and *Efraasia* are set aside as basal taxa. He identified two nested clades among basal sauropodomorphs: *Anchisaurus* plus more derived sauropodomorphs and unnamed 'broad-footed'

genera plus more derived sauropodomorphs, listing four synapomorphies for the first clade and ten for the second (Table 2). 'Broad-footed' prosauropods, in his view, 'are more closely related to sauropods, thus demonstrating the paraphyly of Prosauropoda' (Gauthier 1986, p. 44). Although he followed traditionalists by suggesting that *Riojasaurus* was even more closely linked to Sauropoda, no supporting evidence was cited, and his less inclusive use of Sauropoda (camarasaurids plus titanosaurians) was not followed by later authors (Text-fig. 2A).

Later authors have set aside most of Gauthier's synapomorphies, presumably because some are plesiomorphic, others no longer apply to basal sauropodomorph nodes, and several are size comparisons without comparative ratios. At most five (or 25 per cent) of these synapomorphies (Table 2: 1, 3, 5, 10, 13) are present in modified form in the matrix of Yates (2004). In summary, Gauthier (1986) provided a valuable initial cladistic interpretation of the traditional argument for prosauropod paraphyly, and a few of the synapomorphies he articulated have remained relevant to subsequent analyses.

Yates (2003a, 2004). Yates (2003*a*) scored 164 characters in 18 ingroups, 14 of which are non-sauropod sauropodomorphs ('prosauropods') (Text-fig. 2B). Other ingroups include the basal sauropod *Vulcanodon*, Eusauropoda and two theropod taxa. The taxonomic scope of the analysis therefore is more inclusive than analyses limited to basal sauropodomorph relationships (e.g. Galton 1990; Sereno 1999*a*; Benton *et al.* 2000) and includes character data for Sauropodomorpha and Saurischia (Text-fig. 2B–C). Additive binary coding was used to eliminate all but one multistate character, and only opposing character states were listed rather than the character followed by its character states (Table 1).

Re-analysis of Yates' dataset yields results similar to those reported (Text-fig. 2B). There are five minimum-length trees of 350 steps (351 reported) with very limited branch support for nodes within Sauropodomorpha. To collapse basal nodes linking *Saturnalia*, *Thecodontosaurus* and *Efraasia* with all other sauropodomorphs, six (seven reported), four and two additional steps are needed, respectively. All other nodes within Sauropodomorpha collapse with a single additional step. The dataset supports 'prosauropod' paraphyly, especially if all basal sauropodomorphs are considered; Yates reported that 24 extra steps were needed to place all non-sauropod basal sauropodomorphs within a single clade. If one imposes less severe constraints to include only what are here called 'core prosauropods' (*Plateosaurus, Massospondylus, Lufengosaurus, Yunnanosaurus* and *Riojasaurus*), only ten steps are required. Ten steps, nevertheless, constitute a very significant difference. Yates (2003*a*) further pointed

TABLE 1. Profile of phylogenetic analyses that consider relationships at the base of Sauropodomorpha. When synapomorphies alone are given as character evidence (e.g. Gauthier 1986), the characters themselves and their primitive states are missing and thus are tallied below as absent (–).

	Authors	Matrix				Character documentation				
No.	Analysis	Printed	e-Archived	No. ingroups	No. characters	Character listed	Character states listed	Original author cited	Character description (%)	Character figured (%)
1	Gauthier (1986)	–	–	5	20	–	+, –	–	0	0
2	Sereno (1989)	–	–	3	13	–	+, –	–	0	0
3	Galton (1989)	–	–	13	19	–	+, –	–	0	50
4	Galton (1990)	–	–	13	49	–	+, –	–	0	50
5	Sereno (1999a)	–	+	9	32	+	+	–	0	0
6	Benton et al. (2000)	+	–	12	49	+	+	–	0	0
7	Yates (2003a)	+	+	15	164	–	+	+	19	12
8	Yates and Kitching (2003)	–	+	18	212	+	+	–	–	0
9	Yates (2004)	+	–	17	205	+	+	+	19	5
10	Galton and Upchurch (2004)	–	+	23	137	+	+	+	–	5
11	Pol (2004)	+	–	47	277	+	+	–	–	0

out that a prosauropod clade reappears if basal taxa (*Saturnalia*, *Thecodontosaurus*) and derived taxa ('*Euskelosaurus*', *Melanorosaurus*, *Blikanasaurus*) are removed. Branch support for this clade (three steps), however, is not substantial even under pruned circumstances. In summary, Yates (2003a) made a strong case for prosauropod paraphyly, if one accepts his dataset.

Yates (2003a) provided a critique of the 19 'prosauropod' synapomorphies identified by Sereno (1999a; Table 3), reporting the following results: three were omitted; six could not be scored in 'prosauropod' outgroups; seven were diagnosed as more-inclusive or less-inclusive clades; and only three 'unambiguously support "prosauropod" monophyly' (2003a, p. 30). His phylogenetic results depend to a large extent on how this set of characters was rescored, recoded or discarded. Using the original numbering scheme for these characters (Table 3), Yates' evaluation is reconsidered below. The term 'core prosauropods' will be used for the following five genera: *Plateosaurus*, *Massospondylus*, *Lufengosaurus*, *Yunnanosaurus* and *Riojasaurus*.

Although Yates claimed to have omitted characters 1, 3 and 9 from his analysis, character 3 was included (numbered 12) with a citation to Sereno (1999a). Yates modified character 3 to read 'Development of a secondary *internal* wall of the antorbital fossa' (emphasis added) and scored it as a saurischian synapomorphy. The character, however, involves the *external*, or secondary, wall of the antorbital fossa rather than the primitive internal wall. Several archosaur clades partially enclose the antorbital cavity with a lateral (external) secondary wall. Among dinosaurs this occurs notably in ornithischians and core prosauropods, in which a secondary vertical wall encloses

the ventral margin of the fossa (Text-fig. 4). This appears to be poorly developed in the fragmentary maxilla available for *Thecodontosaurus* (Yates 2003a, fig. 3) and in *Ammosaurus* (= *Anchisaurus*; YPM 1883). In the latter, differences between opposing sides suggests that some plastic deformation might have flattened the left maxilla. The secondary wall is well developed in 'core prosauropods' but entirely absent in sauropods (e.g. *Riojasaurus*, Wilson and Sereno 1998; *Shunosaurus*, Chatterjee and Zheng 2002).

Yates noted correctly that he omitted characters 1 and 9, but he also omitted characters 13 and 16. Character 1 involves the presence of a keratinous beak covering the anterior margin of the premaxilla. Without doubt, this character is difficult to observe in most taxa, is possibly correlated with character 6, and was scored as present by Sereno (1999a) only in several taxa examined closely (*Plateosaurus*, *Massospondylus* and *Riojasaurus*). Yates (2003a, p. 27) described the feature as 'parasagittal ridges': I had described it as a raised platform (Sereno 1997, pp. 451–452). The derived condition was first observed in *Riojasaurus* (Wilson and Sereno 1998, fig. 36A) and later in mature individuals of other core prosauropods. The details supporting this inference are presented below. Although there are various neurovascular foramina on the premaxillae of basal sauropods, no comparable premaxillary attachment area has been described.

Character 9, the degree of deflection of the deltopectoral crest, was discarded by Yates because of the potential for correlation with body size, difficulty in measuring the angle when the crest is reduced and ease of post-mortem distortion. Without specific justification or evidence, it

TEXT-FIG. 3. Qualitative (A) and quantitative (B–E) phylogenetic hypotheses for basal Sauropodomorpha that support 'prosauropod' monophyly. For ease of comparison, terminal taxa outside Sauropodomorpha are excluded and sauropods are collapsed to a single taxon. A, cladogram from Galton (1990). B, consensus cladogram from analysis of 32 characters (after Sereno 1999*a*). C, single most-parsimonious cladogram from analysis of 49 characters (after Benton *et al.* 2000). D, consensus cladogram from analysis of 137 characters (after Galton and Upchurch 2004). E, consensus cladogram from analysis of the same dataset but with modified scores for *Lufengosaurus huenei* (after Barrett *et al.* 2005). Unmarked branch points = 1–4 unambiguous synapomorphies; circled branch points = 5–9 unambiguous synapomorphies; solid nodes = 10 or more unambiguous synapomorphies.

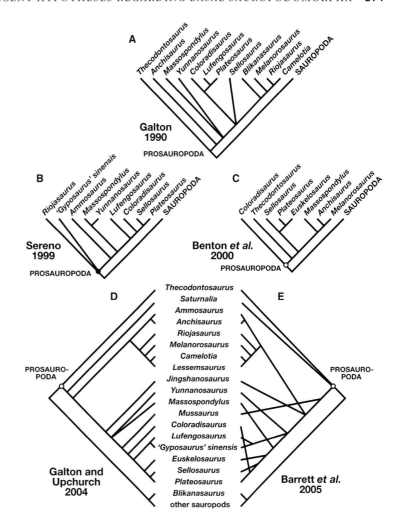

seems these criteria could be applied to any character. *Saturnalia* (Langer *et al.* 2007) and *Thecodontosaurus* (Benton *et al.* 2000) are both small sauropodomorphs, the former clearly showing the derived condition present as in core prosauropods and the latter the primitive condition (60 degree deflection or less). In both cases, the authors had more than a single specimen to observe and did not express concern regarding their estimated angle of deflection. *Omeisaurus* is an example of a basal sauropod with a weakly developed crest but one that clearly is not strongly deflected as seen in proximal view (He *et al.* 1988, pl. 14).

Character 13, the presence of a protruding proximal heel, or ventral intercondylar process, on the first phalanx of manual digit I characterizes core prosauropods, even when immature (Text-fig. 9D–E), as well as heavier-bodied genera, such as *Jingshanosaurus* (Zhang and Yang 1994, fig. 32A). The heel is associated with a well-developed intercondylar crest that gives the proximal end of the phalanx a rounded apex in ventral view rather than a broadly concave margin (Text-fig. 9E). Contrary to Yates,

all of these features are present in the smaller-bodied *Efraasia* (Galton 1973, fig. 10G–L). Sauropods do not have this process (e.g. *Camarasaurus*; Ostrom and McIntosh 1999, pl. 62). Among dinosaurian herbivores, only in the grasping hand of heterodontosaurids is there comparable development of this articular surface. Even in basal theropods, such as *Herrerasaurus*, the ventral heel is not as well developed; it protrudes proximally but the intercondylar crest is incomplete and the proximoventral margin is broadly convex without a distinct apex (Sereno 1993, figs 14–15). Yates correctly observed that a heel of equivalent prominence is present in *Allosaurus* and other neotheropods, tentatively ascribed the apomorphy to Saurischia, and then omitted it from the analysis. 'Given that the difference between a large and a small protrusion is slight, I prefer not to draw a distinction and simply regard the presence of the protrusion as a derived character that is probably diagnostic of the Saurischia' (Yates 2003*a*, p. 29).

Character 16 identifies a swollen rugose welt on the lateral aspect of the iliac preacetabular process. Yates

TABLE 2. Initial synapomorphies listed for basal nodes within Sauropodomorpha (Gauthier 1986).

Sauropodomorpha	
1	Manual digit I robust with enlarged claw
2	Lanceolate teeth with coarsely serrated crowns
3	Skull small on long neck of ten cervical vertebrae
4	Cervical vertebrae longer than most trunk vertebrae
5	Hind limb subequal to, or shorter than, the trunk
6	Tibia shorter than the femur
Anchisaurus plus more derived sauropodomorphs	
7	More robust digit I and metacarpal 1
8	Anterior caudal vertebrae with wider-based neural spines
9	Ilium with arched dorsal margin
10	Acetabulum completely open
'Broad-footed' genera plus more derived sauropodomorphs	
11	Quadrate condyle set below tooth row
12	Premaxillary internasal process compressed
13	Nares very large
14	Teeth increase in size anteriorly in upper tooth row
15	Proximal caudal centra compressed anteroposteriorly; broad-based neural spines
16	Robust forelimbs with short, broad, stout manus (same for pes to a lesser extent)
17	Manual digit I greatly enlarged
18	Proximal carpals absent (unossified)
19	Acetabulum much larger than femoral head
20	Tibial posterolateral flange rudiment present

TABLE 3. Synapomorphies (unambiguous in bold type) listed for core prosauropods by Sereno (1999a). This analysis did not consider *Thecodontosaurus*, and *Saturnalia* had yet to be described.

1	Premaxillary beak: absent (0); present (1).
2	Premaxilla-maxilla external suture: oblique (0); L-shaped (1).
3	Secondary antorbital fossa wall: absent (0); present (1).
4	Maxillary vascular foramina, form: irregular (0); one directed posteriorly, 5–6 anterior (1).
5	Squamosal ventral process, shape: tab-shaped (0); strap-shaped (1).
6	Dentary tooth 1, position: terminal (0); inset (1).
7	Axial postzygapophyses, length: overhang (0), or flush with (1), the posterior centrum face.
8	Deltopectoral crest, length: less (0), or equal to or more (1), than 50 per cent of the length of the humerus (1).
9	Deltopectoral crest, deflection: 45–60 degrees (0), or 90 degrees (1), to the transverse axis of the distal condyles.
10	Distal carpal 1, size: small (0); large (1).
11	Metacarpal 1, basal articulation: flush with other metacarpals (0); inset into the carpus (1).
12	Metacarpal 1, basal width: less than 50 per cent (0), or more than 65 per cent (1), maximum length.
13	Manual digit I, phalanx 1, proximal heel: absent (0); present (1).
14	Manual digit I-phalanx 1, rotation of axis through distal condyles: rotated slightly ventromedially (0); rotated 45 degrees ventrolaterally (1); rotated 60 degrees ventrolaterally (2).
15	Iliac preacetabular process, shape: blade-shaped (0); subtriangular (1).
16	Iliac preacetabular process, scar: absent (0); present (1).
17	Ischial distal shaft cross-section: ovate (0); subtriangular (1).
18	Metatarsal 2 proximal articular surface: subtriangular or subquadrate (0); hourglass-shaped (1).
19	Metatarsal 4 proximal end, transverse width: subequal (0), three times broader than (1), dorsoventral depth.

misunderstood this character as referring to the dorsal margin or blunt end of the preacetabular process, describing the scar as the site of attachment for a 'cartilaginous cap' (Yates 2003a, p. 30). The character was then omitted from the analysis. This attachment welt is a subtle textural feature, comparable in scale with the ambiens process on the pubis or anterior trochanter of the fibula. Drawings of basal

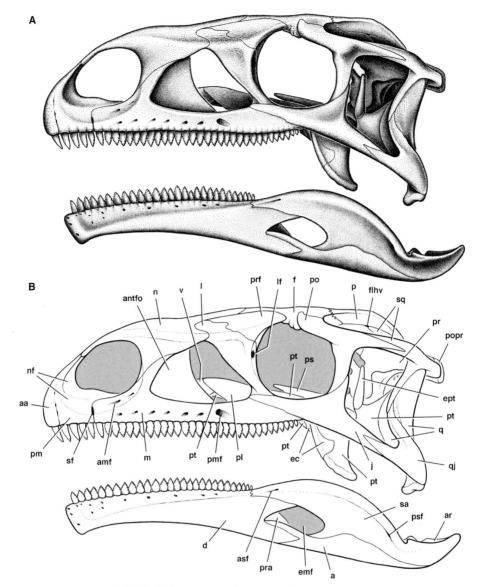

TEXT-FIG. 4. A–B, reconstruction of the skull of *Plateosaurus longiceps* in lateral view, based especially on SMNS 12949, 12950, 13200 and AMNH 6810. Abbreviations: a, angular; aa, attachment area for keratinous upper bill; amf, anterior maxillary foramen; antfo, antorbital fossa; ar, articular; asf, anterior surangular foramen; d, dentary; ec, ectopterygoid; emf, external mandibular fenestra; ept, epipterygoid; f, frontal; flhv, foramen for the lateral head vein; j, jugal; l, lachrymal; lf, lachrymal foramen; m, maxilla; n, nasal; nf, narial fossa; p, parietal; pl, palatine; pm, premaxilla; pmf, posterior maxillary foramen; po, postorbital; popr, paroccipital process; pr, prootic; pra, prearticular; prf, prefrontal; ps, parasphenoid; psf, posterior surangular foramen; pt, pterygoid; q, quadrate; qj, quadratojugal; sa, surangular; sf, subnarial foramen; sq, squamosal; v, vomer.

sauropodomorph ilia in the literature often do not distinguish the feature well. A similar swelling and rugosity has been described in *Saturnalia* (Langer *et al.* 2007) and attributed to the origin of the iliofemoralis cranialis. A swelling is present in a similar position in the herrerasaurid *Staurikosaurus* but absent in its close cousin *Herrerasaurus*. The attachment welt is widespread among basal sauropodomorphs but absent in *Ammosaurus* and sauropods.

Character 19, which identified the extreme relative width of the proximal end of metatarsal 4 in core prosauropods as derived, was modified by Yates to allow other saurischians to be scored with the derived condition. The original character identified a width three times dorsoventral depth as the derived condition; Yates suggested that such a dimension only characterized *Riojasaurus* and reduced the ratio to twice dorsoventral depth. The base of metacarpal 4 in core prosauropods, nevertheless, is proportionately very broad as originally coded (e.g. *Massospondylus*, Cooper 1981, fig. 76; *Plateosaurus*, von Huene 1926, pl. 6).

Characters 2, 4, 7 and 12 were scored as polymorphic in Sauropoda or present in Ornithischia, reducing or eliminating unambiguous support for the monophyly of core prosauropods. Character 4 identifies a stereotypical pattern of neurovascular foramina on the maxilla, the posteriormost of which is largest and exits posterolaterally. Basal sauropods, such as *Shunosaurus* and *Omeisaurus*, have an irregular pattern of foramina and do not have a noticeable posterolaterally directed neurovascular foramen. Many neosauropods have a pneumatic opening on the maxilla identified as the preantorbital fenestra (Wilson and Sereno 1998). Typically developed as a slit-shaped opening or larger fenestra, it opens anterolaterally. To eliminate character 4 as a factor favouring core prosauropod monophyly, Yates ignored the absence of the derived condition in basal sauropods, likened the posterior neurovascular foramen to the preantorbital fenestra, and scored the condition for Sauropoda as polymorphic.

Character 12 highlights the broad proximal width of metacarpal 1 in core prosauropods (proximal width greater than 65 per cent metacarpal length). Yates remarked, 'The only basal sauropods with well-preserved forefeet are those of euhelopodids, most of which have stout, "prosauropod"-like first metacarpals', citing *Shunosaurus* and *Hudeisaurus* (Yates 2003a, p. 29). *Shunosaurus*, does have the proportionately shortest hand of any basal (non-neosauropod) sauropod (proximal width approximately 60–80 per cent metacarpal length). *Hudeisaurus*, however, has a much longer metacarpal 1 (proximal width less than 30 per cent metacarpal length), as does *Omeisaurus* (proximal width about 50 per cent metacarpal length), which was included among 'euhelopodids' (Upchurch 1995). If *Shunosaurus* is situated alone as the most basal sauropod with data available for the manus, then the character state for Sauropoda would be ambiguous (or polymorphic).

Characters 10 and 11 also illustrate selective scoring that effectively undermines these and several of the remaining characters as synapomorphies for a larger group of core prosauropods. Character 10 highlights the size of distal carpal 1 (Table 3). No specific relative measure was originally given; this carpal is by far the largest in the carpus, equalling or exceeding the width of the base of the broadened first metacarpal (Text-fig. 10). Yates recoded the character (number 88) relative to distal carpal 2 (greater than 120 per cent of distal carpal 2). *Ammosaurus* (= *Anchisaurus*, YPM 1883) is an important taxon, but only the posterior aspect of the carpal of interest is exposed. Although it would have been larger if fully exposed in anterior view, even as exposed it is nearly as broad as the base of the enlarged metacarpal 1 as in other core prosauropods (Text-fig. 10). Also like these basal sauropodomorphs, distal carpal 1 is inset medially from the medial margin of metacarpal 1 (Text-fig. 9). In Galton's figure of the manus and carpal, however, an outline of the posterior contour of this carpal is centred directly over metacarpal 1 with no medial inset (1976, figs 17–18). In another specimen from the same region, an enlarged, medially inset distal carpal 1 is visible overlapping a small distal carpal 2 (erroneously figured as a single carpal; YPM 2125; Galton 1976, fig. 32). Yates, nonetheless, scored the condition in *Ammosaurus* (= *Anchisaurus*) as unknown, presumably because the enlarged distal carpal 1 in YPM 1883 is the only carpal exposed. Yates then scored both Neotheropoda and Sauropoda as derived, the former apparently based mainly on *Allosaurus* (rather than several other theropods with unfused equal-sized distal carpals 1 and 2) and the latter based on *Shunosaurus*. Three disc-shaped carpals are preserved in *Shunosaurus*, the first slightly larger than the second. Neither closely resembles the condition in core prosauropods (Zhang 1988, fig. 48). A few small carpal bones were reported in *Omeisaurus* as well, another basal sauropod. As a result of the character state scores outlined above, the distinctive enlarged, medially inset, distal carpal 1 of core prosauropods was optimized on Yates' tree as a synapomorphy uniting Neotheropoda and Sauropodomorpha (excluding Herrerasauridae).

Regarding the proximal inset of metacarpal 1 into the carpus (character 11), Yates scored several basal sauropodomorphs as primitive (with flush metacarpal bases), including *Thecodontosaurus*, *Ammosaurus* (= *Anchisaurus*), *Efraasia* and *Riojasaurus*. Earlier in this paper I described the inset position of metacarpal 1 in *Ammosaurus*. Well-preserved specimens of metacarpal 1 in *Riojasaurus* also strongly suggest this bone was inset into the carpus. The single specimen referred to *Thecodontosaurus* that preserves the carpus and metacarpus, on the other hand, is partially disarticulated and not particularly well preserved (Benton *et al.* 2000). It was scored as having the primitive condition with metacarpal bases flush against the distal carpals. As a result of the character state scores outlined above, an inset metacarpal 1 was optimized on Yates' trees as an autapomorphy of *Plateosaurus* and a synapomorphy for *Lufengosaurus*, *Yunnanosaurus* and *Massospondylus*.

Yates and Kitching (2003) and Yates (2004) performed additional analyses (Text-fig. 2D). Yates (2004) scored 205 characters in 17 ingroups, only ten of which are commonly understood as 'prosauropods'. Other ingroups include the basal sauropods *Vulcanodon*, *Kotasaurus*, *Shunosaurus*, *Barapasaurus*, *Omeisaurus* and Neosauropoda. Compared with Yates (2003a), the taxonomic scope of ingroups is both trimmed and extended; two basal sauropodomorphs ('*Euskelosaurus*', *Yunnanosaurus*) and all ingroups outside Sauropodomorpha have been trimmed, whereas Eusauropoda is split into five terminal taxa. Building on the dataset in Yates (2004), Yates and Kitching (2003) employed 212 characters in 19 ingroups, two of which were new to the analysis (*Antetonitrus*, *Isanosaurus*). Compared with Yates (2004), one character was removed and eight added for a total of 212 characters (Text-fig. 2C).

How this dataset corresponds to that in Yates (2003*a*) is not clear, because the character data now includes ordered and unordered multistate characters, and the results are quite different (Text-fig. 2B–C). Substantial character support is now present for a clade of core prosauropods that in the previous dataset required ten additional steps to compose. This major phylogenetic difference is presumably a byproduct of variation in character selection, coding and scoring. Several additional characters were added, some of which link *Anchisaurus* and Sauropoda (Yates 2004), such as wrinkled enamel. Interpretation of the cranial morphology of this taxon, however, remains controversial (Fedak 2005).

Pol (2004). Pol (2004) and Pol and Powell (2005) have recently presented a broad-scale phylogenetic analysis of basal Sauropodomorpha, with 24 non-sauropod sauropodomorphs, 13 sauropods and 16 taxa outside Sauropodomorpha scored for 277 characters (Text-fig. 2C). The results are broadly consistent with Yates and Kitching (2003) and Yates (2003*a*, 2004) insofar as *Saturnalia*, *Thecodontosaurus* and *Efraasia* are in successively less inclusive basal positions and *Anchisaurus*, *Antetonitrus*, *Melanorosaurus* and *Blikanasaurus* are most closely related to Sauropoda. Several genera in between these extremes, which include the five core prosauropods, are depicted as successive stem taxa, more closely resembling the results of Yates (2003*a*). Branch support is less than five for all of the depicted nodes, even when the most poorly known, unstable taxa are removed.

How the data of Pol (2004) and Yates (2004) compare in terms of characters used and how those characters are coded and scored is not known. Pol (2004) compiled 469 characters from the literature. He rejected 192 (41 per cent) and modified most of the remaining 277 characters (59 per cent). Character selection is as important as character scoring in the evaluation of data, especially when considering cladistic hypotheses like these with relatively with low levels of branch support. Further commentary awaits formal publication of Pol's dataset.

Prosauropod monophyly

Prosauropod monophyly (Text-fig. 3) was proposed independently in the same year in qualitative analyses by Sereno (1989) and Galton (1989, 1990). Quantitative versions of these analyses eventually would appear (Sereno 1999*a*; Galton and Upchurch 2004). In addition, Benton *et al.* (2000) drew the same conclusions in considering the phylogenetic placement of *Thecodontosaurus*.

Sereno (1989, 1999). Sereno (1989) listed 11 synapomorphies to unite prosauropods or a larger clade composed of prosauropods plus 'segnosaurs' (now referred to as therizinosauroids and clearly situated among coelurosaurian theropods). Nine of the 11 features were later coded in the first quantitative analysis of basal sauropodomorphs (Sereno 1999*a*). Using Sauropoda and Theropoda as successive outgroups, Sereno (1999*a*) scored 32 characters in nine of the best-known 'prosauropod' genera (Tables 1, 3). Nineteen characters, or approximately 60 per cent of the character data, supported a 'prosauropod' clade without any homoplasy. The consensus of six minimum-length trees (34 steps, CI = 0·97, RI = 0·98, character 27 ordered) shows *Riojasaurus* as the most basal taxon and *Plateosaurus* the most nested (Text-fig. 3B). The published cladogram simplified these results, showing only six of nine ingroup genera (Sereno 1999*a*, fig. 2). The remaining 13 characters provided little branch support among prosauropods; fewer than five supported any particular node, and all nodes except *Efraasia* + *Plateosaurus* collapse with two additional steps. Furthermore, this last node actually unites two species of the genus *Plateosaurus*, as the material upon which *Efraasia* was scored has more recently been referred to *Plateosaurus gracilis* (Yates, 2003*b*).

Other less complete taxa, such as *Thecodontosaurus*, *Melanorosaurus*, *Mussaurus* or *Blikanasaurus*, were excluded because few, if any, of the characters included in the analysis could be scored. Interpretation of the morphology of *Thecodontosaurus*, the most complete specimen of which is now referred to a new species, *T. caducus* (Yates 2003*a*), is complicated by its immaturity and small size, raising some concern over whether primitive features are merely growth related or allometric correlates of small body size, such as the proportionately smaller deltopectoral crest. The partial preservation and disarticulation of available skull elements has also complicated their interpretation and reconstruction (Kermack 1984; Yates 2003*a*). There is little chance that this situation will change, as the best material was recovered long ago from fissure-fill deposits. *Melanorosaurus readi*, on the other hand, may soon be represented by a much better skeleton with a skull (Yates 2005, 2007).

Although the 1999 dataset strongly favoured prosauropod monophyly, many less complete taxa were not considered, besides those described recently (*Saturnalia*, *Antetonitrus*). Sauropoda, in fact, was originally logged as an outgroup, although the results are the same if

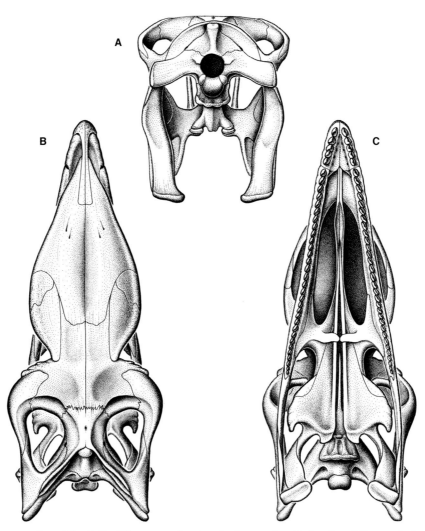

TEXT-FIG. 5. Reconstruction of the skull of *Plateosaurus longiceps* in A, posterior, B, dorsal, and C, ventral views. Based especially on SMNS 12949, 12950, 13200 and AMNH 6810. See Text-figure 6 for labelling.

Sauropoda is transferred to the ingroup. To imply on the basis of this analysis therefore that all, or nearly all, taxa commonly considered 'prosauropods' comprise a monophyletic clade is an overstatement (Sereno 1997, 1999a). The analysis provided some initial evidence in support of the monophyly of a core of prosauropod genera.

Galton (1989, 1990) and Galton and Upchurch (2004).
Galton (1989, p. 82; 1990, p. 321) listed eight derived features in support of 'prosauropod' monophyly (Text-fig. 2A). Only one of these characters, the 'twisted' phalanx 1 of manual digit I, was used in support of prosauropod monophyly by Sereno (1989, 1999a) or the subsequent analysis by Galton and Upchurch (2004: Text-fig. 3D). The remaining features originally listed by Galton are either present in sauropodomorph outgroups

(e.g. diminutive size of manual digits IV and V) or were relocated to less inclusive nodes within Sauropodomorpha by later authors.

Galton and Upchurch scored 137 binary characters (ten of which are uninformative) in 18 'prosauropod' genera as well as in *Blikanasaurus* and four sauropods (Table 1). The data yield two minimum-length trees of 260 steps (279 reported) that include a monophyletic Prosauropoda that excludes only the stocky-limbed genus *Blikanasaurus* (Text-fig. 3D). The dramatic increase in character number is due in part to the incorporation of data for the nodes Sauropodomorpha and Sauropoda and nodes within Sauropoda, which account for approximately 40 unique synapomorphies on minimum-length trees. Character data in Galton and Upchurch (2004) broadly overlap that in Sereno (1999a), and the resulting support at basal nodes is very similar. Seventeen of 19 synapomorphies

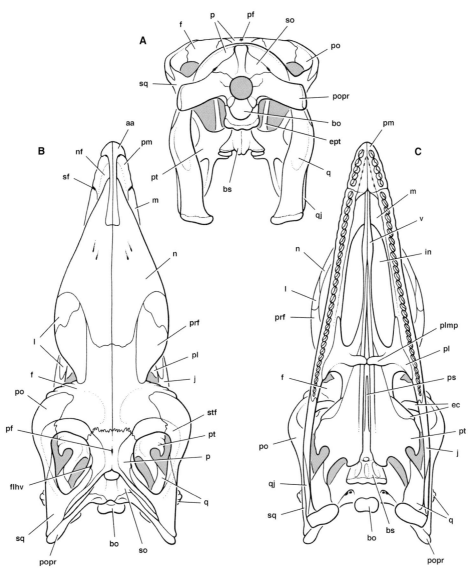

TEXT-FIG. 6.   Reconstruction of the skull of *Plateosaurus longiceps* in A, posterior, B, dorsal, and C, ventral views. Abbreviations: aa, attachment area for keratinous upper bill; bo, basioccipital; bs, basisphenoid; ec, ectopterygoid; ept, epipterygoid; f, frontal; flhv, foramen for the lateral head vein; in, internal nares; j, jugal; l, lachrymal; m, maxilla; n, nasal; nf, narial fossa; p, parietal; pf, parietal foramen; pl, palatine; pm, premaxilla; plmp, palatine medial process; po, postorbital; popr, paroccipital process; prf, prefrontal; ps, parasphenoid; pt, pterygoid; q, quadrate; qj, quadratojugal; sf, subnarial foramen; so, supraoccipital; sq, squamosal; stf, supratemporal fossa; v, vomer.

that supported prosauropod monophyly in Sereno (1999*a*) are present at similar basal nodes in Galton and Upchurch (2004). A single additional step is sufficient to collapse all structure outside Sauropoda, because several terminal taxa are very poorly known taxa. If ingroups are limited to those considered by Sereno (1999*a*: Text-fig. 3B), however, there are two strongly supported nodes (more than ten synapomorphies) within Sauropodomorpha: Prosauropoda and Sauropoda. This is not an artefact arising after removal of homoplastic taxa; these synapomorphies are present at basal nodes in minimum-length trees with all taxa included. There is very little structure

within Prosauropoda, in contrast, which is clearly revealed by the range of relationships that were altered after adjusting character state scores for *Lufengosaurus* (Barrett *et al.* 2005: Text-fig. 3E).

Benton et al. *(2000).* Benton *et al.* (2000) scored 49 characters in nine 'prosauropods' and four sauropods. The dataset, which includes four uninformative characters, yields 42 minimum-length trees of 84 steps (83 reported) with no resolution of relationships among basal sauropodomorphs. By removing *Riojasaurus*, Benton *et al.* (2000) were able to obtain a single tree of

78 steps (77 reported), now with five uninformative characters. Prosauropod monophyly was weakly supported. Yates (2003a) remarked that this monophyletic clade collapses when a single character state for sauropods is corrected. One additional step, in fact, collapses all resolution except a subclade of three well-known sauropods. There are two fundamental reasons for this. First, fully one-half of the characters are not informative for 'prosauropod' relationships (five are uninformative and 20 constitute unique synapomorphies for Sauropodomorpha, Sauropoda or nodes within Sauropoda). Second, approximately 70 per cent of character state scores are unknown for several 'prosauropod' genera, namely *Coloradisaurus*, '*Euskelosaurus*' and *Melanorosaurus*, leaving them poorly constrained. With this degree of missing data, it is difficult to justify the removal of *Riojasaurus* with the aim of obtaining meaningful resolution; *Riojasaurus* is scored for 98 per cent of the character data (all but 1 character).

The character that was rendered uninformative by removal of *Riojasaurus* involves the large size of distal carpal 1 (character 25). Besides *Massospondylus*, *Riojasaurus* was the only ingroup taxon scored with the derived condition (size twice that of other distal carpals), despite the fact that elsewhere in the paper *Thecodontosaurus* was clearly shown with the derived condition (Benton *et al.* 2000, fig. 12A). *Plateosaurus* was also shown with a large distal carpal 1 in a paper by one of the coauthors (Galton and Cluver 1976, fig. 7M). This character was one of five that appeared as prosauropod synapomorphies in the analysis of Sereno (1999a). Only two of these (manual phalanx I-1 twisted, subtriangular preacetabular process) continue to support 'prosauropod' monophyly in Benton *et al.* (2000), given differences in how they were scored. Although Benton *et al.* (2000) might have gone to press before Sereno (1999a) was available, they cited his use of the inset position of metacarpal 1, a character used for the first time in the 1999 analysis.

KEY QUESTIONS

Morphology

The most parsimonious scheme has ramifications for morphology. Some notable features will either be verified as unique 'prosauropod' synapomorphies or viewed as outstanding instances of characters that evolved and then were reversed in the line leading to sauropods. Two are discussed below.

Keratinous beak, tooth retraction. Sereno (1999a) cited as two separate characters the presence of a keratinous sheath on the anterior end of the premaxilla and the

retraction of the first dentary tooth from the anterior end of the dentary (Table 3; Text-figs 4–8). Here I describe them together as possibly associated with the presence of a narrow keratinous beak on upper and lower jaws.

I became aware of the derived morphology of the premaxilla while examining the skull of *Riojasaurus*, in which the attachment area on the premaxilla is raised as a distinct platform (Wilson and Sereno 1998, fig. 36A). A similar raised platform is now known in another undescribed specimen from Argentina pertaining to a different genus (R. Martinez, pers. comm. 2006). More subtle expression of the character is present in other basal sauropodomorphs, such as *Massospondylus* (Gow *et al.* 1990, fig. 9) and a basal sauropod from the Kayenta Formation (Crompton and Attridge 1986). As in *Plateosaurus* the attachment surface is covered with fine pores much like the surface of other keratin-covered jaw bones among dinosaurs, such as the predentary or rostral (Text-figs 4–7). One to three nutrient foramina enter the premaxilla along the posterior border of this region, which is located above the first premaxillary crown or between first and second crowns. It appears from these features that a keratinous sheath covered the anterior end of the premaxillae, which was scored only in these few taxa with exceptional preservation (Sereno 1999a). Although there are various pits and neurovascular foramina on the premaxillae of basal sauropods, no comparable premaxillary attachment area has been described.

Retraction of the first dentary tooth is easier to verify than fine points of premaxillary form. The retraction of the position of the first dentary tooth may well be related to a keratinous lower beak. The condition in *Saturnalia* has not been described in detail and may not be determinable (scored as absent by Yates 2003a), whereas in *Thecodontosaurus* some retraction is present (Benton *et al.* 2000). In *Plateosaurus* a flat platform is located in the position usually occupied by the first tooth (Text-fig. 8A). A shallow trough is present between the dentaries anterior to the symphysis, which also may have functioned for attachment, and a sizeable pair of nutrient foramina always seems to be present and inset a short distance from the leading edge of the dentary (*Plateosaurus*, *Massospondylus*, *Thecodontosaurus*; Text-fig. 8).

A platform of similar form and position is present in other basal sauropodomorphs, such as a basal sauropodomorph from the Kayenta Formation. A keratinous beak shaped like an ornithischian predentary was previously inferred to have been present in this specimen (Crompton and Attridge 1986, fig. 17.6). Based in large part on a preservational artefact that misaligned upper and lower jaws, they suggested that the lower jaw in this specimen is unusually short, that the tips of maxillary crowns inserted into neurovascular foramina on the dentary, and that a

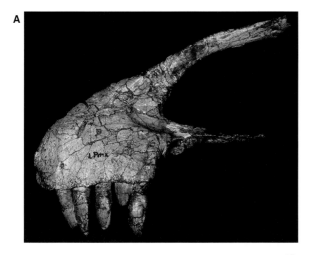

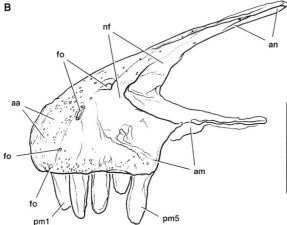

TEXT-FIG. 7. Left premaxilla of *Plateosaurus longiceps* (AMNH 6810) in lateral view. Abbreviations: aa, attachment area for keratinous upper bill; am, articular surface for the maxilla; an, articular surface for the nasal; fo, foramen; pm1–5, premaxillary teeth; nf, narial fossa. Scale bar in B represents 3 cm.

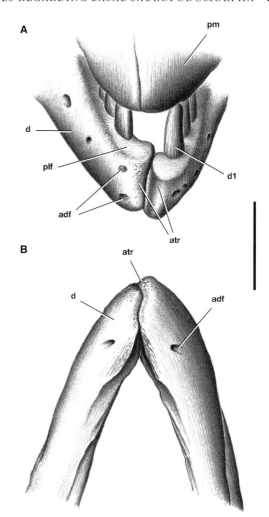

TEXT-FIG. 8. Anterior end of the skull of *Plateosaurus longiceps* (SMNS 12949) in A, anterodorsal, and B ventral views. Abbreviations: adf, anterior dentary foramen; atr, attachment trough; d, dentary; d1, dentary tooth 1; plf, platform; pm, premaxilla. Scale bar represents 2 cm.

keratinous sheath covered other neurovascular foramina and projected anteriorly to complete the lower jaw. Although I cannot support these conclusions for several reasons, Crompton and Attridge (1986) also reported the presence of a high-angle wear facet on a premaxillary crown. Wear facets are unknown elsewhere among basal sauropodomorphs. They have been reported thus far only in isolated teeth referred without justification to *Yunnanosaurus* (Galton 1985) and in fragmentary material improperly referred to Sauropodomorpha (Flynn *et al.* 1999).

In the Kayenta skull, all four premaxillary crowns are preserved on the right side and the second and third are preserved on the left side. A high-angle wear facet is present on the first crown on the right side and second crown on the left side. The facet cuts through the thin

enamel that covers the anteromedial edge of the crown. The crowns are canted anteromedially, so the wear facets were made by a structure passing inside the anterior end of the premaxillary tooth rows. The anteriormost dentary crowns do not have compensatory wear and were not positioned far enough anteriorly to account for the premaxillary facets. The reduced denticulation of the anterior premaxillary teeth also resemble ornithischian premaxillary teeth that oppose a lower keratinous bill.

The inference of an anterior beak in basal sauropodomorphs, thus, is based on the form of the anterior ends of the premaxilla and dentary and the presence of high-angle wear facets on the inside of the anteriormost premaxillary crowns. Sauropods show no indication of this morphology. The lower tooth row in sauropods extends to the midline and the muzzle is often squared. Tooth retraction is a common phenomenon within Tetrapoda, but with rare exception it is progressive and irreversible. These features at the anterior end of the snout either characterize a basal sauropodomorph clade or were reversed without trace in the line leading to Sauropoda.

Carpal-metacarpal complex. Distal carpal 1 is enlarged and overlaps distal carpal 2 in an arrangement unique to basal sauropodomorphs as originally described by Broom (1911: Text-fig. 9A–C). Metacarpal 1 is inset into the carpus and develops a synovial joint on the lateral side of its base, against which rests the smaller distal carpal 2 (Text-figs 9B–C, 10B). In this configuration, distal carpal 1 completely overlaps the proximal surface of distal carpal 2, which effectively separates distal carpal 1 and metacarpal 2. The distal carpals and metacarpals of basal sauropodomorphs show very little variation in form and arrangement, the telltale clues for which are present in synovial joints on the base of metacarpal 1 (Text-fig. 9B–C). The presence of this configuration, in other words, seems possible to infer with only metacarpal 1 at hand (e.g. *Antetonitrus*: Yates and Kitching 2003). The complex is well preserved in articulation in *Massospondylus*, which documents the dramatic size decrease from distal carpal 1 to distal carpal 3 and concomitant change in shape from flattened hemispheroid to elongate hemispheroid to small pyramid, respectively (Broom 1911; Cooper 1981). *Lufengosaurus* shows exactly this condition (Young 1947, figs 4–8). Other basal sauropodomorphs with carpus and metacarpus in natural articulation show the derived configuration, including cf. *Ammosaurus* (YPM 1225), an unnamed taxon from the Navajo Sandstone in western North America (Galton 1976; Irmis 2005), *Plateosaurus* from Europe (von Huene 1932) and Greenland (Galton 2001), *Lufengosaurus* (Young 1941), and *Mussaurus* (Casamiquela 1980; Pol 2004). Whether this is also the case in *Saturnalia* (Langer 2003), *Thecodontosaurus* (Benton *et al.* 2000) or in the more derived *Jingshanosaurus* (Zhang and Yang 1994) remains uncertain.

Misinformation is present in the literature for several taxa. In *Lufengosaurus*, for example, Young (1947) described all of the distal carpals and metacarpals but reassembled the manus with proximal ends of the metacarpals aligned. Young described a pair of distal articular fossae on the enlarged distal carpal 1 as accommodating metacarpals 1 and 2, a condition common among theropods. These fossae, to the contrary, are fitted to metacarpal 1 and distal carpal 2, as has long been documented in *Massospondylus* (Text-fig. 9B–C). This configuration is also present in YPM 1225 from the Connecticut River valley (cf. *Ammosaurus major*), but distal carpals 1 and 2 were drawn as a single element (Galton 1976, fig. 32). Casamiquela (1980) documented and Pol (2004) described this configuration in *Mussaurus*,

but the character was omitted in his phylogenetic analysis (Pol 2004). Santa Luca (1980, fig. 15) figured a similar condition in *Heterodontosaurus* and this has been refigured elsewhere (Langer and Benton 2006, fig. 8). His interpretation was based on the right manus, in which digit I is slightly disarticulated distally. The fully articulated left manus shows the base of the metacarpals in alignment. The inset position of metacarpal 1 and its lateral articulation with distal carpal 2 appears to constitute a unique configuration among dinosaurs.

Sauropods do not exhibit any aspect of this carpalmetacarpal complex. That absence is difficult to discount as a necessary correlative of using the hand in obligate quadrupedal locomotion, not least because some basal sauropodomorphs of considerable size have been interpreted as facultative quadrupeds. *Antetonitrus*, which appears to have maintained this complex within a particularly stout manus, has long front limbs (humerus 90 per cent femoral length) and may well have been a facultative quadruped (Yates and Kitching 2003). To discount all evidence from sauropods as too transformed to score is problematic, when many basal species retain an ossified distal carpus as well as five metacarpals. The unusual configuration of carpus and metacarpus and rotated manual digit I either characterizes a basal sauropodomorph clade or, as now appears more likely, evolved only to be lost without trace in the line leading to Sauropoda.

Phylogeny

New fossil remains described in the last decade, especially *Saturnalia tupiniquim* (Langer 2003; Langer *et al.* 2007), *Mussaurus* (Pol 2004), *Jingshanosaurus xinwaensis* (Zhang and Yang 1995) and *Antetonitrus ingenipes* (Yates and Kitching 2003), have broadened the morphologically narrow core of basal sauropodomorphs that had been central to early phylogenetic hypotheses. In recent years broader sampling of taxa and characters have generated hypotheses for basal sauropodomorphs that are both monophyletic (Galton and Upchurch 2004) and paraphyletic (Yates 2003a, 2004; Yates and Kitching 2003; Pol 2004). The nodes in these hypotheses en route to Sauropoda typically have branch support of only one or two steps. Only one node in one analysis is supported by as many as five unambiguous synapomorphies (Text-fig. 2C), a node uniting sauropodomorphs above *Thecodontosaurus*. I try to frame below what seems reasonable to conclude at this juncture about the phylogenetic relationships of basal sauropodomorphs and where major questions remain.

New basal and derived species. *Saturnalia* is clearly basal to other known sauropodomorphs, given the primitive morphology of its ilium, ischium, pubis, femur and metatarsus, the description of which has only recently become available (Langer 2003) (Text-fig. 11A). In *Saturnalia* these bones are easily distinguished from comparable bones in *Thecodontosaurus*, *Efraasia* and core prosauropods. The ilium, for example, has a broad flange backing the acetabulum (Langer 2003), the femur has a trochanteric shelf, and the proximal end of metatarsal 4 is not strongly compressed. These are primitive conditions unknown elsewhere among basal sauropodomorphs.

Thecodontosaurus is more derived but only positioned tentatively here, given some uncertainty in the association of disparate and often immature specimens (Benton *et al.* 2000; Yates 2003*a*).

Mussaurus patagonicus (Casamiquela 1980; Pol 2004; Pol and Powell 2005) and *Jingshanosaurus xinwaensis* (Zhang and Yang 1994) exhibit notable derived features in the skull. One or both exhibit a skull with anteroposteriorly short proportions, external nares retracted relative to the antorbital opening, a dentary that expands

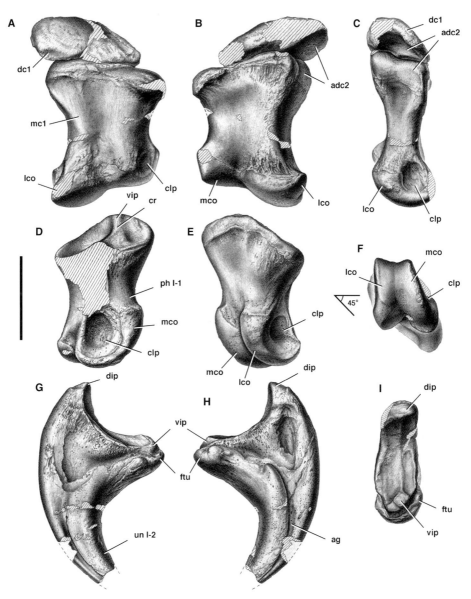

TEXT-FIG. 9.   Right distal carpal 1 and manual digit I of *Massospondylus carinatus* (AMNH 5624). A–C, right distal carpal 1 and metacarpal 1 in dorsal, ventral, and lateral views. D–F, right phalanx 1 of manual digit I in dorsal, ventral and distal views. G–I, right ungual of manual digit I in dorsal, ventral and proximal views. Abbreviations: adc2, articular surface for distal carpal 2; ag, attachment groove; clp, collateral ligament pit; cr, crest; dc1, distal carpal 1; dip, dorsal intercondylar process; ftu, flexor tubercle; lco, lateral condyle; mc1, metacarpal 1; mco, medial condyle; ph I-1, phalanx 1 of manual digit I; un I-2, ungual of manual digit II; vip, ventral intercondylar process. Angle in F is a measurement of the degree of torsion in the shaft of the phalanx. Scale bar represents 2 cm.

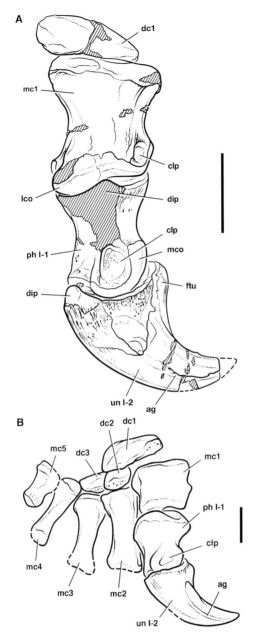

TEXT-FIG. 10. Right carpus and manus of *Massospondylus carinatus*. A, right distal carpal 1 and manual digit I in dorsal view (AMNH 5624). B, right carpus and manus in dorsal view (UCR9558: after Cooper 1981, fig. 35). Abbreviations: ag, attachment groove; clp, collateral ligament pit; dc1–3, distal carpal 1–3; dip, dorsal intercondylar process; ftu, flexor tubercle; lco, lateral condyle; mc1–5, metacarpal 1–5; mco, medial condyle; ph I-1, phalanx 1 of manual digit I; un I-2, ungual of manual digit II. Scale bars represent 2 cm.

than core prosauropods, although more detailed information on each is needed.

Antetonitrus is clearly more derived than other basal sauropodomorphs (Yates and Kitching 2003: Text-fig. 11A). Based on a partial disarticulated skeleton, *Antetonitrus* has a broadened scapular blade, unusually long forelimbs relative to its hind limbs, proportionately short manus and pes, and especially robust metatarsal 1. *Blikanasaurus* (Galton and Van Heerden 1998), represented by the distal portion of a hind limb, has been recognized by many as a sister group to sauropods, but such fragmentary material severely limits confidence in such phylogenetic conclusions. *Antetonitrus*, on the other hand, is represented by several key axial and appendicular elements that help paint a picture of an intermediate form between better known 'core prosauropods' and basal sauropods like *Vulcanodon*.

'Core prosauropods'. These basal sauropodomorphs have been known for some time from near complete skeletons and include *Plateosaurus longiceps*, *Massospondylus carinatus*, *Lufengosaurus huenei*, *Yunnanosaurus huangi* and *Riojasaurus incertus*. I exclude *Thecodontosaurus caducus*, *Efraasia diagnostica* and *Ammosaurus major* on the grounds that their skeletal remains are either relatively incomplete, immature, variously interpreted, poorly preserved, or some combination of the foregoing. Furthermore, given the geological setting where they were discovered, it is unlikely that more complete specimens will be found. I also exclude *Melanorosaurus readi* because of the extremely limited material that has, until recently, limited its interpretation.

The aim here is not to demote the importance of the four species cited above, but rather to highlight a core group of five well-documented species on which to rest the traditional taxon Prosauropoda. Yates and Kitching (2003) revised Prosauropoda in a very similar manner, uniting four of the five taxa listed above (they did not consider *Yunnanosaurus*). Some taxonomists, to be sure, have continued to list every single non-sauropod sauropodomorph under Prosauropoda, even the most derived (e.g. *Blikanasaurus*: Galton and Van Heerden 1998). Yet, if it is possible to unite the five species cited above as a monophyletic clade, salvaging the name Prosauropoda seems appropriate.

Data documentation and comparison. Phylogeneticists increasingly realize the need to document characters and character states better with supporting information. Explanatory notes, specimen documentation and images, for example, are likely to be increasingly attached to the cells of a taxon-character matrix (Pol 2004). The maturity of specimens must also be more carefully considered, because some characters change with age. Yates and Kitching (2003, p. 1755) downplayed growth as a con-

anteriorly and textured enamel on some crowns. Their postcranial skeleton, however, is similar to that in other basal sauropodomorphs and characterized by robust short forelimbs, an inset metacarpal 1 and relatively slender metatarsals. They may occupy a position more advanced

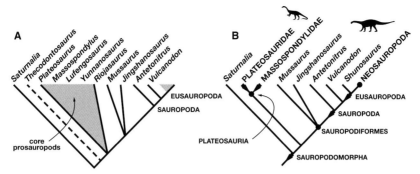

TEXT-FIG. 11. A, summary cladogram depicting a conservative interpretation of the current status of basal sauropodomorph phylogeny. Relations among core prosauropods and between them and other basal sauropodomorphs remain unresolved. B, summary cladogram of suggested phylogenetic taxonomy. Should core prosauropods be united as a clade to the exclusion of Sauropodiformes, the taxon Prosauropoda would be available (see Table 4 for definitions). Dots indicate node-based taxa and arrows indicate stem-based taxa.

founding factor, suggesting that the 'appendicular skeleton of sauropodomorphs experiences little, if any, allometric changes with growth.' The proportions of metacarpal 1 and degree of torsion in phalanx I-1, nevertheless, are good examples (Table 3, character 14). The basal width of metacarpal 1 changes in proportion to length by at least 25 per cent from subadult to adult, from a width considerably less than length to one that exceeds length (Text-fig. 10). Torsion in phalanx I-1 measures approximately 45 degrees in subadult *Massospondylus* (Text-fig. 9F) but apparently increases to 60 degrees in adults (Text-fig. 10B; Broom 1911; Cooper 1981). Sereno (1999a) and Yates (2003a) scored *Massospondylus* as derived (60 degrees or more). Yates (2003a) scored *Thecodontosaurus* as primitive (torsion of less than 50 degrees), although the degree of torsion so far has been described only as similar to that 'in all other prosauropods' (Benton *et al.* 2000, p. 92). Documenting the maturity of specimens and age-related variation may well be an important factor to consider for the many proportional characters in basal sauropodomorph data.

Although increased data documentation will constitute a significant improvement, other challenges rank as more important to achieve a measure of synthesis in phylogenetic resolution. We must attempt to compose morphological characters with more uniformity and then compare character data in more rigorous ways (Sereno, in manuscript). As outlined above, each major analysis of basal sauropodomorph relationships has assembled a unique dataset that is then analysed and compared *a posteriori* to the results obtained from previous datasets. Character comparisons were done selectively, rather than systematically. In this way, we understand the similarities and differences of results, but often have little idea regarding the underlying root causes. The datasets themselves must be compared. Root causes for differing phylogenetic results are few in number and include most importantly (1) character selection and (2) character scoring. As discussed above, there are striking differences in the characters used and how similar characters have been scored in different analyses. These differences can be logged and quantified with indices as intuitive as the consistency index is to better understand *a posteriori* results.

PHYLOGENETIC TAXONOMY

The phylogenetic definitions listed in Table 4 are summarized below. The aim is to provide heuristic definitions that are historically consistent and maximize stability of taxonomic content in the face of known phylogenetic uncertainty (Sereno 2005a). Further historical information regarding previous definitions and usage is available online (http://www.taxonsearch.org; Sereno 2005b, in manuscript).

Sauropodomorpha

Sauropodomorpha was coined as both a node- and stem-based group by Salgado *et al.* (1997) and Upchurch (1997), respectively. Sereno (1998) used a node-based definition, because it allowed the formation of a node-stem triplet for what was regarded as a basal split of Sauropodomorpha into Prosauropoda and Sauropoda (Galton 1989; Sereno 1989, 1997, 1999a, b). Support for this dichotomy, nevertheless, has been seriously eroded by the discovery of basal and derived genera that have relocated synapomorphies and increased homoplasy as outlined above. Given the phylogenetic uncertainty that currently exists, there is no sense in erecting a node-stem triplet at the base of Sauropodomorpha. The taxonomy of basal dinosaurs, in my opinion, is better served by adopting a stem-based definition of Sauropodomorpha as

TABLE 4. Phylogenetic definitions for basal Sauropodomorpha as recommended in this paper. For background, see Sereno (in manuscript; also http://www.taxonsearch.org).

Taxon	Phylogenetic definition	Definitional type
SAUROPODOMORPHA von Huene, 1932	The most inclusive clade containing *Saltasaurus loricatus* Bonaparte and Powell, 1980 but not *Passer domesticus* (Linnaeus, 1758), *Triceratops horridus* Marsh,1889*b*	stem
SAUROPODIFORMES Sereno, 2005b	The least inclusive clade containing *Mussaurus patagonicus* Bonaparte and Vince, 1979 and *Saltasaurus loricatus* Bonaparte and Powell, 1980	node
SAUROPODA Marsh, 1878	The most inclusive clade containing *Saltasaurus loricatus* Bonaparte and Powell, 1980 but not *Jingshanosaurus xinwaensis* Zhang and Yang, 1994, *Mussaurus patagonicus* Bonaparte and Vince, 1979	stem
EUSAUROPODA Upchurch, 1995	The most inclusive clade containing *Saltasaurus loricatus* Bonaparte and Powell, 1980 but not *Vulcanodon karibaensis* Raath, 1972	stem
PROSAUROPODA von Huene, 1920	The most inclusive clade containing *Plateosaurus engelhardti* Meyer, 1837, *Massospondylus carinatus* Owen, 1854, *Lufengosaurus hunei* Young, 1941, *Yunnanosaurus huangi* Young, 1942, and *Riojasaurus incertus* Bonaparte, 1969 but not *Saltasaurus loricatus* Bonaparte and Powell, 1980	stem
PLATEOSAURIA Tornier, 1913	The least inclusive clade containing *Massospondylus carinatus* Owen, 1854 and *Plateosaurus engelhardti* Meyer, 1837 but excluding *Saltasaurus loricatus* Bonaparte and Powell, 1980	node
PLATEOSAURIDAE Marsh, 1895	The most inclusive clade containing *Plateosaurus engelhardti* Meyer, 1837 but not *Massospondylus carinatus* Owen, 1854, *Saltasaurus loricatus* Bonaparte and Powell, 1980	stem
MASSOSPONDYLIDAE von Huene, 1914*b*	The most inclusive clade containing *Massospondylus carinatus* Owen, 1854 but not *Plateosaurus engelhardti* Meyer, 1837, *Saltasaurus loricatus* Bonaparte and Powell, 1980	stem

initially proposed by Upchurch (1997) and later modified by Upchurch *et al.* (2004). The definition recommended here, thus, is a first-order revision of that in Upchurch (1997) using more deeply nested specifiers (Table 4). In this way, the taxonomic content of the clade is stabilized under any arrangement of basal taxa.

Sauropodiformes

Description of the emerging phylogenetic pattern leading to Sauropoda would benefit by having an appropriate taxon name for a clade that unties basal sauropodomorphs more advanced than 'core prosauropods' but less advanced than taxa that might be regarded as basal sauropods. *Mussaurus patagonicus* and *Jingshanosaurus xinwaensis* are known from very complete material including skulls and share several significant synapomorphies with sauropods (Casamiquela 1980; Zhang and Yang 1994; Pol 2004; Pol and Powell 2005). They are destined to become two of the better-known advanced basal sauropodomorphs. Sauropodiformes Sereno, 2005*b* ('in the form of a sauropod') therefore has been proposed as

a node-based taxon anchored by definition to *Mussaurus patagonicus*, *Jingshanosaurus xinwaensis* and the derived sauropod *Saltasaurus loricatus*. A node-based Sauropodiformes is more useful as an anchor outside Sauropoda than a stem-based definition, which for stability would require citation of a broad range of basal sauropodomorphs as external specifiers.

Sauropoda

Sauropoda was initially defined as both node- and stem-based by Salgado *et al.* (1997) and McIntosh (1997), respectively. The node-based definition focused on the basal sauropod *Vulcanodon*. The stem-based definition was initially constructed when at least 'core prosauropods' were regarded as monophyletic. A stem-based Sauropoda has been adopted by several subsequent authors (e.g. Sereno 1998; Wilson and Sereno 1998; Upchurch *et al.* 2004) but should be retooled to exclude the most derived of basal sauropodomorphs. Otherwise, as Yates and others have observed, Sauropoda will incorporate many taxa formerly considered prosauropods, a marked departure from its traditional taxonomic content. The active definition, thus, is a first-order revision of the stem-based definition of McIntosh (1997), Wilson and Sereno (1998) and Sereno (1998). Two external specifiers were selected (*Mussaurus patagonicus, Jingshanosaurus xinwaensis*) for their completeness and efficacy in limiting the taxonomic content of Sauropoda to a less inclusive clade than Sauropodiformes.

Prosauropoda, Plateosauria, Plateosauridae and Massospondylidae

The definition recommended here for Prosauropoda, a first-order revision of the original definition (Upchurch 1997), includes five species viewed as 'core prosauropods' as internal specifiers and *Saltasaurus loricatus* as an external specifier. Should 'core prosauropods' prove to be paraphyletic, the taxon Prosauropoda would not be applicable. Other taxa, such as Plateosauria, Plateosauridae and Massospondylidae (Table 4), are available for less inclusive clades that include the well-known genera *Plateosaurus* and *Massospondylus*. Although the content of these clades would not be particularly stable in some current analyses, the definitions are an attempt to provide a useful taxonomic framework based on pre-existing taxa that is anchored upon well-known species.

Acknowledgements. I thank C. Abraczinskas for executing final drafts of the figures and photographs and for drawings from original specimens, R. Masek and R. Vodden for fossil preparation, M. Langer, J. Wilson and A. Yates for reviewing the manuscript, and E. Gaffney (AMNH), J. Ostrom (YPM), R. Wild (SMNS) and Zhao Xi-Jin (IVPP) for access to fossil material in their care. This research was funded by The David and Lucile Packard Foundation and the National Geographic Society.

REFERENCES

ATTRIDGE, J., CROMPTON, A. W. and JENKINS, F. A. Jr 1985. The southern African Liassic prosauropod *Massospondylus* discovered in North America. *Journal of Vertebrate Paleontology*, **5**, 128–132.

BAI ZI-QI, YANG JIE and WANG GUO-HUI 1990. *Yimenosaurus*, a new genus of Prosauropoda from Yimen County, Yunnan Province. *Yuxiwenbo (Yuxi Culture and Scholarship)*, **1**, 14–23. [In Chinese].

BAKKER, R. T. and GALTON, P. M. 1974. Dinosaur monophyly and a new class of vertebrates. *Nature*, **248**, 168–172.

BARRETT, P. M., UPCHURCH, P. and WANG XIAO-LIN 2005. Cranial osteology of *Lufengosaurus huenei* Young (Dinosauria: Prosauropoda) from the Lower Jurassic of Yunnan, People's Republic of China. *Journal of Vertebrate Paleontology*, **25**, 806–822.

BENTON, M. J. 1986. The Late Triassic reptile *Teratosaurus* – a rauisuchian, not a dinosaur. *Palaeontology*, **29**, 293–301.

—— JUUL, L., STORRS, G. W. and GALTON, P. M. 2000. Anatomy and systematics of the prosauropod dinosaur *Thecodontosaurus antiquus* from the Upper Triassic of southwest England. *Journal of Vertebrate Paleontology*, **20**, 77–108.

BONAPARTE, J. F. 1969. Dos nuevos 'faunas' de reptiles Triásicos de Argentina. 283–306. *In* INTERNATIONAL UNION OF GEOLOGICAL SCIENCES (ed.). *Gondwana Stratigraphy (First International Gondwana Symposium)*. United Nations Educational, Scientific and Cultural Organization, Paris, 1173 pp.

—— 1972. Los tetrápodos del sector superior de la Formación Los Colorados, La Rioja, Argentina (Triásico Superior). 1 Parte. *Opera Lilloana*, **22**, 1–183.

—— 1978. *Coloradia brevis* n. g. et n. sp. (Saurischia-Prosauropoda), dinosaurio Plateosauridae de la Formación Los Colorados, Triásico Superior de La rioja, Argentina. *Ameghiniana*, **15**, 327–332.

—— 1986. The early radiation and phylogenetic relationships of the Jurassic sauropod dinosaurs, based on vertebral anatomy. 247–258. *In* PADIAN, K. (ed.). *The beginning of the age of dinosaurs*. Cambridge University Press, Cambridge, 378 pp.

—— 1999. Evolución de las vértebras presacras en Sauropodomorpha. *Ameghiniana*, **36**, 115–187.

—— and POWELL, J. E. 1980. A continental assemblage of tetrapods from the Upper Cretaceous beds of El Brete, northwestern Argentina (Sauropoda, Coelurosauria, Carnosauria, Aves). *Mémoires de la Société Géologique de France, Nouvelle Série*, **139**, 19–28.

—— and PUMARES, J. A. 1995. Notas sobre el primer craneo de *Riojasaurus incertus* (Dinosauria, Prosauropoda, Melanorosauridae) del Triásico Supérior de La Rioja, Argentina. *Ameghiniana*, **32**, 341–349.

—— and VINCE, M. 1979. El hallazgo del primer nido de dinosaurios triásicos (Saurischia, Prosauropoda), Triásico Superior de Patagonia, Argentina. *Ameghiniana*, **16**, 173–182.

BROOM, R. 1906. On the South African dinosaur (*Hortalotarsus*). *Transactions of the South African Philosophical Society*, **16**, 201–206.

—— 1911. On the dinosaurs of the Stormberg, South Africa. *Annals of the South African Museum*, **7**, 291–307.

CARROLL, R. L. and GALTON, P. M. 1977. 'Modern' lizard from the Upper Triassic of China. *Nature*, **266**, 252–255.

CASAMIQUELA, R. M. 1980. La presencia del genero *Plateosaurus* (Prosauropoda) en el Triásico Superior de la Formación El Tranquilo, Patagonia. *Actas del Segundo Congreso Argentino de Paleontología y Bioestratigrafía y Primero Congreso Latinoamericano de Paleontología*, **1**, 143–158.

CHARIG, A. J., ATTRIDGE, J. and CROMPTON, A. W. 1965. On the origin of the sauropods and the classification of the Saurischia. *Proceedings of the Linnean Society of London*, **176**, 197–221.

CHATTERJEE, S. and ZHENG ZHONG 2002. Cranial anatomy of *Shunosaurus*, a basal sauropod dinosaur from the Middle Jurassic of China. *Zoological Journal of the Linnean Society*, **136**, 145–169.

COLBERT, E. H. 1964. Relationships of saurischian dinosaurs. *American Museum Novitiates*, **2181**, 1–24.

COOPER, M. R. 1981. The prosauropod dinosaur *Massospondylus carinatus* Owen from Zimbabwe: its biology, mode of life and phylogenetic significance. *Occasional Papers of the National Museum of Rhodesia, Series B, Natural Sciences*, **6**, 689–840.

—— 1984. A reassessment of *Vulcanodon karibaensis* Raath (Dinosauria: Saurischia) and the origin of the Sauropoda. *Palaeontologia Africana*, **25**, 203–231.

CROMPTON, A. W. and ATTRIDGE, J. 1986. Masticatory apparatus of the larger herbivores during Late Triassic and Early Jurassic times. 223–236. *In* PADIAN, K. (ed.). *The beginning of the age of dinosaurs*. Cambridge University Press, Cambridge, 378 pp.

CRUICKSHANK, A. R. I. 1975. The origin of sauropod dinosaurs. *South African Journal of Science*, **71**, 89–90.

DUTUIT, J.-M. 1972. Decouvert d'un dinosaure ornithischien dans de Trias superieur de l'Atlas occidental marocian. *Comptes Rendus de l'Académie des Sciences de Paris, Series D*, **275**, 2841–2844.

EVANS, S. E. and MILNER, A. R. 1989. *Fulengia*, a supposed early lizard reinterpreted as a prosauropod dinosaur. *Palaeontology*, **32**, 223–230.

FEDAK, T. 2005. Two heads are better than one: considering *Anchisaurus* as a small sauropod. *Journal of Vertebrate Paleontology*, **25** (Supplement to No. 3), 56A.

FLYNN, J. J., PARRISH, J. M., RAKOTOSAMIMANANA, B., SIMPSON, W. F., WHATLEY, R. L. and WYSS, A. R. 1999. A Triassic fauna from Madagascar, including early dinosaurs. *Science*, **286**, 763–765.

GALTON, P. M. 1971. The prosauropod dinosaur *Ammosaurus*, the crocodile *Protosuchus*, and their bearing on the age of the Navajo Sandstone of northeastern Arizona. *Journal of Paleontology*, **45**, 781–795.

—— 1973. On the anatomy and relationships of *Efraasia diagnostica* (Huene) n. gen., a prosauropod dinosaur (Reptilia: Saurischia) from the Upper Triassic of Germany. *Paläontologische Zeitschrift*, **47**, 229–255.

—— 1976. Prosauropod dinosaurs (Reptilia: Saurischia) of North America. *Postilla*, **169**, 1–98.

—— 1984. Cranial anatomy of the prosauropod dinosaur *Plateosaurus* from the Knollenmergel (Middle Keuper, Upper Triassic) of Germany. I – Two complete skulls from Trossingen/Württ. with comments on the diet. *Geologica et Palaeontologica*, **18**, 139–172.

—— 1985. Diet of prosauropod dinosaurs from the Late Triassic and Early Jurassic. *Lethaia*, **15**, 105–123.

—— 1989. Prosauropoda, the basal Sauropodomorpha. 80–84. *In* PADIAN, K. and CHURE, D. J. (eds). *The age of dinosaurs*. Short Courses in Palaeontology, **2**, 210 pp.

—— 1990. Basal Sauropodomorpha – Prosauropoda. 320–344. *In* WEISHAMPEL, D. B., DODSON, P. and OSMÓLSKA, H. (eds). *The Dinosauria*. University of California Press, Berkeley, CA, 733 pp.

—— 1998. Saurischian dinosaurs from the Upper Triassic of England: *Camelotia* (Prosauropoda, Melanorosauridae) and *Avalonianus* (Theropoda, ?Carnosauria). *Palaeontographica Abteilung A*, **250**, 155–172.

—— 2000. The prosauropod dinosaur *Plateosaurus* Meyer, 1837 (Saurischia: Sauropodomorpha). I. The syntypes of *P. engelhardti* Meyer, 1837 (Upper Triassic, Germany), with notes on other European prosauropods with 'distally straight' femora. *Neues Jahrbuch für Geologie und Pälaontologie, Abhandlungen*, **216**, 233–275.

—— 2001. The prosauropod dinosaur *Plateosaurus* Meyer, 1837 (Saurischia: Sauropodomorpha; Upper Triassic). II. Notes on the referred species. *Revue de Paléobiologie*, **20**, 435–502.

—— and CLUVER, M. A. 1976. *Anchisaurus capensis* (Broom) and a revision of the Anchisauridae (Reptilia: Saurischia). *Annals of the South African Museum*, **69**, 121–159.

—— and UPCHURCH, P. 2004. Prosauropoda. 232–258. *In* WEISHAMPEL, D. B., DODSON, P. and OSMÓLSKA, H. (eds). *The Dinosauria*. Second edition. University of California Press, Berkeley, CA, 861 pp.

—— and VAN HEERDEN, J. 1998. Anatomy of the prosauropod dinosaur *Blikanasaurus cromptoni* (Upper Triassic, South Africa), with notes on other tetrapods from the lower Elliot Formation. *Paläontologische Zeitschrift*, **72**, 163–177.

GAUFFRE, F.-X. 1993. The most recent Melanorosauridae (Saurischia, Prosauropoda), Lower Jurassic of Lesotho, with remarks on the prosauropod phylogeny. *Neues Jahrbuch für Geologie und Paläontologie, Monatschefte*, **1993**, 648–654.

GAUTHIER, J. 1986. Saurischian monophyly and the origin of birds. 1–55. *In* PADIAN, K. (ed.). *The origin of birds and the evolution of flight*. Memoirs of the California Academy of Science, **8**, 98 pp.

GOW, C. E., KITCHING, J. W. and RAATH, M. A. 1990. Skulls of the prosauropod dinosaur *Massospondylus carinatus* Owen in the collections of the Bernard Price Institute for Palaeontological Research. *Palaeontologica Africana*, **27**, 45–58.

HAMMER, W. R. and HICKERSON, W. J. 1994. A crested theropod dinosaur from Antarctica. *Science*, **264**, 828–830.

HE XIN-LU, LI KUI and CAI KAI-JI 1988. *The Middle Jurassic dinosaur fauna from Dashanpu, Zigong, Sichuan: sauropod dinosaurs (2)*. Omeisaurus tianfuensis. Sichuan Publishing House of Science and Technology, Chengdu, 143 pp., 20 pls. [In Chinese, English summary].

HITCHCOCK, E. 1855. Shark remains from the Coal formation of Illinois, and bones and tracks from the Connecticut River sandstone. *American Journal of Science, Series 2*, **20**, 416–417.

HITCHCOCK, E. Jr 1865. *A supplement to the ichnology of New England*. Wright and Potter, Boston, 90 pp.

HUENE, F. VON 1908. Die dinosaurier der europäischen Triasformation mit Berücksichtigung der aussereuropäischen Vorkommnisse. *Geologische und Paläeontologische Abhandlungen Supplement*, **1**, 1–419.

—— 1914a. Das natürliche System der Saurischia. *Zentralblatt für Mineralogie, Geologie und Paläontologie, B*, **1914**, 154–158.

—— 1914b. Saurischia et Ornithischia triadica ('Dinosauria' triadica). *Fossilum Catalogus (Animalia)*, **4**, 1–21.

—— 1920. Bemerkungen zur Systematik und Stammesgeschicht einiger Reptilien. *Zeitschrift für Induktive Abstammungs und Vererburgslehre*, **22**, 209–212.

—— 1926. Vollständige Osteologie eines Plateosauriden aus dem schwäbischen Trias. *Geologische und Paläontologische Abhandlungen, Neue Folge*, **15**, 129–179.

—— 1929. Kurze Übersicht über die Saurischia und ihre natürlichen Zusammenhänge. *Paläontologische Zeitschrift*, **11**, 269–273.

—— 1932. Die fossile Reptil-Ordnung Saurischia, ihre Entwicklung und Geschichte. *Monographien zur Geologie und Paläontologie*, **4**, 1–361.

—— 1956. *Paläontologie and Phylogenie der Niederen Tetrapoden*. Gustav Fisher, Jena, 716 pp.

IRMIS, R. B. 2005. A review of the vertebrate fauna of the Lower Jurassic Navajo Sandstone in Arizona. 55–71. *In* McCORD, R. D. (ed.). *Vertebrate paleontology of Arizona*. Bulletin of the Mesa Southwest Museum, **11**, 180 pp.

JENKINS, F. A., SHUBIN, N. H., AMARAL, W. A., GATESY, S. M., SCHAFF, C. R., CLEMMENSEN, L. B., DOWNS, W. R., DAVIDSON, A. R., BONDE, N. and OSBÆCK, F. 1995. Late Triassic continental vertebrates and depositional environments of the Fleming Fjord Formation, Jameson Land, East Greenland. *Meddelelser om Grønland, Geoscience*, **32**, 1–25.

KERMACK, D. 1984. New prosauropod material from South Wales. *Zoological Journal of the Linnean Society*, **82**, 101–117.

KIRKLAND, J. I., ZANNO, L. E., SAMPSON, S. D., CLARK, J. M. and DEBLIEUX, D. D. 2005. A primitive therizinosauroid dinosaur from the Early Cretaceous of Utah. *Nature*, **435**, 84–87.

KUTTY, T. S. 1969. Some contributions to the stratigraphy of the upper Gondwana formations of the Prahita-Godavari Valley, central India. *Journal of the Geological Society of India*, **10**, 33–48.

LANGER, M. C. 2003. The pelvic and hind limb anatomy of the stem-sauropodomorph *Saturnalia tupiniquim* (Late Triassic, Brazil). *PaleoBios*, **23**, 1–30.

—— and BENTON, M. J. 2006. Early dinosaurs: a phylogenetic study. *Journal of Systematic Palaeontology*, **4**, 1–50.

—— ABDALA, F., RICHTER, M. and BENTON, M. J. 1999. A sauropodomorph dinosaur from the Upper Triassic (Carnian) of southern Brazil. *Comptes Rendus de l'Académie des Sciences de Paris, Sciences de la Terre et des Planètes*, **329**, 511–517.

—— FRANÇA, M. A. G. and GABRIEL, S. 2007. The pectoral girdle and forelimb anatomy of the stem-sauropodomorph *Saturnalia tupiniquim* (Upper Triassic, Brazil). 113–137. *In* BARRETT, P. M. and BATTEN, D. J. (eds). *Evolution and palaeobiology of early sauropodomorph dinosaurs*. Special Papers in Palaeontology, **77**, 289 pp.

LEAL, L. A., AZEVEDO, S. A. K., KELLNER, A. W. A. and DA ROSA, A. A. S. 2004. A new early dinosaur (Sauropodomorpha) from the Caturrita Formation (Late Triassic), Parana Basin, Brazil. *Zootaxa*, **690**, 1–24.

LINNAEUS, C. 1758. *Systema naturae*. Lucae, Stockholm, 376 pp.

LOEWEN, M., SERTICH, J. J., SAMPSON, S. D. and GETTY, M. 2005. Unusual preservation of a new sauropodomorph from the Navajo Sandstone of Utah. *Journal of Vertebrate Paleontology*, **25** (Supplement to No. 3), 84A.

LULL, R. S. 1915. Triassic life of the Connecticut Valley. *Bulletin of the Connecticut Geology and Natural History Survey*, **24**, 1–285.

—— 1953. Triassic life of the Connecticut Valley. Revised edition. *Bulletin of the Connecticut Geology and Natural History Survey*, **81**, 1–331.

MARSH, O. C. 1878. Principal characters of American Jurassic dinosaurs. Part 1. *American Journal of Science, Series 3*, **16**, 411–416.

—— 1882. Classification of the Dinosauria. *American Journal of Science, Series 3*, **23**, 81–86.

—— 1884. Principal characters of American Jurassic dinosaurs. Part VIII. The Order Theropoda. *American Journal of Science, Series 3*, **27**, 329–341.

—— 1885. Names of extinct reptiles. *American Journal of Science, Series 3*, **29**, 169.

—— 1889*a*. Notice of new American dinosaurs. *American Journal of Science, Series 3*, **37**, 331–336.

—— 1889*b*. Notice of gigantic horned Dinosauria from the Cretaceous. *American Journal of Science, Series 3*, **38**, 173–175.

—— 1891. Notice of new vertebrate fossils. *American Journal of Science, Series 3*, **42**, 265–269.

—— 1892. Notes on Triassic Dinosauria. *American Journal of Science, Series 3*, **43**, 543–546.

—— 1893. Restoration of *Anchisaurus*. *American Journal of Science, Series 3*, **45**, 169–170.

—— 1895. On the affinities and classification of the dinosaurian reptiles. *American Journal of Science, Series 3*, **50**, 483–498.

McINTOSH, J. S. 1997. Sauropoda. 654–658. *In* CURRIE, P. J. and PADIAN, K. (eds). *Encyclopedia of dinosaurs*. Academic Press, San Diego, 869 pp.

MEYER, H. VON 1837. Mitteilung an Prof. Bronn (*Plateosaurus engelhardti*). *Neues Jahrbuch für Mineralogie, Geologie und Paläontologie*, **1837**, 817.

OLSEN, P. E., SHUBIN, N. H. and ANDERS, M. H. 1987. New Early Jurassic tetrapod assemblages constrain Triassic-Jurassic tetrapod extinction event. *Science*, **237**, 1025–1029.

OSTROM, J. H. 1969. The case of the missing specimen. *Discovery*, **5**, 50–51.

—— and McINTOSH, J. S. 1999. *Marsh's dinosaurs: the collections from Como Bluff*. Yale University Press, New Haven, 388 pp.

OWEN, R. 1854. *Descriptive catalogue of the fossil organic remains of Reptilia and Pisces contained in the Museum of the Royal College of Surgeons of England*. Taylor and Francis, London, 184 pp.

PAUL, G. S. 1984. The segnosaurian dinosaurs: relics of the prosauropod ornithischian transition? *Journal of Vertebrate Paleontology*, **4**, 507–515.

POL, D. 2004. Phylogenetic relationships of basal Sauropodomorpha. Unpublished PhD dissertation, Columbia University, NY, 303 pp.

—— and POWELL, J. E. 2005. Anatomy and phylogenetic relationships of *Mussaurus patagonicus* (Dinosauria, Sauropodomorpha) from the Late Triassic of Patagonia. 208–209. *In* KELLNER, A. W. A., HENRIQUES, D. D. R. and RODRIGUES, T. (eds). *Boletim de resumos, II Congresso Latin-Americano de Paleontologia de Vertebrados*. Museu Nacional UFRJ, Rio de Janeiro, 285 pp.

RAATH, M. A. 1972. Fossil vertebrate studies in Rhodesia: a new dinosaur (Reptilia, Saurischia) from the near the Trias-Jurassic boundary. *Arnoldia*, **5** (30), 1–37.

RILEY, H. and STUTCHBURY, S. 1836. A description of various remains of three distinct saurian animals, recently discovered in the Magnesian Conglomerate near Bristol. *Proceedings of the Geological Society of London*, **2**, 397–399.

ROGERS, R. R., SWISHER, C. C. III, SERENO, P. C., MONETTA, A. M. and MARTINEZ, R. N. 1993. The Ischigualasto tetrapod assemblage (Late Triassic, Argentina) and ^{40}Ar/^{39}Ar dating of dinosaur origins. *Science*, **260**, 794–797.

ROMER, A. S. 1956. *Osteology of the reptiles*. University of Chicago Press, Chicago, IL, 772 pp.

—— 1966. *Vertebrate paleontology*. Third edition. University of Chicago Press, Chicago, IL, 468 pp.

—— 1968. *Notes and comments on vertebrate paleontology*. University of Chicago Press, Chicago, IL, 304 pp.

SALGADO, L., CORIA, R. A. and CALVO, J. O. 1997. Evolution of titanosaurid sauropods. Part I. Phylogenetic analysis based on the postcranial evidence. *Ameghiniana*, **34**, 3–32.

SANTA LUCA, A. P. 1980. The postcranial skeleton of *Heterodontosaurus tucki* (Reptilia, Ornithischia) from the Stormberg of South Africa. *Annals of the South African Museum*, **79**, 15–211.

SEELEY, H. G. 1895. On the type of the genus *Massospondylus*, and on some vertebrae and limb-bones of *M.* (?) *browni*. *Annals and Magazine of Natural History*, **15**, 102–125.

SERENO, P. C. 1989. Prosauropod monophyly and basal sauropodomorph phylogeny. *Journal of Vertebrate Paleontology*, **9** (Supplement to No. 3), 38A.

—— 1991. *Lesothosaurus*, 'fabrosaurids', and the early evolution of Ornithischia. *Journal of Vertebrate Paleontology*, **11**, 168–197.

—— 1993. The pectoral girdle and forelimb of the basal theropod *Herrerasaurus ischigualastensis*. *Journal of Vertebrate Paleontology*, **13**, 425–450.

—— 1997. The origin and evolution of dinosaurs. *Annual Review of Earth and Planetary Sciences*, **25**, 435–489.

—— 1998. A rationale for phylogenetic definitions, with application to the higher-level taxonomy of Dinosauria. *Neues Jahrbuch für Geologie und Paläontologie, Abhandlungen*, **210**, 41–83.

—— 1999a. The evolution of dinosaurs. *Science*, **284**, 2137–2147.

—— 1999b. A rationale for dinosaurian taxonomy. *Journal of Vertebrate Paleontology*, **19**, 788–790.

—— 2005a. The logical basis of phylogenetic taxonomy. *Systematic Biology*, **54**, 595–619.

—— 2005b. *TaxonSearch*: a relational database for supra-generic taxa and phylogenetic definitions. *PhyloInformatics*, **8**, 1–20.

SUES, H.-D., REISZ, R. R., HINIC, S. and RAATH, M. A. 2004. On the skull of *Massospondylus carinatus* Owen, 1854 (Dinosauria: Sauropodomorpha) from the Elliot and Clarens formations (Lower Jurassic) of South Africa. *Annals of the Carnegie Museum*, **73**, 239–257.

TORNIER, G. 1913. Reptilia (Paläontologie). *Handwörterbuch der Naturwissenschaften*, **8**, 337–376.

UPCHURCH, P. 1995. The evolutionary history of sauropod dinosaurs. *Philosophical Transactions of the Royal Society of London, Series B*, **349**, 365–390.

—— 1997. Sauropodomorpha. 658–660. *In* CURRIE, P. J. and PADIAN, K. (eds). *Encyclopedia of dinosaurs*. Academic Press, San Diego, 869 pp.

—— BARRETT, P. M. and DODSON, P. 2004. Sauropoda. 259–322. *In* WEISHAMPEL, D. B., DODSON, P. and OSMÓLSKA, H. (eds). *The Dinosauria*. Second edition. University of California Press, Berkeley, CA, 861 pp.

VAN HEERDEN, J. 1978. *Herrerasaurus* and the origin of the sauropod dinosaurs. *South African Journal of Science*, **74**, 187–189.

—— 1979. The morphology and taxonomy of *Euskelosaurus* (Reptilia: Saurischia; Late Triassic) from South Africa. *Navorsinge van die Nasionale Museum, Bloemfomtein*, **4**, 21–84.

—— and GALTON, P. M. 1997. The affinities of *Melanorosaurus* – a Late Triassic prosauropod dinosaur from South Africa. *Neues Jahrbuch für Geologie und Paläontologie, Monatschefte*, **1997**, 39–55.

VICKERS-RICH, P., RICH, T. H., McNAMARA, G. C. and MILNER, A. C. 1999. Is *Agrosaurus macgillivrayi* Australia's oldest dinosaur? *Records of the Western Australian Museum, Supplement*, **57**, 191–200.

WILSON, J. A. and SERENO, P. C. 1998. Higher-level phylogeny of sauropod dinosaurs. *Memoir of the Society of Vertebrate Paleontology*, **5**, 1–68.

YANG ZHONG-JIAN (YOUNG CHUNG-CHIEN). 1982. A new ornithopod from Lufeng, Yunnan. 29–35. *In* ZHOU MING-ZHEN (ed.). *Selected works of Yang Zhong-Jian*. Science Press, Beijing, 219 pp. [In Chinese].

YATES, A. M. 2003a. A new species of the primitive dinosaur *Thecodontosaurus* (Saurischia: Sauropodomorpha) and its implications for the systematics of early dinosaurs. *Journal of Systematic Palaeontology*, **1**, 1–42.

—— 2003b. The species taxonomy of the sauropodomorph dinosaurs from the Löwenstein Formation (Norian, Late Triassic) of Germany. *Palaeontology*, **46**, 317–337.

—— 2004. *Anchisaurus polyzelus* (Hitchcock): the smallest known sauropod dinosaur and the evolution of gigantism among sauropodomorph dinosaurs. *Postilla*, **230**, 1–58.

—— 2005. The skull of the Triassic sauropodomorph, *Melanorosaurus readi*, from South Africa and the definition of the Sauropoda. *Journal of Vertebrate Paleontology*, **25**, 132A.

—— 2007. The first complete skull of the Triassic dinosaur *Melanorosaurus* Haughton (Sauropodomorpha: Anchisauria). 9–55. *In* BARRETT, P. M. and BATTEN, D. J. (eds). *Evolution and palaeobiology of early sauropodomorph dinosaurs*. Special Papers in Palaeontology, **77**, 289 pp.

—— and KITCHING, J. W. 2003. The earliest known sauropod dinosaur and the first step towards sauropod locomotion. *Proceedings of the Royal Society of London, Series B*, **270**, 1753–1758.

YOUNG CHUNG-CHIEN 1941. *Gyposaurus sinensis* (sp. nov.), a new Prosauropoda from the Upper Triassic Beds at Lufeng, Yunnan. *Bulletin of the Geological Society of China*, **21**, 205–253.

—— 1942. *Yunnanosaurus huangi* (gen. et sp. nov.), a new Pro-sauropoda from the Red Beds at Lufeng, Yunnan. *Bulletin of the Geological Society of China*, **22**, 63–104.

—— 1947. On *Lufengosaurus magnus* Young (sp. nov.) and additional finds of *Lufengosaurus huenei* Young. *Palaeontologia Sinica, New Series*, **12**, 1–53.

—— 1948. Further notes on *Gyposaurus sinensis* Young. *Bulletin of the Geological Society of China*, **28**, 91–103.

—— 1951. The Lufeng saurischian fauna in China. *Palaeontologica Sinica*, **13**, 1–96.

ZHANG YI-HONG 1988. *The Middle Jurassic dinosaurian fauna from Dashanpu, Zigong, Sichuan. Vol. III. Sauropod dinosaurs (I).* Sichuan Scientific and. Technological Publishing House, Chengdu, Sichuan, 89 pp. [In Chinese, English summary].

—— and YANG ZHAO-LONG 1994. *A complete osteology of Prosauropoda in Lufeng Basin, Yunnan, China.* Jingshanosaurus. Yunnan Science and Technology Publishing House, Kunming, 100 pp. [In Chinese, English summary].

NOTE ADDED IN PROOF

Yates (2006) has recently published another analysis of basal sauropodomorphs based on 353 characters in 41 ingroup taxa, 19 of which are traditional basal (non-sauropod) sauropodomorphs. How these character data differ from those in previous analyses (Yates 2003*a*, 2004; Yates and Kitching 2003; Text-fig. 2B–C) is not indicated, although the results exhibit significant differences. *Yunnanosaurus* is positioned closer to sauropods than other core prosauropods, unlike the most parsimonious hypothesis in Yates (2003*a*), and *Plateosaurus* and *Riojasaurus* are positioned basal to many other core prosauropods, unlike the shortest trees in Yates and Kitching (2003) and Yates (2004).

Two new suprageneric taxa of questionable utility were coined and defined. Riojasauridae was erected to include *Riojasaurus incertus* and *Eucnemasaurus fortis*, a new genus and species based on fragmentary postcranial bones; Massopoda was erected for all basal sauropodomorphs that are closer to sauropods than to *Plateosaurus engelhardti*. Plateosauria was used without any qualifying conditions, unlike the present recommendation (Table 4; Sereno *et al.* 2005); as a result many taxa including Sauropoda are subsumed, and the taxon bears little resemblance to its historical usage (Text-fig. 1A). Finally, Sauropoda was redefined using *Melanorosaurus readi* as an external specifier rather than *Jingshanosaurus xinwaensis* and *Mussaurus patagonicus* (Sereno *et al.* 2005; Table 4). Although a reasonable alternative definition, the positioning of *Melanorosaurus readi* may depend in good measure on the status of newly referred material.

YATES, A. M. 2006. Solving a dinosaurian puzzle: the identity of *Aliwalia rex* Galton. *Historical Biology*, **2006**, 1–31.